IMAGE-GUIDED NEUROSURGERY

ELSEVIER

ELSEVIER science & technology books

Companion Web Site:

http://booksite.elsevier.com/9780128008706

Image-Guided Neurosurgery

Edited by Alexandra J. Golby

Resources for Professors:

- All figures from the book available as PowerPoint slides

ELSEVIER

ACADEMIC PRESS

IMAGE-GUIDED NEUROSURGERY

Edited by

ALEXANDRA J. GOLBY

Associate Professor of Neurosurgery and Radiology Harvard Medical School

Associate Surgeon
Director of Image-guided Neurosurgery
Clinical Co-Director AMIGO

Co-Director Clinical Functional MRI Brigham and
Women's Hospital Department of Neurosurgery
Boston, MA, USA

ELSEVIER

AMSTERDAM • BOSTON • HEIDELBERG • LONDON
NEW YORK • OXFORD • PARIS • SAN DIEGO
SAN FRANCISCO • SINGAPORE • SYDNEY • TOKYO
Academic Press is an imprint of Elsevier

AP

Academic Press is an imprint of Elsevier
125, London Wall, EC2Y 5AS
525 B Street, Suite 1800, San Diego, CA 92101-4495, USA
225 Wyman Street, Waltham, MA 02451, USA
The Boulevard, Langford Lane, Kidlington, Oxford OX5 1GB, UK

ISBN: 978-0-12-800870-6

British Library Cataloguing-in-Publication Data
A catalogue record for this book is available from the British Library

Library of Congress Cataloging-in-Publication Data
A catalog record for this book is available from the Library of Congress

For information on all Academic Press publications
visit our website at http://store.elsevier.com/

Publisher: Mica Haley
Acquisition Editor: Melanie Tucker
Editorial Project Manager: Kristi Anderson
Production Project Manager: Melissa Read
Designer: Alan Studholme

DEDICATION

This book is dedicated to my mentor, colleague, and friend Ferenc Jolesz. Ferenc welcomed me to Brigham and Women's Hospital during the height of the world's first experiences with intra operative MRI. I was just starting my career and he provided me a most fertile environment in which to collaborate and develop innovative neurosurgical approaches.

Ferenc saw the broadest possible perspective of image guided therapy. Having trained first as a neurosurgeon and then as neuroradiologist, his soft spot was often in developing techniques for neurosurgery while taking full advantage of radiological innovations. This bias made us natural allies, and together we started our journey of understanding the complexity of the brain and of brain pathologies. Over the years, in addition to the MRT program and its follow-on, the AMIGO suite, he was powerfully involved in high intensity focused ultrasound, development of biomarkers for imaging, robotics for neurosurgery, and functional and structural mapping of the brain; I discussed nearly every topic represented in this book frequently with Ferenc. He introduced me personally to many of the authors of the chapters. He wrote the Foreword whilst recuperating from one of many interventions he had soldiered through for over two decades, his force undiminished, and as always, directed towards technical innovation and guiding and mentoring our close-knit team with interests across the whole spectrum of image guided therapies. Ferenc was a very special leader who will leave an indelible mark in neurosurgery as well as many other interventional specialties. This book brings some of that together in one place. I hope that it will serve as a key reference for many. It would have been Ferenc's greatest wish to see these ideas and technologies disseminated throughout the world.

CONTENTS

FOREWORD

This book is the first in a decade to tackle the subject of image guidance in neurosurgery and to incorporate all current aspects and future potential of this wide-ranging and fast-advancing field. There are increasing numbers of users of image-guided technology not only in neurosurgery but also in other surgical fields and new clinical applications are constantly evolving. This book will help the reader to understand all the current proven advantages and future perspectives across the full range of image-guided techniques in neurosurgery and at the same time become familiar with the wide variety of already existing and only foreseen clinical applications.

Since its introduction in the early 1990s intraoperative MRI (iMRI) has substantively changed modern neurosurgery. The vision and inspiration of iMRI was the direct consequence of the widespread acceptance of MRI as the primary method of diagnostic neuroimaging. Since then iMRI has followed the progress of MRI technology and has grown with its advances. Before iMRI was introduced in neurosurgery it became obvious that there is a new and better way to guide and control interventions like biopsies and thermal ablations by MRI. That recognition of the power of MRI for nondiagnostic application is the origin of interventional MRI. The discovery of MRI-guided interstitial laser surgery and its potential to treat brain tumors was the initial motivation for the development of the first genuine iMRI designed almost exclusively for neurosurgical use by General Electric in collaboration with Brigham and Women's Hospital investigators.

At the time this groundbreaking new iMRI was installed, preoperative MRI-based computerized navigational methods were already customary in neurosurgery. However, their inability to maintain accuracy during surgeries in the presence of substantial brain shifts and deformations remained unresolved. It was recognized early that iMRI could provide a possible solution for the inherent problem of preoperative imaging-based navigation. Serial imaging during brain surgeries provides anatomically correct updates for navigation. The first iMRI system therefore integrated navigation with continuous multioblique image plane selection and also provided monitoring and control for thermal ablations using temperature-sensitive MRI. In addition to open craniotomies, biopsies, minimally invasive laser procedures, and endoscopies were performed in an open configuration environment with two neurosurgeons having full access to the head without the need for moving the patient. All these novel technological features and the optimal workflow environment made this system ideal for neurosurgical use, but the high cost and relatively low image quality precluded its broad adoption.

In the first decade of iMRI, several open and more or less closed magnet configurations with increasing field strength were tested for neurosurgical use by multiple investigators. In the end, closed magnets with higher field strength were the final choice of most academic neurosurgical centers. The reason for this choice has been the improvement of image quality achieved with advanced high-field MRI, which now dominates diagnostic neuroimaging. It is difficult to use a lower quality image during surgery compared with the preoperative diagnostic images that show much more detail. High-resolution imaging is particularly important for intraoperative guidance when visualization of intricate details is even more essential than in the diagnostic workup. Unfortunately, this requirement for better image quality and resolution has led to a major compromise. Higher field closed configuration magnets with superb image quality have become the choice of most neurosurgeons but the cost is a very inconvenient environment requiring a complicated workflow that limits what can be accomplished during a surgical session. Requirements of the closed bore include moving the patient deep into the bore of the magnet or moving the magnet to the patient on the operating room table in order to obtain images. More importantly, with this approach, one must give up one of the main advantages of iMRI: to provide imaging updates by serial imaging for compensation of brain shifts. The result of this compromise is usually only a single imaging session at the end of surgery that may identify residual tumor for additional resection if needed. This compromise is now generally accepted as a standard feature of routine iMRI that in most cases involves only two imaging sessions, at the beginning and at the end of surgeries, instead of the originally used and more appropriate serial imaging. Updated navigation and the resulting sustained intraoperative accuracy became the casualties of improved image quality and resolution.

The other original source of the general idea of interventional and iMRI is the use of temperature-sensitive MRI to monitor and control thermal ablations of brain tumors and other diseases. This minimally invasive method has since become a major new direction in modern neurosurgery. MRI-guided interstitial laser surgery is now, 20 years after its introduction, available commercially and there are several successful clinical applications not only for malignant but also for benign brain tumors and for non-neoplastic diseases like epilepsy. MRI-guided Focused Ultrasound (FUS) is a noninvasive ablation method with the potential to change not only brain tumor surgery and functional neurosurgery, but also other related fields of clinical neurosciences. FUS does not use ionizing radiation, is repeatable, and can be monitored and controlled in real time unlike all currently practiced radiation therapy. Neuromodulation by FUS can have several applications in neurology and psychiatry; targeted drug delivery by FUS opening of the blood—brain barrier could be a real game changer in chemotherapy and in other branches of neuropharmacology. All of these currently discovered and far-reaching advances are the results of the introduction of MRI in neurosurgery and in therapy.

Early results from all the pioneers of iMRI demonstrated that this novel approach could improve the completeness of glioma resections and help to perform more extensive tumor removal without causing new neurological deficits. First, it was proven mostly for low-grade tumors that surgeries using iMRI could lengthen survival; later clear benefits were demonstrated for higher-grade tumors, too. Since then a substantial number of investigations indicate a clear benefit of gross total tumor resection on overall- and recurrence-free survival in patients with both low- and high-grade gliomas. These findings strongly motivate further advancing of the field. It was also demonstrated that in some benign tumor surgeries, like those for pituitary adenoma, iMRI might provide important imaging clues that can make the surgical interventions more complete, more successful and safer. IMRI was also introduced into endoscopic sinus surgeries with potential use in skull-base surgery.

The last 10 years have brought even more monumental changes. The introduction of MRI methods for depicting functional anatomy using advanced MRI techniques like functional MRI (fMRI) and Diffusion Tensor Imaging (DTI) were embraced not only by basic neuroscientists but also by some clinical neurosurgeons. Comprehensive imaging that integrates all the anatomical and functional information and relates these to the actual location of tumors can provide the surgeon with the wealth of information that was unavailable and unimaginable in traditional neurosurgery. These imaging data now can be used for surgical planning before and for guidance during surgery. This combination of preoperative and intraoperative imaging data can demonstrate a less well-known advantage of iMRI: helping to execute the preoperative surgical plan. Surgeons relying on the complex information provided by advance neuroimaging are able to remove more tumors and avoid complication after careful preoperative preparation and intraoperative presentation of the plan. Computerized image-processing methods like image fusion with nonrigid registration and more advanced navigation techniques can facilitate the process of intraoperative decision-making. Updated imaging can include intraoperative fiber tracking combined with electrophysiology in awake patients.

Intraoperative decision-making that requires the knowledge of tumor extent can be enhanced by intraoperative methods that may provide real-time biomarkers for tumor. Areas under investigation include mass spectrometry, Raman spectroscopy, and probes detecting radioactive molecular tumor detecting agents. These methods are potentially more sensitive and specific than MRI and can be localized and registered to the MR images. Using the combination of preoperative functional anatomical data, applying those for model generation for surgical planning, utilizing all features of intraoperative imaging and navigation and exploiting additional surgical decision-making tools, it is possible to further improve surgical outcomes for malignant brain tumor surgeries beyond the current results.

Proving efficacy, however, is challenging. Total resection of malignant brain tumors is an unworkable goal since in most cases tumor infiltrates normal brain and it is impossible to resect without causing new neurological deficits. Each individual case is different in the extent of tumor invasion and by location in relationship to functionally critical regions. Also, currently we have no optimal specific MRI method to distinguish edema from infiltrative tumor and the sensitivity of MRI is insufficient to produce an accurate map of tumor extent. The most significant for influencing efficacy and outcome among them is the mandate for the neurosurgeon to prevent new neurologic deficit as this impacts quality of life and has also been found to impact survival. Data supports that the greatest impact of iMRI on extent of resection is in tumors in noneloquent regions. The combination of tumor map and correctly registered fMRI-DTI data can provide an optimal surgical plan that can be correctly executed only by using iMRI by updating the anatomic image during surgery. However, if the definition of success is the removal of maximal amount of tumor without new functional defect, the definition of success and outcome is different and not based on survival alone.

iMRI has continued to improve glioma surgery but more exciting applications are ahead in other areas of neurosurgery. iMRI has been applied for vascular, spinal, and skull-base treatments, too. Most of these newer applications require advanced 3T MRI platforms for faster and more flexible image acquisitions than those available in the early stages. The full potential of iMRI can only be reached if advanced MRI is complemented with other imaging modalities in an intraoperative setting. For vascular interventions and surgeries X-ray fluoroscopy and angiography is necessary, for spine procedures MRI is extremely helpful but not without X-ray CT that provides details of the bone; similarly skull-base surgeries require both MRI and CT guidance. The potential advantages of molecular imaging are only conceivable if PET/CT or optical imaging is available in the surgical environment. Ultrasound can be an essential real-time monitoring device of brain shift, replacing the need for multiple serial MR images during surgeries. Development of an integrated navigational system based on the combination of US and MRI is necessary to manage the brain shift challenge that every neurosurgeon is faced with. This clinically well-justified solution also requires further improvement of multimodality nonlinear registration methods.

The Brigham and Women's Hospital Advanced Multimodality Image Guided Operating room (AMIGO) is the first implementation of this multimodality concept. MRI is the primary intraoperative imaging modality that can be supplemented with any other imaging method that is necessary for a given clinical procedure. AMIGO is a testbed not only for new neurosurgical procedures but also for other surgical and interventional applications. The success of iMRI in neurosurgery initiated and motivated the acceptance and use of intraoperative MR imaging and integrated navigational guidance in other fields. Neurosurgeons who pioneered stereotactic surgery, frameless navigation and iMRI have inspired their colleagues in other fields to embrace modern advanced image-guided surgical technologies in their particular fields. In AMIGO, which is a translational component of an active multidisciplinary program, the interaction and cross-fertilization between neurosurgical and other projects is an assurance for further progress of the entire image-guided therapy field.

Future improvements of iMRI may result not only from further progress of imaging technologies, but also from devices and tools that are guided by MRI. Among those are endoscopes, especially flexible neuroendoscopes, endovascular catheters and devices, and robotic surgical assistants. MRI tracking and MRI-based control of these devices may have a substantial role in the future of neurosurgery. It has been anticipated since the introduction of iMRI that new surgical approaches will be developed by innovative neurosurgeons equipped with iMRI technology. These developments continue to emerge. Further innovations will rest on the development of MRI-compatible devices like endoscopes, catheters, electrodes, and robots. Integration of the MR imaging methods with various therapy devices and robots can transform open neurosurgical procedures into minimally invasive image-guided surgeries by changing surgical techniques and approaches leading to new treatments for tumors, vascular abnormalities, and other diseases of the brain and spine.

In the last two decades, iMRI has been advancing and improving. It has moved through the stages of discovery, acceptance, and routine clinical use in neurosurgery and now it spreads into other surgical fields. It is still a not fully developed and mature technology, with presently only a limited number of clinical applications in which efficacy is already proven. Nevertheless, it is anticipated that, in combination with other advanced imaging, image processing and navigational technologies, and after integration with therapy delivery devices, its usefulness and effectiveness will be further verified. This book represents an important step in that direction.

Ferenc A. Jolesz, MD
B. Leonard Holman Professor of Radiology
Director, Division of MRI and National Center for Image Guided Therapy
Department of Radiology
Brigham and Women's Hospital
Harvard Medical School
Boston, MA, USA

LIST OF CONTRIBUTORS

John R. Adler, Jr.
Varian Medical Systems, Inc., Palo Alto, CA, USA; Department of Neurosurgery, Stanford University, Stanford, CA, USA

Nathalie Y. Agar
Department of Neurosurgery and Department of Radiology, Harvard Medical School, Brigham and Women's Hospital, Boston, MA, USA

Costas D. Arvanitis
Department of Radiology, Brigham and Women's Hospital, Harvard Medical School, Boston, MA, USA

Wael Asaad
Department of Neurosurgery, Alpert Medical School of Brown University and Department of Neurosurgery, Brigham and Women's Hospital, Harvard Medical School, Boston, MA, USA

Wenya Linda Bi
Department of Neurosurgery, Brigham and Women's Hospital, Harvard Medical School, Boston, MA, USA

Nicole M. Petrovich Brennan
Functional MRI Laboratory, Department of Radiology, Memorial Sloan Kettering Cancer Center, New York, NY, USA

David Calligaris
Department of Neurosurgery and Department of Radiology, Harvard Medical School, Brigham and Women's Hospital, Boston, MA, USA

Barbara Carl
Department of Neurosurgery, University Marburg, Marburg, Germany

Xiaolei Chen
Department of Neurosurgery, Chinese PLA General Hospital, Beijing, China

Tyler Cole
Stanford University School of Medicine, Stanford, CA, USA

G. Rees Cosgrove
Department of Neurosurgery, Alpert Medical School of Brown University and Department of Neurosurgery, Brigham and Women's Hospital, Harvard Medical School, Boston, MA, USA

Rose Du
Department of Neurosurgery, Brigham and Women's Hospital, Harvard Medical School, Boston, MA, USA

Ian F. Dunn
Department of Neurosurgery, Brigham and Women's Hospital, Harvard Medical School, Boston, MA, USA

Nabgha Farhat
Surgical Planning Lab, Department of Radiology, Brigham and Women's Hospital, Harvard Medical School, Boston, MA, USA

Robert L. Galloway, Jr.
Center for Technology-Guided Surgery, Vanderbilt University, Nashville, TN, USA

Yakov Gologorsky
Department of Neurosurgery, Mount Sinai School of Medicine, New York, NY, USA

Bradley A. Gross
Department of Neurosurgery, Brigham and Women's Hospital, Harvard Medical School, Boston, MA, USA

Robert E. Gross
Department of Neurosurgery, Emory University School of Medicine, Atlanta, GA, USA; Department of Neurology, Emory University School of Medicine, Atlanta, GA, USA; Interventional MRI Program, Emory University Hospital, Atlanta, GA, USA; Coulter Department of Biomedical Engineering, Georgia Institute of Technology, Atlanta, GA, USA

Nathan C. Himes
Department of Radiology, Brigham and Women's Hospital, Harvard Medical School, Boston, MA, USA

Andrei I. Holodny
Functional MRI Laboratory, Department of Radiology, Memorial Sloan Kettering Cancer Center, New York, NY, USA

Ferenc A. Jolesz
Department of Neurosurgery and Department of Radiology, Harvard Medical School, Brigham and Women's Hospital, Boston, MA, USA

Tina Kapur
Surgical Planning Lab, Department of Radiology, Brigham and Women's Hospital, Harvard Medical School, Boston, MA, USA

Rajiv Khajuria
Department of Neurosurgery, Brigham and Women's Hospital, Harvard Medical School, Boston, MA, USA

Ron Kikinis
Surgical Planning Lab, Department of Radiology, Brigham and Women's Hospital, Harvard Medical School, Boston, MA, USA

Sanju Lama
Department of Clinical Neurosciences and Hotchkiss Brain Institute, University of Calgary, Calgary, Alberta, Canada

Edward R. Laws, Jr.
Department of Neurosurgery, Brigham and Women's Hospital, Harvard Medical School, Boston, MA, USA

Fa-Ke Lu
Department of Neurosurgery and Department of Radiology, Harvard Medical School, Brigham and Women's Hospital, Boston, MA, USA

Yi Lu
Department of Neurosurgery, Brigham and Women's Hospital, Boston, MA, USA

Nathan McDannold
Department of Radiology, Brigham and Women's Hospital, Harvard Medical School, Boston, MA, USA

Ziev Moses
Department of Neurosurgery, Brigham and Women's Hospital, Boston, MA, USA

Christopher Nimsky
Department of Neurosurgery, University of Marburg, Marburg, Germany

Olutayo I. Olubiyi
Department of Neurosurgery and Department of Radiology, Harvard Medical School, Brigham and Women's Hospital, Boston, MA, USA

Kyung K. Peck
Functional MRI Laboratory, Department of Radiology, Memorial Sloan Kettering Cancer Center, New York, NY, USA

Sonia Pujol
Department of Radiology, Brigham and Women's Hospital, Harvard Medical School, Boston, MA, USA

Ashley Ralston
Section of Neurosurgery, The University of Chicago, Chicago, IL, USA

R. Mark Richardson
Epilepsy and Movement Disorders Surgery, University of Pittsburgh, Pittsburgh, PA, USA

David W. Roberts
Geisel School of Medicine at Dartmouth, Section of Neurosurgery, Dartmouth-Hitchcock Medical Center, Lebanon, NH, USA

Agam Sharda
Varian Medical Systems, Inc., Palo Alto, CA, USA

Nina Shevzov-Zebrun
Functional MRI Laboratory, Department of Radiology, Memorial Sloan Kettering Cancer Center, New York, NY, USA

Garnette R. Sutherland
Department of Clinical Neurosciences and Hotchkiss Brain Institute, University of Calgary, Calgary, Alberta, Canada

Jack K. Tung
Coulter Department of Biomedical Engineering, Georgia Institute of Technology, Atlanta, GA, USA

Viren Vasudeva
Department of Neurosurgery, Brigham and Women's Hospital, Boston, MA, USA

Peter C. Warnke
Section of Neurosurgery, The University of Chicago, Chicago, IL, USA

Jon T. Willie
Department of Neurology, Emory University School of Medicine, Atlanta, GA, USA; Department of Neurosurgery, Emory University School of Medicine, Atlanta, GA, USA; Interventional MRI Program, Emory University Hospital, Atlanta, GA, USA

Geoffrey Young
Department of Radiology, Brigham and Women's Hospital, Harvard Medical School, Boston, MA, USA

CHAPTER 1

Introduction and Historical Perspectives on Image-Guided Surgery

Robert L. Galloway, Jr.
Center for Technology-Guided Surgery, Vanderbilt University, Nashville, TN, USA

While a number of dry reviews exist of the rise of image-guided neurosurgery (IGN), mine included,[1–4] I find a discussion of why IGN came into being somewhat lacking. So, consider the brain as a surgical target. It is the only organ entirely encased in bone and it lacks redundancy. That is, the function of the tissue at one location is not replicated in other locations. Contrast this to a liver or a kidney. In addition, the value of any given section of brain is different from other similarly sized section of brain. Highly valued sections of brain tissue are referred to as "eloquent areas" and they can change from patient to patient depending on hand dominance, patient age, and existing disease. Thus, in considering the rise of IGN, one has to consider both the surgical target—tumor, vascular anomaly, or seizure foci—and the *path* to that target. A lack of guidance can lead to significant collateral damage of previously healthy functional tissue on the way to the surgical target. Rough guidelines embodied in the idea of standard approaches were an attempt to bring collective wisdom into the selection of surgical path. Implicit in these approaches is the thinking "bad things happen less often if you go this way" or "there is less damage to the average person's anatomy if you use this approach."

It was that generalization of anatomy which inspired Sir Victor Horsley and Robert Clarke to develop the first stereotactic frame,[5] but also ultimately limited its value. Horsley and Clarke were interested in mapping brain function and (rightly) believed that function which lay in one area of the brain of a subject, such as dogs, cats and/or monkeys, would be reflected in similar locations in other subjects of the same species. They developed a gridding system in the form of a glass plate which they would lay over a sliced section of a freshly sacrificed animal of the same species and use those locations to provide coordinates in the living animal. An "electrolytic" needle was used to make small lesions. The estimated lesion location and the observed response of the subject were recorded. In addition, Clarke had developed a microtome for slicing excised brains and Horsley produced photographic albums of 1 to 2 mm thick cut sections in transverse, sagittal and coronal planes, presaging modern tomography by 70 years. In addition, they gave rise to the concept of a device existing external to the cranium being used to guide tools into the interior of the cranium.

A. Golby (Ed): Image-Guided Neurosurgery
DOI: http://dx.doi.org/10.1016/B978-0-12-800870-6.00001-7

The ultimate limitation of the work by Horsley and Clarke lies in the presumption that all subjects of similar size and same species will have corresponding internal and external anatomy. While Horsley and Clarke were able to produce lesions and occasionally get similar effects in their work, the variability of their results showed that the exterior of a subject did not correctly predict the interior of the subject. Their careers and standing in medical and scientific circles was such that, even given the variability of their animal results, the use of their frame in humans was proposed.[6]

With the exterior of the subject failing to predict the inside, stereotaxy had to wait for developments that allowed patient-specific information to be obtained. The first candidate was X-ray, but simple radiographs of the head resulted in skull images with no perceivable soft-tissue contrast. Dandy's work on ventriculography[7] (later called pneumoencephalography) allowed the delineation of the ventricles and therefore intuition of the location of space-occupying objects such as tumors or subdural hematomas. Moniz[8] and others developed cerebral angiography in the 1930s and one of the other major structures, the cerebral vasculature and malformations thereof, could then be seen in radiographs.

STEREOTACTIC FRAMES

By the 1940s radiography had progressed to the point that most of the subcerebral structures could be approximated if not seen. That, and injuries resulting from World War II, inspired Spiegel and Wycis[9] to return to the concepts of stereotaxy. They developed a frame, affixed to the patient's head with plaster and imaged using pneumoencephalography. However, for the most part, their targets were not cancer or vascular volumes but electrophysiological in nature:

> *This apparatus is being used for psychosurgery. In a series of patients studied in collaboration with H. Freed, lesions have been placed in the region of the medial nucleus of the thalamus (medial thalamotomy) in order to reduce the emotional reactivity by a procedure much less drastic than frontal lobotomy. The results so far obtained are promising. Further applications of the stereotaxic technic are under study, e.g. interruption of the spinothalamic tract in certain types of pain or phantom limb; production of pallidal lesions in involuntary movements; electrocoagulation of the Gasserian ganglion in trigeminal neuralgia; and withdrawal of fluid from pathological cavities, cystic tumors.*

So their image-driven targeting did not have to be particularly fine. Their intent was to lesion problematic neural sites and they could use the placed electrode to refine their position.

Whether inspired by Spiegel and Wycis or driven by the same problems and opportunities, the late 1940s and early 1950s saw an explosion of stereotactic systems and techniques. These include systems by Leksell,[10] Talairach,[11] Reichert,[12] and others. Most of these techniques were targeted toward electrophysiological

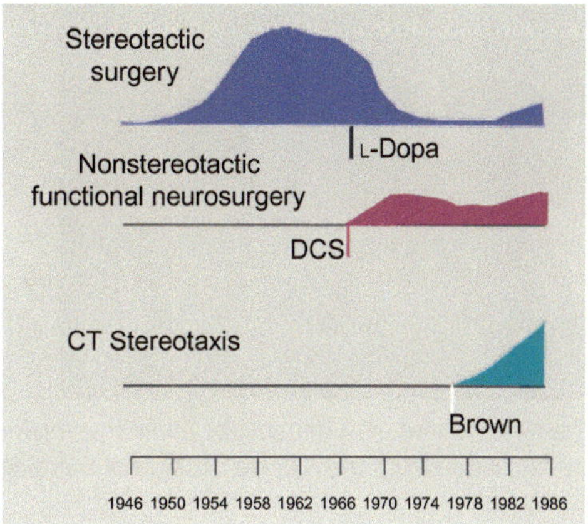

Figure 1.1 The rise and fall of stereotactic surgery cases. *Adapted from Gildenberg.*[13]

procedures using electrophysiological measures for refinement of position. The rise (and subsequent fall) of stereotactic procedures was elegantly documented by Gildenberg[13] (see Figure 1.1).

Since the majority of stereotactic cases were ultimately electrophysiological in nature, the development of antiseizure medications such as L-Dopa allowed diseases such as Parkinson's disease to be treated without surgery—thus, the rapid decline in stereotactic cases. But the development and commercialization of a new form of imaging changed stereotaxy.

ADVENT OF VOLUMETRIC TOMOGRAPHY

While it is difficult to define who exactly "invented" CT scanning, there is little question that the first commercially available system was invented in large part by Godfrey Hounsfield and developed by EMI (Electric and Music Industries) in 1972. With that development the imaging uncertainty of X-ray based imaging went from being the thickness of the patient's head to millimeters in plane and a centimeter across planes. Contrast agents still had value but it was the improvement in the third dimension of visualization that was critical.

The availability of three-dimensional data was not lost on the stereotactic community but there was a lack of clarity as to how to make use of it. This was addressed when Russell Brown[14] developed the simple but elegant N-localizer system. This system is shown in Figure 1.2.

Because the X and Y location of any target could be visualized and the height above the base ring calculated, each image held the target coordinates. By having two lateral and one AP sets of N-bars, the orientation of the image plane could also be determined from any tomographic image in which the N-bars were visible.

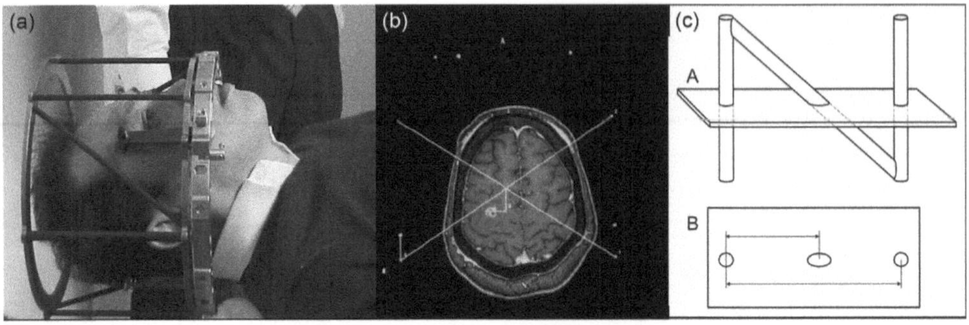

Figure 1.2 (a) Stereotactic frame mounted on a patient. (b) An image showing the N-bars on an MRI. (c) Height of the crossing can be calculated from the relative position of the diagonal bar to the end bars.

It did not take the innovators in stereotactic surgery long to embrace the new technology. Leksell[15] and Mundinger[16] both published papers on using the new technology. Gildenberg[17] even developed a technique for creating quasi-AP and Lateral images from CT so that surgeons with old calculation methods could use the images. But it was two surgeons with access to greater computational intensity who began to see the real value. By using the N-bars they could confirm the orientation and spacing of the CT slices. That allowed them to trust the volumetric nature of the scan and to use it not merely for neuroelectrical interventions but for volumetric resections. This work was led by Shelden[18,19] and Kelly.[20,21]

The development and commercialization of magnetic resonance imagers made MRI available to neurosurgeons in the early 1980s. MRI was of great interest to neurosurgeons due to its higher soft tissue contrast. In particular it allowed clear demarcation of the gray/white matter junction and improved visualization of lesion margins at the cost of lower resolution and the potential for geometric distortion.[22] Again, the presence of the N-bars allowed quantization of any distortion.

It was in the 1985–1990 time frame that several factors came together to facilitate the move from stereotaxy to image guidance. The presence of CT and MRI reduced the spatial uncertainty in the task of intracranial surgery. The volumetric nature of CT and MRI had led to an expansion of stereotaxy from almost solely electrophysiological surgery to volumetric resections of tumors[23] and vascular structures.[24] The IBM AT personal computer was released in 1984 and provided the disk storage and addressable memory space necessary to manage medical images. The open architecture of the PC encouraged the development of plug-in boards, which allowed devices such as articulated arms to interact with the device. In an often-overlooked paper, Columbo et al.[25] demonstrated that if one trusted the tomogram's orthography then one only needed three reference points to locate any target point in the volume. The years 1986 and 1987 brought new algorithms[26–28] for the closed form determination of the rotation between homologous points in two three-dimensional representations. These methodologies reduced the time uncertainties in finding a least-square solution to the rotation aspect of the transformation necessary to map physical space into image space.

IMAGE GUIDANCE

Stereotaxy was the process of finding a point in images or, with tomography and the N-bars, an image; and making physical adjustments to an external mechanism to guide the surgeon to that point. With all of the precursors coming together, laboratories across the world made a very similar leap. Could the stereotactic process be reversed? Is it possible to track a mechanical device in a three-dimensional space and show its location in image space? While four groups, two in the United States, one in Japan, and one in Germany, all independently made that leap, the Dartmouth group was the first to press in 1986 with an ultrasonically tracked surgical microscope[29] (see Figure 1.3).

The "outriggers" visible on the microscope held spark-gap sonic sources. They would fire in sequence and be received by detectors rigidly fixed in the room. By measuring the time between the spark and the reception the distance between spark and detectors could be resolved by time of flight measurements. Since the microscope was rigid, if three or more spark gaps could be detected the microscope could be localized and tracked.

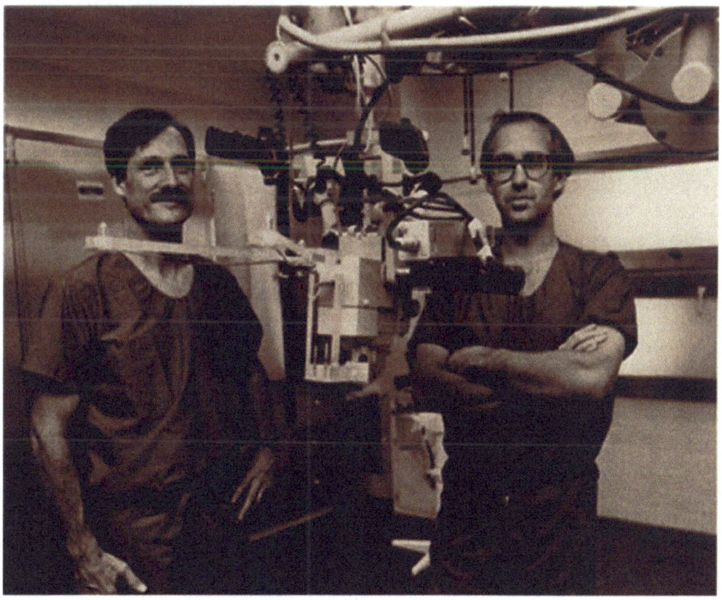

Figure 1.3 John Strobehn (left) and David Roberts (right) with the Dartmouth ultrasonically tracked microscope. *Figure courtesy of Dr. David Roberts.*

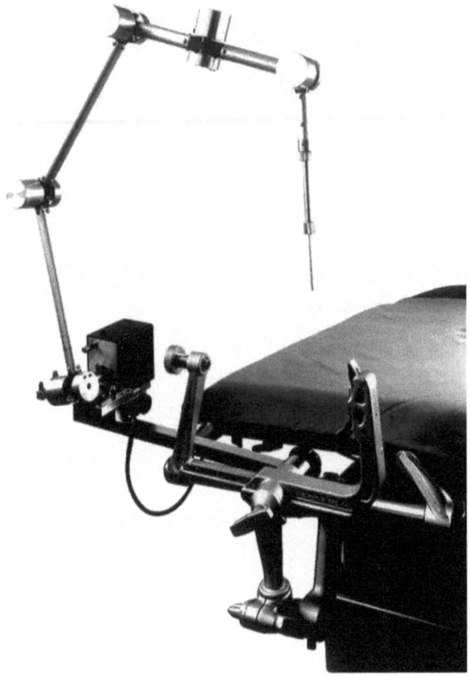

Figure 1.4 The Tokyo Neuronavigator. *Figure courtesy of Dr. Eiju Watanabe.*

In 1987, the second published group, from Tokyo Police Hospital, adapted an industrial articulated arm to create a "Neuronavigator"[30] (see Figure 1.4).

One type of articulated arm is a "revolute" arm, in that the link lengths are fixed and the angles sensed. From there the location of the tip could be calculated in its frame of reference. The Neuronavigator was not purely revolute. The last link length could be extended to one of several standard lengths and held in that position by a ball-indent.

Our group at Vanderbilt University had the next publications.[31,32] The Vanderbilt Mark I was a revolute articulated arm, this one custom-designed for the surgical task (Figure 1.5).

Virtually coincident with the Vanderbilt system, a group at the University of Aachen showed an articulated arm system.[33,34] The arm was purely revolute but demonstrated an innovative suspension system to remove some of the weight carried by the surgeon using the arm. The original arm is shown in Figure 1.6.

In addition to having revolute arms, both the Vanderbilt and Aachen groups showed the 2 over 2 display, which has become the standard in image guidance. This is shown in Figure 1.7.

It should also be noted that these systems emerged from surgeon/engineer teams: Roberts/Strobehn, Watanabe/Kosugi, Maciunas/Galloway and Schlöndroff,

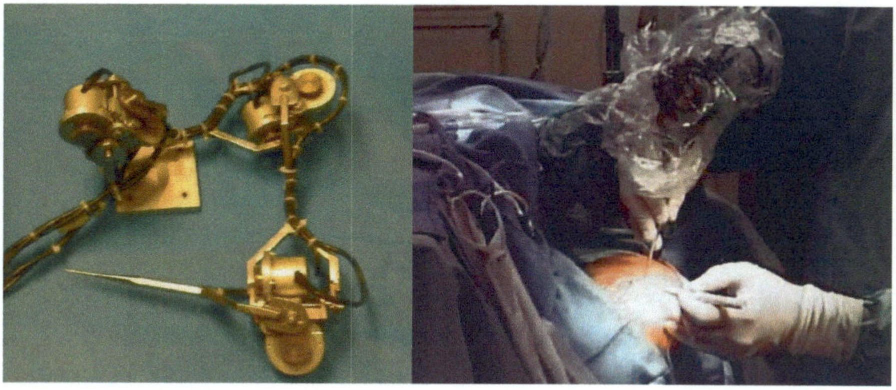

Figure 1.5 The Vanderbilt Mark I articulated arm.

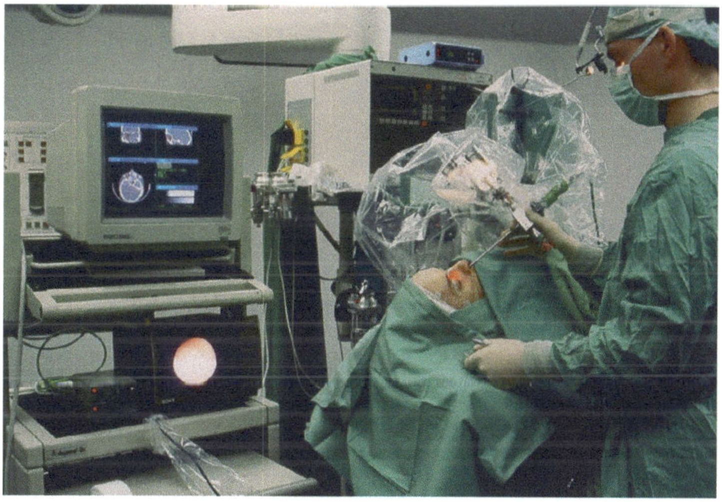

Figure 1.6 The University of Aachen surgical guidance arm. *Image courtesy of Dr. Ralph Mösges.*

and Mösges/Meyer-Ebrecht. Beyond those four, there were groups across the world developing guidance systems. These include a magnetic localizer technique,[35] an optical system,[36] and a sonic localizer.[37]

It did not take long for commercial systems to emerge. The first was from a Canadian medical image processing firm, ISG Technologies, Inc., which used the Faro commercial articulated arm to develop a product called the Viewing Wand.[38,39] Following closely on the heels of the ISG system, Smith et al.[40] came out with a system that was the progenitor of the present Stealth Station.

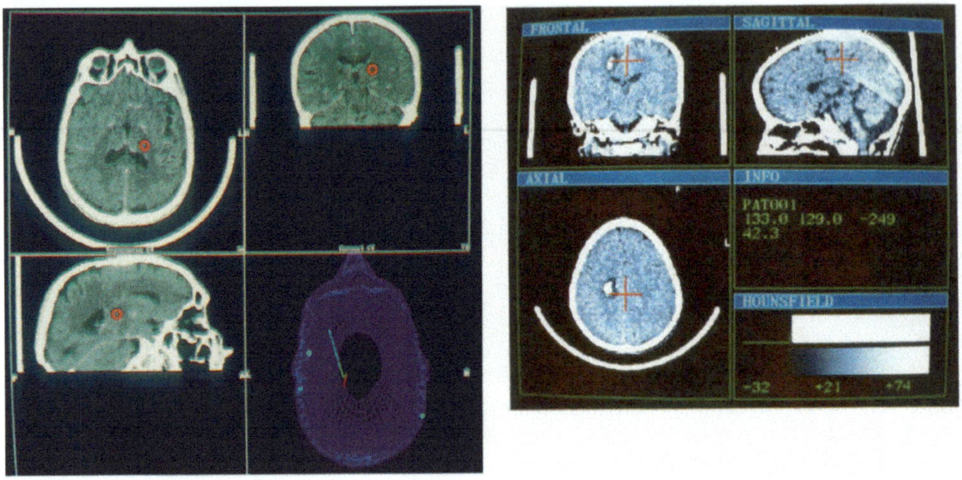

Figure 1.7 (Left) The Vanderbilt 2 × 2 display, (right) Aachen 2 × 2 display. *Aachen display image courtesy of Dr. Ralph Mösges.*

With the acceptance of image guidance as a neurosurgical tool, there was a rapid movement to virtually all forms of intracranial interventions. These include epilepsy surgery,[41] vascular abnormalities,[42] pituitary surgeries,[43] and ventricular surgeries.[44] Concomitant with the spread to multiple surgical procedures there was a generalized move away from articulated arms. Suspension such as that pioneered in Aachen and internal counterbalances and springs could greatly reduce the weight the surgeon had to support, but they could not reduce the mass of the arm. It had inertia, requiring a push to get it moving in a direction desired by the surgeon and an active braking action to stop it at the desired target. This led to triangulation systems being examined as physical space localizers.

OPTICAL LOCALIZATION

There were two primary triangulation methodologies: sonic and optical. The Dartmouth sonic system was the pioneer in image-guided surgery[37] and one of the early commercial systems, Picker's ViStar system,[45] was also sonic. But sonic systems faced a challenge in that they were predominantly time-of-flight devices. While the time between transmission and reception could be measured accurately and precisely, the speed of sound in the operating room was not a constant. This led to inaccuracies in converting time to distance. In addition, the speed of sound meant that the localization of each source took tens of milliseconds and between localizations there had to be a delay to allow confounding reverberations to die out. Even with the slow motions dictated by the care taken in neurosurgery, it was difficult to get the excess of

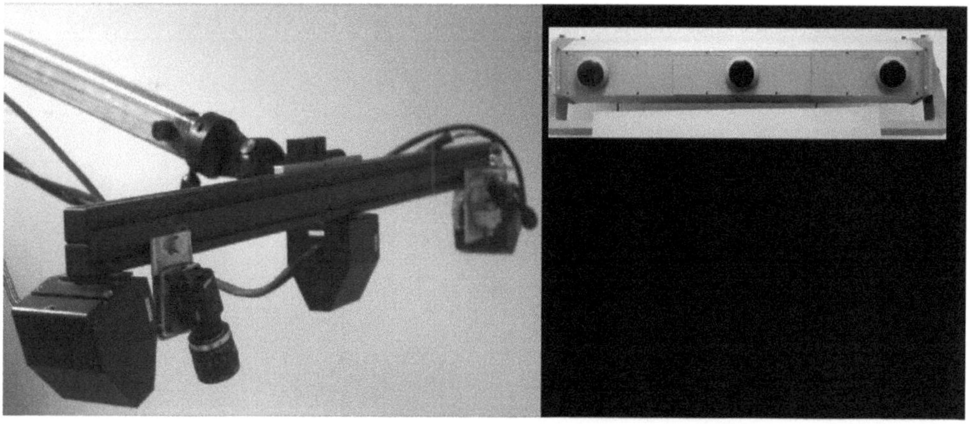

Figure 1.8 (Left) The Pixsys Flashpoint with an attached video camera, (right) The Optotrak 3020.

localizations which allows attenuation of bipolar noise. A comparison of optical versus sonic localizers for surgery was provided by Bucholz et al.[46]

There were three major optical triangulation localizers. Two used commercially available optical tracking devices and the third used distortion-corrected video imaging. That third system, dubbed VISLAN,[47] used structured light and two video cameras. The structured light allowed the establishment of correspondence between structures seen in the two cameras. The other two approaches used flashing infrared LED (IRED) sources and three linear sensors. Since only one source was illuminated at any given moment, there was no difficulty in establishing correspondence between the three sensors. The two optical systems were from Northern Digital Inc. and Pixsys (Figure 1.8). The Northern Digital system, dubbed the Optotrak 3020, was highly accurate, fast, and could localize a large number of sources in a data frame. However, its cost and size limited its acceptance in operating rooms.[36,17] The other commercially available optical tracking system was the Flashpoint from Pixsys Inc. Several research groups incorporated this localizer into their work.[48,49] The original StealthStation also used a Flashpoint as its localizer.[50]

Both systems used IREDs as their transmitters. Examples of the probes are shown in Figure 1.9. The faster read-out of optical location by the Optotrak allowed the placement of an excess number of IREDs on each tool. Please see Figures 1.9b and 1.9c. Since only three are required for device tracking, the "excess" provides three advantages. First, the use of more than three IREDs allows for the mathematical overdetermination of tool location and the improved robustness available in a least-square error solution. Second, by having IREDs wrapping completely around the tool, that tool may be tracked from almost any angle. Finally, a surplus of sources means that the localization is not dependent on any given IRED. Thus, if an IRED is obscured by blood or anything else in the surgical field, a localization can still be performed.

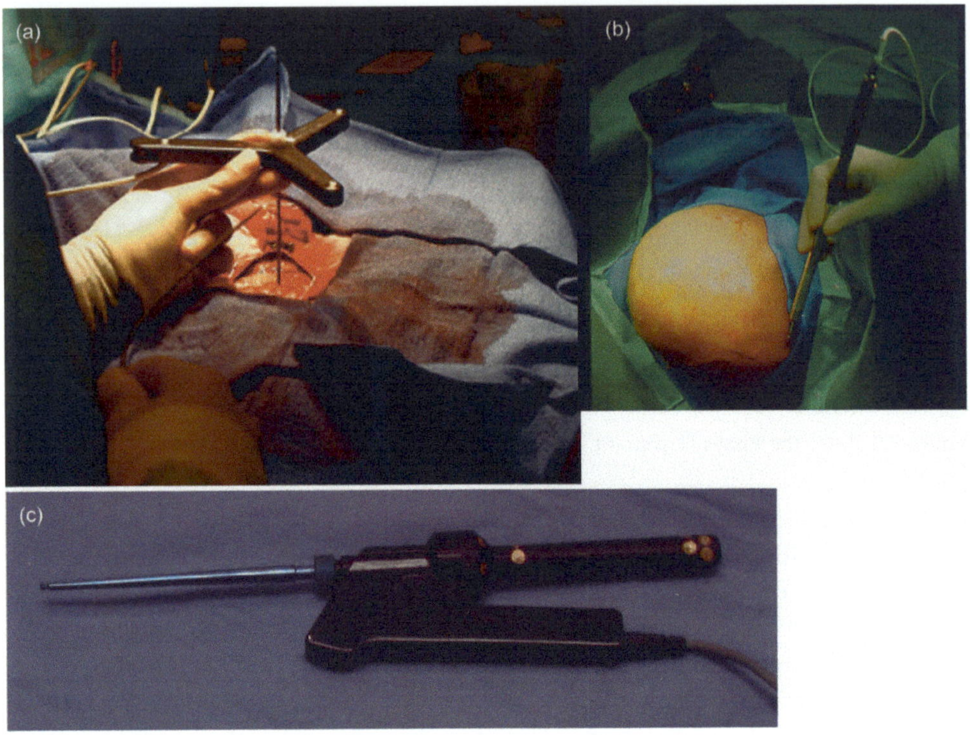

Figure 1.9 (a) A Flashpoint tool used to guide a biopsy needle; (b) an Optotrak tool being used for fiducial registration; and (c) a commercial Optotrak tool.

REGISTRATION

Stereotactic frames had one significant advantage over image-guided systems. Because the frames could be seen in the images, then the marriage of frame space to image space was very straightforward. With the rise of free-hand localization, determining the mathematical relationship between image space and physical space, a process known as **registration** became critical. If the relationship could be described with one translation vector and one rotation matrix, the process was known as a **rigid registration**.

It was noted earlier that a critical step in the development of IGN was the development of closed-form solutions to create a transformation matrix for rigid registration. While the two primary approaches, Horn[27] and Arun,[28] used different mathematical constructs, they both require lists of homologous corresponding points in both spaces. Such homologous points are called fiducials, a map-making term for trusted point. These points can be anatomic landmarks (also known as an intrinsic fiducial)[51,52] or external objects (extrinsic fiducial markers).[32,53] It should be noted that the mathematical point is the fiducial whether derived from anatomy or computed from the geometry of a marker.

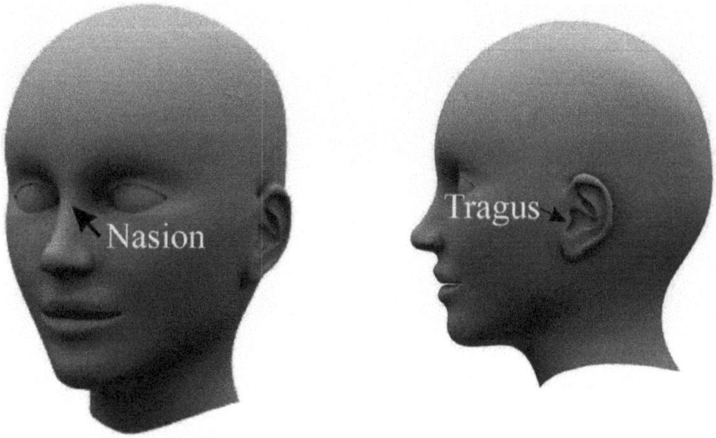

Figure 1.10 The most common intrinsic points used as fiducials in image-guided neurosurgery were the point of maximum inflection on the nose (the nasion) and the small eminences in the ear (the tragi).

The most commonly used intrinsic points for image-space to physical space neurosurgery were the nasion and the tragi (see Figure 1.10).

The primary difficulty of using intrinsic fiducials is that the body rarely comes to a point. Even calculable points like the maximum inflection of the nose might be distinct localizations such as a physical space tracker and an image. While the appeal of a *retrospective* technique, that is one that allows point location without intervention before a scan, is undeniable, it quickly became obvious that a *prospective approach*, one in which imageable objects are placed on the patient before the scan, provided much greater accuracy.

The first extrinsic fiducials were often objects of convenience. This included radio-absorptive beads[53] or surgical staples[32] for use in CT. Vitamin E capsules were used as extrinsic fiducials in MRI[32,54] but really only provided a good T1 signal and there were issues with chemical shift.

Removable fiducial systems that bound to the upper teeth, for example,[55] appeared in a number of different incarnations but failed to gain wide usage.

The three most commonly used fiducial systems were commercial. The first was developed by Zinreich and colleagues.[56] It consisted of a spongy material impregnated with CT and MR contrast materials and was marketed by IZI. A peel-off adhesive back allowed it to be attached to the skin. Examples of these markers are shown in Figure 1.11.

The other two fiducial systems were both designed to be implanted into the outer table of the skull. Their bone implantation gave them much less potential motion with skin motion relative to the skin surface-mounted markers, but it did so at the cost of increased invasiveness and surgeon time. The two organizations were: Johnson and

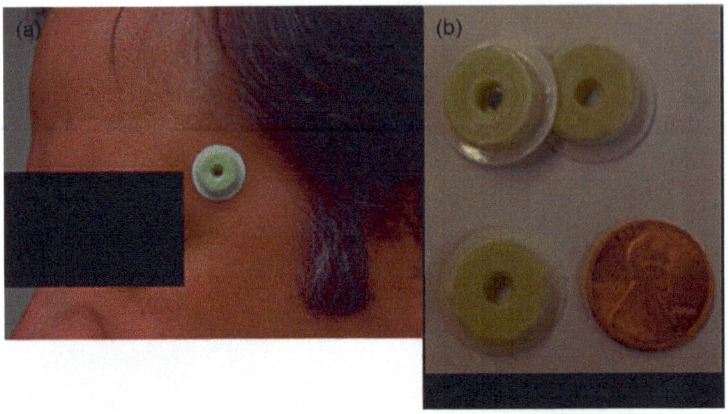

Figure 1.11 Example of IZI skin-mounted marker on a patient (a) and loose (b).

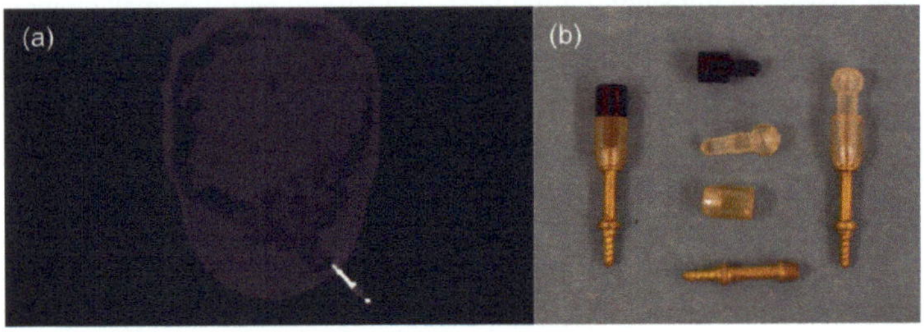

Figure 1.12 The Leibinger skull-mounted fiducial. Panel (a) shows a CT of a phantom with an implanted marker. Panel (b) shows the localization cap and the reservoir for imaging contrast agents. The J&J markers consisted of a surgical plastic post with an imaging cap containing a fluid visible in both MRI and CT images and a localization cap which accepted a ball tip probe. The J&J marker system is shown in Figure 1.13a with a drawing of the localizing process shown in Figure 1.13b.

Johnson (J&J), specifically Codman and Shurtleff working with researchers at Vanderbilt University[57]; and Howmedica Leibinger. The exact timeline of invention and embodiment of the markers is a bit murky and has been the subject of legal actions but the markers' designs in the late 1990s were strikingly different. The Leibinger markers were based on a tall, thin metal screw with separate tops to be placed on the mounting screw for MR, CT, and physical location (please see Figures 1.12 and 1.13).

The J&J markers were designed to appear in multiple pixels and multiple slices. This allowed the determination of the markers center, the fiducial, to be made using multiple pixel locations. Such a technique allowed for localization of the fiducial to a finer resolution than the images themselves.

Several authors including Ammarati et al.[58] performed comparisons of skin-mounted to bone-implanted fiducials. While the bone-implanted fiducials provided not-unexpected better accuracy, the value of that improvement was overwhelmed by the ease of use of the stick-on markers.

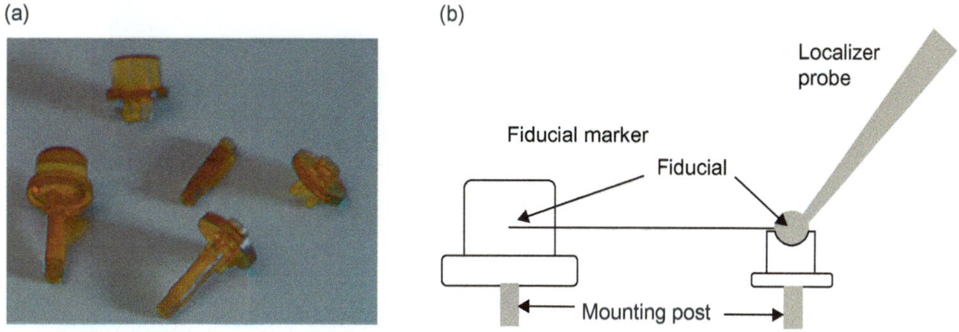

Figure 1.13 The Johnson and Johnson fiducial markers. (a) The skull-mounted posts, the caps filled with CT and MRI contrast agents as well as the physical localization cap; (b) How the physical location cap holds the center of the ball tip at the equivalent point as the middle of the imaging marker.

Throughout the development of image-guided surgery there was a continuing desire to use intrinsic properties of the patient to perform registrations. Not only would this reduce cost, it allowed the registration to be retrospective, that is not requiring an additional scan after fiducial placement. However, as mentioned above, trying to find reliable, accurate intrinsic fiducials is difficult. So, there were attempts at using fiducials combined with other physical space attributes such as surfaces to perform registrations. An example of this is the work by Maurer et al.[59]

A large multicenter trial looking at *image to image* registration[60] provided several insights into the general registration problem. Surface-based registrations are sensitive to rotational symmetries, that is, a rounded surface will fit equally well to the top as on the side of a bowling ball. In addition, a descriptive set of mathematics looking at point based registration[61–65] emerged, allowing for the quantitative assessment of the quality of a point-based registration. No such mathematical development has been demonstrated for other registration methodologies.

BEYOND POINTS

One pioneering surface-based approach was the VISLAN from Guy's Hospital.[47] As described previously, it used distortion-corrected stereo video cameras for tracking. For registration it could provide a surface description by projecting a structured light source onto the surface of the patient's head. The stereo cameras used the nature of the structured light to solve the correspondence problem to localize a cloud of surface points.

A second noncontact surface acquisition technique for surface-based registration was the Medtronic FAZER. This device used a tracked laser triangulation system to localize points in the tracker space. The handheld device broadcast a laser spot and detected the location of that point with a sensor mounted in the handheld unit. The

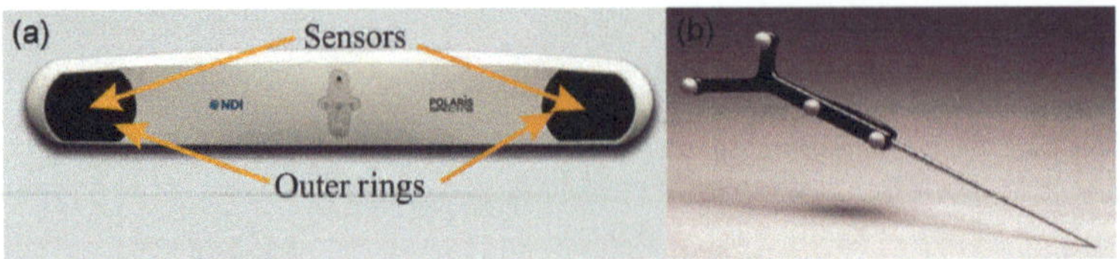

Figure 1.14 The Polaris localization system (a) and tracked probe (b).

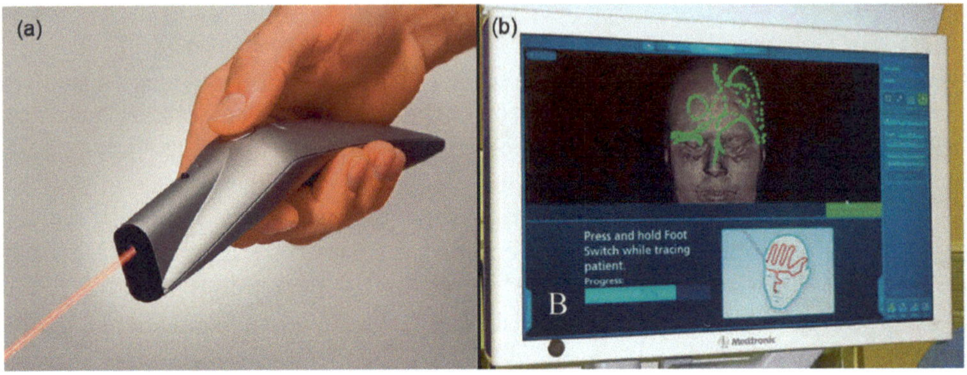

Figure 1.15 The Brainlab z-touch. (a) A hand-held system used to create surface maps; (b) shows the result of the Medtronic Fazer. Both systems produce sparse surfaces.

laser was swept as a point-by-point acquisition over the surface of the head to acquire a point cloud for registration.

Central to understanding the performance of the next surface registration is knowledge of the most often used localizer of the time, the Northern Digital Polaris. The Polaris, shown in Figure 1.14, transmits a flash of IR light from the outer rings seen to the left and right of Figure 1.14a. The flash is reflected by the retroreflective balls seen in Figure 1.14b and is detected by the sensors on the transmission unit.

The Brainlab z-touch® is a hand-held laser system that mixes visible and IR light. The IR beam is selected to be the same frequency of light as the Polaris's broadcast frequency. By suppressing the Polaris's IR flash, the system will detect the reflection of the IR source from the hand-held unit. This allows the surgeon to sweep the beam over the patient's head and acquire a cloud of surface points for registration already in frame of the Polaris device.[66] This device is shown in Figure 1.15a.

However, while these techniques are useful especially if the surgeon is careful to acquire points that demonstrate high curvature change—such as both sides of the nose and the orbital rims—they are sparse with uneven coverage (please see Figure 1.15b).

Another approach was to use a laser range scanner (LRS) to obtain the surface (see Figure 1.16). Early examples of this are can be found in refs[66,67]. The LRS provided dense, regularly spaced points of the surface in a few seconds. The problem with this

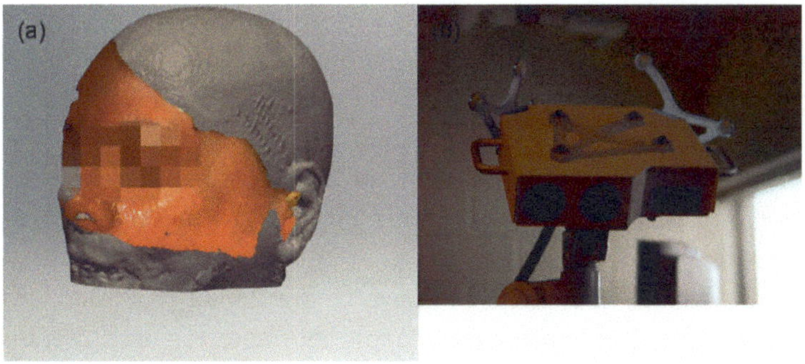

Figure 1.16 Face surface (a) acquired with LRS for registration; (b) LRS Unit from Pathfinder Therapeutics Inc. *Figure 1.16b is courtesy of Dr. Amber Simpson.*

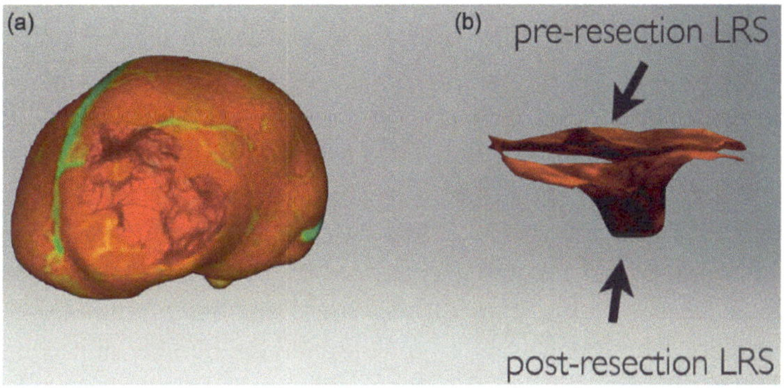

Figure 1.17 An LRS-generated cortical surface (a) matched to a preoperative contrast-enhanced MRI image volume; panel (b) shows LRS images both before and after a resection. *Figure courtesy of Dr. Michael Miga.*

was that the laser points alone provided no information from which they arose; was this head, neck, or OR table? By adding a co-registered video image it was possible to texture map the surface and easily trim the LRS cloud to be only the desired surface.

With the development of a more precise LRS, Miga and colleagues[66,68] (see Figure 1.17) have refined the registration process to encompass the cortical surface and now the bottom and sides of the postsurgical resection cavity.[69] With that, the LRS moves beyond simply a registration methodology and more into an intraoperative guidance role.

INTRAOPERATIVE IMAGING

Imaging during surgery is common although, until relatively recently, it was mainly two-dimensional. Endoscopy, fluoroscopy, and ultrasound all are common in general surgery but have a smaller role in intracranial neurosurgery.[70] Manwaring et al.[71]

pioneered both image-guided endoscopic neurosurgery as well as magnetic tracking. Because endoscopes require a transparent medium in which to work, most neurosurgical applications focused on ventricular and cystic procedures.[72]

Ultrasound has great appeal in that it often allows direct real-time visualization of a tumor, ventricle, vascular malformation, or hemorrhage.[73] However, it is a low signal-to-noise slice imaging technique with an inconsistent slice thickness. Early on in the development of IGN, it was paired with MRI to retain the advantages of real-time imaging but to minimize the issues of low signal to noise.[74] As ultrasound transducers got smaller they became more applicable both for structural imaging[75] and registration data.[76]

In the early 2000s intra or perioperative tomography arose. This included intraoperative MRI (both low field and high field) and interoperative CT. Some research teams used standard MRI scanners with tracks or mobile OR tables (for example, Hall et al.[77]). Another approach was to place the patient into a specially adapted MR unit, which would allow the surgical gesture and imaging to be performed without moving either the patient or the magnet. This approach led to the GE "double donut" in which the patient was placed between two 3 Tesla (3T) MR sources. Between the sources there was a sufficient gap to allow the surgeon to work. Imaging of the surgical volume was performed in a 0.5T fringe field established by the two 3T sources.[78] While clinically appealing, the large magnets required considerable investment in the system, in OR redesign, in staff safety training, and in the surgical equipment and tools brought into the OR. In addition, the average surgical time limited the number of patients who could be treated in such a room, requiring that the costs be amortized over a smaller base.

Another method developed low field systems using permanent magnets in the 0.12–0.2 Tesla range. These were designed to mount below the OR table and swing up on demand during the surgery to provide updated imaging. Initially marketed in the United States by an Israeli company (Odin),[79] the low field meant that images had low signal to noise. However, the system was only marginally disruptive to the surgical process and required fewer changes in conventional equipment and tools.

A more complete discussion of intraoperative MRI can be found in[80] and in Chapter 7 of this volume.

MRI, with its ability to distinguish gray matter from white matter, is the diagnostic imaging of choice in the brain. However, CT can distinguish the surface of the brain, the ventricles, tumors and, with the use of contrast, can image blood vessels. It does so at four times the resolution of most MRIs and images can be made much more quickly.[81] While CT intrudes less on the surgical process than MRI, there are radiation dose considerations to the patient and OR personnel.

There are two major reasons for intraoperative tomography during an image-guided neurosurgical procedure. The first is to test the completeness of any resection during the process, presumably to increase the percentages of gross total resections.

The second reason is unique to image-guided surgery. In the presence of a space-occupying lesion in a rigid skull, intracranial pressure (ICP) rises. For the past 35 years[82] a high molecular weight alcohol, mannitol, has been used to move water from the brain and into the bloodstream. When coupled with a diuretic such as Lasix, water is removed from both the brain and the surrounding vasculature. This temporarily shrinks the brain, reducing the ICP and reducing the chance of the brain herniating out of the skull when it is opened. While this process is common to all intracranial interventions, it has an additional implication in IGN. For the majority of intracranial surgeries the patient is rotated so that the skull flap is the highest point relative to gravity. Thus, when the flap is opened, the space created by the reduction in fluid volume occurs at the surface of the brain. This is a material sag, often called a "brain shift," and was first quantified by Kelly et al.[83] Moreover, the surgical gesture itself including resection of the lesion, egress of CSF, and swelling can cause further brain displacements with complex and unpredictable directions of shift. This phenomenon is especially deleterious in IGN because it means that the brain that underwent preoperative imaging changed shape after imaging—thus the interest in obtaining new tomograms during surgery. However, the cost and complexity of intraoperative tomography has given rise to another method of addressing perioperative brain changes.

DEFORMATION CORRECTION

Intraoperative tomography can provide three-dimensional information about the present state of the brain during surgery (Figure 1.18). However, even before the current concerns of the high cost of medical care, ownership of such a system was beyond the reach of most hospitals. On the other hand, if the deformation can be measured[84,85] biomechanical models can be created,[86,87] which estimate the deformation that the brain has undergone. While such a technique cannot make the error go to zero, it can dramatically reduce targeting error[88] and it has the great advantage that it can be distributed for reasonable costs.

WORK UNDERWAY

In 1993 at a conference of the American Association of Neurologic Surgeons, an early image-guided system was being demonstrated during a workshop. A neurosurgeon watching the demonstration said, "That's very nice but why would I need that? I'm never lost." Since that time, IGN has become the standard of care in intracranial neurosurgery. However, as evidenced by this book, there is still work to be done. Clear studies that document the value of IGN must be done, especially in light of the developing evidence-based focus of reimbursement.

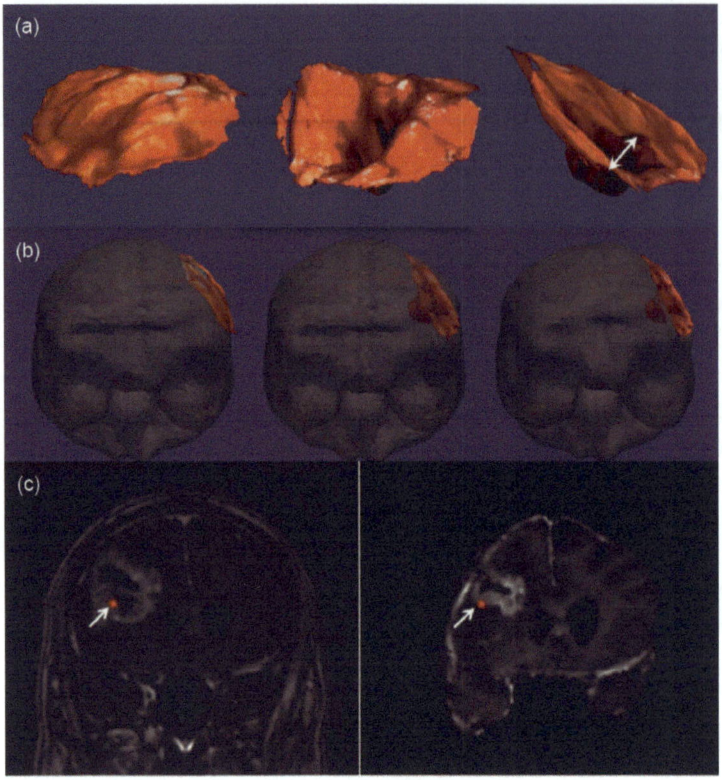

Figure 1.18 (a) *Left, middle,* pre-post laser range scanned cortical surfaces with shift shown *right,* (b) *Left, middle* preop model with registered cortical surface before and after shift respectively, with corrected model shown *right,* and (c) the red dot is location of swabbed point on bed of resected tumor cavity surface with *left* uncorrected, and *right* corrected. In this case, an alternative visualization was used whereby rather than showing a resection hole in *right* corrected image, tumor was interpolated to fit predicted resection cavity. *Figure courtesy of Dr. Michael Miga.*

Careful delineation of tumor margins must be defined and, if possible, linked back to preoperative imaging. This may be accomplished through a large clinical trial with multiple stereotactically localized biopsies, a new intraoperative imaging modality, or via a detectable gold standard biomarker.

Deformation modeling has been shown to be helpful and viable within the surgical timeframe. However, as of this writing there are no commercial IGN systems that incorporate any model-correction algorithms.

The increasing functionality of magnetic tracking will allow nonlinear approaches to structures of surgical interest near eloquent areas, but such methodologies still have to be proven to be as accurate and robust as the state-of-the-art optical systems.

Robotic systems can provide extraordinary precision; however, their accuracy must still be demonstrated in the case of a specific patient with unique brain structure.

Lastly, I began by describing the brain in terms of its spatial distribution of critical function and lack of redundancy. During the almost 30 years of IGN, we continue to guide based on anatomic decisions with only a minor knowledge of the function of the tissue being resected, disturbed, or displaced. True guidance will not be image-guided but rather information-guided.

DEDICATION

This chapter is dedicated to the memory of Robert Joseph Maciunas, MD: gentleman, scholar, neurosurgical innovator.

REFERENCES

1. Grunert P, Darabi K, Espinosa J, Filippi R. Computer-aided navigation in neurosurgery. *Neurosurg Rev.* 2003;26(2):73–99.
2. Galloway RL. The process and development of image-guided procedures. *Annu Rev Biomed Eng.* 2001;3:83–108.
3. Galloway RL, Peters TM. Overview and history of image-guided interventions. In: Peters TM, Cleary K, eds. *Image-Guided Interventions.* New York, NY: Springer; 2008:1–21.
4. Venbrux AC, Brozyna JM, Chandra S, Chen HK, Tran GD, Gagarin DA. A brief history of image-guided therapy. In: Mauro MA, Murphy KPJ, Thompson KR, Venbrux AC, Morgan RA, eds. *Image-Guided Interventions.* Philadelphia, PA: Elsevier Saunders; 2008:1–10.
5. Horsley V, Clarke RH. The structure and functions of the cerebellum examined by a new method. *Brain.* 1908;31:45–124.
6. Gildenberg PL. Stereotactic surgery: present and past. In: Heilbrun MP, ed. *Stereotactic Neurosurgery.* Baltimore, MD: Lippincott Williams & Wilkins; 1988.
7. Dandy WE. Ventriculography following the injection of air into the cerebral ventricles. *Ann Surg.* 1918;68(1):5–11.
8. Sassard R, O'Leary JP. Egas Moniz: pioneer of cerebral angiography. *Am Surg.* 1998;64(11): 1116–1117.
9. Spiegel EA, Wycis HT, Marks M, Lee A. Stereotactic apparatus for operations on the human brain. *Science.* 1947;106:349–350.
10. Leksell L. Stereotaxic apparatus for intracerebral surgery. *Acta Chir Scand.* 1949;99:229–233.
11. Talairach J, Hecaen M, David M, Monnier M, Ajuriaguerra J. Recherches sur la coagulation therapeutique des structures sous-corticales chez l'homme. *Rev Neurol.* 1949;81:4–24.
12. Riechert T, Wolff M. Ueber ein neues Zielgeraet zur intrakraniellen elektrischen Abteilung und Ausschaltung. *Arch Pyschiatr Z Neurol.* 1951;186:225–230.
13. Gildenberg PL. Whatever happened to stereotactic surgery? *Neurosurgery.* 1987;20(6):983–987.
14. Brown RA. A computerized tomography-computer graphics approach to stereotaxic localization. *J Neurosurg.* 1979;50(6):715–720.
15. Leksell L, Jernberg B. Stereotaxis and tomography. A technical note. *Acta Neurochir.* 1980;52:1–7.
16. Mundinger F. CT-stereotactic biopsy of brain tumors. In: Voth D, Gutjahr P, Langmaid C, eds. *Tumors of the Central Nervous System in Infancy and Childhood.* Berlin, Germany: Springer; 1982.
17. Gildenberg PL, Kaufman HH, Murthy KS. Calculation of the stereotactic coordinates from the computed tomographic scan. *Neurosurgery.* 1982;10:580–586.
18. Shelden CH, McCann G, Jacques S, et al. Development of a computerized microstereotaxic method for localization and removal of minute CNS lesions under direct 3-D vision. *J Neurosurg.* 1980;52 (1):21–27.
19. Jacques S, Hunter Shelden CH, McCann G, Freshwater DB, Rand R. Computerized three-dimensional stereotaxic removal of small central nervous system lesions in patients. *J Neurosurg.* 1980;53(6):816–820.

20. Goerss S, Kelly PJ, Kall B, Alker GJ. A computed tomographic stereotactic adaptation system. *Neurosurgery.* 1982;10:375–379.
21. Alker G, Kelly PJ. An overview of CT-based stereotactic systems for the localization of intracranial lesions. *Comput Radiol.* 1984;8:193–196.
22. Kondziolka D, Dempsey PK, Lunsford LD, et al. A comparison between magnetic resonance imaging and computed tomography for stereotactic coordinate determination. *Neurosurgery.* 1992;30(1): 402–407.
23. Apuzzo ML, Sabshin JK. Computed tomographic guidance stereotaxis in the management of intracranial mass lesions. *Neurosurgery.* 1983;12(3):277–285.
24. Wild AM, Xuereb JH, Marks PV, Gleave JR. Computerized tomographic stereotaxy in the management of 200 consecutive intracranial mass lesions: analysis of indications, benefits and outcome. *Br J Neurosurg.* 1990;4(5):407–415.
25. Colombo F, Angrilli F, Zanardo A, Pinna V, Benedetti A. A new method for utilizing CT data in stereotactic surgery: measurement and transformation technique. *Acta Neurochir.* 1981;57(3–4):195–203.
26. Faugeras OD, Hebert M. The representation, recognition, and locating of 3D Objects. *Int J Robot Res.* 1986;5:27–52.
27. Horn BKP. Closed-form solution of absolute orientation using unit Quaternions. *JOSA.* 1987;4: 629–642.
28. Arun KS, Huang TS, Blostein SD. Least square fitting of two 3-D point sets. *IEEE Trans Pattern Anal Mach Intell.* 1987;9:698–700.
29. Roberts DW, Strohbehn JW, Hatch JF, Murray W, Kettenberger H. A frameless stereotaxic integration of computerized tomographic imaging and the operating microscope. *J Neurosurg.* 1986;65: 545–549.
30. Watanabe E, Watanabe T, Manaka S, Mayanagi Y, Takakura K. Three-dimensional digitizer (Neuronavigator): new equipment for computed tomography-guided stereotaxic surgery. *Surg Neurol.* 1987;27:543–547.
31. Galloway RL, Edwards C, Haden GL, Maciunas RJ. An interactive, image-guided articulated arm for laser surgery. *Proceedings of the Strategic Defense Initiative Organization's Fourth Annual Meeting on Medical Free-Electron Lasers.* Dallas, TX. September 22–24, 1989:15.
32. Galloway RL, Maciunas RJ, Edwards CA. Interactive, image-guided neurosurgery. *IEEE Trans BME.* 1992;39(12):1126–1231.
33. Schlöndroff G, Mösges R, Meyer-Ebrecht D., et al. CAS (computer assisted surgery) Ein neuartiges Verfahren in der Kopfund Halschirurgie HNO, 1989;37(1989):187–189.
34. Adams L, Krybus W, Meyer-Ebrecht D, et al. Computer assisted surgery. *IEEE Comput Graph Appl.* 1990;10(1990):43–49.
35. Kato A, Yoshimine T, Hayakawa T, et al. A frameless, armless navigational system for computer-assisted neurosurgery. *J Neurosurg.* 1991;74(5):845–849.
36. Zamorano LJ, Nolte L, Kadi AM, Jiang Z. Interactive intraoperative localization using an infrared-based system. *Neurol Res.* 1993;15(5):290–298.
37. Barnett GH, Kormos DW, Steiner CP, Weisenberger J. Intraoperative localization using an armless, frameless stereotactic wand. *J Neurosurg.* 1993;78(3):510–514.
38. Drake JM, Rutka JT, Hoffman HJ. ISG viewing wand system. *Neurosurgery.* 1994;34(6):1094–1097.
39. Sandeman DR, Patel N, Chandler C, Nelson RJ, Coakham HB, Griffith HB. Advances in image-directed neurosurgery: preliminary experience with the ISG viewing wand compared with the Leksell G frame. *Br J Neurosurg.* 1994;8(5):529–544.
40. Smith KR, Frank KJ, Bucholz RD. The NeuroStation—a highly accurate, minimally invasive solution to frameless stereotactic neurosurgery. *Comput Med Imaging Graph.* 1994;18(4):247–256.
41. Olivier A, Germano IM, Cukiert A, Peters TM. Frameless stereotaxy for surgery of the epilepsies: preliminary experience. *J Neurosurg.* 1994;81(4):629–633.
42. Peters TM, Davey B, Munger P, Comeau R, Evans A, Olivier A. Three-dimensional multimodal image-guidance for neurosurgery. *IEEE Trans Med Imaging.* 1996;15(2):121–128.
43. Sandeman D, Moufid A. Interactive image-guided pituitary surgery: an experience of 101 procedures. *Neurochirurgie.* 1998;44(5):331–338.

44. Muacevic A, Müller A. Image-guided endoscopic ventriculostomy with a new frameless armless neuronavigation system. *Comput Aided Surg*. 1999;4(2):87–92.
45. Barnett G, Kormos D, Steiner C. Frameless stereotaxy using a sonic digitizing wand: development and adaptation to the Picker ViStar medical imaging system. In: Maciunas RJ, ed. *Interactive, Image-Guided Neurosurgery*. Park Ridge, IL: AANS Publications; 1993.
46. Bucholz RD, Smith KR. A comparison of sonic digitizers versus light emitting diode-based localization. In: Maciunas R, ed. *Interactive Image-Guided Neurosurgery*. Park Ridge, IL: AANS Publications; 1993:179–200.
47. Galloway RL, Maciunas RJ, Bass WA, Carpini W. Optical localization for interactive, image-guided neurosurgery. *Proc. SPIE Medical Imaging 1994: Image Capture, Formatting, and Display*. 2164:137–145.
48. Grimson E, Leventon M, Ettinger G, et al. Clinical experience with a high precision image-guided neurosurgery system. *Lect Notes Comput Sci*. 1998;1496:63–73.
49. Ryan MJ, Erickson RK, Levin DN, et al. Frameless stereotaxy with real-time tracking of patient head movement and retrospective patient—image registration. *J Neurosurg*. 1996;85(2):287–292.
50. Kaus M, Steinmeier R, Sporer T, Ganslandt O, Fahlbusch R. Technical accuracy of a neuronavigation system measured with a high-precision mechanical micromanipulator. *Neurosurgery*. 1997;41(6):1431–1437.
51. Hill DL, Hawkes DJ, Crossman JE, et al. Registration of MR and CT images for skull base surgery using point-like anatomical features. *Br J Radiol*. 1991;64(767):1030–1035.
52. Drake JM, Prudencio J, Holowaka S, Rutka JT, Hoffman HJ, Humphreys RP. Frameless stereotaxy in children. *Pediatr Neurosurg*. 1994;20(2):152–159.
53. Friets EM, Strohbehn JW, Hatch JF, Roberts DW. A frameless stereotaxic operating microscope for neurosurgery. *IEEE Trans Biomed Eng*. 1989;36(6):608–617.
54. Masamune K, Masutani Y, Nakajima S, et al. Three-dimensional slice image overlay system with accurate depth perception for surgery. *MICCAI 2000 Lect Notes Comput Sci Vol*. 2000;1935:395–402.
55. Howard III MA, Dobbs MB, Simonson TM, LaVelle WE, Granner MA. A noninvasive, reattachable skull fiducial marker system: technical note. *J Neurosurg*. 1995;83(2):372–376.
56. Bare RO, Zinreich J, Zinreich ES. Radiographic multi-modality skin markers. US Patent #5469847 A.
57. Maurer CR, Fitzpatrick JM, Wang MY, Galloway RL, Maciunas RJ, Allen GS. Registration of head volume images using implantable fiducial markers. *IEEE TMI*. 1997;16(4):447–462.
58. Ammirati M, Gross JD, Ammirati G, Dugan S. Comparison of registration accuracy of skin- and bone-implanted fiducials for frameless stereotaxis of the brain: a prospective study. *Skull Base*. 2002;12(3):125–130.
59. Maurer CR, Maciunas RJ, Fitzpatrick JM. Registration of head CT images to physical space using a weighted combination of points and surfaces [image-guided surgery]. *IEEE-TMI*. 1998;17(5):753–761.
60. West J, Fitzpatrick JM, Wang MY, et al. Comparison and evaluation of retrospective intermodality image registration techniques. *J Comput Assist Tomogr*. 1997;21:554–566.
61. Fitzpatrick JM, West JB, Maurer Jr. CR. Predicting error in rigid-body point-based registration. *IEEE Trans Med Imaging*. 1998;17(5):694–702.
62. Andrew D, Wiles AD, Likholyot A, Frantz DD, Peters TM. A statistical model for point-based target registration error with anisotropic fiducial localizer error. *IEEE-TMI*. 2008;27(3):378–390.
63. Schicho K, Figl M, Seemann R, et al. Comparison of laser surface scanning and fiducial marker-based registration in frameless stereotaxy: technical note. *J Neurosurg*. 2007;106(4):704–709.
64. Pfisterer WK, Papadopoulos S, Drumm DA, Smith K, Preul MC. Fiducial versus nonfiducial neuronavigation registration assessment and considerations of accuracy. *Neurosurgery*. 2008;62(3 Suppl 1):201–207.
65. Schlaier J, Warnat J, Brawanski A. Registration accuracy and practicability of laser-directed surface matching. *Comput Aided Surg*. 2002;7(5):284–290.
66. Miga MI, Sinha TK, Cash DM, Galloway RL, Weil RJ. Cortical surface registration for image-guided neurosurgery using laser-range scanning. *IEEE Trans Med Imaging*. 2003;22(8):973–985.
67. Sinha TK, Miga MI, Cash DM, Weil RJ. Intraoperative cortical surface characterization using laser range scanning: preliminary results. *Neurosurgery*. 2006;59(4 Suppl 2):ONS368–ONS376.

68. Cao A, Thompson RC, Dumpuri P, et al. Laser range scanning for image-guided neurosurgery: investigation of image-to-physical space registrations. *Med Phys.* 2008;35:1593.
69. Sinha TK, Dawant BM, Duay V, et al. A method to track cortical surface deformations using a laser range scanner. *IEEE-TMI.* 2005;24(6):767–781.
70. Galloway Jr RL, Berger MS, Bass WA, et al. Registered intraoperative information: electrophysiology, ultrasound and endoscopy. In: Maciunas R, ed. *Interactive Image-Guided Neurosurgery.* Park Ridge, IL: AANS Publications; 1993:247–258.
71. Manwaring KH, Manwaring ML, Moss SD. Magnetic field guided endoscopic dissection through a burr hole may avoid more invasive craniotomies. *Acta Neurochir.* 1994;61:34–39.
72. Hopf NJ, Perneczky A. Endoscopic neurosurgery and endoscope-assisted microneurosurgery for the treatment of intracranial cysts. *Neurosurgery.* 1998;43(6):1330–1336.
73. Sutcliffe JC. The value of intraoperative ultrasound in neurosurgery. *BJN.* 1991;5(2):169–178.
74. Comeau RM, Fenster A, Peters TM. Integrated MR and ultrasound imaging for improved image guidance in neurosurgery. *Proc. SPIE* 3338, *Medical Imaging* 1998: *Image Processing*, 747.
75. Unsgaard G, Selbekk T, Müller TB, et al. Ability of navigated 3D ultrasound to delineate gliomas and metastases—comparison of image interpretations with histopathology. *Acta Neurochir.* 2005;147 (12):1259–1269.
76. Ji S, Wu Z, Hartov A, Roberts DW, Paulsen KD. Mutual-information-based image to patient re-registration using intraoperative ultrasound in image-guided neurosurgery. *Med Phys.* 2008; 35:4612.
77. Hall WA, Martin AJ, Liu H, Nussbaum ES, Maxwell RE, Truwit CL. Brain biopsy using high-field strength interventional magnetic resonance imaging. *Neurosurgery.* 1999;44(4):807–813.
78. Alexander 3rd(1) E, Moriarty TM, Kikinis R, Black P, Jolesz FM. The present and future role of intraoperative MRI in neurosurgical procedures. *Stereotact Funct Neurosurg.* 1997;68(1–4 Pt 1): 10–17.
79. Hadani M, Spiegelman R, Feldman Z, Berkenstadt H, Ram Z. Novel, compact, intraoperative magnetic resonance imaging-guided system for conventional neurosurgical operating rooms. *Neurosurgery.* 2001;48(4):799–809.
80. Mislow JM, Golby AJ, Black PM. Origins of intraoperative MRI. *Neurosurg Clin N Am.* 2009; 20(2):137–146.
81. Gumprecht H, Lumenta CB. Intraoperative imaging using a mobile computed tomography scanner. *Minim Invasive Neurosurg.* 2003;46(6):317–322.
82. Ferrer E, Vila F, Isamaat F. Mannitol response and histogram analysis in raised ICP. Intracranial Pressure IV Springer 1980:647–652.
83. Kelly PJ, Kall B, Goerss S, Earnest FI. Computer-assisted stereotaxic laser resection of intra-axial brain neoplasms. *J Neurosurg.* 1986;64:427–439.
84. Hill DLG, Maurer CR, Maciunas RJ, Barwise JA, Fitzpatrick JM, Wang MY. Measurement of intraoperative brain surface deformation under a craniotomy. *Neurosurgery.* 1998;43:514–528.
85. Roberts DW, Hartov A, Kennedy FE, Miga MI, Paulsen KD. Intraoperative brain shift and deformation: a quantitative analysis of cortical displacement in 28 cases. *Neurosurgery.* 1998;43:749–760.
86. Paulsen KD, Miga MI, Kennedy FE, Hoopes PJ, Hartov A, Roberts DW. A computational model for tracking subsurface tissue deformation during stereotactic neurosurgery. *IEEE Trans Biomed Eng.* 1999;46:213–225.
87. Warfield SK, Talos F, Tei A, Bharatha A, Nabavi A, Ferrant M, et al. Real-time registration of volumetric brain MRI by biomechanical simulation of deformation during image guided neurosurgery. *Comput Vis Sci.* 2002;5(1):3–11.
88. Dumpuri P, Thompson RC, Dawant BM, Cao A, Miga MI. An atlas-based method to compensate for brain shift: preliminary results. *Med Image Anal.* 2007;11(2):128–145.

SECTION I

Methods

CHAPTER 2

Background on Imaging Structural Imaging

Nathan C. Himes and Geoffrey Young
Department of Radiology, Brigham and Women's Hospital, Harvard Medical School, Boston, MA, USA

INTRODUCTION

The goals of imaging for neurosurgical guidance are many and often involve developing a differential diagnosis, predicting the grade of brain tumors, evaluating safety of resection, establishing optimal targets for biopsy, planning operative approach, and extent of resection and monitoring disease following intervention, among many others. Cross-sectional imaging is the primary method of imaging the brain and spinal cord. The two primary modalities used are computed tomography (CT) and magnetic resonance imaging (MRI), with MRI providing the most sensitive and specific results because its superior contrast resolution allows accurate differentiation of diverse tissues and pathologies. This chapter will briefly introduce the basic image acquisition techniques in MR and CT to illustrate common applications in tumor surgery and outline the principles that guide interpretation of images in this application.

BASIC MAGNETIC RESONANCE IMAGING

Magnetic resonance imaging has been in clinical use since the early 1980s when its introduction revolutionized central nervous system diagnosis, surgical guidance, and monitoring. Most clinical MRI is based on acquisition of signal from mobile free water protons within the body. Some of the basic properties of a proton include mass, a positive electric charge, and spin.[1] Since a proton has an electric charge its spin produces a small, but detectable, magnetic field. Free water is the largest reservoir of protons in the body, followed by fat. When the body is placed into the large external magnetic field of the MRI magnet, a few more mobile protons per million will align parallel to the main external magnetic field of the scanner than align opposite the field because the parallel alignment is a lower energy state. When this slight excess of low-energy state protons absorbs radio-frequency (RF) non ionizing radiation broadcast by the "transmit coils" in the MRI scanner, the protons enter a transient higher energy state. As they relax again to the low-energy state these protons release the absorbed energy as photons at the same RF. This "signal" broadcast from the patient's mobile protons forms that basis of the MR image. The percentage of protons aligning with the external magnetic field is determined

A. Golby (Ed): Image-Guided Neurosurgery
DOI: http://dx.doi.org/10.1016/B978-0-12-800870-6.00002-9

in part by the field strength of the MRI magnet. Therefore, higher field strength scanners, such as clinical scanners with 3-Tesla magnets, can provide a higher signal-to-noise ratio by magnetizing and subsequently exciting more protons.

A rigorous understanding of MR physics must be based on quantum mechanics because on the individual proton level, *spin* is an intrinsic quantum mechanical property of the proton nucleus. This gives spin-1/2 nuclei, such as protons, certain unintuitive fundamental properties. First, protons can only spin at one precisely defined frequency determined only by the proton gyromagnetic ratio and the applied external magnetic field. Equally importantly, protons can only occupy two orientations in a magnetic field: the low energy "spin down" position parallel to the applied field and the high energy "spin up" position opposite to the applied magnetic field. The transition from the low energy quantum state to the high energy quantum state requires absorption of a photon of frequency corresponding to its precession frequency and the return from the high-energy state to the low-energy state requires emission of a photon at the same frequency. This leads to the concept that protons can transfer spin to other protons, known as *spin exchange*. The aggregate of a large number of protons with these unique quantum mechanical properties gives rise to a net magnetic moment that can be thought of in classical physics terms. This more intuitive formulation can allow a useful understanding of many of the basic processes that underlie MRI.

Just as a spinning top precesses about its axis in the earth's gravitational field, the net magnetic moment of an aggregate of proton spins can be thought of as a dipole that precesses about the axis of the main external magnetic field of the magnet. The frequency of the precession is directly proportional to the strength of the magnetic field and is referred to as the Larmor or resonance frequency.[2–4] If an electromagnetic RF pulse is applied at or near this resonance frequency, the proton resonance leads to absorption of this energy. As individual protons absorb quanta of photon energy from this RF pulse and transition to the quantized higher energy state, the net proton magnetization of an aggregate of these protons can be pictured as a vector which tips out of the longitudinal or *Z*-axis of the main magnetic field and into the transverse or *X-Y* plane. It is the resulting component of the magnetic field of this aggregate of excited protons precessing within the transverse plane that can be detected as the "MRI signal." Although signal from every tissue voxel has the same resonant frequency, pulse sequences can be created to modulate the detected signal by one of a number of mechanisms, the most important of which are the two fundamental physical spin relaxation processes. Once the RF transmitter is turned off, two independent mechanisms of relaxation begin to happen simultaneously. The excited protons begin to return to the original *Z*-axis orientation and in that process exchange the absorbed RF energy with the surrounding molecular bond lattice ("spin-lattice relaxation" characterized by an equation with time constant T1), and the excited protons begin to dephase from each other within the transverse plane ("spin-spin dephasing" characterized by a different equation with time constant T2). Each of these mechanisms of relaxation to the ground state is described by a separate exponential function.

Longitudinal relaxation

Following the end of the excitation pulse, as the excited protons return from the high-energy state opposite to the main magnetic field to the low-energy state parallel to the field, the protons transfer energy into the surrounding tissue spin-lattice. The time constant T1 in the equation describing this relaxation is intrinsic to the aggregate of mobile protons in each tissue, and determined in large part by the local macromolecular environment of the free water protons, especially the presence of protein and lipid within each voxel. Images can be created to reflect the intensity of the signal from the fraction of protons in a given voxel that have returned to the low-energy state at any time between the end of the excitation pulse and the time when all protons in a sample have relaxed. Such images are said to be *weighted* by the T1 intrinsic to each tissue voxel. Images created so the principle weighting that determines the image contrast in tissue T1 are called *T1-weighted images*. Image contrast between tissues of different T1 is called *T1 contrast*.

Transverse dephasing

Starting at the same moment, at the end of the excitation pulse, an additional completely independent physical relaxation process starts in which nearby mobile proton spins in each voxel, which initially all precess "in phase" ("in sync") together, exchange magnetization with each other. As the spins exchange magnetization with each other, different protons accumulate different amount of spin and end up out of phase with each other (*spin—spin dephasing*). Since the aggregate of the transverse plane magnetization of the in-phase spins comprises the precessing bulk magnetic field vector which induces the alternating "signal" detected in the MRI receive coil, the detected signal intensity decays as the spins get out of phase with each other. The time constant in the equation describing relaxation by this mechanism is called T2. T2 is an intrinsic property of the mixture of protons in each tissue and voxel and largely reflects the proportion of mobile free water protons in each voxel. Images created in which the intensity of signal from each voxel reflects the fraction of protons that remain in-phase with each other in that voxel are called *T2-weighted images*. Image contrast between different voxels with different net T2 is called *T2 contrast*. T2-weighted images are exquisitely sensitive to the amount of tissue water in each voxel and hence to the presence of tissue edema.

Magnetic susceptibility effects (T2* decay)

Pure T2 relaxation is defined by transient random interaction between protons that causes cumulative loss in phase coherence resulting in a decay of signal detected from each voxel. In addition, different but chemically identical protons in tissue encounter slight differences in their local magnetic fields. These differences are due to differences in proximity to ferromagnetic (e.g., iron) or paramagnetic (e.g., magnesium, lithium, oxygen) materials, which strengthen the external applied field, or diamagnetic materials (e.g., calcium, copper and many other elements), which slightly weaken the applied field, as well as slight imperfections in the homogeneity of the applied magnetic field itself. The differences in the local field experienced by different protons are generally constant or "fixed" during the duration of the MRI scan. These *fixed field inhomogeneity* effects cause some protons in a given voxel to precess faster and other protons to precess slower. As a result of these tiny differences in rotational frequencies, the protons accumulate differences in phase from each other (become out of sync), resulting in a loss of the net transverse magnetization in each voxel and hence decay of the signal detected in that voxel. Each tissue has an intrinsic magnetic susceptibility related to the mix of paramagnetic and diamagnetic elements within each tissue voxel. Tissue voxels also can be subject to differences in magnetic susceptibility arising from differences in extrinsic ferromagnetic paramagnetic, diamagnetic materials (e.g., steel, iron, blood products, air) near that tissue voxel. Because, unlike spin-spin exchange, these fixed field inhomogeneity effects alter the resonant frequency of the protons slightly, they also distort the resonant frequency gradient based anatomic localization of the signal from voxels near to air/tissue interfaces, surrounding metal, or even blood products. When T2-weighted images are created that depict signal arising from each voxel without correcting for fixed field inhomogeneity (generally by use of gradient echo sequences) these are called *T2*-weighted images*. The asterisk denotes that the image contrast is modulated by uncorrected magnetic susceptibility effects as well as the underlying spin-spin relaxation effect that dominates spin-echo T2-weighted images. T2* image contrast is exploited to the fullest in susceptibility-weighted imaging to detect minute amounts of blood products, calcium, etc. and in BOLD functional MRI (see Chapter 3).

Proton density

In every MRI image, the signal obtained from each voxel of tissue is proportional to the total number of detectable mobile protons within that voxel called the *proton density* (PD) or *spin density*. Images created to minimize the amount of T1-weighting (by allowing full longitudinal magnetization recovery), the amount of T2-weighting (by allowing little time for spin-spin dephasing) and to correct for susceptibility effect related signal loss, are called proton density-weighted images. In general, PD images provide the highest SNR of any images, but the least tissue contrast.

MR pulse sequence

The MR signal emitted by a group of protons is referred to as an *echo*. To create an echo, a series of precisely calibrated RF excitation and refocusing pulses and time-varying magnetic field gradients referred to as an MR *pulse sequence* is applied to the sample. The echo can be weighted to reflect predominantly T1, T2, T2*, PD, or a combination of these physical properties of the imaged protons. Different timing of the period between application of the excitation pulses (repetition time or TR) and the period between the excitation pulse and the measurement of the echo (echo time or TE) are the primary determinants of the weighting of an image. There are others, such as the amount of energy instilled with the excitation RF pulse, also referred to as flip angle; however, these are beyond the scope of an introductory explanation. T1-weighted sequences have a relatively short TE and intermediate TR, T2-weighted sequences have a relatively intermediate TE and long TR, and PD-weighted images have a relatively long TR and short TE. A long TR minimizes the effects of longitudinal relaxation and a short TE minimizes the effects of transverse relaxation. In reality, all MRI sequences reflect a combination of PD, T1 relaxation and T2 relaxation effects, but judicious selection of TR and TE creates images that principally reflect one or a combination of these properties.

There are two basic types of MR pulse sequences in common use: gradient echo and spin-echo sequences.[5] A basic gradient echo pulse sequence consists of an excitation RF pulse followed by application of dephasing and then rephasing magnetic field gradients that generate the echo. The RF excitation pulse typically has a smaller angle (less than 90°) that decreases the time required for magnetization recovery in the z-axis and therefore allows use of a shorter TR and hence faster image acquisition than spin-echo sequences that require a refocusing pulse. Gradient echo sequences are also very sensitive to the field inhomogeneity that causes T2* dephasing. This makes them sensitive for detection of paramagnetic components of hemorrhage but also very sensitive to artifact from metal that can obliterate signal from adjacent protons and create anatomic distortion. Spin echo-sequences reduce the effects of local field inhomogeneity by introducing a 180° "refocusing" RF pulse between the excitation pulse and the echo. This corrects for the effects of imperfections in the main magnetic field or gradients, bulk paramagnetic or ferromagnetic materials in or adjacent to the tissue and paramagnetic or ferromagnetic tissue components. Spin-echo sequences can produce true T2-weighted sequences and generally are less sensitive to artifact from metal or hemorrhage and subject to less anatomic distortion but require longer acquisition times and application of more RF to the patient. Fast spin-echo sequences accelerate image acquisition by applying an "echo-train" of multiple 180° refocusing pulses and collecting multiple echoes after each 90° excitation pulse. Since each echo in the train is acquired at a different TE, fast spin-echo imaging results in blurring of the image contrast and spatial resolution as well as high RF exposures. One of the innumerable other techniques for acceleration of imaging is echo planar imaging with long trains of gradient refocused echoes used in diffusion imaging, perfusion imaging, and BOLD functional imaging. Another is parallel imaging with simultaneous acquisition from multiple small surface coils, which is now ubiquitous at 3T. A third is half Fourier acquisition single-shot fast spin-echo (HASTE or SSFSE) imaging which produces a heavily T2-weighted image, often used clinically in the brain for evaluation in ventricular size in pediatric patients or other patients who cannot hold still in the MRI magnet[6] as well as for MRI-based near-real-time guidance of biopsy needles or deep brain stimulator electrodes.[7–11] Every acceleration technique adds additional artifacts, effects on image contrast and potential for spatial anatomic distortion in the images.

MRI equipment

Although permanent ferromagnets are still in widespread use for low field MRI scanners with field strength <0.5T, most hospitals now use intermediate field (1–2T, most commonly 1.5T), high field (3–4.5T), and ultrahigh field (7–9.4T) superconducting magnet-based MRI systems. The magnetic field from a superconducting magnet is produced from very high currents that flow continuously within multiple loops of wire made from superconducting materials that, when cooled below their critical temperature (near absolute zero) by liquid helium, lose all resistance to current flow. Therefore, once an MRI magnet is cooled at initial installation, current is applied and the power supply is disconnected, the current will continue to flow indefinitely, creating a magnetic field that is always "on"—even when the power is "off."

In addition to the large loops of wire used to create the main magnetic field, there are multiple additional wire loops with much smaller currents running through them that are used for other essential components of MRI. The most important of these are the gradient coils used for spatial localization of the MRI signal. By applying well-timed currents within the *x*, *y*, or *z* plane aligned gradient coils, rapid changes can be made in the strength of the local magnetic field at each location in the scanner. This allows slightly different field strength to be applied to each voxel of imaged tissue at each moment of time, altering the resonance frequencies of the protons in that voxel that in turn allows the signal detected from the protons in each voxel to be assigned to the correct spatial location of that voxel within the tissue. The peak current amplitude, the linearity, and maximum rate at which the gradient coils can be turned on or off (slew rate), determined in part by the need to dissipate the heat produced by resistance in the coils, has a very significant impact on the speed of image acquisition, the achievable spatial resolution and the anatomic distortion in the resulting images.

In a typical MR pulse sequence, one set of gradient coils is used to select the area of the patient excited by the 90° pulse, and the other two to localize the signal from each individual voxel within that slice by a combination of phase-encoding and frequency encoding. The gradient amplitudes are used to map the digitized signal intensity data from an MR acquisition into a mathematical matrix referred to as *k-space*. Each data point in k-space contributes to the entire image. Data in the central portion of k-space provides the majority of the overall signal, and contrast and data on the edges of k-space provide the majority of the spatial resolution information. The k-space data is then transformed into image data, typically by applying a multidimensional Fourier transform analysis to extract from the compound RF signal the components of the derived signal from each tissue voxel based on the unique combination of frequency and phase encoded into the protons in that voxel by the temporal and spatial magnetic field gradients specified in the pulse sequence. These signal intensities are mapped as grayscale values in the resulting image displayed on the computer monitor.

In addition to the gradient coils used for spatially localizing the MRI signal, there are RF coils used for transmitting RF energy into the patient and for receiving the signal coming out of the patient. These are generally divided into volume coils that provide a uniform field for transmitting signal into the patient and surface coils for receiving signal. Much modern 3T imaging uses volume coils, often with the body coil built into the magnet system, to achieve homogeneous excitation and localized surface receiver coils—or commonly a phased array of such coils—placed directly over the anatomy of interest to provide very high signal to noise. This is sometimes called "receive only" coil imaging, to distinguish it from "transmit receive" imaging in which the localized coil is used for both transmit and receive.

MRI sequences in clinical use

The MRI sequences used most commonly in standard of care brain tumor imaging include: 2D FSE T2-weighted image (T2WI), 2D FSE Fluid Attenuated Inversion Recovery (FLAIR) T2-weighted, spin-echo echo planar diffusion-weighted imaging (DWI), conventional or susceptibility-weighted gradient echo T2* (GRE or SWI), dynamic susceptibility contrast (DSC) gradient or spin-echo echo planar perfusion (DSC or PWI), 2D T1-weighted images (T1WI), postcontrast 2D T1-weighted and pre- and/or postcontrast Axial 3D T1-weighted Spoiled Gradient Echo (SPGR or MPRAGE®) images (Figure 2.1). 2D sequences can be acquired in any plane but are most commonly acquired in the axial plane. 3D sequences can be acquired in any plane and when implemented to produce near isotropic voxels allow reformatted images to be produced in the other two planes or in any oblique plane, particularly important for treatment planning. Each of these sequences adds value in the evaluation of brain tumor, infarct and other pathologies. In certain cases dynamic contrast-enhanced (DCE) T1-weighted imaging may be performed to allow quantitative assessment of tissue permeability. Similarly, spectroscopic imaging is occasionally useful to allow assessment of the concentrations of certain macromolecules present high concentration in tissue by detection of their bound protons. These techniques are more time consuming to perform than the standard sequences and require meticulous attention to acquisition and postprocessing and so are not routinely used outside of research settings.

Because of its long T2, CSF produces high signal intensity on T2-weighted sequences and hence T2WI provides good contrast between the CSF spaces and the brain parenchyma and between different structures in the brain with different water

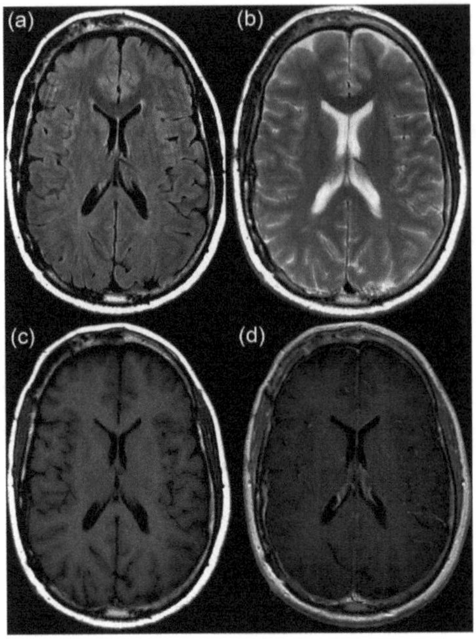

Figure 2.1 The four standard anatomical sequences used in brain MRI include (a) FLAIR T2-weighted, (b) T2-weighted, (c) T1-weighted, and (d) delayed post-contrast T1-weighted images.

content. T2WI can be used to distinguish extra-axial from intra-axial peripheral lesions by depicting a CSF cleft between the lesion and the brain parenchyma and displaying displacement of the cortical gray matter and pial vessels between the lesion and the white matter. In addition, highly cellular tumors often appear hypointense on T2 weighted sequences, similar to the normal white matter in signal intensity.

FLAIR T2WI are produced by a T2-weighted sequence containing an inversion recovery RF pulse before the excitation pulse that is timed to suppress signal from normal CSF. The inversion pulse also slightly suppresses signal from the normal white matter and results in images that are exquisitely sensitive to abnormal brain parenchymal water content in edema associated with tumor, infarction, infection, trauma, necrosis, etc. The suppression of CSF allows more sensitive detection of periventricular and cortical lesions than standard T2-weighted sequences. The FLAIR pulse suppresses CSF by exciting the protons at the point after the inversion pulse when half of the CSF protons have relaxed. As such the FLAIR sequence accentuates signal from protons with T1 shorter than normal CSF and so allows detection of abnormal signal within the subarachnoid spaces related to subarachnoid hemorrhage or from pus in meningitis.

Since T1WI signal intensity reflects the degree of recovery of longitudinal proton magnetization between excitation pulses, on T1 weighted images, subcutaneous fat is bright, lipid dense white matter is brighter than gray matter, and CSF and abnormal tissue water will appear dark. Lesions containing hemorrhage, fat, melanin, certain hydration states of

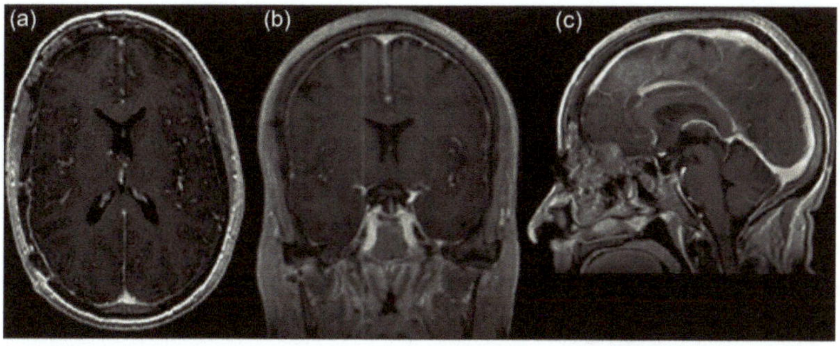

Figure 2.2 3D T1-weighted spoiled gradient echo (SPGR® or MPRAGE®) post-contrast sequences are often used for surgical planning because for this application the ability to reformat the data in any plane with minimal loss of spatial resolution is a critical advantage that outweighs the drawbacks of relatively low in-plane spatial resolution on the source images and long acquisition times. These slight motion and truncation artifacts related to the long acquisition time and fast imaging are most evident on the axial source images (a), but in combination with the slightly anisotropic voxel size contribute to the slight but acceptable blurring evident on the coronal (b) and sagittal (c) reformatted images.

calcium and concentrated protein have a short T1 and appear hyperintense on T1WI. For this reason, tumor or fibrosis replacing the normal short T1 marrow fat in the marrow of the skull and cervical spinal marrow can be detected on T1WI. Most importantly, because extravascular gadolinium-based contrast agents (GBCA) shorten the T1 of adjacent brain parenchymal water protons, the combination of precontrast and postcontrast T1WI allows detection of local blood–brain-barrier (BBB) impairment in tumor, infection, inflammation, and subacute infarction, among other pathologies. Because 3D imaging allows production of volumetric near-isotropic images suitable for reformatting in multiple planes and use in surgical navigation software, 3D T1-weighted post-contrast sequences are the mainstay of surgical guidance for contrast-enhancing tumors. To allow acquisition during the time the patient can lie still, gradient echo sequences are used. In order to produce gradient echo images principally weighted to reflect tissue T1, "spoiler" gradients are added during each repetition to dephase residual transverse magnetization. The resulting images are known as *spoiled gradient echo* (SPGR)T1WI (Figure 2.2). A slight variation of the SPGR T1WI sequence involves addition of an inversion recovery spin preparation pulse to accentuate gray-white tissue contrast. The resulting pulse sequence is called magnetization prepared rapid gradient echo (MP RAGE) T1WI.[12]

Diffusion-weighted imaging

In DWI, large *diffusion-sensitizing gradients* are typically applied to single-shot echo-planar T2WI to amplify the dephasing related signal loss due to Brownian (i.e., thermal) diffusion of water protons in the tissues. This results in images in which the net signal intensity of each voxel reflects the fraction of water protons in that voxel that encounter a microscopic barrier to diffusion during the pulse sequence. In practice water molecules diffuse roughly 10 μm during a typical DWI pulse sequence. On that spatial scale, the principal tissue barriers to diffusion in the brain are cellular and intracellular membranes. As such the apparent diffusion coefficient (ADC) calculated in each voxel is related to the extracellular volume fraction in that voxel principally determined by tissue cellularity, cell size, and cellular swelling.[13,14] Adjusting strength and timing of the diffusion sensitizing gradients can alter the sensitivity of DWI sequences. The magnitude of these gradients, and resulting degree of diffusion weighting in the DWI images, is characterized by the *b factor.* In clinical brain imaging, b factors range from 800 to 1500 s/mm^2. Because axon and myelin sheath membranes in white matter are anisotropic—meaning water diffusivity along the fiber tracts is much higher than water diffusivity perpendicular to the fiber tracts—the diffusion sensitizing gradients are applied in multiple diffusion directions (minimally three) during each acquisition.[15,16] The gradients are also applied with multiple different b factors (typically 0, 500 and 1000 s/mm^2) to allow accurate fitting of the ADC curve. A combined DWI image—sometimes called a *trace* or *average* image—roughly representing a directionally weighted average diffusion-weighted image is produced from these multiple whole brain acquisitions, which is interpreted clinically along with corresponding ADC maps (see Figure 2.3). ADC maps generated from the diffusion-weighted MRI sequence depict high ADC values characteristic of tissue with fewer membrane barriers to diffusion as brighter pixels and low ADC values characteristic of tissue with more membrane barriers to diffusion as darker pixels. The use of single-shot EPI techniques permits whole brain image acquisition in less than 100 ms and complete DWI of the brain in as little as 75 seconds. This produces a sequence that is very insensitive to motion artifact, but very sensitive to magnetic susceptibility artifact from blood products, bone, calcifications and tissue–air interfaces. As such it is critical to interpret DWI in the context of the other anatomic MRI sequences.

High signal on DWI with corresponding dark pixels on the ADC maps allows exquisitely sensitive detection of early cerebral ischemia and small infarcts (Figure 2.4). In arterial infarction, cytotoxic edema results in cell swelling and membrane damage that increases the proportion of membrane associated water within less than 20 minutes of onset of ischemia hours to days before it is evident on CT or other MRI sequences.[17] Because the underlying sequence is T2*-weighted, care must be taken to distinguish hyperintensity on DWI representing infarction or hypercellularity from hyperintensity related to "T2 shine-through effect" in which high signal is seen on the ADC map, or susceptibility artifact in which a typical pattern of susceptibility related bright or dark pixels are seen on the B0 images. The arterial infarction abnormality on DWI persists for approximately 7–14 days with pseudonormalization of the ADC map at roughly one week.[18–20] Chronic infarcts characterized pathologically by astrogliosis and encephalomalacia, like gliosis of other etiologies and edema, have high diffusivity seen as bright pixels on the ADC maps.

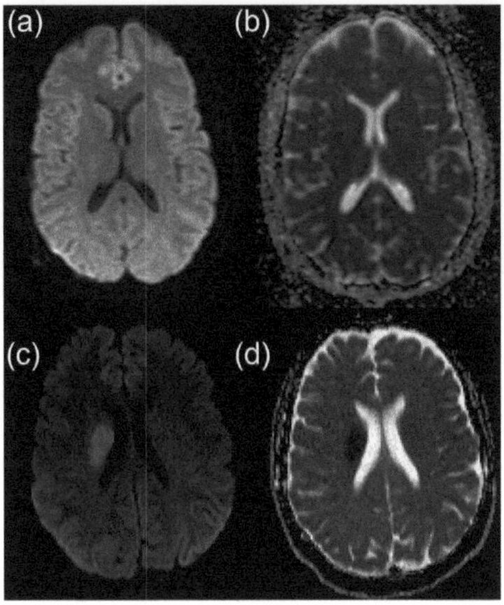

Figure 2.3 Diffusion-weighted sequences are especially sensitive for the detection of early ischemia. The combined (aka "trace") diffusion-weighted image (DWI) (a) has underlying T2*-weighting and it therefore must be analyzed in conjunction with its corresponding calculated apparent diffusion coefficient (ADC) map (b) as in this normal patient. With acute ischemia, there will be high signal intensity on the combined DWI image (c) with corresponding dark pixels on the ADC map (d) reflecting cytotoxic edema, as is the case in this infarct in the right MCA territory.

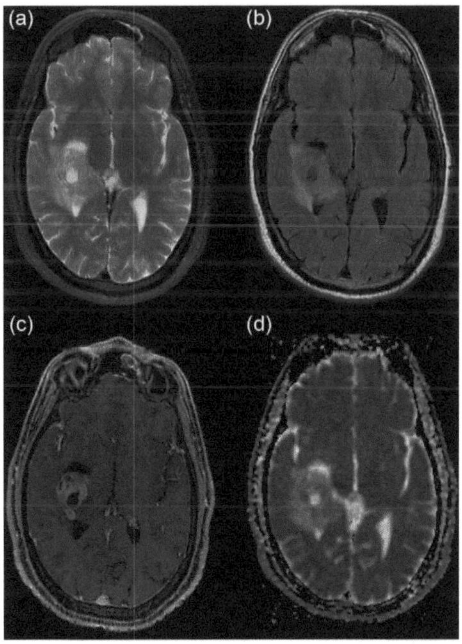

Figure 2.4 Typical appearing glioblastoma in the right temporal lobe on (a) T2 weighted, (b) FLAIR T2 weighted, (c) 3D Axial T1 MPRAGE®, and (d) ADC map. Note the intermediate signal intensity on T2-weighted images and intermediate diffusivity on ADC map within the peripheral solid component of the mass consistent with high cellularity and the heterogeneous enhancement on post-contrast imaging with areas of non-enhancing necrosis centrally.

DWI is also useful in the evaluation of brain tumors. Studies have shown that ADC values are inversely related to tumor cellularity,[13,14] which can be useful in the differentiation of highly cellular tumors from infiltrative tumor, vasogenic edema and radiation-induced necrosis. In addition, ADC maps can aid biopsy guidance in heterogeneous tumors by localizing the areas of highest cellularity, which likely represent higher-grade tumor. DWI is extremely useful in immediate postoperative MRI to distinguish infarction at the margin of the surgical resection or elsewhere from residual tumor since subacute infarction may demonstrate enhancement, nodularity indistinguishable from residual or recurrent tumor on later follow-up MRI. On later follow-up, DWI is useful for detection of cerebral abscess in which profound low diffusivity is seen, related to the presence of viable white cells in pus. Similarly, DWI MRI can be very valuable in differentiating an epidermoid cyst, which has very low diffusivity, related to exfoliated squames within the cyst, from other extra-axial cystic lesions. In both cases care must be taken to correlate with other pulse sequences since cellular acute or subacute hematoma has very low diffusivity until the red blood cells lyse. Finally, longitudinal follow-up DWI can be useful in following tumor treatment response and detecting recurrence because cytotoxic radiation and chemotherapy-induced cell death increase ADC within a given area of tumor.[21–25]

Typical brain white matter has an ADC value of around 0.7×10^{-3} mm^2/s. Typical values for the highly cellular portions of glioblastoma multiforme are slightly higher whereas typical values for radiation necrosis changes tend to be slightly lower, in the range of 0.5×10^{-3} mm^2/s. An inverse correlation between minimum ADC (ADC_{min}) and tumor cellularity has been verified by histology in a wide variety of tumors, including high- and low-grade glioma, lymphoma, medulloblastoma, meningioma, and metastases.[26–30] Within meningioma, a lower ADC value has been demonstrated in atypical and malignant versus typical subtypes, but the overlap of the two groups precludes the use of ADC for definite differentiation in individual patients.[31] In glioma, however, a number of groups have found that ADC_{min} values below a cutoff in the range of 1.7 to 2.5 can be used to distinguish high-grade glioma from low-grade glioma.[32,33] Again, overlap between tumor grades mandates that ADC_{min} values be combined with other advanced and conventional MRI data to reliably distinguish high-grade from low-grade.[32,34–36]

T2* dynamic susceptibility contrast perfusion imaging

Similar to DWI, dynamic susceptibility contrast (DSC) capillary perfusion imaging is based on EPI SE-EPI or GE-EPI T2*-weighted rapid imaging of the whole brain. In DSC the whole brain is imaged every 1–2 seconds during the first pass of a bolus of intravenously injected paramagnetic contrast agent, most commonly gadolinium-based contrast agents.[37–39] As the concentrated bolus of paramagnetic contrast agent passes through the capillary bed, it causes transient susceptibility related dephasing of protons in the nearby extravascular space that leads to a transient decrease in signal intensity on the T2*-weighted images.[40] With a standard dose of contrast agent (0.1 mmol/kg body weight), a signal loss of approximately 15–35% observed in normal white matter provides a very robust and reproducible measure of capillary perfusion. Given certain critical assumptions, the application of a kinetic model based on the nondiffusable tracer theory[41] to the time versus signal intensity (TIC) curve allows for estimation of quantitative maps of cerebral blood flow (CBF), cerebral blood volume (CBV) and mean transit time (MTT) typically displayed as color maps.[42,43] Inspection of the shape of the TIC is essential during DSC data quality assurance and interpretation to assess for inadequate bolus timing, bolus concentration, bolus volume, susceptibility and other artifacts or high first pass capillary leak, any of which can render the resulting maps unreliable. It is important to note that there is usually no way to detect these artifacts on the maps themselves so unreliable data can only be detected by inspecting the TIC. CBV is the most robust measure and most widely used for evaluation of brain tumors because of its direct relationship to neocapillary size and density. All three maps are used in the evaluation of stroke. Although relative CBV is proportional to the area under the contrast agent concentration versus time curve[44] absolute arterial concentration cannot be determined by DSC. Normal gray matter has a relative CBV approximately 2.7 times that of normal white matter. As such, the blood volumes of normal-appearing gray and white matter are used as internal references for visual comparisons and measurements of normal appearing white matter CBV are used for normalized CBV (nCBV) region-of-interest measurements.[45]

The extensive literature establishes the utility of DSC MRI in evaluation of ischemic and infarcted tissues, preoperative differential diagnosis of intracranial masses, preoperative tumor grade estimation, operative resection planning, and in follow-up of therapeutic response.

Perfusion imaging is important in allowing identification of areas of reduced perfusion that are at risk of infarction versus areas that are already infarcted. Distinguishing at-risk (ischemic penumbra) and infarcted tissues allows for the identification of potentially salvageable tissue that could guide treatment decisions in individual patients. Among the most widespread interpretation of PWI in stroke imaging is based on the hypothesis that the combination of prolonged MTT and normal CBV in brain tissue outside of the area of low diffusivity seen on DWI is a marker of tissue at risk.[46–53]

In GE-EPI imaging of tumors, the rough estimate of first pass leakage derived from the shape of the TIC has been reported to provide a clue to the differential diagnosis of an intracranial mass.[54] Microvessels within tumors of extra-axial and nonglial origin—meningioma, choroid plexus papilloma, metastases, lymphoma, and so forth—do not form a BBB and therefore a large fraction of the bolus leaks into the extravascular space during the first pass.[34,55,56] Because glioma microvessels form a BBB that is impaired but not absent, the TIC returns toward the baseline in these tumors, although not as much as in normal brain. The difference between these patterns can contribute significantly to the discrimination of tumor types in cases of peripherally located enhancing tumors when the differential diagnosis includes meningioma and peripheral GBM and in periventricular enhancing lesions when the differential diagnosis includes choroid plexus papilloma/carcinoma and GBM.

Perfusion MRI can also be used to measure relative CBV and vascular permeability within brain tumors, which can be directly correlated with histopathologic evidence of angiogenesis.[57] In areas of infiltrative and low-grade cellular glioma supplied by native vessels, hypoxia can lead to secretion of vasoactive substances by glioma and host immune cells that induce the expression of aquaporins and suppress the expression endothelial tight junction proteins, resulting in varying degrees of impairment of the BBB.[54,58] In GBM, on the other hand, new formation of dense beds of characteristically tortuous and structurally abnormal neocapillaries produces the extremely high local tissue blood volume. In these neocapillaries, deficiency or absence of basal lamina and pericytes and reduced endothelial expression of occludins and other cell surface proteins result in large endothelial gaps or fenestrations and leaky intercellular tight junctions that together produce markedly increased capillary permeability.[59,60] These two central features of tumor neovasculature are the focus of the two types of microvascular imaging methods for brain tumors: DSC MRI to measure nCBV and DCE T1-weighted "permeability" techniques used to estimate impairment of the BBB by monitoring passage of the contrast into the extravascular space during early recirculation phases.

In brain tumors, the increased microvascular density produces increased nCBV on DSC PWI. Normalized cerebral blood volume measurements have been shown to correlate with tumor grade and histological findings of increased tumor vascularity with higher nCBV values in high-grade gliomas compared to low-grade gliomas.[57,61–66] Using an nCBV value ratio threshold of 1.75 resulted in 95.0% sensitivity and 57.5% specificity in predicting high-grade gliomas.[64] And tumors with a baseline nCBV ratio of <1.75 had a significantly longer time to progression than those with nCBV >1.75.[67] Relative CBV values may also be useful in predicting tumors undergoing malignant transformation before changes are seen in conventional contrast-enhanced MRI sequences.[68] Maximum rCBV has been demonstrated to be very helpful in preoperative planning to ensure biopsy, resection or ablation of the highest-grade portion of a heterogeneous tumor.[69–71] Therefore, the rCBV map should be used in conjunction with the conventional, particularly contrast-enhanced, MRI sequences to guide the biopsy of a tumor in order to try to target the region that likely represents the highest grade. Relative CBV maps may be particularly helpful in nonenhancing tumors to guide sampling of the areas of presumed increased tumor vascularity. In treated glioma, normalized CBV >1.47 has been shown to have an 81.5% sensitivity and 77.8% specificity to differentiate pseudoprogression from true early progression,[72] with higher values indicating true progression (Figure 2.5). Perfusion imaging can also aid in

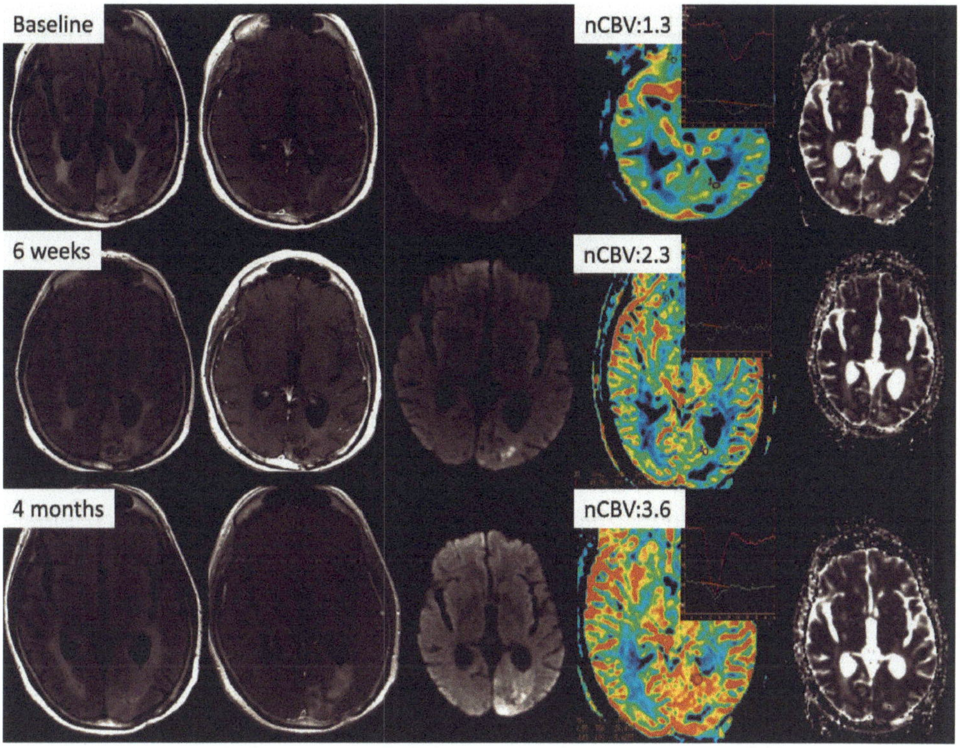

Figure 2.5 Glioblastoma demonstrating increasing tumor edema, mass effect, contrast enhancement, cellularity and blood volume consistent with high grade local tumor progression.

the differentiation of delayed radiation induced necrosis from recurrent or residual tumor. If there is substantial elevation of nCBV, this is predictive of recurrent tumor rather than delayed radiation induced necrosis.[73] In addition, with the recent use of antiangiogenesis cancer therapies that directly diminish tumor vascularity, DSC MRI has been shown to demonstrate changes in nCBV of the tumor during treatment and can be an indicator of response to this therapy.[74]

DSC MRI may be useful in differentiating solitary metastasis from a primary glioma on the basis of increased nCBV in the peritumoral area defined on conventional MRI sequences as the region of nonenhancing T2 signal abnormality adjacent to the enhancing portion of the mass. The measured relative cerebral blood volumes in the peritumoral region in high-grade gliomas and metastases were 1.31 ± 0.97 (mean $\pm$ SD) and 0.39 ± 0.19, respectively. The difference was statistically significant ($P < 0.001$)[75].

In short, the use of DSC MRI adds sensitivity and specificity in glioma assessment and clinical decision-making. However, the importance of correlation with other anatomic and advanced sequences cannot be overstated. Increased nCBV is not synonymous with malignancy: a number of intracranial neoplasms, especially meningiomas, choroid plexus papillomas, and low-grade oligodendrogliomas, have been shown to have high nCBV.

T1 dynamic contrast enhanced permeability imaging

Dynamic contrast enhanced (DCE) permeability imaging is a semiquantitative adaptation of gadolinium-enhanced imaging, in which fast spoiled gradient echo T1-weighted images of the whole brain are obtained continuously every few seconds starting before the arrival of the contrast bolus until roughly 3 minutes after injection. As the GBCA contrast recirculates, more leaks out of the capillaries into the interstitium over time at a rate determined largely by the local permeability of the capillary BBB. The increasing concentration of GBCA in the extravascular space shortens the T1 of the adjacent extravascular free water until a steady state is reached. Time-intensity curves are plotted that represent the resulting gradual increase in signal intensity of each perfused voxel on T1WI. Because the change in longitudinal relaxation rate R_1 ($R_1 = 1/T_1$) is proportional to the change in tissue concentration of GBCA, the TIC can be used to estimate rate of change of the extravascular GBCA concentration. In order to improve the reliability of this estimate T1-mapping is generally performed immediately before DCE imaging. This change in concentration can be used to derive a large number of parameters related to leakage of GBCA from the intravascular to the extravascular compartment of the brain and hence to BBB impairment.[76–78] Higher-grade brain tumors tend to have a more permeable BBB because the dysplastic high volume capillaries induced by secretion of VEGF and other cytokines have large gaps in their walls. This fact underlies the familiar enhancement of higher-grade tumors on delayed contrast-enhanced TWI.[79] Of the derived semiquantitative measures of permeability, the most widely used is the net forward volume transfer constant (K^{trans}) derived from a two-compartment pharmacokinetic modeling equation. The permeability surface area product (PS) is also another commonly used measure of tumor vascular leakiness.

Tumor permeability correlates reproducibly with increasing tumor grade when a reliable acquisition and postprocessing technique is used.[80–83] Tumor permeability can also aid targeting of higher grade regions within a tumor for biopsy to help distinguish between recurrent glioma and radiation necrosis and to track response to VEGF inhibitors.[84]

In short, DCE permeability imaging is a very promising technique, but has yet to achieve widespread clinical use because the acquisition requires an additional 5+ minutes of imaging time including T1-mapping, because the quantitative estimates of permeability parameters rely on complex pharmacokinetic modeling and vary substantially depending on the model and software used, and the information provided is not truly independent of that provided by standard of care delayed contrast-enhanced T1WI.

Proton MR spectroscopy

Proton MR spectroscopy (MRS) assays the intensity of resonance from a number of chemically distinct proton species present at high concentration in the brain. Brain metabolites have concentrations in the range of mmol/l or less, whereas protons in water and lipids are present in exponentially higher concentrations and therefore sequences must be optimized to suppress the free water signal from protons in water by more than four orders of magnitude in order to detect the much smaller signal from the metabolites of interest. Each proton species has a slightly different nuclear magnetic resonance frequency related to differences in the covalent bond structure of the parent molecule and resultant differences in shielding of the proton nucleus as well as spin exchange (coupling) between neighboring nuclei. This chemical shift of the proton resonance frequency is displayed in a graphical format with the *x*-axis representing the differences in proton resonance frequency in units of parts per million (ppm) of the resonance frequency of a standard reference compound, tetramethylsilane (TMS). PPM is used rather than hertz in order to produce spectra that are comparable across different magnetic field strengths. Because clinical MRS is not directly quantifiable, the *y*-axis represents arbitrary units of signal intensity scaled relative to the highest peak.[85,86] In addition to the normal graphical format of each voxel, MRS data can be used to produce color "metabolite maps" that depict the spatial distribution of the different peak heights, areas, or peak ratios that can be derived from the spectra. Currently, clinical MRS sequences are typically obtained with pulse sequences referred to as Point Resolved Spectroscopy Sequence (PRESS) or Stimulated Echo Acquisition Mode (STEAM), although more advanced sequences are becoming available. There are two basic approaches to spatial localization in MR spectroscopy: single voxel techniques and multi-voxel techniques. In routine use, single voxels of roughly 1 cm^3 can be acquired in 5 minutes or less. Multivoxel techniques can be acquired as either two- or three-dimensional (3D) arrays of contiguous voxels and are usually acquired with a PRESS sequence.

The primary metabolites detected by clinical MRS include branch chain amino acids (0.9–1.0 ppm), lipid (0.9–1.5 ppm), lactate (doublet at 1.31 ppm), alanine (1.5 ppm), *n*-acetyl aspartate (NAA; primarily at 2.01 ppm), glutamate/glutamine (2.35–2.45 ppm), choline (3.21 ppm), creatine (3.02 ppm and 3.91 ppm), and myoinositol (3.56 and 4.05 ppm). Creatine produces two resonant peaks because it contains two chemically distinct species of protons. The lipid and amino acid peaks are broad because each contains a large number of different molecules with similar but not identical bond structures and hence numerous protons in similar but not identical chemical environments. Overlap between the spectral ranges of the amino acids, lactate, and lipid peaks can be resolved in many cases with a combination of short and intermediate or intermediate and long echo time spectra because the distinct protons have distinct T2 relaxation times.

Short TE (usually <35 ms) spectra have better SNR. Because short TE spectra detect protons with T2 too short to be detectable on long TE spectra, short TE spectra can also detect more compounds than longer echo time spectra. On the other hand, short TE spectra typically suffer from more contamination by water and lipid signal and therefore have a more variable baseline. Longer echo time spectra (e.g., 130–150 ms or 260–300 ms), despite slightly lower SNR, are insensitive to lipids and macromolecules and therefore are more robust and constitute the default for clinical MRS. Typically short echo time spectra can be obtained if there is a need to detect compounds such as glutamate/glutamine or myoinositol which are only observed with short TE spectra.

A combination of 135 ms and 270 ms TE spectra can be used to confirm the presence of lactate. The three identical lactate methyl group protons are detected on MRS at 1.3 ppm exchange spin with the adjacent two identical methine protons through the shared electrons cloud of the molecular covalent bond. This "indirect spin-spin coupling" is often referred to as *J-coupling* (or *scalar coupling* or *through-bond coupling* to distinguish it from *through space coupling* or *dipolar coupling*). Because of J-coupling, the lactate methyl group protons can occupy two slightly different energy states with roughly 7.35 Hz different resonant frequency. Thus each lactate containing voxel has a mix of protons resonating at two slightly different frequencies which cycle in and out of phase with each other roughly every 1/7.35 second (i.e., 135 ms). As such, at 1.5T the combination of the two longer echo times (typically 135 ms and 270 ms) can be used to confirm the presence of lactate by observing an inverted 1.3 ppm doublet when the methyl proton pools are in negative in-phase on TE = 135 ms spectra and upright doublet when they are in positive in-phase at TE = 270 ms spectra. The underlying J-coupling phenomenon is a ubiquitous interaction of protons bound to macromolecules. J-coupling is also seen prominently in MRS of alanine and amino-acid resonances, is responsible for the low signal intensity of fat on conventional SE T2WI, and underlies many of the more advanced spectral editing and correlation spectroscopy techniques such as COSY, TOCY, etc. Special attention to this phenomenon is needed to correctly identify the lactate, amino acid and alanine peaks at 3T because of anomalous J-coupling effects.[87]

Brain metabolite resonances

NAA is manufactured in neurons and is routinely used as an indicator of neuronal injury. It is decreased in any condition that decreases the number or impairs the normal metabolic homeostasis of neurons, including tumors, infarctions, demyelination, infection, radiation injury, necrosis, etc. Choline is a marker of cell membrane turnover. High choline is seen in any condition that causes either formation of new cells or injury to cells—especially formation of or injury to glial cells—including tumors, demyelination, and other pathological processes. Although the choline/NAA ratio is high in brain tumors, and very high ratios in tumor suggest higher-grade tumor, increased Cho/NAA ratio is not specific for tumor. Thus, while Cho/NAA can be useful for biopsy targeting in known glioma, it is rarely helpful in differential diagnosis. Any process that disrupts brain aerobic glycolysis can result in lactate formation, and all processes that produce necrosis release lipid and decrease creatine.[88–90] Because the normal concentrations of these metabolites vary by anatomic location and because the relative signal detected from a given concentration of each metabolite varies with echo time chosen for a spectroscopy sequence, reference to spectra of normal-appearing voxels is critical for clinical interpretation of MRS.[91–95] More interesting metabolites present at lower concentrations and/or having a NMR resonant frequency overlapping one of the major resonances can be measured by using two spectroscopic dimension correlation (2-D correlation single-voxel MRS) techniques employing a second spin-evolution dimension. These 2-dimensional spectroscopy techniques should not be confused with one-dimensional spectroscopy performed in multiple simultaneous voxels in two or three spatial dimensions (2D or 3D multivoxel MRSI). An exciting application of correlation spectroscopy currently under development is 2-hydroxyglutarate (2HG) imaging in glioma designed to detect the metabolic consequence of the IDH-1 mutation.[96] Although these techniques have been known for more than half a century, and have been used in in vivo research for decades, data acquisition in clinical tumor patients has only become possible in the last few years because of hardware improvements, and remains invalidated, technically challenging, and very time consuming. Efforts at clinical validation are underway and the technical barriers to routine use may decrease substantially as 7+ Tesla imaging becomes available in the near future.

Magnetic resonance spectroscopy in differential diagnosis

As noted, the individual metabolites and ratios that comprise the typical pattern of glioma spectra—high choline and low or absent NAA peaks, with lipid and lactate peaks often seen in GBM—have been extensively investigated and failed to demonstrate value in differential diagnosis between tumor types or differentiation of tumor from nonneoplastic processes such as demyelination, ischemia, and gliosis.[97–100] With this understanding, there are a limited number of niche applications in which MRS can be useful in differential diagnosis. The spectra of extra-axial tumors such as meningioma generally reveals high choline related to membrane formation and no NAA because the tumors contain no neurons. While this may be indistinguishable from high-grade glioma, a high alanine peak observed in a subset of meningioma can be strongly suggestive of meningioma. It should be noted that the low levels of alanine seen in up to 80% of meningioma is not specific because it is detected in similar frequency in metastases and schwannoma.[101] A second niche in which MRS adds value to differential diagnosis is the distinction of abscess from rim-enhancing tumor by demonstrating branch chain amino acids within the contents of the cyst, a finding that is essentially diagnostic of the presence of activated polymorphonuclear leukocytes, and thus of bacterial or, less likely, parasitic infection.[102,103] Of note, this is true only if the voxel is completely contained within the cyst lumen since partial volume averaging of brain parenchyma will result in false positive detection of AA resonances in uninfected tumoral or other cysts. As such this is useful only in cystic lesions more than roughly 1.5 cm in diameter, and meticulous attention to the prescription of the voxel is required in all three spatial dimensions to prevent inadvertent inclusion of parenchyma.

Magnetic resonance spectroscopy in preoperative glioma grading and operative guidance

Within known or very strongly suspected glioma, qualitative or quantitative detection of high choline/NAA peak height ratios is predictive of the presence of high-grade tumor.[33,104] Similarly, the presence of lipid/lactate in untreated glioma suggests the presence of necrotic grade IV tumor.[33,105] Although there is considerable overlap between high- and low-grade tumor spectra, meticulously acquired spectra revealing choline/NAA ratios above 1.5 can improve the accuracy of anatomic MRI prediction of tumor grade.[106,107] Although more technically robust and cost-effective perfusion and permeability techniques have significantly reduced the use of MRS for tumor grade estimation, lesions suggestive of oligodendroglioma remain a notable exception because high blood volume is seen even in low-grade oligodendroglioma, reducing the usefulness of perfusion imaging for grading.[108–111] In addition the use of MRS to target biopsies to areas with high choline/NAA ratios has been reported to increase the accuracy of tumor biopsy by targeting areas of metabolically active tumor within areas of heterogeneous glioma, thus reducing the false-negative rate.[112,113] Similar methods have been used to guide stereotactic radiosurgery.[114–116] Finally, a recent report showed that glioma may decrease whole-brain NAA 30% more than can be explained by the visible tumor burden, suggesting that decreased whole-brain NAA may reflect the global burden of infiltrative tumor.[117] Because infiltration is a feature of glioma that cannot be detected reliably with current techniques, the significance of whole-brain NAA deserves further exploration as a marker of poor prognosis and diffuse tumor spread. Another recently reported MRS technique for detection of tumor infiltration that deserves further study is the use of CH_2/CH_3 ratios within the normal brain lipid pool to assay for tumor burden.[118]

Magnetic resonance spectroscopy in assessment of treatment response

Because delayed radiation necrosis is also characterized by the presence of lactate/lipid peaks, the presence of these peaks alone is not useful in the distinction of radiation necrosis from tumor recurrence.[119] Complete absence of NAA and choline peaks on multivoxel spectroscopy containing internal reference voxels or serial MRSI documenting progressive decrease in NAA and choline peaks combined with lactate/lipid peaks should suggest necrosis, particularly when corroborated by a rising ADC and low blood volume. Conversely, a significant increase in choline plus a decrease in NAA over time, with a consequent increase in the choline/NAA ratio or in derived statistics such as the choline/NAA ratio R value, is a sensitive indicator of tumor recurrence when seen in the appropriate anatomic imaging context.[113,114,120–122] Overall, serial MRSI under carefully controlled conditions has been shown to be a useful adjunct to conventional imaging for discrimination of high-grade focal brain tumor recurrence from delayed radiation necrosis in the hands of a few research groups, especially when combined with other imaging data (Figure 2.6). Unfortunately, because the spatial variation in choline, NAA, lactate, and lipid peaks within an individual tumor is often much greater than the change in these peaks over time, slight differences between scans in voxel placement, registration or acquisition technique can render the assessment of longitudinal change unreliable. Experience has shown that with currently available commercial MRI hardware and software, reliable clinical MRSI requires the direct supervision of each MRS data acquisition and postprocessing by a trained spectroscopist under direct physician supervision. In the United States, where MRS and MRSI are not reimbursable at present, these personnel costs impose an unsupportable burden that prevents the routine use of serial MRSI brain tumor monitoring at most centers.

Magnetic resonance spectroscopy summary

Selection of an appropriate area of interest and voxel size is critical to produce useful information while avoiding artifacts from partial volume averaging of calvarial marrow fat and susceptibility from bone or metal. Moreover, because the information provided by spectroscopy is generally not specific enough to be useful when only a single time point is available, serial comparison of change in spectra over time is critical for accurate interpretation. This need for serial comparison compounds the data acquisition problem because it requires a high degree of reproducibility in voxel selection over serial scans. As a result, the spectroscopy groups that have had the greatest success using MRS and MRSI in neurooncology have found it necessary to develop significant additional human resources for monitoring data acquisition and processing beyond what is generally available in the routine clinical MR setting. At the present time, MRSI is not reimbursable, so the number of institutions with the financial resources to develop effective spectroscopy laboratories remains small, even among the major academic centers.

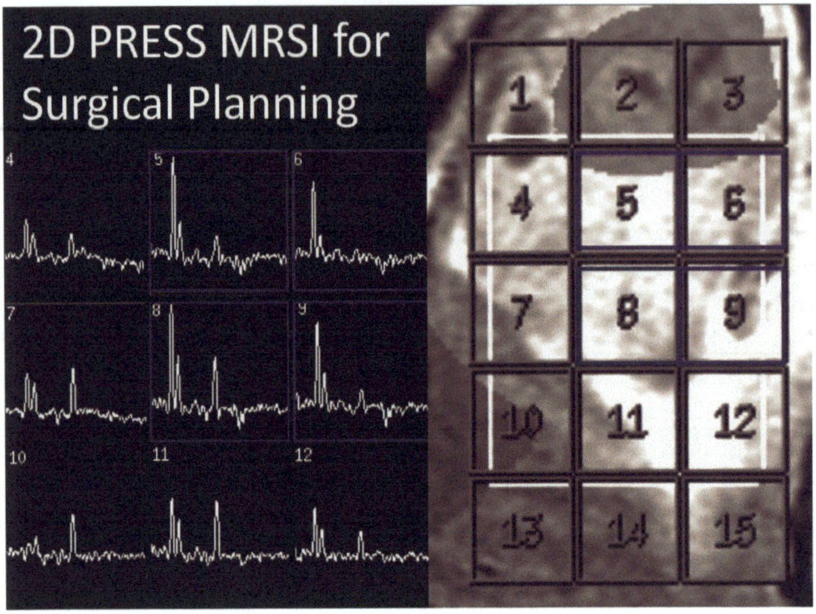

Figure 2.6 2D PRESS MRSI demonstrating high Cho/NAA ratio in voxels 5, 6, and 9 suggesting that biopsy in this location will increase the chance of obtaining the correct tumor grade.

CT/DYNAMIC CTA/CT PERFUSION

Computed tomography (CT) is a technique that mathematically constructs digital cross-sectional images of the body reflecting attenuation of narrow beam ionizing X-radiation emitted by a rotating X-ray tube. Because X-ray attenuation is related almost exclusively to photon–electron interactions, CT of the brain reflects only one underlying contrast mechanism: electron density. While this limitation means that CT provides far less soft tissue contrast than MRI, CT acquisition is fast, safe, and requires no patient safety screening. When adequate technique is used, CT provides high sensitivity to hemorrhage, herniation and hydrocephalus, which would require immediate surgical intervention. Because the relationship between X-ray photon attenuation is quite linear and the transmission detection is not subject to the variation in scaling introduced by RF coupling in MRI, CT is much more easily quantifiable than MRI. The gray-scale values in CT are called Hounsfield Units (HU). By definition, the HU for water is 0 with HU ranging from -1000 (air) to $+3,095$ for most 12-bit CT scanners. Some important HU ranges include acute hemorrhage that in a patient with normal hematocrit is on the order of 60 HU (less if hyperacute), CSF at approximately 15 HU, gray matter at 37–41 HU, white matter at 30–34 HU, adipose tissue at −80 to −30 HU, and bone ranging from 700–3000 HU. The minimal difference in attenuation between gray and white matter should make clear why subtle abnormalities easily detected on MRI are often difficult or impossible to detect by CT. On the other hand, the linear path of X-ray photons through tissue means that CT is not subject to the anatomic distortions encountered in MRI. This constitutes a major advantage of CT for surgical planning. In many cases superior contrast MRI images are registered ("fused") to anatomically precise and undistorted CT images in third-party surgical guidance software systems for this reason.

CT images are "reconstructed" from massive source data files constituting temporal series of photon detection counts from each detector array and row. This has typically been performed by an operation known as *filtered back projection*, but in recent years more complex model based and iterative reconstruction methods have been introduced to try to reduce radiation dose. Each image reconstruction employs a selected *reconstruction kernel* that specifies the degree of spatial averaging applied to smooth out image noise. Typically for brain imaging, the user will be provided with two images: a bone kernel reconstruction producing higher spatial resolution in the image allowing evaluation for thin fractures and other bone detail at the cost of increased noise, and a soft tissue kernel reconstruction with less noise and hence better brain tissue contrast at cost of lower spatial resolution. In addition, CT images can be reconstructed at different slice thicknesses with the same trade-off of spatial resolution vs. noise. Once the image is reconstructed with a certain slice thickness and kernel, the user can choose to view the images with different window level and width settings which highlight differences between HU of pixels but cannot alter the underlying voxel size or spatial resolution determined by the field of view and kernel selected at the time of image reconstruction. In addition to soft tissue and bone windows, it is important to review the brain parenchyma using "stroke" windows (40 window width, 40 window level) for high contrast between gray and white matter and also "subdural" windows (approximately 350 window width, 90 window level) to evaluate for subtle extra-axial hemorrhage.

CT angiography (CTA) and CT venography (CTV) are methods for depicting and evaluating the arterial and venous anatomy in a less invasive way than standard fluoroscopic angiography (Figure 2.7). Because of the higher contrast resolution of CT, compared to fluoroscopy, contrast injected by a venous access can be used to depict the arterial anatomy on CTA as opposed to arterial access needed for fluoroscopy. This has led to a substantial increase in the use of CTA for initial evaluation of stroke patients and for patients with arterial venous malformations, among other pathologies, while leaving the more invasive fluoroscope angiograms for cases requiring treatment after initial evaluation or cases in which higher resolution provided by conventional angiogram is needed (Figure 2.8). More advanced CT scanners with up to 320

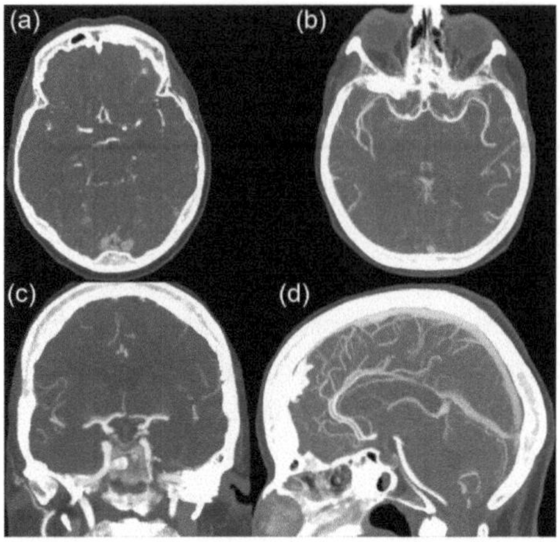

Figure 2.7 CT angiography (CTA) is often very useful for surgical planning because it depicts the relationship of a lesion to the adjacent arteries, veins, dural sinuses, and bones without spatial distortion and can easily be fused with MRI images for surgical navigation. CTA source images (a) are often reconstructed with overlapping thick section maximum intensity projection images that can be depicted as a long segment of the vessels with preserved edge information in all three orthogonal planes: axial (b), coronal (c), and sagittal (d).

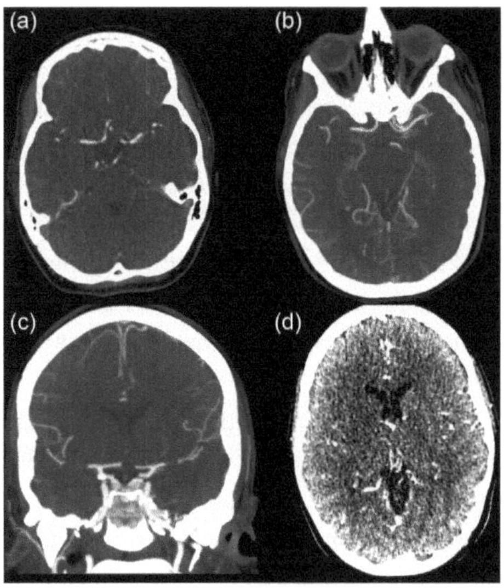

Figure 2.8 CTA depicting an acute stroke with focal cut-off of the right M1 MCA segment on source images (a) and overlapping thick section maximum intensity projection images in both axial (b) and coronal (c) planes. Using a narrow window centered around the vessels to display the source images (d) allows detection of paucity of distal branch vessels in the MCA territory and subtle effacement of the gray-white matter differentiation in the right basal ganglia and right insular cortex strongly supporting the diagnosis of acute or early sub-acute occlusive MCA infarct.

detector rows can acquire up to 16 cm FOV volumetric data during a single rotation of the gantry, allowing for dynamic volume CT with high temporal resolution displaying both anatomical and functional data, including multiple different phases of blood flow in the intracranial arterial and venous systems and producing a virtual CT digital subtracted angiogram with both high spatial and temporal resolution. These studies are often very useful for surgical planning because they depict the relationship of a lesion to the adjacent arteries, veins, dural sinuses, and bones in a way that can easily be fused with MRI images for surgical navigation. In most cases, this has obviated the need for more invasive conventional angiograms prior to surgical intervention.

In addition, whole brain CT perfusion imaging can be performed by an analogous method to MR perfusion imaging by acquiring intermittent whole brain images during bolus intravenous injection of iodinated contrast in order to assess capillary relative cerebral blood volume, mean transit time, relative cerebral blood flow, and other physiological parameters. CT perfusion is most often indicated for the acute evaluation of a potential stroke patient, allowing for quick assessment of the vasculature with the CTA portion and depicting brain "at risk" for further infarction with the CT perfusion portion. Although early iterations of CT perfusion technique required radiation exposures up to 5–6 times that of a standard head CT,[123,124] techniques implemented in all CT scanners sold since 2008 allow CT perfusion to be acquired with radiation dose similar to or less than a standard head CT.[125] Barring failures of due diligence and quality assurance that have resulted in substantial radiation overdoses, radiation dose in modern CT perfusion should not be a significant consideration.[125]

SAFETY OF CT AND MRI

While a full discussion of the risks of each modality is beyond the scope of this chapter, recent media disinformation about radiation risk and increasing attention to MRI-related accidents calls for a brief discussion of basic principles. In MRI, the energy transmitted into the body is in the frequency spectrum between FM radio waves and microwaves with frequencies in the range of 64–128 MHz. The photon energies encountered at these frequencies are far too low to strip electrons from atoms, and so MRI does not create free radicals in tissue the way CT X-ray photons do. However, the "non-ionizing" photon energies used in MRI are roughly matched to the covalent bond vibrational modes of water, fat, and other organic molecules and so significant amounts of energy are deposited in tissue as heat in the same way as a microwave oven heats food. Current knowledge of radiation biology and DNA repair raises legitimate doubt as to whether low-energy ionizing radiation from CT causes any increase in cancer, and to date substantial investigation has failed to convincingly demonstrate direct evidence of causality, suggesting that such risks, if real, are extremely low. The theoretical possibility that CT radiation may contribute to cancer is largely derived from inappropriate epidemiological extrapolation from flawed high dose data using a linear no-threshold (LNT) model. Expert opinion concurs with a recent review which concludes that the "LNT [model] appears to be scientifically invalid in the low-dose range" and that for both this reason and because of the low inherent risks, epidemiological extrapolation cannot accurately predict cancer risk, if any, associated with low-dose radiation exposure.[126] Despite this dearth of solid evidence, the purported evidence of CT radiation producing cancer has received significant attention in the media and from the FDA, leading to the perception that MRI is safer than CT. Close attention to the evidence reveals a different picture. A number of recent reports document that, like CT, MRI creates DNA damage, likely similar in magnitude to that produced by CT.[127,128] The significance of this is as unclear for CT as for MRI. Extensive investigation in cells, tissues, and living organisms has demonstrated no convincing evidence of long-term adverse tissue effects from MRI, and 30 years of clinical CT and MRI have not produced convincing evidence of any cancer, other significant disease or delayed or cumulative adverse biological effect among CT or MRI patients.[129]

While there are no known serious patient injuries or deaths attributable to CT, MRI is associated with certain known hazards.[129,130] Estimating the frequency of patient injury is difficult because of widespread systematic underreporting of MRI related injury caused by fears of legal liability and damage to institutional reputations. Ongoing analysis of government incident report databases suggest that the several dozen deaths and many hundreds of injuries known from the published literature and FDA databases underreport the actual incidence by more than 50- to 100-fold and that the rate of such events is increasing faster than the rate of MRI utilization.[131] This plausible estimate suggests that MRI has resulted in possibly more than a hundred thousand injuries and at least many hundreds and more likely thousands of patient deaths. The increase in reported incidents is likely in part due to increasing ascertainment related to increasing attention to patient safety and new regulatory reporting rules, but also likely due to the increasing prevalence of implanted medical devices, many of which pose special risks in MRI.[130]

The majority of MRI patient injuries are burns of the skin or subcutaneous fat. These have been reported in patients with almost any metal-containing object: drug patches, tattoos, metallic clothing, bullets, central lines, pulse oximeters, ECG, and EEG leads and myriad implanted devices. Burns also occur less frequently from the wires and coils of the MRI scanner and rarely without obvious cause. These are mostly minor, but third and fourth degree burns have occurred leading to digit and extremity amputation, and deep tissue burns of the brain have been reported in patients with deep brain stimulators and transcranial pressure monitors.[132–136] The majority of known patient deaths caused by MRI relate to inappropriate or inadvertent scanning of patient with cardiac pacemakers and resultant lead heating and/or accidental stimulation. In addition to heating, the extremely powerful magnetic field can produce hazardous magnetic torques in patients with unsafe implanted devices and can produce missile hazards if an unsafe metal object is brought into the scanner room. Notable examples include patients who have died when magnetic torque on an MRI-unsafe cerebral aneurysm clip avulsed the middle cerebral artery and patients who have been seriously injured or killed by blunt head trauma when MRI-unsafe gas cylinders brought into the room flew into the center of the bore.[137–139] The neurosurgeon as referring physician plays a critical role in helping to prevent such catastrophes and more minor patient injuries by clearly documenting the precise manufacturer, model and serial number of every medical device (aneurysm clip, VP shunt, DBS, VNS, etc.) prominently in the medical record at the time of implantation, asking patients about implanted devices when referring to MRI and alerting the MRI technologists to the presence of any known devices. Nevertheless, given that hundreds of millions—reliable estimates suggest roughly 800 million—MRI scans and at least the same number of CT scans have been performed in patients,[140] physicians and patients should be reassured that when standard precautions are observed, the known risks of each of these modalities is extremely low.

In sum, it can be said with confidence that after more than 30 years of widespread human use there is no convincing direct evidence linking either standard diagnostic CT or MRI to cancer or other significant human injury or disease. If such risks are real for either modality, they are extremely low. As such, for any adult patient suspected of having disease significant enough to merit neuroimaging evaluation, let alone consideration of neurosurgery, the risk from the underlying disease and surgery or other therapies are generally far greater—orders of magnitude greater in most cases—than any plausible risk from properly performed medical imaging with CT or MRI. Thus, it seems appropriate that in general when MRI and CT imaging is performed in accordance with standard of care patient safety and quality assurance practices, both are regarded by the international medical and scientific community as insignificant risk procedures for which informed consent is neither indicated nor required. Therefore, the choice of imaging modality and technique in neuroimaging should be driven by the unique and complementary capabilities of each modality to contribute to accurate diagnosis and optimal surgical planning and guidance and not by undue concern about the very low risks inherent to these modalities.

Contrast agent risks

The use of contrast agents in MRI and CT introduces both risks and benefits distinct from the modalities themselves. Although safe and appropriate use of contrast agents is an extremely complex topic, it may be valuable to introduce a few underlying principles. The following is not intended as a detailed guide to clinical decision making about contrast agent use. If further detail is desired, the reader is urged to consult the *ACR Manual on Contrast Media*, and relevant FDA drug labeling.[141]

Although a number of advanced imaging contrast agents are under development and beginning niche clinical use, the most common CT and MRI contrast agents both consist of a heavy metal bound to a large chelating molecule. In CT the metal is usually iodine and the agents are termed *iodinated contrast agents* (ICA). In MRI the metal is gadolinium and the agents are termed *gadolinium based contrast agents* (GBCA). In CT the iodine electrons absorb X-rays. In MRI the paramagentic properties of the gadolinium when present at low concentration shorten the T1 of the adjacent water protons, increasing the signal intensity detected on T1WI and when present in high concentrations shorten the T2* of nearby water protons on GRE sequences. These are used both for imaging of the intravascular compartment in CTA, MRA, CTV, MRV, and CTP and MRP and for detecting areas of abnormal capillary permeability related to the presence of tumor or inflammation, infection, radiation or mechanical injury, etc. These agents add vital information for surgical planning which significantly outweighs the attendant risks in the vast majority of patients. It should be noted that the paramagnetic properties of GBCA combined with the exquisite tissue contrast of MRI mean that much greater sensitivity to abnormal capillary permeability is achieved with much smaller doses.

The most immediate risk of both ICA and GBCA is anaphylactic or allergic contrast reaction. This is much less common with modern low-osmolar nonionic ICA than older ionic CT contrast agents, and even less common in MRI than CT, but can be life-threatening when it occurs.[142] Although a number of risk factors are known, the most important risk factor for contrast agent related anaphylaxis is a history of prior allergic reaction to the same type (ICA or GBCA) and same formulation of contrast agent. When this is known, detailed documentation in the medical record and clear communication to the radiologists and radiology staff are essential elements of patient safety. Although both ICA and GBCA are principally excreted by the kidneys and both are intrinsically nephrotoxic, this is rarely clinically relevant for GBCA because of the much smaller doses used for conventional imaging. In CT, on the other hand, the nephrotoxicity of ICA is significant in patients with pre-existing acute or chronic renal insufficiency and dehydration and is a significant contributor to acute renal failure in inpatients. In such patients hydration can have a protective effect and is essential if ICA use cannot be avoided. Similarly, while both GBCA and ICA create an osmotic challenge, this is rarely a significant issue with the typical small doses of GBCA used in MRI but can produce osmotoxic reactions and life-threatening fluid shifts/intravascular fluid overload in CT patients with impaired renal or cardiac function or preexisting fluid overload.

In contrast to these immediate effects, GBCA are now known to produce nephrogenic systemic fibrosis (NSF), a progressive, incurable and often crippling or fatal systemic fibrosing syndrome in MRI patients with severe renal insufficiency or renal failure. To date this has only been identified in patients with pre-existing renal impairment.[143] Since the identification of NSF, GBCA use has been regarded as generally contraindicated in patients with chronic kidney disease and estimated glomerular filtration rates (eGFR) less than 30 or acute renal impairment, especially in patients on hemodialysis. Unfortunately, hemodialysis is not thought to be effective in preventing or treating NSF. NSF is felt to be related to toxicity of free gadolinium ions that dissociate from the chelating molecule during the prolonged plasma circulation times encountered in patients with renal failure. Consistent with this, certain newer classes of GBCA with much lower dissociation constants appear to have a better safety profile in regard to NSF.[144] The critical element in preventing NSF is to identify patients with acute renal impairment or eGFR <30 chronically. In such patients use of GBCA should be avoided if at all possible.

Finally three recent reports in the literature document long-term accumulation of either chelated or dechelated Gd in the cerebellar and deep nuclei of patients with normal renal function who have received multiple MRI scans using one of the two FDA approved linear nonionic GBCA.[145–147] This basic finding is consistent with established biochemical principles and in vitro, animal and human tissue literature documenting some release of free Gd and long-term accumulation in tissue. To date the accumulation in the cerebellar outlet nuclei and deep nuclei is not known to have any pathophysiologic or pathologic correlate, but these locations are worrisome because of their involvement both in a host of neurodegenerative conditions including neurodegeneration related to heavy metal toxicity. The most widely used standard GBCA and especially the newer macrocyclic agents have thermodynamic Gd-binding dissociation constants that are many orders of magnitude lower than the linear nonionic GBCA. Because dechelation is substantially affected by ion exchange with zinc, copper, calcium and iron (aka, transmetallation) thermodynamic stability is not the whole story, but in vivo and in vitro testing has demonstrated 4–10-fold higher levels of dechelation and free Gd recovery from animal and human tissue with the linear nonionic agents compared to the more common agents and the newer macrocyclic agents.[148,149,144] While this data is still preliminary, it may be prudent for the moment to avoid use of the lower affinity agents whenever possible, especially in patients with a long life expectancy and potential to need multiple contrast-enhanced MRI scans. Although use of GBCA is not known to be associated with adverse effects on the fetus of pregnant patients, the recent data on dechelation also adds support to the longstanding practice of avoiding use of GBCA in pregnant patients unless absolutely necessary.

CONCLUSION

MRI and CT neuroimaging play a critical role in diagnosis, preoperative assessment, surgical planning, and posttherapeutic monitoring of the neurosurgical patient. Optimal use of the different MRI and CT techniques can be a substantial determinate of surgical safety, success and quality of care and requires a detailed understanding of the physical principles, application, interpretation and strengths and weaknesses of each technique outlined here.

REFERENCES

1. Bloch F. Nuclear induction. *Phys Rev.* 1946;70:460–474.
2. Bitar R, Leung G, Perng R, et al. MR pulse sequences: what every radiologist wants to know but is afraid to ask. *Radiographics.* 2006;26:513–537.
3. Mikulis DJ, Roberts TP. Neuro MR: protocols. *J Magn Reson Imaging.* 2007;26:838–847.
4. Roberts TP, Mikulis D. Neuro MR: principles. *J Magn Reson Imaging.* 2007;26:823–837.
5. Jackson EF, Ginsberg LE, Schomer DF, Leeds NE. A review of MRI pulse sequences and techniques in neuroimaging. *Surg Neurol.* 1997;47:185–199.
6. Iskandar BJ, Sansone JM, Medow J, Rowley HA. The use of quick-brain magnetic resonance imaging in the evaluation of shunt-treated hydrocephalus. *J Neurosurg.* 2004;101:147–151.
7. Hall WA, Liu H, Martin AJ, Truwit CL. Comparison of stereotactic brain biopsy to interventional magnetic-resonance-imaging-guided brain biopsy. *Stereotact Funct Neurosurg.* 1999;73:148–153.
8. Hall WA, Martin AJ, Liu H, Nussbaum ES, Maxwell RE, Truwit CL. Brain biopsy using high-field strength interventional magnetic resonance imaging. *Neurosurgery.* 1999;44:807–813:discussion 13–4.
9. Liu H, Hall WA, Truwit CL. Remotely-controlled approach for stereotactic neurobiopsy. *Comput Aided Surg.* 2002;7:237–247.
10. Martin AJ, Hall WA, Roark C, Starr PA, Larson PS, Truwit CL. Minimally invasive precision brain access using prospective stereotaxy and a trajectory guide. *J Magn Reson Imaging.* 2008;27:737–743.
11. Martin AJ, Larson PS, Ostrem JL, et al. Placement of deep brain stimulator electrodes using real-time high-field interventional magnetic resonance imaging. *Magn Reson Med.* 2005;54:1107–1114.
12. Brant-Zawadzki M, Gillan GD, Nitz WR. MP RAGE: a three-dimensional, T1-weighted, gradient-echo sequence—initial experience in the brain. *Radiology.* 1992;182:769–775.
13. Mardor Y, Pfeffer R, Spiegelmann R, et al. Early detection of response to radiation therapy in patients with brain malignancies using conventional and high b-value diffusion-weighted magnetic resonance imaging. *J Clin Oncol.* 2003;21:1094–1100.
14. Provenzale JM, Mukundan S, Barboriak DP. Diffusion-weighted and perfusion MR imaging for brain tumor characterization and assessment of treatment response. *Radiology.* 2006;239:632–649.
15. Ulug AM, Beauchamp Jr. N, Bryan RN, van Zijl PC. Absolute quantitation of diffusion constants in human stroke. *Stroke.* 1997;28:483–490.
16. Chong J, Lu D, Aragao F, et al. Diffusion-weighted MR of acute cerebral infarction: comparison of data processing methods. *Am J Neuroradiol.* 1998;19:1733–1739.
17. Gass A, Ay H, Szabo K, Koroshetz WJ. Diffusion-weighted MRI for the "small stuff": the details of acute cerebral ischaemia. *Lancet Neurol.* 2004;3:39–45.
18. Schulz UG, Briley D, Meagher T, Molyneux A, Rothwell PM. Abnormalities on diffusion weighted magnetic resonance imaging performed several weeks after a minor stroke or transient ischaemic attack. *J Neurol Neurosurg Psychiatry.* 2003;74:734–738.

19. Augustin M, Bammer R, Simbrunner J, Stollberger R, Hartung HP, Fazekas F. Diffusion-weighted imaging of patients with subacute cerebral ischemia: comparison with conventional and contrast-enhanced MR imaging. *Am J Neuroradiol.* 2000;21:1596–1602.
20. Geijer B, Lindgren A, Brockstedt S, Stahlberg F, Holtas S. Persistent high signal on diffusion-weighted MRI in the late stages of small cortical and lacunar ischaemic lesions. *Neuroradiology.* 2001;43:115–122.
21. Chenevert TL, Stegman LD, Taylor JM, et al. Diffusion magnetic resonance imaging: an early surrogate marker of therapeutic efficacy in brain tumors. *J Natl Cancer Inst.* 2000;92:2029–2036.
22. Chenevert TL, McKeever PE, Ross BD. Monitoring early response of experimental brain tumors to therapy using diffusion magnetic resonance imaging. *Clin Cancer Res.* 1997;3:1457–1466.
23. Hein PA, Eskey CJ, Dunn JF, Hug EB. Diffusion-weighted imaging in the follow-up of treated high-grade gliomas: tumor recurrence versus radiation injury. *Am J Neuroradiol.* 2004;25:201–209.
24. Chan YL, Yeung DK, Leung SF, Chan PN. Diffusion-weighted magnetic resonance imaging in radiation-induced cerebral necrosis. Apparent diffusion coefficient in lesion components. *J Comput Assist Tomogr.* 2003;27:674–680.
25. Tsui EY, Chan JH, Ramsey RG, et al. Late temporal lobe necrosis in patients with nasopharyngeal carcinoma: evaluation with combined multi-section diffusion weighted and perfusion weighted MR imaging. *Eur J Radiol.* 2001;39:133–138.
26. Guo AC, Cummings TJ, Dash RC, Provenzale JM. Lymphomas and high-grade astrocytomas: comparison of water diffusibility and histologic characteristics. *Radiology.* 2002;224:177–183.
27. Okamoto K, Ito J, Ishikawa K, Sakai K, Tokiguchi S. Diffusion-weighted echo-planar MR imaging in differential diagnosis of brain tumors and tumor-like conditions. *Eur Radiol.* 2000;10:1342–1350.
28. Toh CH, Chen YL, Hsieh TC, Jung SM, Wong HF, Ng SH. Glioblastoma multiforme with diffusion-weighted magnetic resonance imaging characteristics mimicking primary brain lymphoma. Case report. *J Neurosurg.* 2006;105:132–135.
29. Krabbe K, Gideon P, Wagn P, Hansen U, Thomsen C, Madsen F. MR diffusion imaging of human intracranial tumours. *Neuroradiology.* 1997;39:483–489.
30. Kotsenas AL, Roth TC, Manness WK, Faerber EN. Abnormal diffusion-weighted MRI in medulloblastoma: does it reflect small cell histology? *Pediatric Radiol.* 1999;29:524–526.
31. Filippi CG, Edgar MA, Ulug AM, Prowda JC, Heier LA, Zimmerman RD. Appearance of meningiomas on diffusion-weighted images: correlating diffusion constants with histopathologic findings. *Am J Neuroradiol.* 2001;22:65–72.
32. Sugahara T, Korogi Y, Kochi M, et al. Usefulness of diffusion-weighted MRI with echo-planar technique in the evaluation of cellularity in gliomas. *J Magn Reson Imaging.* 1999;9:53–60.
33. Catalaa I, Henry R, Dillon WP, et al. Perfusion, diffusion and spectroscopy values in newly diagnosed cerebral gliomas. *NMR Biomed.* 2006;19:463–475.
34. Yang S, Law M, Zagzag D, et al. Dynamic contrast-enhanced perfusion MR imaging measurements of endothelial permeability: differentiation between atypical and typical meningiomas. *Am J Neuroradiol.* 2003;24:1554–1559.
35. Bulakbasi N, Kocaoglu M, Ors F, Tayfun C, Ucoz T. Combination of single-voxel proton MR spectroscopy and apparent diffusion coefficient calculation in the evaluation of common brain tumors. *Am J Neuroradiol.* 2003;24:225–233.
36. Calli C, Kitis O, Yunten N, Yurtseven T, Islekel S, Akalin T. Perfusion and diffusion MR imaging in enhancing malignant cerebral tumors. *Eur J Radiol.* 2006;58:394–403.
37. Rosen BR, Belliveau JW, Aronen HJ, et al. Susceptibility contrast imaging of cerebral blood volume: human experience. *Magn Reson Med.* 1991;22:293–299:discussion 300–303.
38. Rosen BR, Belliveau JW, Buchbinder BR, et al. Contrast agents and cerebral hemodynamics. *Magn Reson Med.* 1991;19:285–292.
39. Rosen BR, Belliveau JW, Vevea JM, Brady TJ. Perfusion imaging with NMR contrast agents. *Magn Reson Med.* 1990;14:249–265.
40. Villringer A, Rosen BR, Belliveau JW, et al. Dynamic imaging with lanthanide chelates in normal brain: contrast due to magnetic susceptibility effects. *Magn Reson Med.* 1988;6:164–174.

41. Meier P, Zierler KL. On the theory of the indicator-dilution method for measurement of blood flow and volume. *J Appl Physiol.* 1954;6:731–744.
42. Calamante F, Thomas DL, Pell GS, Wiersma J, Turner R. Measuring cerebral blood flow using magnetic resonance imaging techniques. *J Cereb Blood Flow Metab.* 1999;19:701–735.
43. Zaharchuk G. Theoretical basis of hemodynamic MR imaging techniques to measure cerebral blood volume, cerebral blood flow, and permeability. *Am J Neuroradiol.* 2007;28:1850–1858.
44. Perkio J, Aronen HJ, Kangasmaki A, et al. Evaluation of four postprocessing methods for determination of cerebral blood volume and mean transit time by dynamic susceptibility contrast imaging. *Magn Reson Med.* 2002;47:973–981.
45. Nakagawa T, Tanaka R, Takeuchi S, Takeda N. Haemodynamic evaluation of cerebral gliomas using XeCT. *Acta Neurochir.* 1998;140:223–233:discussion 33–4.
46. Rivers CS, Wardlaw JM, Armitage PA, et al. Do acute diffusion- and perfusion-weighted MRI lesions identify final infarct volume in ischemic stroke? *Stroke.* 2006;37:98–104.
47. Parsons MW, Barber PA, Chalk J, et al. Diffusion- and perfusion-weighted MRI response to thrombolysis in stroke. *Ann Neurol.* 2002;51:28–37.
48. Schlaug G, Benfield A, Baird AE, et al. The ischemic penumbra: operationally defined by diffusion and perfusion MRI. *Neurology.* 1999;53:1528–1537.
49. Neumann-Haefelin T, Wittsack HJ, Wenserski F, et al. Diffusion- and perfusion-weighted MRI. The DWI/PWI mismatch region in acute stroke. *Stroke.* 1999;30:1591–1597.
50. Grandin CB, Duprez TP, Smith AM, et al. Usefulness of magnetic resonance-derived quantitative measurements of cerebral blood flow and volume in prediction of infarct growth in hyperacute stroke. *Stroke.* 2001;32:1147–1153.
51. Parsons MW, Yang Q, Barber PA, et al. Perfusion magnetic resonance imaging maps in hyperacute stroke: relative cerebral blood flow most accurately identifies tissue destined to infarct. *Stroke.* 2001;32:1581–1587.
52. Rohl L, Ostergaard L, Simonsen CZ, et al. Viability thresholds of ischemic penumbra of hyperacute stroke defined by perfusion-weighted MRI and apparent diffusion coefficient. *Stroke.* 2001;32:1140–1146.
53. Thijs VN, Adami A, Neumann-Haefelin T, Moseley ME, Marks MP, Albers GW. Relationship between severity of MR perfusion deficit and DWI lesion evolution. *Neurology.* 2001;57: 1205–1211.
54. Young GS. Advanced MRI of adult brain tumors. *Neurol Clin.* 2007;25:947–973:viii.
55. Hartmann M, Heiland S, Harting I, et al. Distinguishing of primary cerebral lymphoma from high-grade glioma with perfusion-weighted magnetic resonance imaging. *Neurosci Lett.* 2003;338: 119–122.
56. Rollin N, Guyotat J, Streichenberger N, Honnorat J, Tran Minh VA, Cotton F. Clinical relevance of diffusion and perfusion magnetic resonance imaging in assessing intra-axial brain tumors. *Neuroradiology.* 2006;48:150–159.
57. Shiroishi MS, Castellazzi G, Boxerman JL, et al. Principles of T*-weighted dynamic susceptibility contrast MRI technique in brain tumor imaging. *J Magn Reson Imaging.* 2014.
58. Manoonkitiwongsa PS, Schultz RL, Whitter EF, Lyden PD. Contraindications of VEGF-based therapeutic angiogenesis: effects on macrophage density and histology of normal and ischemic brains. *Vascul Pharmacol.* 2006;44:316–325.
59. Liebner S, Fischmann A, Rascher G, et al. Claudin-1 and claudin-5 expression and tight junction morphology are altered in blood vessels of human glioblastoma multiforme. *Acta neuropathol.* 2000; 100:323–331.
60. Davies DC. Blood-brain barrier breakdown in septic encephalopathy and brain tumours. *J Anat.* 2002;200:639–646.
61. Maeda M, Itoh S, Kimura H, et al. Tumor vascularity in the brain: evaluation with dynamic susceptibility-contrast MR imaging. *Radiology.* 1993;189:233–238.
62. Sugahara T, Korogi Y, Kochi M, Ushio Y, Takahashi M. Perfusion-sensitive MR imaging of gliomas: comparison between gradient-echo and spin-echo echo-planar imaging techniques. *Am J Neuroradiol.* 2001;22:1306–1315.

63. Aronen HJ, Gazit IE, Louis DN, et al. Cerebral blood volume maps of gliomas: comparison with tumor grade and histologic findings. *Radiology.* 1994;191:41–51.
64. Law M, Yang S, Wang H, et al. Glioma grading: sensitivity, specificity, and predictive values of perfusion MR imaging and proton MR spectroscopic imaging compared with conventional MR imaging. *Am J Neuroradiol.* 2003;24:1989–1998.
65. Morita N, Wang S, Chawla S, Poptani H, Melhem ER. Dynamic susceptibility contrast perfusion weighted imaging in grading of nonenhancing astrocytomas. *J Magn Reson Imaging.* 2010;32: 803–808.
66. Donahue KM, Krouwer HG, Rand SD, et al. Utility of simultaneously acquired gradient-echo and spin-echo cerebral blood volume and morphology maps in brain tumor patients. *Magn Reson Med.* 2000;43:845–853.
67. Law M, Oh S, Babb JS, et al. Low-grade gliomas: dynamic susceptibility-weighted contrast-enhanced perfusion MR imaging—prediction of patient clinical response. *Radiology.* 2006;238:658–667.
68. Danchaivijitr N, Waldman AD, Tozer DJ, et al. Low-grade gliomas: do changes in rCBV measurements at longitudinal perfusion-weighted MR imaging predict malignant transformation? *Radiology.* 2008;247:170–178.
69. Chaskis C, Stadnik T, Michotte A, Van Rompaey K, D'Haens J. Prognostic value of perfusion-weighted imaging in brain glioma: a prospective study. *Acta neurochir.* 2006;148:277–285:discussion 85.
70. Lupo JM, Cha S, Chang SM, Nelson SJ. Dynamic susceptibility-weighted perfusion imaging of high-grade gliomas: characterization of spatial heterogeneity. *Am J Neuroradiol.* 2005;26:1446–1454.
71. Maia Jr. AC, Malheiros SM, da Rocha AJ, et al. Stereotactic biopsy guidance in adults with supratentorial nonenhancing gliomas: role of perfusion-weighted magnetic resonance imaging. *J Neurosurg.* 2004;101:970–976.
72. Kong DS, Kim ST, Kim EH, et al. Diagnostic dilemma of pseudoprogression in the treatment of newly diagnosed glioblastomas: the role of assessing relative cerebral blood flow volume and oxygen-6-methylguanine-DNA methyltransferase promoter methylation status. *Am J Neuroradiol.* 2011;32:382–387.
73. Mitsuya K, Nakasu Y, Horiguchi S, et al. Perfusion weighted magnetic resonance imaging to distinguish the recurrence of metastatic brain tumors from radiation necrosis after stereotactic radiosurgery. *J Neurooncol.* 2010;99:81–88.
74. Pechman KR, Donohoe DL, Bedekar DP, Kurpad SN, Hoffmann RG, Schmainda KM. Characterization of bevacizumab dose response relationship in U87 brain tumors using magnetic resonance imaging measures of enhancing tumor volume and relative cerebral blood volume. *J Neurooncol.* 2011;105:233–239.
75. Law M, Cha S, Knopp EA, Johnson G, Arnett J, Litt AW. High-grade gliomas and solitary metastases: differentiation by using perfusion and proton spectroscopic MR imaging. *Radiology.* 2002; 222:715–721.
76. Jain R. Measurements of tumor vascular leakiness using DCE in brain tumors: clinical applications. *NMR Biomed.* 2013;26:1042–1049.
77. Tofts PS. Modeling tracer kinetics in dynamic Gd-DTPA MR imaging. *J Magn Reson Imaging.* 1997; 7:91–101.
78. Sourbron SP, Buckley DL. Tracer kinetic modelling in MRI: estimating perfusion and capillary permeability. *Phys Med Biol.* 2012;57:R1–33.
79. Ginsberg LE, Fuller GN, Hashmi M, Leeds NE, Schomer DF. The significance of lack of MR contrast enhancement of supratentorial brain tumors in adults: histopathological evaluation of a series. *Surg Neurol.* 1998;49:436–440.
80. Roberts HC, Roberts TP, Ley S, Dillon WP, Brasch RC. Quantitative estimation of microvascular permeability in human brain tumors: correlation of dynamic Gd-DTPA-enhanced MR imaging with histopathologic grading. *Acad Radiol.* 2002;9(Suppl 1):S151–S155.
81. Uematsu H, Maeda M, Sadato N, et al. Vascular permeability: quantitative measurement with double-echo dynamic MR imaging—theory and clinical application. *Radiology.* 2000;214:912–917.

82. Roberts HC, Roberts TP, Brasch RC, Dillon WP. Quantitative measurement of microvascular permeability in human brain tumors achieved using dynamic contrast-enhanced MR imaging: correlation with histologic grade. *Am J Neuroradiol.* 2000;21:891–899.
83. Provenzale JM, Wang GR, Brenner T, Petrella JR, Sorensen AG. Comparison of permeability in high-grade and low-grade brain tumors using dynamic susceptibility contrast MR imaging. *Am J Roentgenol.* 2002;178:711–716.
84. Batchelor TT, Sorensen AG, di Tomaso E, et al. AZD2171, a pan-VEGF receptor tyrosine kinase inhibitor, normalizes tumor vasculature and alleviates edema in glioblastoma patients. *Cancer Cell.* 2007;11:83–95.
85. Marshall I, Wardlaw J, Cannon J, Slattery J, Sellar RJ. Reproducibility of metabolite peak areas in 1H MRS of brain. *Magn Reson Imaging.* 1996;14:281–292.
86. Calvar JA. Accurate 1H tumor spectra quantification from acquisitions without water suppression. *Magn Reson Imaging.* 2006;24:1271–1279.
87. Lange T, Dydak U, Roberts TP, Rowley HA, Bjeljac M, Boesiger P. Pitfalls in lactate measurements at 3T. *Am J Neuroradiol.* 2006;27:895–901.
88. Birken DL, Oldendorf WH. N-acetyl-L-aspartic acid: a literature review of a compound prominent in 1H-NMR spectroscopic studies of brain. *Neurosci Biobehav Rev.* 1989;13:23–31.
89. Moffett JR, Ross B, Arun P, Madhavarao CN, Namboodiri AM. N-Acetylaspartate in the CNS: from neurodiagnostics to neurobiology. *Prog Neurobiol.* 2007;81:89–131.
90. Wyss M, Kaddurah-Daouk R. Creatine and creatinine metabolism. *Physiol Rev.* 2000;80: 1107–1213.
91. Babb SM, Ke Y, Lange N, Kaufman MJ, Renshaw PF, Cohen BM. Oral choline increases choline metabolites in human brain. *Psychiatry Res.* 2004;130:1–9.
92. Brief EE, Whittall KP, Li DK, MacKay A. Proton T1 relaxation times of cerebral metabolites differ within and between regions of normal human brain. *NMR Biomed.* 2003;16:503–509.
93. Christiansen P, Toft P, Larsson HB, Stubgaard M, Henriksen O. The concentration of N-acetyl aspartate, creatine + phosphocreatine, and choline in different parts of the brain in adulthood and senium. *Magn Reson Imaging.* 1993;11:799–806.
94. Degaonkar MN, Pomper MG, Barker PB. Quantitative proton magnetic resonance spectroscopic imaging: regional variations in the corpus callosum and cortical gray matter. *J Magn Reson Imaging.* 2005;22:175–179.
95. Kent C. Regulatory enzymes of phosphatidylcholine biosynthesis: a personal perspective. *Biochim Biophys Acta.* 2005;1733:53–66.
96. Pope WB, Prins RM, Albert TM, et al. Non-invasive detection of 2-hydroxyglutarate and other metabolites in IDH1 mutant glioma patients using magnetic resonance spectroscopy. *J Neurooncol.* 2012;107:197–205.
97. Del Sole A, Falini A, Ravasi L, et al. Anatomical and biochemical investigation of primary brain tumours. *Eur J Nucl Med.* 2001;28:1851–1872.
98. Delorme S, Weber MA. Applications of MRS in the evaluation of focal malignant brain lesions. *Cancer Imaging.* 2006;6:95–99.
99. Gajewicz W, Papierz W, Szymczak W, Goraj B. The use of proton MRS in the differential diagnosis of brain tumors and tumor-like processes. *Med Sci Monit.* 2003;9:MT97–105.
100. Preul MC, Caramanos Z, Collins DL, et al. Accurate, noninvasive diagnosis of human brain tumors by using proton magnetic resonance spectroscopy. *Nat Med.* 1996;2:323–325.
101. Cho YD, Choi GH, Lee SP, Kim JK. (1)H-MRS metabolic patterns for distinguishing between meningiomas and other brain tumors. *Magn Reson Imaging.* 2003;21:663–672.
102. Lai PH, Ho JT, Chen WL, et al. Brain abscess and necrotic brain tumor: discrimination with proton MR spectroscopy and diffusion-weighted imaging. *Am J Neuroradiol.* 2002;23:1369–1377.
103. Mishra AM, Gupta RK, Jaggi RS, et al. Role of diffusion-weighted imaging and in vivo proton magnetic resonance spectroscopy in the differential diagnosis of ring-enhancing intracranial cystic mass lesions. *J Comput Assist Tomogr.* 2004;28:540–547.

104. Devos A, Lukas L, Suykens JA, et al. Classification of brain tumours using short echo time 1H MR spectra. *J Magn Reson.* 2004;170:164–175.
105. Li X, Vigneron DB, Cha S, et al. Relationship of MR-derived lactate, mobile lipids, and relative blood volume for gliomas in vivo. *Am J Neuroradiol.* 2005;26:760–769.
106. Chen J, Huang SL, Li T, Chen XL. In vivo research in astrocytoma cell proliferation with 1H-magnetic resonance spectroscopy: correlation with histopathology and immunohistochemistry. *Neuroradiology.* 2006;48:312–318.
107. Fayed N, Morales H, Modrego PJ, Pina MA. Contrast/noise ratio on conventional MRI and choline/creatine ratio on proton MRI spectroscopy accurately discriminate low-grade from high-grade cerebral gliomas. *Acad Radiol.* 2006;13:728–737.
108. Jenkinson MD, Smith TS, Joyce K, et al. MRS of oligodendroglial tumors: correlation with histopathology and genetic subtypes. *Neurology.* 2005;64:2085–2089.
109. Lev MH, Ozsunar Y, Henson JW, et al. Glial tumor grading and outcome prediction using dynamic spin-echo MR susceptibility mapping compared with conventional contrast-enhanced MR: confounding effect of elevated rCBV of oligodendrogliomas [corrected]. *Am J Neuroradiol.* 2004;25:214–221.
110. White ML, Zhang Y, Kirby P, Ryken TC. Can tumor contrast enhancement be used as a criterion for differentiating tumor grades of oligodendrogliomas? *Am J Neuroradiol.* 2005;26:784–790.
111. Xu M, See SJ, Ng WH, et al. Comparison of magnetic resonance spectroscopy and perfusion-weighted imaging in presurgical grading of oligodendroglial tumors. *Neurosurgery.* 2005;56:919–926.
112. Gajewicz W, Grzelak P, Gorska-Chrzastek M, Zawirski M, Kusmierek J, Stefanczyk L. The usefulness of fused MRI and SPECT images for the voxel positioning in proton magnetic resonance spectroscopy and planning the biopsy of brain tumors: presentation of the method. *Neurol Neurochir Pol.* 2006;40:284–290.
113. Hall WA, Martin A, Liu H, Truwit CL. Improving diagnostic yield in brain biopsy: coupling spectroscopic targeting with real-time needle placement. *J Magn Reson Imaging.* 2001;13:12–15.
114. Graves EE, Nelson SJ, Vigneron DB, et al. A preliminary study of the prognostic value of proton magnetic resonance spectroscopic imaging in gamma knife radiosurgery of recurrent malignant gliomas. *Neurosurgery.* 2000;46:319–326.
115. Graves EE, Pirzkall A, Nelson SJ, Larson D, Verhey L. Registration of magnetic resonance spectroscopic imaging to computed tomography for radiotherapy treatment planning. *Med Phys.* 2001;28:2489–2496.
116. Payne GS, Leach MO. Applications of magnetic resonance spectroscopy in radiotherapy treatment planning. *Br J Radiol.* 2006;79(Spec No 1):S16–S26.
117. Cohen BA, Knopp EA, Rusinek H, Babb JS, Zagzag D, Gonen O. Assessing global invasion of newly diagnosed glial tumors with whole-brain proton MR spectroscopy. *Am J Neuroradiol.* 2005;26:2170–2177.
118. Matulewicz L, Sokol M, Wydmanski J, Hawrylewicz L. Could lipid CH2/CH3 analysis by in vivo 1H MRS help in differentiation of tumor recurrence and post-radiation effects? *Folia Neuropathol.* 2006;44:116–124.
119. Chan YL, Yeung DK, Leung SF, Cao G. Proton magnetic resonance spectroscopy of late delayed radiation-induced injury of the brain. *J Magn Reson Imaging.* 1999;10:130–137.
120. Graves EE, Nelson SJ, Vigneron DB, et al. Serial proton MR spectroscopic imaging of recurrent malignant gliomas after gamma knife radiosurgery. *Am J Neuroradiol.* 2001;22:613–624.
121. Hollingworth W, Medina LS, Lenkinski RE, et al. A systematic literature review of magnetic resonance spectroscopy for the characterization of brain tumors. *Am J Neuroradiol.* 2006;27:1404–1411.
122. Plotkin M, Eisenacher J, Bruhn H, et al. 123I-IMT SPECT and 1H MR-spectroscopy at 3.0 T in the differential diagnosis of recurrent or residual gliomas: a comparative study. *J Neurooncol.* 2004;70:49–58.
123. Cohnen M, Wittsack HJ, Assadi S, et al. Radiation exposure of patients in comprehensive computed tomography of the head in acute stroke. *Am J Neuroradiol.* 2006;27:1741–1745.

124. Imanishi Y, Fukui A, Niimi H, et al. Radiation-induced temporary hair loss as a radiation damage only occurring in patients who had the combination of MDCT and DSA. *Eur Radiol.* 2005;15:41–46.
125. Wintermark M, Lev MH. FDA investigates the safety of brain perfusion CT. *Am J Neuroradiol.* 2010;31:2–3.
126. Averbeck D. Does scientific evidence support a change from the LNT model for low-dose radiation risk extrapolation? *Health Phys.* 2009;97:493–504.
127. Knuuti J, Saraste A, Kallio M, Minn H. Is cardiac magnetic resonance imaging causing DNA damage? *Eur Heart J.* 2013;34(30):2337–2339.
128. Tanaka R, Yumoto T, Shiba N, et al. Overheated and melted intracranial pressure transducer as cause of thermal brain injury during magnetic resonance imaging: case report. *J Neurosurg.* 2012; 117(6):1100–1109.
129. Shellock FG, Crues JV. MR procedures: biologic effects, safety, and patient care. *Radiology.* 2004; 232(3):635–652:Review.
130. Expert Panel on MR Safety, Kanal E, Barkovich AJ, Bell C, et al. ACR guidance document on MR safe practices: 2013. *J Magn Reson Imaging.* 2013;37(3):501–530.
131. Gilk, T. Analysis of reported MRI accidents/injuries: type, frequency and severity—Unpublished data. Presented at 2014 ISMRM Workshop on MRI Safety. Washington, DC; 2014.
132. Henderson JM, Tkach J, Phillips MM, et al. Permanent neurological deficit related to magnetic resonance imaging in a patient with implanted deep brain stimulation electrodes for Parkinson's disease: case report. *Neurosurgery.* 2005;57(5):pE1063.
133. Fiechter M, Stehli J, Fuchs TA, et al. Impact of cardiac magnetic resonance imaging on human lymphocyte DNA integrity. *Eur Heart J.* 2013;34(30):2340–2345.
134. Haik J, Daniel S, Tessone A, Orenstein A, Winkler E. MRI induced fourth-degree burn in an extremity, leading to amputation. *Burns.* 2009;35(2):294–296.
135. Dempsey MF, Condon B. Thermal injuries associated with MRI. *Clin Radiol.* 2001;56(6): 457–465:Review.
136. Jones S, Jaffe W, Alvi R. Burns associated with electrocardiographic monitoring during magnetic resonance imaging. *Burns.* 1996;22(5):420–421.
137. Klucznik RP, Carrier DA, Pyka R, Haid RW. Placement of a ferromagnetic intracerebral aneurysm clip in a magnetic field with a fatal outcome. *Radiology.* 1993;187(3):855–856.
138. Chaljub G, Kramer LA, Johnson 3rd RF, et al. Projectile cylinder accidents resulting from the presence of ferromagnetic nitrous oxide or oxygen tanks in the MR suite. *Am J Roentgenol.* 2001;177 (1):27–30.
139. Chen D.W. Boy, 6, dies of skull injury during MRI; oxygen tank becomes fatal missile in hospital. *New York Times*; July 31, 2001. p B1, 5.
140. Schenck, J. B0 fields for MRI: overview. Presented at 2014 ISMRM Workshop on MRI Safety, Washington, DC, Sept 5 2014.
141. ACR Committee on Drugs and Contrast Media. ACR Manual on Contrast Media Version 9: 2013. American College of Radiology 2013. Available from: <www.acr.org>.
142. Prince MR, Zhang H, Zou Z, Staron RB, Brill PW. Incidence of immediate gadolinium contrast media reactions. *Am J Roentgenol.* 2011;196(2).
143. Idée JM, Port M, Dencausse A, Lancelot E, Corot C. Involvement of gadolinium chelates in the mechanism of nephrogenic systemic fibrosis: an update. *Radiol Clin North Am.* 2009;47(5): 855–869.
144. Idée JM, Port M, Robic C, Medina C, Sabatou M, Corot C. Role of thermodynamic and kinetic parameters in gadolinium chelate stability. *J Magn Reson Imaging.* 2009;30 (6):1249–1258.
145. Errante Y, Cirimele V, Mallio CA, et al. Progressive increase of T1 signal intensity of the dentate nucleus on unenhanced magnetic resonance images is associated with cumulative doses of intravenously administered gadodiamide in patients with normal renal function, suggesting dechelation. *Invest Radiol.* 2014;49(10):685–690.

146. Kanda T, Ishii K, Kawaguchi H, et al. High signal intensity in the dentate nucleus and globus pallidus on unenhanced T1-weighted MR images: relationship with increasing cumulative dose of a gadolinium-based contrast material. *Radiology.* 2014;270(3):834–841.
147. Roccatagliata L, Vuolo L, Bonzano L, et al. Multiple sclerosis: hyperintense dentate nucleus on unenhanced T1-weighted MR images is associated with the secondary progressive subtype. *Radiology.* 2009;251(2):503–510.
148. Sherry AD, Caravan P, Lenkinski RE. Primer on gadolinium chemistry. *J Magn Reson Imaging.* 2009; 30(6):1240–1248.
149. Tweedle M, Kanal E, Muller E. Considerations in the selection of new gadolinium based contrast agent. *Appl Radiol.* 2014;(Suppl):1–11.

CHAPTER 3

Advanced Functional Imaging: fMRI, PET, and MEG

Nina Shevzov-Zebrun, Nicole M. Petrovich Brennan, Kyung K. Peck, and Andrei I. Holodny
Functional MRI Laboratory, Department of Radiology, Memorial Sloan Kettering Cancer Center, New York, NY, USA

INTRODUCTION

Functional imaging is an important tool for clinicians, psychologists, and other research scientists. Techniques such as positron emission tomography (PET), magnetoencephalography (MEG), and functional magnetic resonance imaging (fMRI) not only help provide a better understanding of the anatomy, physiology, and connectivity of specific areas of the brain, but also allow for more effective and precise characterization and localization of brain lesions. Prior to the advent of functional imaging, neurosurgeons relied upon invasive functional mapping with the intracarotid amytal test or electrocorticography during craniotomy, often in awake patients. As a result, operations—especially those involving deep-seated lesions—entailed much more estimation, approximation, and thus room for error and complication.[1] Neurosurgeons are now able to use such functional imaging methods to help guide surgery and intraoperative mapping and to make more informed, patient-specific choices about the most effective way to perform tumor resections or other surgeries. Functional imaging also has applications beyond tumor resection, playing a role in evaluation and treatment of Parkinson's and Alzheimer's diseases, epilepsy, and Huntington's disease. Overall, PET, MEG, and fMRI can each help clinicians formulate more personalized, efficient treatment plans.[2] This chapter reviews the physiological basis, clinical uses, and current research-related applications of PET, MEG, and fMRI as they pertain to neurosurgery.

POSITRON EMISSION TOMOGRAPHY

Overview and physiological basis

PET uses key biological molecules—often their analogs—tagged with a positron emitting radiotracer to offer insight into certain biological processes (such as glucose metabolism) in which the tagged molecules are involved.[3] Although the molecules used range from amino acids to water,[3] the most common molecule used for PET is

A. Golby (Ed): Image-Guided Neurosurgery
DOI: http://dx.doi.org/10.1016/B978-0-12-800870-6.00003-0

fluorodeoxyglucose (FDG), an analog of glucose lacking a hydroxyl group at the C2 position and instead tagged with ^{18}F.[2]

FDG is taken up by cells as normal glucose, and hexokinase phosphorylates the molecule at the C6 position to yield FDG-6-P. Since FDG lacks the C2 hydroxyl group, however, FDG-6-P becomes trapped in cells as it cannot complete the glycolysis pathway as regular glucose would.[2] In this way, PET scans are able to provide a temporal, quantitative measure of glucose metabolism rate in various regions of the body or brain, based on the accumulation of FDG-6-P.[2] Decay of the radiotracer, which in the case of ^{18}F has a half-life of about 110 minutes, results in photon release.[2] PET scanners register this photon emission and thus help to map the usage and distribution of the radiotracer-tagged molecule in the body.[4] FDG is then cleared from the body through the urine.[5] Radiotracer half-life is an important consideration; if the half-life is short, radiopharmaceutical injection and PET scan must occur without a break in between if usable results are to be obtained.

PET imaging allows for the delineation and characterization of lesions largely due to the fact that cancer cells use glucose at an increased rate.[2] FDG therefore becomes trapped more significantly in cancerous cells than in healthy cells. Even when there is adequate oxygen, tumor cells appear to prefer to harvest energy anaerobically (anaerobic glycolysis) and metabolize the resultant pyruvate molecules to lactic acid.[2] This less efficient process—combined with the fact that glucose transporter proteins, namely GLUT-1, tend to be overexpressed in cancer cells—helps account for the faster rate of glucose uptake in tumor cells.[5] (See Figure 3.1.)

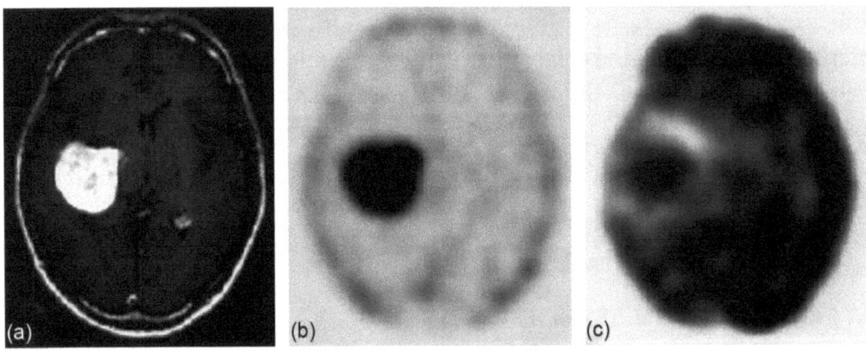

Figure 3.1 A 58-year-old patient with a meningioma on enhanced T1 (a) is shown. An 11 C-choline PET image obtained at 5 min (b) and 50 min (c) shows increased tumor uptake (b) with a tumor to white matter ratio of 29.10. 50 minutes postinjection shows a tumor with a ratio of 1.65. *Reproduced from Zhu A et al.* Seminars in Oncology. *Volume 38, Issue 1. Figure 2.* Metabolic positron emission tomography imaging in cancer detection and therapy response, *Elsevier 2011, with permission.*

Clinical importance and applications

PET has clinical applications in a range of disciplines, though it is used primarily in oncology, neurology, and cardiology.[6] This chapter will review the importance of PET and some of its uses in oncology and neurology.

First, PET is a widely used and helpful tool for clinicians in the identification, description, and staging of tumors. Unlike fMRI and MEG, which offer useful anatomical information, PET is able to provide a more sensitive molecular-level, physiology-based analysis of lesions/tissues (rather than pure localization) due to its foundation in biochemical processes such as glucose metabolism.[2] This physiological, biochemical aspect of PET (specifically with FDG/glucose metabolism) is a double-edged sword, however; though it allows for better, more complete characterization of lesions,[2] it limits the types of cancers that can be investigated.[4] Only about 10 different cancer types can be imaged with FDG PET.[4] These limitations stem from the fact that glucose uptake is determined by many different outside factors—inflammation, muscle activity, or infection, for example—and thus an observed increase in glucose uptake rate in a particular region may not actually be related to cancerous cell activity.[6,7] In other words, differences in the rate of glycolysis may not necessarily be indicative of the presence of cancer cells.

PET can play a role in determining malignancy in situations in which other tests fail.[3] Furthermore, PET scans can help identify malignant lesions earlier than other imaging techniques.[6] It is important to note that changes in physiology and function occur before anatomical, structural shifts, and thus PET, which relies on physiological processes, allows for earlier evaluation, characterization and diagnosis of any changes seen in those biochemical processes.[6] PET can also be used in tumor restaging, evaluating relapse, and monitoring the progression and development of a lesion.[4]

Second, PET allows for earlier assessment of response to treatment.[2] It has been shown that a decrease in FDG PET signal (signal measured before and after the course of treatment has begun) is significantly associated with increased survival rate.[4] The ability to evaluate a therapy plan early in its course not only indicates that FDG PET could be helpful in drug development and evaluation of the efficacy of novel drugs,[4] but also helps clinicians to create more efficient treatment plans.[8] With an earlier indicator of whether or not a particular therapy—chemotherapy, radiotherapy, —is working well in a particular patient, ineffective treatment plans need not be prolonged unnecessarily, and better, more patient-specific oncologic treatment decisions can be made.[2]

Third, PET can be used as a surgical planning tool. In the past, without presurgical insight into the location of eloquent cortices, neurosurgeons relied on intraoperative mapping techniques such as direct cortical stimulation. Unguided direct cortical stimulation and craniotomies, however, may bring about various complications. If mapping reveals any anatomical abnormalities or unexpected relationships between the lesion and eloquent cortices, surgery time/time under anesthesia may be prolonged as the neurosurgeon decides on a new plan.[9] The neurosurgeon may even unnecessarily decide against resection due to lack of necessary functional information.[1] The success of direct cortical stimulation also depends upon the patient's ability to perform the task given—if the patient struggles or fails to cooperate, necessary functional information may not be obtained and surgery may be terminated.[9]

Like MEG and fMRI, PET can help provide neurosurgeons with information about the areas surrounding lesions prior to surgery in order to avoid the aforementioned complications associated with "blind," unguided direct cortical stimulation. However, PET is not used much for preoperative planning as the spatial and temporal resolution of fMRI and MEG are superior.[10]

Fourth, PET has applications in epilepsy treatment and surgery planning. Similarly to cancer cells, cells in the epileptogenic zone have a higher rate of glucose uptake during seizures. Between seizure episodes, however, their glucose uptake levels actually fall slightly below that of the healthy cells in that brain area.[6] Because glycolysis rate differs for cells in the epileptogenic zone, FDG PET can be used to identify this region in epileptic patients. The PET results can later be used as part of a multimodality evaluation to help decide whether or not the patient is a good candidate for epilepsy surgery and, if so, how best to plan the procedure.[6] (See Figure 3.2.)

Fifth, PET has played a role in gaining a better understanding of Parkinson's disease and both its motor and nonmotor aspects, thus allowing for more patient-specific case management.[10] ^{18}F-DOPA is a radiopharmaceutical used to study diagnosis, development and physiology of the disease.[10] It helps to identify changes in activity levels of L-stereochemistry aromatic amino acid decarboxylases.[11] ^{18}F-DOPA PET has been used to evaluate and follow disease development and progression, as levels of

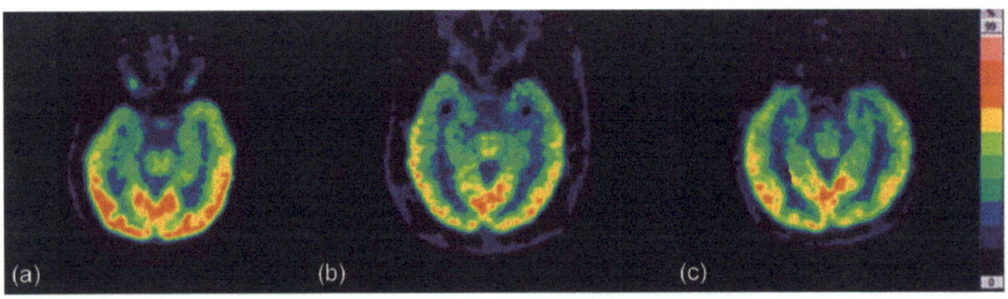

Figure 3.2 PET studies showing right anteromedial unilateral temporal hypometabolism (a), asymmetrical bilateral temporal hypometabolism (b), and symmetrical temporal hypometabolism (c). *Reproduced from Tepmongkol S. et al.* Epilepsy and Behavior. *Figure 1.* Factors affecting bilateral temporal lobe hypometabolism on 18F-FDG PET brain scan in unilateral medial temporal lobe epilepsy, *Elsevier 2013, with permission.*

radiotracer uptake in the putamen are thought to be associated with increased disease severity.[10] Further use of PET in Parkinson's disease treatment and research would be facilitated by new radiopharmaceuticals capable of targeting other biological processes implicated in the disease.[10]

Sixth, FDG PET is used in treatment planning for non-Hodgkin lymphoma (NHL).[12] The information supplied by PET can influence the course of lymphoma treatment and disease management, as decreases in glucose metabolism rates may signify a positive response to a specific treatment (leading to continuation of that treatment or therapy).[12,13] Thus far, most studies and clinical applications have focused on FDG PET use in NHL treatment more broadly and not on its potential uses in primary central nervous system lymphoma (PCNSL), a type of NHL.[13] Relatively recent studies by Mohile and colleagues, however, have suggested that FDG PET is actually more precise and receptive to changes in physiology than conventional staging methods, and may thus prove helpful in PCNSL staging, identification of systemic lymphoma sites, and PCNSL treatment planning.[13]

A major current challenge in the imaging of brain tumors is the similarity in the MR appearance of true tumor progression and the effects of treatment (pseudoprogression).[14] Pseudoprogression is essentially indistinguishable from true tumor progression on routine MR with both presenting with an increase in the volume of enhancement and FLAIR abnormality.[15] This is a crucial clinical difference, since in the former, the patient's tumor is progressing and the treating physician must consider a change in treatment strategy. On the other hand, in the case of pseudoprogression, the treatment is actually having a beneficial effect and should be maintained. Similarly, bevacizumab has been shown to decrease the volume of enhancement in both glial tumors and metastases. This is known as pseudoresponse and also complicates the ability of routine MR to accurately assess the true clinical state of the tumor.[14] FDG-PET has been shown to be useful in differentiating between these two entities.[16,17] In large brain tumors with a homogeneous appearance on routine MR, FDG-PET can occasionally help guide the biopsy to the most metabolically active (and presumably most malignant) part of the tumor.[18]

Limitations

Although PET scans can provide useful temporal, quantitative physiological information about specific brain regions, the imaging technique has certain limitations and pitfalls. First, the procedure requires the use of a radioactive isotope (most commonly ^{18}F), and thus care must be taken to ensure that the patient is properly hydrated to keep FDG concentration in urine at safely low levels.[5]

Second, as previously discussed, the types of situations and cancers that allow for effective FDG PET use are limited. Indeed, due to the multifaceted nature of increased glycolysis rate (infection, inflammation, and muscle activity can all affect this physiological process), determining the nature of differences in glucose uptake rate between cells may present challenges as an elevated signal may not be related specifically to a tumor.[6] In this sense the brain is a "problem area" for PET since the cortex of the brain has a very high glucose metabolism limiting the difference that can occur between glucose uptake rates in cancer versus healthy cells in the brain.[6] In ambiguous cases, use of another imaging technique may be more helpful.[6]

Third, since PET has an inherently lower resolution than MRI, the ability of PET to accurately localize lesions is inferior to MR.[3] Thus a combination technique using PET and CT or PET and MR should allow for better localization of lesions and delineation of where they begin and end while still acquiring the physiological, temporal information from PET.[3]

Last, PET—and the nature of the biochemical processes underlying it—presents the possibility for both false negatives and false positives. A false negative may occur if a tumor is small (usually <1 cm) or slow-growing as a significant increase in glucose uptake rate may not be observed.[6,2] Sometimes—as in some prostate and thyroid carcinoma cases—tumor cells may actually use glucose at a decreased rate.[6] False positives are more common, however, and as previously mentioned can result from other factors such as inflammation, muscle activity, or infection leading to an increase in glycolysis rate.[6]

Patient preparation and procedure (Focus on FDG PET)

The primary goal of FDG PET is to detect increases in glucose metabolism as a way of quantitatively characterizing lesions. Patients must thus be prepped and scan procedures carried out properly in order to help create the most significant contrast possible between cancer and healthy cell glucose uptake rate. In order to identify differences in physiological processes between tumor cells and healthy tissue, the amount of glucose healthy cells take up must be kept at a minimum.[5]

With this goal in mind, several measures should be taken to obtain the most helpful results possible. First, though the patient should be well hydrated so that FDG can be cleared from the body in a safely low concentration, he or she should not eat starting about six hours before the PET scan in order to minimize glucose levels and excess glucose metabolism by healthy cells.[5] Blood glucose level prior to injection generally must be less than 7 mmol/L in order to proceed with the scan.[5] The patient should also not partake in any major exercise starting about six hours before the scan, and should be kept warm both during radiotracer injection and throughout the whole scan.[5] Lastly, any metallic objects should be removed from the patient's body in order to minimize the effects of metallic artifacts.[19]

During FDG injection, the patient should be sitting or reclining and silent so that muscle activity—and thus glucose uptake in muscles *not* targeted—is kept at a minimum. Use of sedatives may be needed to ensure stillness if the lesion is in the head/neck area or if the patient is claustrophobic.[5] In addition, if the lesion is in the brain, the patient should be kept in a quiet, dark room to avoid unwanted brain activity.[5]

Data correction and interpretation

PET allows mapping of differences in physiological, molecular-level functions to specific locations in order to help diagnose and characterize lesions, create or alter a treatment plan, or monitor and stage tumors. In order for PET scan results to aid in the aforementioned tasks, corrections are often applied in order to minimize the effects of artifacts, which can arise from various sources.

The most important correction to raw PET results is attenuation correction.[20] Photon attenuation refers to the fact that some photons released from radiotracer decay may be absorbed or otherwise scattered by surrounding tissue/biological material and therefore go undetected by the PET scanner. The use of CT and PET technologies together not only allows for better combination and integration of anatomical and physiological data, but has also proven to be helpful in attenuation correction.[19] (See Figure 3.3.)

Other factors that can affect PET (and especially PET/CT) results include any metallic implants and respiratory motion. Dental implants, such as fillings, are especially important to consider when imaging a head/neck lesion as they tend to absorb photons readily,[21] and may thus interfere with PET scan results (which rely on detection of photon emission), leading to false negatives or positives.[19] Respiratory artifacts are less pertinent to brain lesions, but movement and muscle activity associated with breathing can prove problematic when imaging lung lesions.[22]

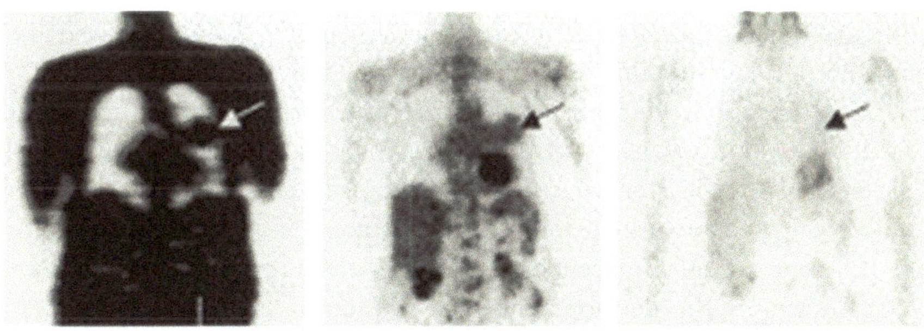

(a) Transmission (attenuation) image (b) FDG uptake image with correction for attenuation (c) FDG uptake image without correction for attenuation

Figure 3.3 *Uptake attenuation correction allows the visualization of a hematoma (arrow). It can be seen in both the transmission image (a) and in the FDG image both with attenuation correction (b) but not in the FDG uptake image without such correction (c). Reproduced from Kinahan P. et al.* Seminars in Nuclear Medicine. *Volume 33, Issue 3. Figure 1.* X-ray-based attenuation correction for positron emission tomography/computed tomography scanners, *Elsevier 2003, with permission.*

Applications and current and future research

Personalized care and therapy evaluation

Current research is revealing possibilities for new, exciting uses and clinical applications of PET in treatment of various types of cancers. First, PET offers a unique opportunity to provide more personalized and thus efficient treatment.[2] As previously mentioned, changes in physiology and function often occur before any anatomical, structural shifts take place.[6] Thus PET—which is based on biochemical processes—can be used for diagnosis, characterization, and evaluation of a lesion earlier than other imaging techniques.[6] This ability to use PET relatively "early on"—detecting increases or decreases in glucose uptake, for instance—allows for evaluation of therapy plans shortly after they have begun.[4] It has been shown in some cancers that a decrease in glucose uptake detected even just days after treatment has commenced is significantly associated with and indicative of increased survival.[4] As a result, FDG PET may prove helpful in the future for evaluating drug trials and new therapies, as PET results could serve as reliable indicators of therapy efficacy well before any visible, more noticeable progress in patient health has been made.[4] As of now, however, there is no consensus on the amount by which glucose uptake rates must decrease in order to signify a true response to the treatment at hand.[2]

Radiopharmaceutical specificity and development

The information gleaned from a PET scan directly corresponds to the type of radiopharmaceutical used and, more specifically, to the biochemical, physiological process in which that molecule (or its analog) typically plays a role. FDG PET, for instance, quantitatively helps to evaluate changes in glucose uptake and metabolism since FDG is initially taken up by cells and phosphorylated as regular glucose. Fluorothymidine (FLT), on the other hand, acts as an analog of thymidine (which is used in DNA synthesis), and thus helps to elucidate areas of increased DNA synthesis and cell proliferation.[2] Since increased glucose metabolism could be caused by certain outside factors and may not necessarily correspond to the presence of cancerous cells and rapid cell proliferation, FLT or other nucleoside-based radiopharmaceuticals have the potential to more specifically map and characterize areas of increased cell division.[2]

Since different radiopharmaceuticals help to detect and map different types of physiological changes, the development of new radiotracer-tagged molecules could help to expand PET use to more types of cancer. Radiopharmaceuticals unrelated to such complex, multifaceted processes as glucose metabolism could help not only ensure a more clear, reliable distinction between cancer and healthy cells, but also allow PET to be used with cancers that do not necessarily exhibit increased glucose uptake.[2] Molecules with apoptosis, hypoxia, and increased levels of protein synthesis as "targets," for instance, have been designed and are currently being researched as all of these physiological processes are in some way implicated in tumor growth and/or cancer therapy,[2] yet are distinct from glucose metabolism.

In addition, the development of radiopharmaceuticals that target proteins either consistently overexpressed or solely expressed in tumor cells is under investigation.[2]

Indeed, several "alternative" radiopharmaceuticals and radiotracers (involved in processes other than glucose metabolism) are becoming more prominently used both in research and the clinical setting. In addition to the aforementioned FLT, a thymidine analog involved in DNA synthesis, ^{11}C-Choline—part of the pathway involved in cell membrane phospholipid synthesis—is a promising radiopharmaceutical in oncology.[11] It can be used to delineate tumors by tracking increased choline phosphorylation/cell membrane synthesis.[11] In addition, *O*-(2-^{18}F-fluoroethyl)-L-tyrosine (^{18}F-FET) PET, has shown promise in brain tumors.[23] Accumulation of ^{18}F-FET in brain tumor cells is presumably linked to high expression of the L-type amino acid transporters (LATs), which are the major transport system for large neutral amino acids.[23] Early reports indicate that ^{18}F-FET may facilitate the differentiation of pseudoprogression from true progression and guide biopsies.[24,25] ^{18}F-DOPA, used in Parkinson's disease diagnosis and evaluation, helps to identify changes in levels of enzyme activity—specifically activity of L-stereochemistry aromatic amino acid decarboxylases.[11]

FUNCTIONAL MAGNETIC RESONANCE IMAGING

Overview and physiological basis

fMRI, like PET and MEG, is a noninvasive brain imaging technique important for offering neurosurgeons, other clinicians, and researchers a better understanding of the functional neuroanatomy of specific areas of the brain. Unlike PET, fMRI does not entail the use of any radioactive isotopes. This technique has a relatively short scan time, is easily repeatable, and has no known risks.[9]

fMRI uses relative changes in cerebral blood flow (CBF), cerebral metabolic rate of oxygen, and cerebral blood volume occurring with increased neuronal activity in order to map task-related brain activity.[9] In response to neuronal activity, both CBF to the active area and oxygen consumption increase. An "overshoot" occurs, however, in the amount of oxyhemoglobin brought to the active area via increased CBF relative to the amount of deoxyhemoglobin present from increased oxygen usage. Since oxyhemoglobin is diamagnetic and deoxyhemoglobin paramagnetic—and since paramagnetic deoxyhemoglobin alone causes a drop in fMRI signal—this imbalance and dilution of a paramagnetic substance by a diamagnetic one actually leads to an increase in fMRI signal on the T2*-weighted images used in fMRI in the active area of the brain.[9] The blood oxygen level dependent (BOLD) signal forms the physiological basis of fMRI.

Clinical importance and applications

Although fMRI has an ever-growing list of clinical applications, fMRI is most commonly used in presurgical planning for tumor resections and offers both patients and neurosurgeons distinct advantages. First, fMRI results help the neurosurgeon avoid "blind" intraoperative direct cortical stimulation and somatosensory-evoked potentials. Unguided direct cortical stimulation, which necessitates a craniotomy, can pose challenges not only due to unforeseen anatomical differences or shifts arising from lesion growth, but also due to lack of patient cooperation in task performance.[9] Patients may have trouble waking, following instructions, or performing the paradigm, thus rendering the intraoperative stimulation nearly useless.[1] In serving as a guide for direct cortical stimulation, therefore, fMRI can help to decrease operation time/time under anesthesia while adding a layer of confirmation in the case of any unforeseen difficulties or anatomical irregularities.

Second, fMRI can help the neurosurgeon decide not only how, but whether to operate at all. If fMRI results suggest, for instance, that the tumor is intimately involved with Broca's area or another key functional area, he or she may decide against surgery, sparing the patient anesthesia, direct cortical stimulation, and a potentially harmful procedure.[1] Without proper information about eloquent cortices and specificities of tumor location, a surgeon may be more likely to be timid in resection or decide against surgery.[1]

If the surgeon does decide to operate, fMRI can help him or her plan the best approach. With preoperative knowledge of the location of important eloquent motor, memory, language or sensory areas, surgeons are better able to maximize resection while minimizing both damage to surrounding functional areas and need for intraoperative testing.

Co-registration of fMRI data to neuronavigational systems using commercially available software allows fMRI to play an even more important, precise role in guiding neurosurgeries. Upon co-registration of fMRI information and higher resolution MRI data, higher quality images can be viewed in the operating room during procedures, guiding surgeons and clarifying the location of lesions and any eloquent cortices in real time.[9]

Last, fMRI can offer insight into any cortical plasticity/reorganization that may have occurred as a result of tumor growth and morphology, leading to more personalized, case-specific treatment. Cortical reorganization loosely refers to the process that occurs when a certain area of the brain loses the ability to perform its function fully and another part of the brain begins to take on that function in an effort to compensate.[1] Since lesions may bring about such shifts in function in regions affected by tumor growth, it is important that a neurosurgeon have the tools necessary to remain aware of any abnormal functional organization as he or she plans the procedure.[26]

Patient preparation and procedure

(Note the following section is general and thus applicable to fMRI and MEG exam preparations and procedures.)

Paradigm selection and design

Prior to discussing paradigm selection, it is important to note a few key features of motor- and language-related neuroanatomy as they directly influence choice and prioritization of task. The motor (and sensory) systems are organized topographically—that is, specific motor functions have corresponding locations on the cortex.[9] The leg and foot are mapped along the interhemispheric fissure, the hand-related region is lateral to the foot, and the face and tongue are represented lateral to the hand. Many different regions are involved in performing and planning motor activities. These include the primary motor cortex (M1), the supplementary motor area (SMA), the superior parietal lobules and the lateral premotor cortex. M1 functions in movement performance, while the SMA has a role in organizing and planning movement.[27] The primary sensorimotor cortex, which is involved in touch and sensing sensory stimulation, is found in the postcentral gyrus.[28] (See Figure 3.4.)

Language function is typically found in the left hemisphere in right-handed individuals.[28] Left-handed people are more likely to be co-dominant or right-hemisphere dominant for language.[29] Broca's area, which is found in the frontal lobe (often left), is involved in speech production, while Wernicke's area, located in the temporal lobe (often left), is implicated in language comprehension and speech planning. (See Figure 3.5.)

Paradigms should be designed with this functional neuroanatomy in mind if the results obtained are to help neurosurgeons and other clinicians offer the most individualized and properly planned care possible. Indeed, paradigms should be chosen based on the location of the lesion and thus possible nearby functional areas—motor, language, sensory, memory, etc.—such that the pros and cons of neurosurgery and

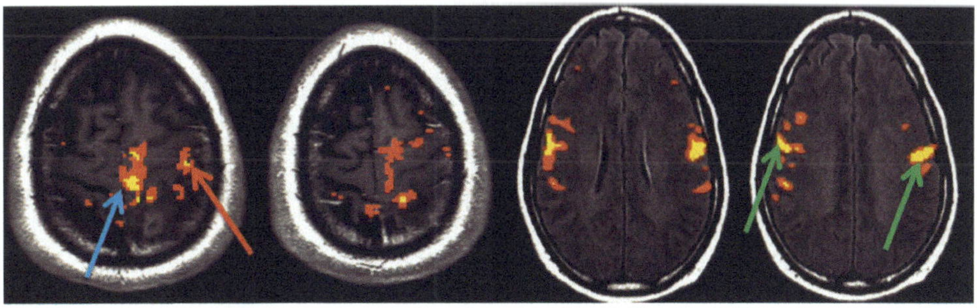

Figure 3.4 fMRI of foot (blue arrow), hand (red arrow), and tongue motor activations (green arrow).

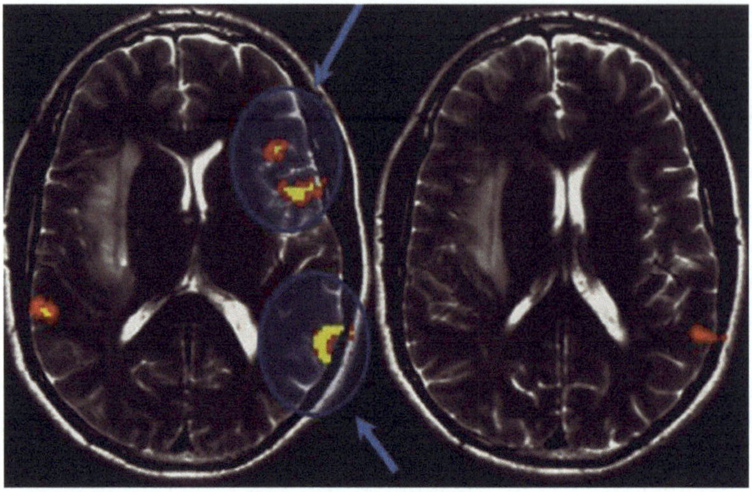

Figure 3.5 An example of a well-lateralized fMRI language map in a patient with a right hemisphere tumor. Broca's area is anterior and Wernicke's area posterior. *Reproduced from Belyaev A, et al. Magnetic Resonance Imaging.* Clinics of North America. *Volume 21, Issue 2. Figure 3.* Clinical Applications of Functional MR Imaging, *Elsevier 2013, with permission.*

other treatment plans can be evaluated in an informed fashion.[9] Paradigm design should also take into account the patient's ability to perform the task with minimal unwanted motion while alone in the MRI scanner. Factors such as age, medical history, and neurologic deficits may affect paradigm choice.[1]

Paradigms are generally either "block" or "event-related," and consist of "ON" states, during which the patient performs the task 3–10 times, and "OFF" states, during which the patient rests. If performing a block paradigm, the patient alternates between ON and OFF states each lasting the same (periodic test delivery) or different (nonperiodic task delivery) amounts of time.[9] Nonperiodic task delivery is thought to minimize the effects of artifacts such as scanner noise and unwanted motion while maximizing signal.[1]

In an event-related paradigm, the patient completes a single, short-lived action (such as a hand motion or fist clench) followed by a longer OFF period (same duration as in block design).[9] In this way, hemodynamic and thus neuronal response to an isolated event can be measured and evaluated. Event-related paradigms are not as statistically robust as those with a block design, however, and thus more images may need to be collected, increasing the time the study takes.[9]

In order to achieve the most ideal, helpful results possible, paradigms should be performed in such a way that minimizes unwanted head and body motion. A finger-tapping paradigm, for instance, should be performed without wrist or upper arm motion, and tongue paradigms should be performed with a closed mouth to avoid head motion and artifact.[28] Proper paradigm explanation prior to the start of the exam—along with other preventative measures such as placing pillows behind the neck—can further help to reduce unwanted motion.[9]

Motor and sensory paradigms

The location of a lesion should be considered when selecting a motor paradigm. If the lesion is close to the interhemispheric fissure, for example, a foot/toe paradigm would be more appropriate than a tongue movement paradigm.[1] Motor paradigms most often include finger, tongue, or toe motion ("tapping") performed in alternating ON and OFF states. Sensory paradigms may involve brushing the patient's foot or hand, and can also help elucidate the location of areas involved in motor activity.[9] Both motor and sensory paradigms follow the ON/OFF state guidelines previously outlined.

Language paradigms

Language paradigms should help both lateralize and identify the regions of the brain involved in language function relative to the lesion in question.[28] Based on the location of the lesion, paradigms can either entail speech production, comprehension/reception, or both. Lesions closer to Broca's area necessitate paradigms involving speech production: patients may be asked to come up with words beginning with a specific letter or with verbs corresponding to a given noun.[1] Patients can also be asked to come up with sets of words falling under a certain category, such as vegetables or animals.[28]

Lesions near Wernicke's area, on the other hand, necessitate the use of reception/comprehension paradigms. Posterior language area activity is much more difficult to "capture" and isolate.[28,1] Patients may be asked to identify silently the subject of pictures shown to them.[1] Alternatively, they may need to answer questions they hear read aloud to them (color of grass, sky, etc.).[28]

It is important to note that, whether speech production or reception paradigms are used, both (and still other) language areas will likely be activated to some extent since language function is complex and requires integration of many brain regions.[28]

Patient preparation for paradigm performance

Proper patient preparation is necessary in order to obtain the most useful results possible. In addition, many of the complications arising during exams can be circumvented if the paradigm and exam procedure are fully explained to the patient beforehand. Extra explanation—and even paradigm modification—might be needed depending on age and any neurologic deficits.[9] Longer preparation time corresponds to more meaningful results.[1] Avoiding false-negative results is particularly important, as the neurosurgeon may not be aware of the location of an important functional area adjacent to the lesion and could proceed with a potentially risky surgery. Patients should thus arrive early enough so that the paradigm and exam procedure can be explained in full. Paradigms and paradigm timing should also be practiced before the actual exam so that any potential problems can be foreseen and avoided with paradigm modification and further repetition of directions.

Data analysis, correction, and interpretation

The overarching goal of fMRI data analysis is to identify voxels in the brain that exhibit statistically significant changes in BOLD signal relating to the paradigm(s) performed and the timing of the ON/OFF states.[30,9] Significant changes are typically small (0.5%–5.0% signal change from initial), and thus false-negative and -positive results should be carefully guarded against with proper data analysis.[31] Commercially available software packages are useful in correcting for various types of artifacts (discussed below) and attaining more reliable results.

Artifacts and limitations

Clinical fMRI has limitations stemming from various artifact sources and physiological processes. First, unwanted motion can lead to movement-related artifacts, resulting in false-negative, false-positive, or otherwise compromised results. Such motion includes voluntary head or body movement as well as smaller motions resulting from patient breathing.[9,31] Unwanted motion artifacts can be minimized through proper patient set up. Patient comfort and stability is thus crucial.

Second, susceptibility artifacts, which lead to signal dropout or disturbance, can affect fMRI results in patients with and without prior surgery.[9] (See Figure 3.6.) Studies have suggested that in patients with prior surgery (who may have residue from skull drills or metallic implants such as titanium plates or staples), signal tampering or dropout and thus a false-negative result is a serious concern.[32] Patient medical history

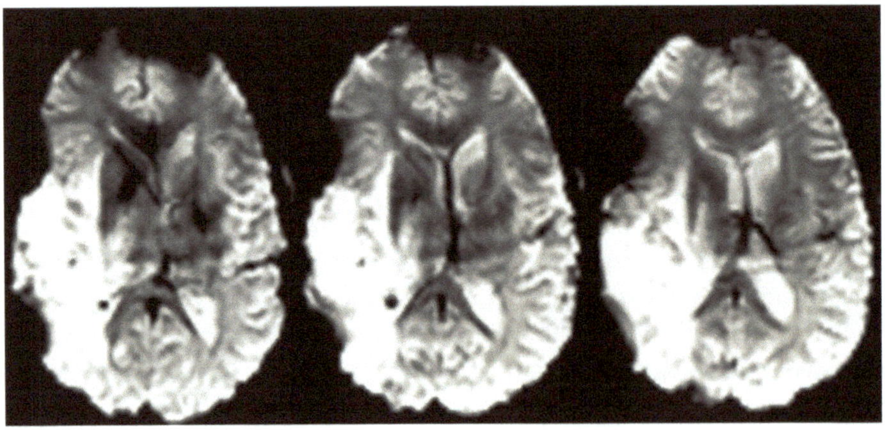

Figure 3.6 T2* image showing drop-out artifact from a previous surgery. Source fMRI images should be inspected in order to assess for risk of false-negative fMRI activation.

should be taken into account when analyzing and interpreting fMRI data. In addition, review of the raw images can demonstrate areas of signal loss.

Even in patients who have not undergone previous surgery, susceptibility artifacts can still affect the reliability of fMRI results. Such artifacts are usually found around moving tissues, cavities, or air–tissue/air–bone interfaces.[9] fMRI signal may be especially weak or remain undetected, for instance, around the temporal lobes (air–bone interface)[1] or eyes (moving tissue).[9] Consistently obtaining reliable BOLD signal-based data is thus more difficult in some areas of the brain, limiting the usefulness of fMRI near those regions.

A third limitation/complication which may arise in interpreting fMRI data stems from the fact that BOLD signal coming both from large draining veins and microvasculature of the activated area may be detected and registered. The distinction between these two "activation" signals is important, since only microvasculature-related BOLD signal (and not signal related to blood flow in large draining veins) is indicative of increased local neuronal activity.[9]

Fourth, tumor neovasculature and its effects on the BOLD response may negatively impact fMRI data. fMRI signal typically increases in active areas due to an overshoot in the amount of diamagnetic oxyhemoglobin present (from increased CBF) relative to the amount of paramagnetic deoxyhemoglobin generated with increased oxygen usage. Tumor neovasculature (in malignant tumors) may have impaired autoregulation, and increased neuronal activity may not cause the same overshoot in CBF and increase in fMRI signal.[9] Such neurovascular decoupling/loss of autoregulation ability can thus lead to false negative results.

Cortical plasticity/reorganization

Last, cortical plasticity/reorganization introduces the possibility for unexpected functional results, which should be recognized and accordingly analyzed in order to properly inform the neurosurgeon. Cortical reorganization occurs when, for some reason (disease, tumor, etc.), a certain part of the brain is no longer able to perform its function fully and, in order to compensate for loss of function at the original location, another area of the brain begins to take on that function.[1,8] Reorganization may be intrahemispheric or it may occur across hemispheres. In patients with tumors affecting the primary motor area, for instance, the SMA may begin to play a role in movement planning and performance, both typically functions of the primary motor cortex.[9] (See Figure 3.7.)

Reorganization may also occur after stroke, and thus fMRI results can help create more individualized stroke therapy and recovery plans. One study demonstrated, for instance, that patients left-dominant for language who suffered strokes affecting Broca's or Wernicke's areas began to show, within a relatively short period of time, language-related activity in the same areas in the *right* hemisphere.[33]

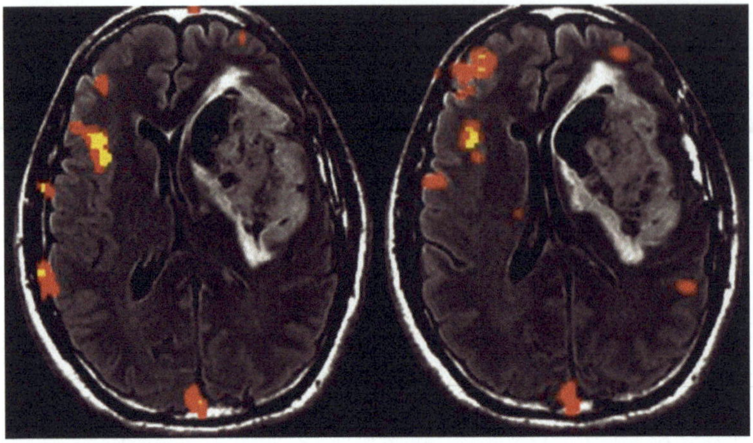

Figure 3.7 47 year old strongly right handed male with a large left insular tumor who presented with headaches and no speech deficit. Functional MRI suggested and direct cortical stimulation confirmed translocation of Broca's area to the right hemisphere.

Since lesion growth may lead to reorganization,[9] and since one of the primary goals of tumor resection is to preserve the function of eloquent cortices, fMRI plays an important role in helping clinicians understand whether or not any functional rearrangement has occurred and if any "new" areas need to be treated with special attention. It is also important to note that reorganization is different than simple movement or relocation of a region due to physical pressure from a growing lesion.[34]

Applications and current and future research

Current research is revealing many new, exciting possible applications for fMRI. Resting-state fMRI, for instance—during which the patient does not perform a paradigm—is showing promising clinical applications and results. The physiological basis of resting-state fMRI stems from spontaneous variations in BOLD signal present even without any voluntary task performance.[35] Since different brain regions have been found to have matching BOLD signal fluctuations, resting-state fMRI can help elucidate the integration and connection between brain areas on a function level—that is, which brain areas' signals "fluctuate" together, forming "resting state networks" (RSNs).[35] A language-related network, for example, encompassing Broca's and Wernicke's (as well as other) areas, has been identified with resting-state fMRI.[28] Knowledge of such RSNs could offer neurosurgeons additional information about critical functional areas—that is, not simply locations of eloquent cortices, but also locations of the specific brain areas with which they interact.[36] (See Figure 3.8.)

Resting-state fMRI could be especially helpful in patients so neurologically or otherwise physically compromised that proper paradigm performance is not possible

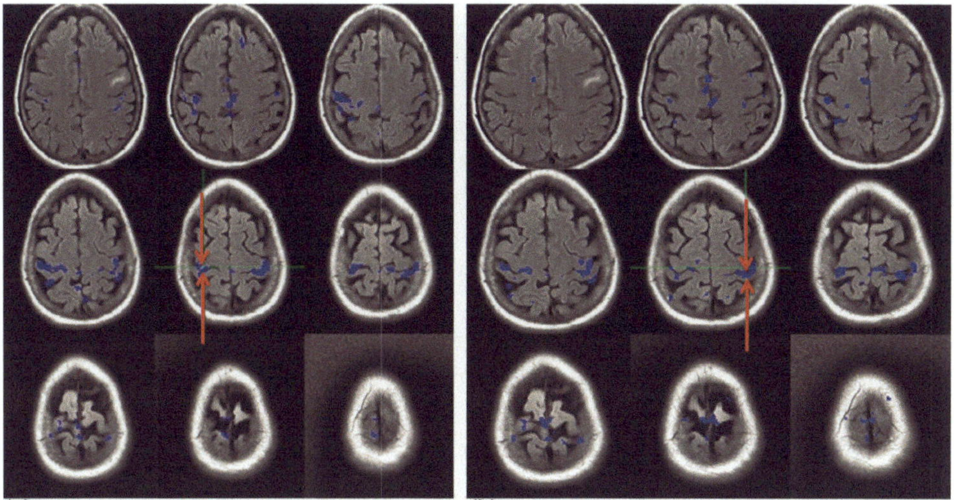

Figure 3.8 Resting-state fMRI connectivity patterns. (a) Seed placed in right side hand motor. (b) Seed placed on left side hand motor. Red arrows indicate the location of the seed placement in the reverse hand omega.

and good paradigm-based fMRI results unachievable. Other advantages include shorter scan time (than paradigm-based fMRI) and the ability to detect numerous neural networks at once without having the patient perform different kinds of tasks to activate different functional areas.[36,37] This technique could also have applications in diagnosis of brain-related disorders in the future.[35]

Resting-state fMRI is still a new technique under research and development, however, and thus has several uncertainties and limitations associated with it. Before RSNs can be detected and mapped consistently, further investigation is needed into the effects on resting-state BOLD signal fluctuations of both medications and natural variation in physiological processes.[36]

fMRI use in diagnosis, evaluation, and treatment of certain brain-related disorders and psychiatric diseases is also being investigated. fMRI can be helpful in understanding of schizophrenia, for instance, as functional brain imaging can elucidate deficiencies in normal neural networks.[38] Future research goals include development of new paradigms for patients with schizophrenia since current paradigms may not be appropriate in such cases (improper or incomplete performance) and may not yield the most reliable results possible.[38]

fMRI has also been used in research relating to Alzheimer's disease. This type of research often involves repeated testing over a long period of time to study disease development. Because fMRI is noninvasive and does not entail the use of radioactive isotopes (as does PET), it is a well-suited technique for such longitudinal Alzheimer's studies.[28]

The use of fMRI in studying other neurodegenerative disorders such as Parkinson's disease is also being investigated. One particularly fascinating and novel use of fMRI—the simultaneous use of fMRI and deep brain stimulation (DBS)—has been successfully applied to rats.[39] This combination technique poses challenges due to possible dangers associated with MRI scanner use on a patient with implanted electrodes.[36] DBS fMRI allows researchers to stimulate specific brain areas while recording and "seeing" the brain activity occurring in response to that stimulation in real time.[39] DBS is currently used in Parkinson's disease treatment. In the future, DBS fMRI in humans could allow for identification of the neural pathways and downstream activation patterns associated with therapy stimulation, paving the way for investigation and discovery of new treatments for Parkinson's disease, based on those same neural networks.[39]

Resting-state fMRI is currently being used as a new tool to help elucidate shifts in neural networks and brain region integration brought about by Parkinson's disease.[40] Interestingly, Parkinson's-related differences in brain connectivity vary with the symptoms examined.[40] Depression, tremor, and indifference/lack of energy, for instance—all different Parkinson's symptoms—are associated with distinct changes in baseline BOLD signal synchronization across brain areas. These changes are thought to come about in response to decreased dopamine levels, and the most affected parts of the brain appear to be sensorimotor regions.[40] In the future, such knowledge could help better to define stages of Parkinson's disease progression and allow for earlier screening and evaluation of patients with high Parkinson's disease risk.[40]

fMRI is also playing an ever-increasing role in the development of more personalized medicine and individualized treatment plans. With a deeper understanding of cortical reorganization and the molecular, biological mechanisms behind it, clinicians will be better prepared to plan approaches to treating patients presenting atypical anatomy.[9] fMRI results can offer invaluable information concerning functional or anatomical irregularities in tumor, stroke, epilepsy, or other patients.

Understanding the body and patient specificity of fMRI results is an important part of designing individualized paradigms and treatment plans. A patient's typical movement habits and experiences (handedness, etc.), as well as medical history, can impact fMRI data.[41] One study showed, for instance, that motor imagery paradigm results differ based on the patient's usual lifestyle and the way he or she would actually choose to perform the action.[41]

Similarly to PET and MEG, fMRI can help to localize the epileptogenic source in epilepsy patients. Such localization is crucial for deciding on and planning surgery. In cases in which the epileptic source cannot be properly identified with a brain imaging technique, an invasive, potentially risky procedure entailing the use of intracranial electrodes may be necessary. But even in these cases, since proper electrode placement is crucial for helpful results, prior knowledge from fMRI or other functional imaging data about the general area likely containing the epileptogenic zone can be useful.[42] EEG-fMRI, a combination technique in which EEG-detected epileptiform abnormalities serve as the events for fMRI analyses, has been shown to be helpful in localization of the epileptogenic zone as part of surgical/treatment planning.[42]

fMRI has also been used in pediatric epilepsy research. A study by Yuan and colleagues aimed at identifying differences in language lateralization patterns between epileptic and healthy children.[43] Results of silent verb-generation paradigms showed a significant increase in frequency of bilateral or right hemisphere dominance among epileptic children (vs. healthy controls). This higher incidence of irregular language lateralization in epileptic children suggests that seizure activity may elicit reorganization of language function—plasticity may help combat seizure-related damage to left hemisphere language areas.[43] Alternatively, the epileptic patients in the study may have had underlying differences in brain organization that contributed to both the development of epilepsy and right hemisphere/bilateral language dominance.[43]

Additionally, prior to this study, a significant database of purely pediatric fMRI data on language localization and lateralization did not exist. Over the course of the study, such a database was gathered and has the potential to be used as a baseline for future studies of children with different brain-related disorders.[43] The data compiled from healthy children revealed an interesting trend—younger children were more regularly right hemisphere dominant or bilaterally dominant for language compared to older children.[43] These results indicate that left hemisphere dominance is likely a gradual process which occurs with maturation.[43]

Indeed, pediatric fMRI, though potentially more challenging to perform and analyze, could have profound clinical benefits and is thus an important area to continue to develop.[44] Paradigm design and patient preparation must be carried out with special consideration when working with children. Paradigm explanation may require more detail, for instance, and a child's ability to cooperate—especially if he or she has a neurological deficit—should be taken into account when practicing the paradigm.[44]

Finally, fMRI has been used to predict naming outcomes in patients undergoing anterior temporal lobectomies as part of treatment for epilepsy.[45] While the Wada test was often used for such prognosticative purposes in the past, the procedure is more complication-prone than fMRI and current research is aimed at establishing fMRI as a reliable, accurate alternative to the Wada test. One study focused on 10 cases in which fMRI (based on silent, picture naming paradigms) and Wada results concerning preoperative language lateralization were discordant.[45] Postoperatively, it was concluded that fMRI was a better predictor of naming outcomes/deficits in seven of the ten patients, Wada was a better predictor in two patients, and the tests equal in predictive power in one case.[45] This study helped to validate fMRI as a reliable—and perhaps even superior—predictor of postoperative naming outcomes in anterior temporal lobectomy patients when compared to the Wada test.[45] Another earlier, similar study by Sabsevitz and colleagues also indicated that preoperative fMRI is a useful tool to estimate the risk of naming deficits after left anterior temporal lobectomies.[46]

MAGNETOENCEPHALOGRAPHY

Overview and physiological basis

MEG detects and measures brain activity by registering weak electromagnetic signals resulting from intracellular neuronal electrical activity.[47,48] In order to boost weak signals from neuron-related magnetic fields, superconducting quantum interference devices (SQUIDs) are used to increase sensitivity.[49]

Unlike PET, MEG does not entail the use of radioactive isotopes. MEG exhibits reasonably good source localization ability and millisecond-level temporal resolution.[50] Such strong source localization is in part due to the fact that MEG signals are not as affected by materials through which they travel (skull, scalp, etc.) unlike EEG signals, which are more disturbed.[51] In addition, it is important to note that MEG more readily registers current sources tangential to the skull surface (rather than radial to it),[50] which indicates that the activity registered comes mostly from sulci (rather than gyri).[51]

MEG is an important functional imaging tool in part because it provides insight into neural networks and the connectivity between multiple brain regions.[52,47] It has been established that significant integration is present between different brain areas on a functional level, and that a particular function most often cannot simply be localized to one region of the brain.[47,53] MEG is thus a powerful, valuable noninvasive imaging tool as it can help elucidate the neural networks and integration underlying complex brain functions such as language.[48] However, the widespread application of MEG has so far been somewhat limited due to the cost of the equipment and the not insubstantial siting requirements.

Clinical importance and applications

MEG has a relatively wide range of clinical applications, though it is used primarily in presurgical planning and localization/mapping of brain functions. Brain imaging techniques are an important part of planning both for tumor resections and other types of neurosurgery. MEG, like fMRI, can help neurosurgeons and other clinicians better understand the anatomy of brain areas surrounding the lesion and navigate eloquent cortices on a patient-by-patient basis, which in turn allows for maximum tumor resection and minimum damage to any nearby language, motor, sensory, or memory functional areas.[9,54]

The general accuracy and reliability of MEG has been validated by other techniques. fMRI, for instance, has shown good concordance with MEG,[31] especially in localization of certain sensory and motor areas based on simple sensory and motor paradigms.[48] Additionally, when the results of MEG and the sodium amobarbital procedure (the Wada test) were compared, the two techniques showed high concordance in determining hemispheric language dominance, suggesting that MEG can be reliably used for this purpose instead of the Wada test, which may have more complications associated with it.[54] In such circumstances, MEG may emerge as a viable noninvasive alternative to the Wada test which requires an invasive angiographic procedure with injections of a short acting barbiturate, sodium amobarbital.

As with fMRI, in order to identify important functional areas surrounding a lesion pre-operatively, patients may need to carry out tasks to activate specific brain regions involved in the functions being investigated. To identify the hand motor area, for example, the patient might be asked to tap his or her fingers.[9] To locate the somatosensory cortex, on the other hand, the patient's toes and fingers may be stimulated by an outside electric source.[51] Brain mapping using MEG can also be used for identification of language-related areas. Since MEG can help elucidate key neural networks, this technique may be especially powerful and helpful in language lateralization and localization since numerous brain regions are implicated in language function.[48]

In addition to functional brain mapping and presurgical planning in tumor patients, MEG can play a role in epilepsy surgery and treatment. Resective surgery is often one of the most effective treatments for medically refractory epilepsy,[49] and thus it is important for surgeons to have tools capable of identifying both the epileptogenic area and any surrounding functional areas. In cases in which MRI may not adequately outline the epileptogenic zone, MEG can help neurosurgeons more reliably identify the epileptic source as well as better understand the anatomy of surrounding functional areas.[49] MEG can also play a role in identifying the most optimal locations for intracranial electrodes, which are costly and often used as part of epilepsy surgery planning.[49,55] The strong source localization ability of MEG—again stemming partially from the fact that MEG signals are not drastically disrupted or changed as they pass through the skull, scalp or other tissues—is especially important in epilepsy therapy since the specific epileptic source may not be previously known.[48,56]

MEG also has pediatric applications, specifically in the treatment of epilepsy.[57] In one study of children with refractory nonlesional extratemporal epilepsy, MEG and intracranial electrode testing results both mapped the somatosensory area for the hand to the same area of the brain.[57] These results suggest not only that MEG should be considered an important part of pediatric epilepsy surgery planning,[57] but also that MEG could potentially replace intracranial electrode usage in epilepsy evaluation and surgery in selected cases.

MEG also has clinical applications in the realm of stroke treatment and stroke recovery evaluation. Stroke may impact the blood oxygen level dependent (BOLD) effect on which fMRI is based, whereas MEG signals remain unaltered.[48] MEG may thus be a better option in evaluating brain functionality and possible reorganization after a stroke.[58]

Additionally, MEG is involved in the diagnosis of neurocognitive disorders insofar as it exhibits good source localization, features millisecond-level temporal resolution,[50] and can illustrate and provide insight into neural networks. A recent auditory processing-related study by Larson and colleagues, for instance, suggested that MEG, when used with proper auditory paradigms, could help diagnose and analyze central auditory processing disorder (CAPD).[52] In cases in which CAPD is present along with another disorder (such as ADHD), MEG could also be used to better understand which (and how) specific brain regions and neural networks separately contribute to each disorder.[52]

MEG is also playing an increasing role in evaluation and treatment of patients with neurodegenerative disorders such as Alzheimer's and Parkinson's diseases. Diagnosis of Alzheimer's and identification of the level of disease progression (both of which necessitate a distinction from simple aging) may be challenging and even debatable since different criteria for diagnosis exist.[59] The identification of distinct, reliable "biomarkers" or changes in neurophysiological processes associated with Alzheimer's could thus facilitate disease diagnosis and evaluation.[59] As MEG can detect changes in neuronal activity based on electromagnetic signals from those cells, this imaging technique has the potential to identify irregularities and inadequacies in a region-specific manner, providing an alternate test to help in Alzheimer's diagnosis and disease progression evaluation.[59]

Patient preparation and procedure

In order to help ensure meaningful results, patients must be adequately prepared before the exam commences. Any metal on the patient's clothing or body should be removed.[60] For cases in which the patient does not need to perform a task, preparation for the exam is minimal and less explanation is needed. If, however, the patient must perform a paradigm, preparation is more complex and paradigm design must be taken into consideration.

For an overview of patient preparation and paradigm design/delivery, refer to the corresponding heading in the fMRI section.

Applications and current and future research

Current research is revealing new clinical applications for MEG. MEG has been used in research relating to cortical plasticity, language function and processing, neurodevelopment, and even social interaction and communication.[48] The neuroscience behind humans' recognition of each other—the specific brain activity that occurs when one person sees another—has been studied, for instance, using MEG. MEG results helped to reveal that primary motor cortex activity can occur simply when one person observes another performing some action. [57]

MEG has also been used to examine and compare brain activity resulting from different types (or lack) of eye contact.[61] One study showed that brain activity increased as a result of a viewer encountering direct eye contact or even averted gaze as opposed to closed eyes.[61]

MEG has proved helpful not only in examining the neuroscience of adult human interaction, but has also been used in research concerning the prenatal period and neonates, offering insights into functional brain development. A study by Draganova and colleagues utilized MEG to record both fetal neural responses to sounds registered through mothers' abdomens and neonate responses to similar sounds after birth.[62] Since the ability to perceive and respond to sound is necessary for proper speech development, such studies could be important for better understanding the developmental basis of deficits appearing at a later age.[62]

Studies have also demonstrated that MEG is useful in memory function lateralization as well as language function lateralization.[48] Indeed, the Wada test and MEG have shown high concordance in determining hemispheric language dominance.[54] Although more research into the ability of MEG to investigate and determine memory lateralization is needed, MEG may prove a reliable, less risky alternative to the Wada test in general.[48]

As previously noted, MEG is playing an ever-increasing role in the diagnosis, evaluation, and treatment of neurodegenerative disorders (Alzheimer's, Parkinson's, etc.) and is also a helpful tool to better understand the basis of such diseases. MEG has been used, for instance, to investigate irregularities in neural networks underlying Parkinson's disease progression. Recent studies have documented that, while Parkinson's patients with dementia exhibit decreased functional connectivity, patients without dementia actually show increases in some neural network connections, namely in the lower alpha frequency band.[63] Changes in beta and theta bands only appeared in more severe Parkinson's disease patients.[63] Detection of irregularities and disruptions in the interactions between specific brain regions in Parkinson's patients may be helpful in future investigations of the direct causes of and physiological mechanisms behind the disease.[63]

The use of MEG in detecting signs of various neurodevelopmental/autism spectrum disorders in young children is also currently under investigation.[64] Initially, whole-head MEG systems were designed for adults; if used in children, results were not as reliable or accurate due to the larger physical distance between the child's brain and the MEG machinery.[64] With the goal of early identification of irregular neuronal activity patterns indicative of development of autism spectrum disorders, a child/infant MEG system was designed.[64] This system, Artemis 123, is used for children three and under, and proper use of the system may require new paradigms since young children may not be able to perform typical, complex paradigms for extended periods of time.[64] Such a child-geared system could have great benefits, detecting indicators of development of neurodevelopmental conditions and allowing for more feasible early, protective intervention.[64]

CONCLUSION

PET, MEG, and fMRI are noninvasive functional imaging techniques that provide researchers, surgeons, and other clinicians with important anatomical, physiological, and neural-network related information pertaining to specific areas of the brain. Each imaging technique has its strengths and limitations in particular clinical situations. PET uses analogs of biological molecules to provide temporal, quantitative measures of various biochemical processes (such as glucose metabolism) in the brain, offering key insights into physiological changes occurring with tumors or other brain-related disorders/conditions.[2] PET, however, does not have strong localization power, and PET scans may not provide the most useful anatomical data.[3] PET also entails the administration of radioactive isotopes and is somewhat limited in its uses, especially in the types of cancer it can image.

MEG and fMRI, on the other hand, provide high signal to noise and are helpful in localization of brain function and detection of important neural networks. fMRI is based on the BOLD effect and thus depends on neurovascular coupling with high spatial resolution but lower temporal resolution. MEG has lower spatial resolution and very high temporal resolution based on measuring brain activity through registering weak electromagnetic signals resulting from intracellular neuronal electrical activity.[21,48] Both techniques can involve the use of paradigms to activate specific functional brain areas and neural networks.

PET, MEG, and fMRI increasingly provide crucial information to help neurosurgeons and other clinicians make more informed decisions even before any invasive procedures are performed. These techniques can also help guide intraoperative mapping and neurosurgery in real time. Personalized medicine is a goal for the future, and PET, MET, and fMRI can play important roles in developing individualized treatment plans not only for tumor/cancer patients, but also in other neurologic and psychiatric diseases. Current research is investigating such novel uses and applications for each technique.

REFERENCES

1. Bogomolny DL, Petrovich NM, Hou BL, Peck KK, Kim MJ, Holodny AI. Functional MRI in the brain tumor patient. *Top Magn Reson Imaging*. 2004;15(5):325–335.
2. Farwell MD, Pryma DA, Mankoff DA. PET/CT imaging in cancer: current applications and future directions. *Cancer*. 2014;120(22):3433–3445.

3. Griffeth LK. Use of PET/CT scanning in cancer patients: technical and practical considerations. *Proc (Bayl Univ Med Cent)*. 2005;18(4):321–330.
4. Kelloff GJ, Hoffman JM, Johnson B, et al. Progress and promise of FDG-PET imaging for cancer patient management and oncologic drug development. *Clin Cancer Res.* 2005; 11(8):2785–2808.
5. Boellaard R, O'Doherty MJ, Weber WA, et al. FDG PET and PET/CT: EANM procedure guidelines for tumour PET imaging: version 1.0. *Eur J Nucl Med Mol Imaging*. 2010;37(1):181–200.
6. Kiston SL, Cuccurullo V, Ciarmiello A, Salvo D, Mansi L. Clinical applications of positron emission tomography (PET) imaging in medicine: oncology, brain diseases and cardiology. *Curr Radiopharm.* 2009;2(4):224–253.
7. Buerkle A, Weber WA. Imaging of tumor glucose utilization with positron emission tomography. *Cancer Metastasis Rev.* 2008;27(4):545–554.
8. Tai YF, Piccini P. Applications of positron emission tomography (PET) in neurology. *J Neurol Neurosurg Psychiatry.* 2004;75(5):669–676.
9. Holodny AI, Shevzov-Zebrun N, Brennan N, Peck KK. Motor and sensory mapping. *Neurosurg Clin N Am.* 2011;22(2):207–218:viii.
10. Loane C, Politis M. Positron emission tomography neuroimaging in Parkinson's disease. *Am J Transl Res.* 2011;3(4):323–341.
11. Miele E, Spinelli GP, Tomao F, et al. Positron Emission Tomography (PET) radiotracers in oncology—utility of 18F-Fluoro-deoxy-glucose (FDG)-PET in the management of patients with non-small-cell lung cancer (NSCLC). *J Exp Clin Cancer Res.* 2008;27:52.
12. Mohile NA, Deangelis LM, Abrey LE. Utility of brain FDG-PET in primary CNS lymphoma. *Clin Adv Hematol Oncol.* 2008;6(11):818–820:840.
13. Mohile NA, Deangelis LM, Abrey LE. The utility of body FDG PET in staging primary central nervous system lymphoma. *Neuro-oncology.* 2008;10(2):223–228.
14. Vogelbaum MA, Jost S, Aghi MK, et al. Application of novel response/progression measures for surgically delivered therapies for gliomas: response Assessment in Neuro-Oncology (RANO) Working Group. *Neurosurgery.* 2012;70(1):234–243:discussion 243–234.
15. Young RJ, Gupta A, Shah AD, et al. Potential utility of conventional MRI signs in diagnosing pseudoprogression in glioblastoma. *Neurology.* 2011;76(22):1918–1924.
16. Omuro A, Beal K, Gutin P, et al. Phase II study of bevacizumab, temozolomide, and hypofractionated stereotactic radiotherapy for newly diagnosed glioblastoma. *Clin Cancer Res.* 2014;20 (19):5023–5031.
17. Basu S, Alavi A. Molecular imaging (PET) of brain tumors. *Neuroimaging Clin N Am.* 2009;19 (4):625–646.
18. Demetriades AK, Almeida AC, Bhangoo RS, Barrington SF. Applications of positron emission tomography in neuro-oncology: a clinical approach. *Surgeon.* 2014;12(3):148–157.
19. Sureshbabu W, Mawlawi O. PET/CT imaging artifacts. *J Nucl Med Technol.* 2005;33(3):156–161: quiz 163–154.
20. Kinahan PE, Hasegawa BH, Beyer T. X-ray-based attenuation correction for positron emission tomography/computed tomography scanners. *Semin Nucl Med.* 2003;33(3):166–179.
21. Schafers KP, Raupach R, Beyer T. Combined 18F-FDG-PET/CT imaging of the head and neck. An approach to metal artifact correction. *Nuklearmedizin.* 2006;45(5):219–222.
22. Nehmeh SA, Erdi YE, Rosenzweig KE, et al. Reduction of respiratory motion artifacts in PET imaging of lung cancer by respiratory correlated dynamic PET: methodology and comparison with respiratory gated PET. *J Nucl Med.* 2003;44(10):1644–1648.
23. Nedergaard MK, Kristoffersen K, Michaelsen SR, et al. The use of longitudinal 18F-FET MicroPET imaging to evaluate response to irinotecan in orthotopic human glioblastoma multiforme xenografts. *PLoS One.* 2014;9(2):e100009.
24. Galldiks N, Dunkl V, Stoffels G, et al. Diagnosis of pseudoprogression in patients with glioblastoma using O-(2-[F]fluoroethyl)-L-tyrosine PET. *Eur J Nucl Med Mol Imaging.* 2014; [Epub ahead of print].

25. Misch M, Guggemos A, Driever PH, et al. F-FET-PET guided surgical biopsy and resection in children and adolescence with brain tumors. *Childs Nerv Syst.* 2014; [Epub ahead of print].
26. Tuntiyatorn L, Wuttiplakorn L, Laohawiriyakamol K. Plasticity of the motor cortex in patients with brain tumors and arteriovenous malformations: a functional MR study. *J Med Assoc Thai.* 2011;94 (9):1134–1140.
27. Peck KK, Bradbury M, Psaty EL, Brennan NP, Holodny AI. Joint activation of the supplementary motor area and presupplementary motor area during simultaneous motor and language functional MRI. *Neuroreport.* 2009;20(5):487–491.
28. Belyaev AS, Peck KK, Brennan NM, Holodny AI. Clinical applications of functional MR imaging. *Magn Reson Imaging Clin N Am.* 2013;21(2):269–278.
29. Knecht S, Drager B, Deppe M, et al. Handedness and hemispheric language dominance in healthy humans. *Brain.* 2000;123(Pt 12):2512–2518.
30. Matthews PM, Honey GD, Bullmore ET. Applications of fMRI in translational medicine and clinical practice. *Nat Rev Neurosci.* 2006;7(9):732–744.
31. Kesavadas C, Thomas B. Clinical applications of functional MRI in epilepsy. *Indian J Radiol Imaging.* 2008;18(3):210–217.
32. Kim MJ, Holodny AI, Hou BL, et al. The effect of prior surgery on blood oxygen level-dependent functional MR imaging in the preoperative assessment of brain tumors. *Am J Neuroradiol.* 2005;26 (8):1980–1985.
33. Thulborn KR, Carpenter PA, Just MA. Plasticity of language-related brain function during recovery from stroke. *Stroke.* 1999;30(4):749–754.
34. Thomas B, Sage C, Eyssen M, Kovacs S, Peeters R, Sunaert S. Brain plasticity and fMRI. *Med Radiol.* 2007;:209–226.
35. Lee MH, Smyser CD, Shimony JS. Resting-state fMRI: a review of methods and clinical applications. *Am J Neuroradiol.* 2013;34(10):1866–1872.
36. Lang S, Duncan N, Northoff G. Resting-state functional magnetic resonance imaging: review of neurosurgical applications. *Neurosurgery.* 2014;74(5):453–464:discussion 464–455.
37. Tomasi D, Volkow ND. Resting functional connectivity of language networks: characterization and reproducibility. *Mol Psychiatry.* 2012;17(8):841–854.
38. Van Hecke J, Gladwin TE, Coremans J, Hulstijn W, Sabbe B. [Functional MRI for schizophrenia: importance of the type of task being scanned]. *Tijdschr Psychiatr.* 2014;56(6):385–393.
39. Younce JR, Albaugh DL, Shih YY. Deep brain stimulation with simultaneous FMRI in rodents. *J Vis Exp.* 2014;84:e51271.
40. Pyatigorskaya N, Gallea C, Garcia-Lorenzo D, Vidailhet M, Lehericy S. A review of the use of magnetic resonance imaging in Parkinson's disease. *Ther Adv Neurol Disord.* 2014;7(4):206–220.
41. Willems RM, Toni I, Hagoort P, Casasanto D. Body-specific motor imagery of hand actions: neural evidence from right- and left-handers. *Front Comput Neurosci.* 2009;3:39.
42. Zijlmans M, Huiskamp G, Hersevoort M, Seppenwoolde JH, van Huffelen AC, Leijten FS. EEG-fMRI in the preoperative work-up for epilepsy surgery. *Brain.* 2007;130(Pt 9):2343–2353.
43. Yuan W, Szaflarski JP, Schmithorst VJ, et al. fMRI shows atypical language lateralization in pediatric epilepsy patients. *Epilepsia.* 2006;47(3):593–600.
44. Wilke M, Pieper T, Lindner K, et al. Clinical functional MRI of the language domain in children with epilepsy. *Hum Brain Mapp.* 2011;32(11):1882–1893.
45. Janecek JK, Swanson SJ, Sabsevitz DS, et al. Naming outcome prediction in patients with discordant Wada and fMRI language lateralization. *Epilepsy Behav.* 2013;27(2):399–403.
46. Sabsevitz DS, Swanson SJ, Hammeke TA, et al. Use of preoperative functional neuroimaging to predict language deficits from epilepsy surgery. *Neurology.* 2003;60(11):1788–1792.
47. Zhang X, Lei X, Wu T, Jiang T. A review of EEG and MEG for brainnetome research. *Cogn Neurodyn.* 2014;8(2):87–98.
48. Hari R, Salmelin R. Magnetoencephalography: from SQUIDs to neuroscience. Neuroimage 20th anniversary special edition. *NeuroImage.* 2012;61(2):386–396.
49. Haneef Z, Chen DK. Functional neuro-imaging as a pre-surgical tool in epilepsy. *Ann Indian Acad Neurol.* 2014;17(Suppl 1):S56–S64.

50. Nevalainen P, Lauronen L, Pihko E. Development of human somatosensory cortical functions—what have we learned from magnetoencephalography: a review. *Front Comput Neurosci.* 2014;8:158.
51. Ray A, Bowyer SM. Clinical applications of magnetoencephalography in epilepsy. *Ann Indian Acad Neurol.* 2010;13(1):14–22.
52. Larson E, Lee AK. Potential use of MEG to understand abnormalities in auditory function in clinical populations. *Front Comput Neurosci.* 2014;8:151.
53. Tognoli E, Kelso JA. Enlarging the scope: grasping brain complexity. *Front Syst Neurosci.* 2014;8:122.
54. Papanicolaou AC, Simos PG, Castillo EM, et al. Magnetocephalography: a noninvasive alternative to the Wada procedure. *J Neurosurg.* 2004;100(5):867–876.
55. Knowlton RC, Razdan SN, Limdi N, et al. Effect of epilepsy magnetic source imaging on intracranial electrode placement. *Ann Neurol.* 2009;65(6):716–723.
56. Burgess RC. How to prepare for your MEG: Cleveland clinic. Accessed 05.01.15. Available from: clevelandclinic.org/epilepsy.
57. Minassian BA, Otsubo H, Weiss S, Elliott I, Rutka JT, Snead 3rd OC. Magnetoencephalographic localization in pediatric epilepsy surgery: comparison with invasive intracranial electroencephalography. *Ann Neurol.* 1999;46(4):627–633.
58. Rossini PM, Altamura C, Ferreri F, et al. Neuroimaging experimental studies on brain plasticity in recovery from stroke. *Eura Medicophys.* 2007;43(2):241–254.
59. Fernandez A, Turrero A, Zuluaga P, et al. MEG delta mapping along the healthy aging-Alzheimer's disease continuum: diagnostic implications. *J Alzheimers Dis.* 2013;35(3):495–507.
60. American Clinical Magnetoencephalography Society. Why is MEG beneficial? A guide for patients and their families. 2015. Available from: <http://www.acmegs.org/what-is-meg>.
61. Taylor MJ, George N, Ducorps A. Magnetoencephalographic evidence of early processing of direction of gaze in humans. *Neurosci Lett.* 2001;316(3):173–177.
62. Draganova R, Eswaran H, Murphy P, Huotilainen M, Lowery C, Preissl H. Sound frequency change detection in fetuses and newborns, a magnetoencephalographic study. *NeuroImage.* 2005;28(2):354–361.
63. Stam CJ. Use of magnetoencephalography (MEG) to study functional brain networks in neurodegenerative disorders. *J Neurol Sci.* 2010;289(1–2):128–134.
64. Roberts TP, Paulson DN, Hirschkoff E, et al. Artemis 123: development of a whole-head infant and young child MEG system. *Front Comput Neurosci.* 2014;8:99.

CHAPTER 4

Imaging White Matter Anatomy for Brain Tumor Surgery

Sonia Pujol
Department of Radiology, Brigham and Women's Hospital, Harvard Medical School, Boston, MA, USA

INTRODUCTION

The cerebral white matter is an intricate network of myelinated fibers that connect gray matter regions and support brain function. A significant challenge in surgical planning of brain tumor resection is the preservation of functional areas and white matter tracts. Magnetic resonance imaging (MRI) helps to depict the spatial relationship of a lesion to eloquent cortical areas. However, conventional structural MRI lacks contrast to delineate white matter fibers. As a result, the mental representation of the trajectory of white matter pathways is traditionally based on anatomical knowledge, but is not able to take into account individual variation and effects of the lesion. The introduction of diffusion MRI and subsequent developments of diffusion tensor imaging (DTI) have opened up the possibility of inferring the architecture of the cerebral white matter in vivo. From DTI data, tractography techniques have enabled the three-dimensional (3D) representation of the trajectory of white matter pathways at the individual patient level. Such techniques provide surgeons critical information on white matter involvement of lesions located in eloquent areas. In particular, the knowledge of the relationship of a tumor with tracts involved in motor, visual or language function is essential to prevent postoperative neurological deficits. Diffusion MRI is an active area of research that involves multidisciplinary teams from MR physics, applied mathematics, computer science and biomedical engineering, and provides clinical researchers and practitioners unprecedented information on the anatomy of white matter pathways in individual patients. This chapter is an introduction to the theoretical foundations of the acquisition and postprocessing of diffusion MRI data, and the current status of the use of tractography in clinical neurosurgery.

THEORETICAL FOUNDATIONS OF DIFFUSION MRI

Diffusion in biological tissue is based on the random displacement of water molecules undergoing constant thermal motion. In 1905, Einstein demonstrated that when the molecules are free to diffuse, the displacement distribution is a Gaussian function

A. Golby (Ed): Image-Guided Neurosurgery
DOI: http://dx.doi.org/10.1016/B978-0-12-800870-6.00004-2

characterized by a diffusion coefficient *D*, and that the mean square displacement is directly proportional to the observation time *t* through the equation.

$$<r^2> = 6Dt \tag{4.1}$$

When water molecules encounter biological barriers such as cell membranes and organelles, the mean square displacement is shorter and the observed diffusion coefficient, or so-called apparent diffusion coefficient (ADC), is smaller than in free water.[1] In gray matter and cerebrospinal fluid (CSF), the diffusion of water molecules is isotropic and independent of the orientation of the tissue. In white matter, the axonal membranes and myelin sheets encasing the axons restrict the displacements.[2] As a result, water molecules diffuse predominantly in the direction parallel to the fiber orientation and the diffusion is anisotropic.

Diffusion MRI combines the physics of diffusion and the physics of MRI to infer the architecture of the brain. This sophisticated technology provides an elegant characterization of the cerebral white matter noninvasively. MRI techniques rely on the intrinsic properties of the hydrogen atoms of water molecules. The nuclei of hydrogen atoms (protons) possess a magnetic moment, often referred to as spin, which is randomly oriented. During an MRI acquisition, the strong external magnetic field B_0 of an MR scanner causes the magnetic nuclear spins to align themselves in the direction of the magnetic field. Radio-frequency (RF) pulses are then added to deviate the magnetic nuclear spins in a direction perpendicular to B_0. When the RF pulse frequency is applied at the Larmor frequency, the spins start rotating in a conical motion, also called precession, with the cone axis aligned with B_0. As the spins coherently precess, they induce a current in the receiver coil positioned next to the patient in the MR scanner. In addition, a set of magnetic field gradients is applied for slice selection and image encoding. These gradients are rapidly turned on and off, and are responsible for the typical noise audible during an MR scan acquisition. The current induced in the receiver coil is then processed to give localized information about the tissues generating an MRI image. The series of RF pulses and magnetic field gradients applied during an MRI acquisition is called a pulse sequence. Diffusion MRI is based on special pulse sequences called diffusion-weighted imaging (DWI). DWI sequences add to standard MRI sequences a series of pairs of bipolar magnetic field gradients called diffusion sensitizing gradients. These pairs of gradients are used to add a phase to the spin precession of the hydrogen nuclei of the water molecules. The first gradient dephases the spins by adding a positive phase, and the second gradient rephases them by adding a negative phase of the opposite value. If water molecules stay at the same location, the negative phase compensates the positive phase, and the sum is zero. If a net movement of the water molecules occurs before the application of the second gradient, the negative phase does not compensate the positive phase, and this results in an attenuation of the signal intensity *S* in the direction of the applied gradient.

In the example of a single pair of bipolar gradients, and under the assumption that the diffusion of water molecules is Gaussian, the attenuation of the signal intensity S can be expressed using the Stejskal and Tanner equation[3]:

$$S = S_0 e^{-bD} \tag{4.2}$$

where S_0 is the signal intensity with the diffusion sensitizing gradients' amplitude set to zero; b (s/mm^2) is the diffusion weighting factor calculated based on the intensity, strength, duration, and temporal separation of the diffusion sensitizing gradients;[1] and D is the ADC (mm^2/s). Figure 4.1 shows an example of DWI acquisitions with b = 0, 1000, and 3000 s/mm^2 in an astrocytoma WHO grade II. From equation (4.2), it can be seen that, that for a given b factor, tissues with different ADC values will cause different signal attenuation. Thus, diffusion-weighted images provide a unique contrast that is sensitive to the diffusion properties of the underlying tissues. The

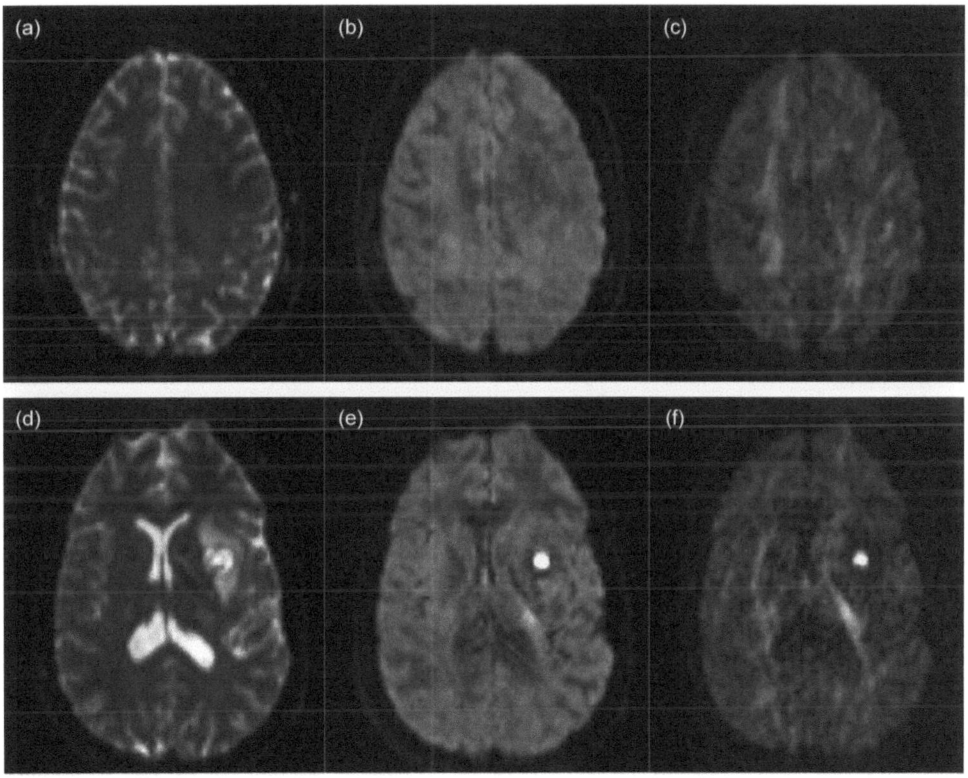

Figure 4.1 Diffusion-weighted images of an astrocytoma WHO grade II case at b = 0 s/mm^2 (a, d), b = 1000 s/mm^2 (b, e), and b = 3000 s/mm^2 (c, f). Images a, b and c are acquired in the centrum semiovale, images d, e and f are acquired at the level of the tumor. At b = 1000 s/mm^2, the white matter shows low signal intensity compared to the gray matter. At b = 3000 s/mm^2, the white matter is relatively more hyperintense than gray matter, and the level of noise is higher than at b = 1000 s/mm^2.

parameters that characterize a DWI sequence include the b factor, number and directions of diffusion sensitizing gradients, and voxel size. The b factor defines the sensitivity of the measurements to the diffusion process: the greater the b value, the stronger the effect of diffusion on the measured signal, and the higher the contrast on the image. Depending on the diffusion model used, b values vary between 800 and 3000 s/mm^2 on clinical scanners. The number of diffusion sensitizing gradients is influenced by the model of diffusion signal chosen. When the model of diffusion is a tensor, described later in this chapter, a minimum of six directions are necessary. However, 30 or more unique sampling orientations are often recommended for a robust estimation of tensor-orientation.[4] Finally, a critical parameter of diffusion-weighted images is the size of the voxel. While the diameter of axons is in the order of 1–30 microns, the size of clinical DWI voxels is typically 2–3 mm. As a consequence, DWI voxels can contain structures with different diffusion properties, which result in partial volume effects. For instance, the presence of both white matter and CSF in the same voxel, or so-called CSF contamination, in structures passing close to the lateral ventricles can introduce a bias in the interpretation of diffusion-derived measurements.[5] Similarly, partial volume effects arising from multiple fiber populations crossing through a voxel have important implications in tractography reconstructions as described later in this chapter.[6] The acquisition of DWI scans with smaller voxel size can help in solving partial volume effects, but this comes at the expense of longer acquisition time and lower signal-to-noise ratio. Postprocessing denoising tools can be applied to improve image quality in current clinical scanners.[7] In addition, the introduction of 7T and 11T MRI scanners offers promising perspective on the acquisition of DWI scans with improved resolution and signal-to-noise ratio.

To date, the main clinical application of diffusion-weighted acquisitions is early detection of acute ischemic stroke as decreased ADC values allow the visualization of ischemic regions within hours of symptom onsets.[8,9] In brain tumor imaging, however, while the tissue characteristics vary significantly between tumor types, the overlap between the range of ADC values precludes its use as a robust parameter for differentiating tumor types.[10] Still, diffusion-weighted scans are gaining increasing importance as an imaging biomarker for tumor detection, staging and treatment response assessment.[11]

DIFFUSION MRI IN PRACTICE: IMAGE ACQUISITION ARTIFACTS

Diffusion MRI uses large and rapidly switched diffusion sensitizing gradients to probe microscopic displacements of water molecules in biological tissues. As a result, the acquisition of diffusion-weighted images is subject to artifacts due to both the strong gradient pulses and patient motion.

Magnetic field inhomogeneities, which are proportional to the strength of the static field B_0, can be important at the interface between bones and air-filled cavities such as the sinuses or the auditory canal. The resulting image distortions are relatively more severe on 3T scanners than on 1.5T scanners, and can be corrected using parallel imaging techniques.[12] Other approaches for correcting the distortions include the use of postprocessing tools based on fieldmap estimation,[13–15] or the registration of EPI distorted images to high resolution structural scans.[16–18] The correction of artifacts caused by magnetic field inhomogeneities is especially critical in neurosurgical interventions using intraoperative MRI scanners as the largest image distortions are caused by the interface between air and brain tissues at the margins of the resection cavity. In such challenging settings, an approach combining fieldmaps and image registration-based correction has demonstrated promising results in reducing geometric distortions.[19]

A second source of artifacts in diffusion-weighted acquisitions is caused by eddy currents that originate from the rapid switching of the diffusion sensitizing gradient pulses. Eddy currents create local magnetic field gradients that can combine with the imaging gradient. As a consequence the gradients experienced by the imaged tissues become different from the gradients programmed in the pulse sequence, and the resulting ghost artifacts visible in the diffusion-weighted images can lead to interpretation errors. Eddy current effects can be reduced using self-shielded gradient coils and proper calibration.[20]

Finally, bulk motion of tissues is a major source of artifacts during diffusion-weighted acquisitions and can lead to heterogeneous misregistration of diffusion-weighted images. In particular, image quality can be severely altered by patient motion, respiratory cycles and cardiac pulsation. Motion-related artifacts can be corrected using cardiac gating and postprocessing image-based registration tools.[21] Alternatively, practical solutions, such as a comfortable head holder, can help reduce the amplitude of the misalignments of diffusion-weighted images.

Magnetic field inhomogeneities, eddy currents, and patient motion create artifacts that increase the difficulty of interpreting diffusion-weighted images. In the clinical setting, the DWI sequence readily available on MRI scanners and most commonly used is the single-shot echo-planar imaging (EPI) sequence.[22] The speed of the EPI acquisitions, typically 100 ms per image, makes the sequence robust to motion. However, EPI scans are still susceptible to magnetic field inhomogeneities and eddy currents.[20] Other alternatives include line scan diffusion imaging (LSDI) sequences which are insensitive to motion and allow diffusion-weighted image acquisitions with reduced image distortions,[23,24] and fast parallel imaging sequences such as sensitivity encoding (SENSE) techniques.[25] Finally, the availability of open-source tools for quality control of DWI shows promise for improving the sensitivity of neuroimaging studies based on clinical diffusion MRI data.[26]

DIFFUSION TENSOR IMAGING

DTI has been introduced to model the anisotropic diffusion of water molecules in the brain.[27] The model uses the diffusion tensor $\boldsymbol{D}$, a 3×3 symmetric matrix that characterizes the covariance of displacement of water molecules in three dimensions:

$$\boldsymbol{D} = \begin{bmatrix} Dxx & Dxy & Dxz \\ Dxy & Dyy & Dyz \\ Dxz & Dyz & Dzz \end{bmatrix} \tag{4.3}$$

The diagonal terms (Dxx, Dyy, Dzz) correspond to the diffusion variances along the three axes of the measurement frame of the MR scanner. The off-diagonal terms (Dxy, Dxz, Dyz) correspond to the correlation between displacements in orthogonal directions. Under the assumption that the displacement distribution of water molecules is Gaussian, equation (4.3) can then be written as:

$$S = S_0 e^{-b:D} = S_0 e^{-(b_{xx}D_{xx}+2b_{xy}D_{xy}+2b_{xz}D_{xz}+b_{yy}D_{yy}+2b_{yz}D_{yz}+b_{zz}D_{zz})} \tag{4.4}$$

where S is the signal attenuation resulting from the application of the gradients. S_0 is the signal intensity without diffusion-sensitizing gradient, $\boldsymbol{D}$ is the diffusion tensor and b is the b matrix. Since $\boldsymbol{D}$ is fully described by six coefficients, a minimum of six diffusion-weighted images acquired with six sensitizing gradients oriented in non-collinear directions are needed in addition to the baseline image S_0. The coefficients of the tensor can be calculated at each voxel by solving the system of linear equations resulting from the diffusion-weighted acquisitions with different gradient directions. Each individual tensor is then diagonalized to compute the three eigenvalues (λ_1, λ_2, λ_3) and three orthogonal eigenvectors (e_1, e_2, e_3). The eigenvectors correspond to the principal axes of the diffusion tensor, with the main eigenvector corresponding to the direction of maximum diffusion. The eigenvalues correspond to the apparent diffusivities along the eigenvectors. The information contained in DTI data can be visualized using diffusion tensor ellipsoids, with the principal axes corresponding to the eigenvectors and the length of the axes proportional to the square root of the eigenvalues.[28] The diffusion tensor ellipsoid represents the isoprobability surface where a molecule of water placed at its center will diffuse. When the eigenvalues are significantly different from each other, the diffusion is anisotropic. If the main eigenvalue λ_1 is much larger than λ_2 and λ_3, the diffusion occurs preferentially in the direction of the eigenvector e_1 and the ellipsoid is cigar shaped. If λ_1 and λ_2 are similar and larger than λ_3, the diffusion occurs in the direction of e_1 and e_2 and the ellipsoid is disk shaped. If λ_1, λ_2, and λ_3 are equivalent, the diffusion is isotropic and the ellipsoid is spherical. With this formalism, in white matter fibers, the diffusion tensor ellipsoids are cigar shaped, with the largest eigenvalue corresponding to the mobility of water molecules in the direction parallel to the fibers. In gray matter or CSF

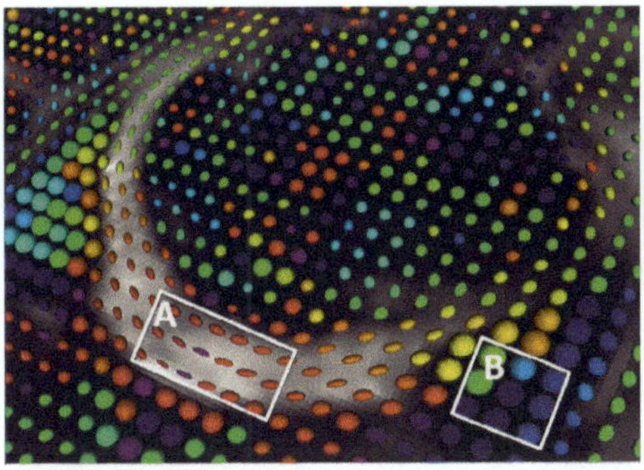

Figure 4.2 Diffusion tensor ellipsoids. The figure represents an axial DTI image overlaid on a T2-weighted slice at the level of the corpus callosum. In the splenium (region A) the diffusion is anisotropic and the tensor ellipsoids are cigar shaped. In the CSF-filled ventricles (region B), the diffusion is isotropic and the tensor ellipsoids are spherical.

diffusion occurs in all directions and the diffusion ellipsoids are spherical. Figure 4.2 shows an axial DTI image at the level of the corpus callosum: the cigar-shaped ellipsoids (region A) represent the anisotropic diffusion in the splenium of the white matter bundle; the spherical ellipsoids (region B) represent the isotropic diffusion in the adjacent lateral ventricle.

The diffusion tensor ellipsoid provides an elegant and intuitive representation of the information contained in a DTI dataset at the voxel scale. Directionally encoded color (DEC) maps of the human brain have been subsequently introduced to represent the directional information contained in diffusion tensor images and visualize the orientation of white matter pathways.[29] In DEC maps, the orientation of fibers is described by the eigenvector associated with the largest eigenvalue and is represented using colors. Fibers oriented in the left–right direction such as the corpus callosum appear in red, fibers oriented in the anterior–posterior direction such as the superior segment of the cingulum in green and fibers oriented in the inferior–superior direction such as the corticospinal tract (CST) in blue. Thus, DEC maps provide another intuitive visualization of the main commissural, association, and projection white matter fiber pathways.

In addition to the color coding of diffusion data, diffusion maps of rotationally invariant anisotropy measures have been introduced to provide quantitative evaluation of the size and shape of the diffusion tensor in different regions of the brain. The two most common measurements are the mean diffusivity (MD) and fractional anisotropy (FA). The MD corresponds to the average of the three eigenvalues, and is relatively homogeneous in the normal brain parenchyma. The FA is based on the normalized variance of the eigenvalues and indicates the degree of anisotropy of the diffusion process. FA values range for 0.0 (isotropic diffusion) to 1.0 (anisotropic diffusion). In the

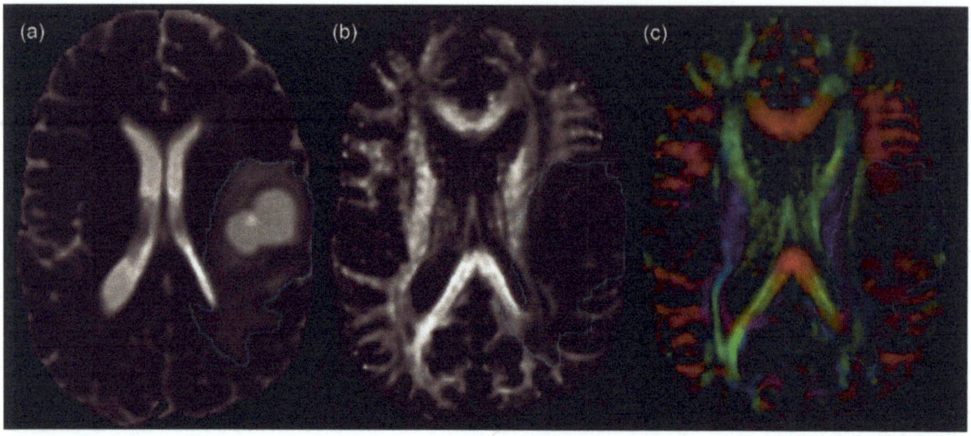

Figure 4.3 Diffusion maps of a patient with WHO grade III astrocytoma. The figure shows the tumor and peritumoral edema in a cross section of mean diffusivity (a), fractional anisotropy (b), and directionally-encoded color (c) map. Peritumoral edema (blue) is outlined in the three axial images.

normal brain, voxels with high anisotropy and a low diffusivity correspond mainly to white matter, voxels with low anisotropy and low diffusivity correspond mainly to gray matter, and voxels with low anisotropy and high diffusivity correspond to CSF.[28] Figure 4.3 shows an example of MD, FA, and DEC maps in a WHO grade III astrocytoma case. During surgical planning, the combination of quantitative information on the change in diffusion patterns with orientation information provided by DEC maps has the potential to provide surgeons clinically relevant information on the location and integrity of peritumoral fibers in the vicinity of a lesion.

In recent years, a number of studies have been conducted to investigate the potential role of DTI-derived measurements in the characterization of the cellularity and anisotropy of tumor tissues. However, results show an overall lack of agreement among the different reports, which may be due to differences in the acquisition and postprocessing protocols. Studies investigating the use of FA and MD values to distinguish between vasogenic and tumor infiltrated edema, for example, have contradictory results, most likely due to the differences in the regions of interest (ROIs) chosen for the evaluation.[30,31] In addition, factors such as the ratio of extracellular space to intracellular space, or the vascularity of the tissues can affect the measurements.[30] Several exploratory works have investigated the use of alternative diffusion maps which include axial and radial diffusivity,[32] gradient of ADC values[33] and DTI-based probability maps.[34] Such maps appear to be promising tools for the evaluation of the diffusion properties of edematous tissues in the vicinity of gliomas. However, the characterization of tumor-altered white matter using DTI-derived still remains challenging due to the heterogeneity of brain tumors, both within and across tumor types, as well as the low resolution of diffusion MRI voxels. As a result, quantitative evaluation of tissue microstructure based on DTI data is not yet used in surgical planning. Further improvement in signal-to-noise ratio, standardization of acquisition parameters, and postprocessing protocols, as well as histological validation will help improve the robustness of DTI metrics and increase their role as noninvasive imaging biomarkers for the evaluation of white matter involvement in gliomas. The recent development of diffusion kurtosis imaging (DKI) allows refinement of the modeling of the diffusion process by quantifying the degree to which diffusion is not Gaussian.[35] Initial findings in the use of DKI metrics to differentiate low-grade and high-grade gliomas showed improved results when compared to classical DTI metrics such as FA and MD values.[36]

HIGH ANGULAR RESOLUTION DIFFUSION IMAGING

While DTI provides a powerful tool to investigate the anisotropic diffusion of water molecules in the brain, the tensor model can only recover a single fiber orientation at each voxel and fails to describe complex anatomical regions where several fiber populations coexist. Technical advances in MR scanners and parallel imaging sequences have enabled the acquisition of diffusion-weighted images with a larger number of diffusion sensitizing gradients, typically in the range of 64–100 directions. As a result, high angular resolution diffusion imaging (HARDI) techniques using a relatively large number of gradient directions and a high b value ($b > 1500\ \mathrm{s/mm^2}$) have been introduced to enable the modeling of multiple fiber populations.[37] In particular, high b values have the potential to solve fiber crossing challenges by providing better contrast between the fast diffusion coefficient of one fiber population and the slow diffusion coefficient of a second fiber population.[38] The design and implementation of HARDI acquisition sequences is an active area of research, and there is no consensus on the optimal b values and number of diffusion sensitizing gradient directions. Studies on healthy subjects suggested a minimum of 45 gradient directions and a b value of $3000\ \mathrm{s/mm^2}$ to properly characterize the diffusion-weighted signal.[39]

Numerous mathematical models have been proposed to describe fiber populations with different orientations within a given voxel from HARDI data.[40] These sophisticated models attempt to estimate the component fibers either discretely or as orientation distribution functions. Examples of such frameworks are based on mixture models such as multitensor fitting and spherical deconvolution. The multitensor model is a mixture of Gaussians based on the assumption that the diffusion signal attenuation is the sum of signal attenuations from several fiber populations, each modeled as a tensor.[37,41] A limitation of the multitensor approach arises from the need to estimate the number of fiber components to be fitted in a given voxel. The spherical deconvolution approach overcomes this limitation by generating estimates of the fiber orientation distribution without the need for prior assumptions regarding the number of fibers populations present in a voxel.[42,43] Other models include the CHARMED framework that characterizes the diffusion signal attenuation in white matter in terms of Gaussian (hindered) diffusion and non-Gaussian (restricted) diffusion contributions,[44] and Q-ball imaging based on the Funk–Randon transform that does not require any assumptions on the diffusion process.[45] Other reconstruction methods focusing on the fiber orientation distribution to represent the orientation of underlying fibers are based on persistent angular structure (PAS-MRI),[46] and diffusion orientation transform (DOT).[47] In recent years, multishell HARDI acquisitions consisting of multiple HARDI shells, each with a different b value, have been investigated. An example of a multishell HARDI scheme is the protocol implemented by the Human Connectome Project consortium.[48] The acquisition protocol consists of 90 gradient directions and five baseline images acquired twice and with three b values ($b = 1000$, 2000, and 3000 s/mm^2). However, such sophisticated sequences require a total scanning time of approximately 1 hour for whole-brain coverage, and thus are not yet compatible with the constraints of neurosurgical care. Similarly, diffusion spectrum imaging (DSI), a model free method based on the measurement of the diffusion spectrum, requires a high number of measurements, typically in the order of 500, high b values and long acquisition time.[49] The recent development of accelerated DSI using compressed sensing shows promising results for the reduction of the acquisition time without loss of critical information.[50] As of today, Q-ball imaging, multishell HARDI, and DSI techniques are mainly used to provide brain connectivity maps in neuroscience studies. Unprecedented efforts are being made through the Human Connectome Project to advance the sensitivity of high angular resolution diffusion imaging acquisitions by improving MRI gradient hardware and acquisition efficiency.[51] It is expected that these new diffusion imaging methods will help to depict regions of crossing white matter pathways.

DIFFUSION MRI TRACTOGRAPHY

Tractography consists of the three-dimensional reconstruction of the trajectory of the brain white matter pathways from diffusion MRI data. The technique is based on the assumption that the principal direction of diffusion at each voxel is aligned with the main orientation of the underlying fibers. Tractography methods can be divided into three categories: deterministic, probabilistic, and global. Deterministic tractography algorithms propagate a single pathway by following the local tensor information at each voxel.[52–55] Starting from a seed point, the technique reconstructs the trajectory of white matter fibers by propagating a line that follows an estimated tract direction until a termination criteria is reached. Streamline tracking algorithms use the main eigenvector as the estimated tract direction.[52,53] Tensor deflection (TEND) approaches use the entire diffusion tensor and have shown improved performance in crossing regions with low anisotropy where the major eigenvector direction is not well defined.[56] The most common termination criteria are based on an anisotropy threshold or a local tract curvature. The anisotropy threshold, which is often based on FA values in the range of 0.1 to 0.2, aims at stopping the tractography reconstruction when the algorithm reaches CSF or gray matter regions. The angular threshold aims at preventing sharp turns which are not consistent with known anatomy during the reconstruction of tracts. Seeding from anatomical ROIs permits the identification of major white matter pathways,[57] and DTI tractography atlases provide a useful resource for defining ROIs for a specific tract in healthy subjects.[58] While deterministic approaches can produce anatomically faithful reconstructions of major pathways, the techniques propagate a unique trajectory for a given seeding point. As a consequence, the algorithms fail to resolve regions of crossing or kissing fibers. In addition, since the propagation is based solely on the local diffusion information, errors in the estimation of the principal direction of diffusion can accumulate due to image noise or image distortion. Thus, a small error at one point in the tracking process can lead to false anatomical reconstructions. These limitations have led to the development of probabilistic approaches that compute the distribution of fiber pathways emanating from a given seed point, and assign a confidence level to each specific trajectory.[59–64] Probabilistic streamline methods rely on the same tract tracing technique as in the deterministic approach, but instead of choosing a unique direction of propagation at each voxel, the algorithms propagate a large number of pathways chosen from the distribution of possible fiber orientations. The results can be represented as probability maps in which each voxel is assigned the number or percentage of reconstructed tracts that pass through it. Figure 4.4 shows an example of probabilistic reconstruction of the pyramidal tract (PT) in a healthy subject.[65] While such maps provide useful information on the reproducibility of the tracking process given the data, the probability values derived from probabilistic tracking do not contain information on the anatomical accuracy of the reconstruction.[66] Still, probabilistic tractography can provide useful information to the neurosurgeon regarding the uncertainty associated with the reconstructed pathways. Recent work on the combination of probabilistic tracking and the Q-ball diffusion model showed better sensitivity and anatomical accuracy than deterministic approaches.[67] Still, deterministic and probabilistic tractography are both local tractography techniques, and a perturbation in a single voxel can corrupt the tract tracing process. This limitation has motivated the development of global tractography methods based on the global reconstruction of the geometry of the complete white matter.[68] In such approaches, fiber bundles have been modeled using small segments optimized to minimize an energy term that represents a trade-off between the information on tissue microstructure provided by the diffusion data and prior knowledge of the curvature of plausible fascicles.[69,70] Other global tractography approaches have modeled the tracts as shortest distance

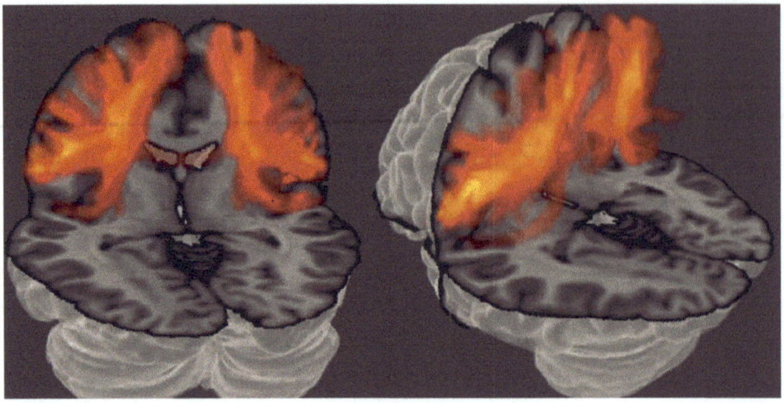

Figure 4.4 Probabilistic tractography reconstruction of the pyramidal tract. Light yellow regions represent the locations where the tract is most likely to be found. Dark red regions represent lower probabilities that the tract runs through a given voxel. From Eickhoff et al.[65], reproduced with permission from Society of Neuroscience.

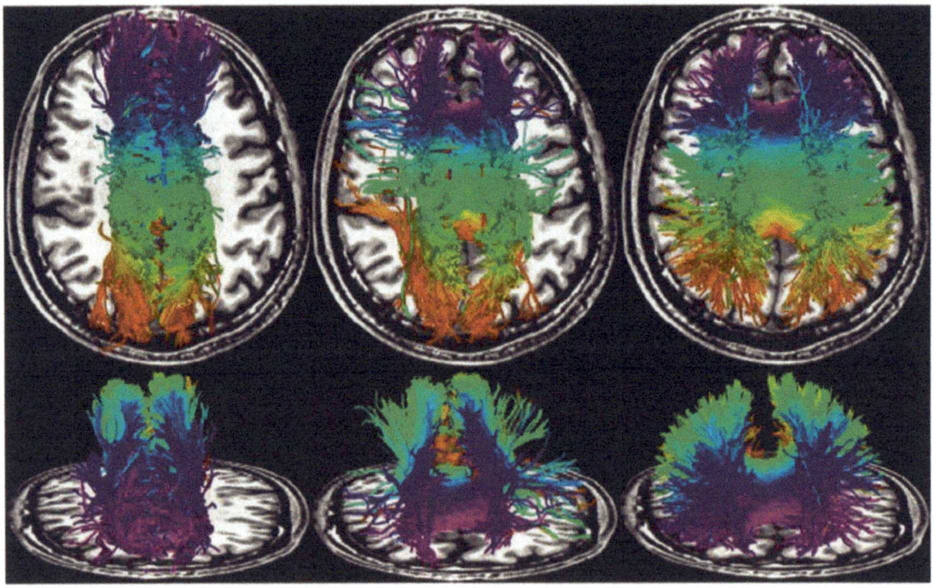

Figure 4.5 Tractography reconstruction of the corpus callosum by local and global methods. Left: tensor-based streamline; middle: Q-ball-based streamline; right: global spin glass. From Mangin et al.[68], reproduced with permission from Elsevier.

paths for a metric derived from the water diffusion profiles.[71,72] Figure 4.5 shows a tractography reconstruction of the corpus callosum by local and global methods. While global tractography methods overcome the problem posed by local perturbations in the diffusion data, the long computation time, and the difficulty in specifying prior knowledge criteria have hindered their transfer to the clinics. Recent efforts to make global tractography more practical have demonstrated promising results for clinical applications.[73,74]

Tractography reconstructions of motor, language, and vision pathways

As tractography methods have been used to investigate major projection, commissural, and association white matter pathways and infer group differences in neuroscience studies, the possibility of visualizing the trajectory of tracts associated with motor, vision, and language function at the individual patient level have gained an increasing interest in the neurosurgical community.

Motor pathways

The vast majority of tractography studies for neurosurgery focus on the PT due to its critical role in voluntary movements of the limbs. The PT connects the motor cortex to the brain stem and spinal cord, and includes both CSTs and corticobulbar tracts (CBTs). CST fibers arise in the primary motor cortex, converge through the corona radiata, and course through the posterior limb of the internal capsule, cerebral peduncles, basis pontis, and medullary pyramid. CBT fibers arise from the lateral part of the motor cortex, converge into the corona radiata, continue through the genu of the internal capsule and the cerebral peduncles, and terminate at the cranial motor nuclei. CST fibers run with CBT fibers, and cannot be distinguished from one another in DTI color maps.

Early reports on the CST reconstructions in brain tumor patients demonstrate the potential of tractography to depict the spatial relationship between the tract and a tumor.[54,75] A clinical study on the impact of using DTI-based neuronavigation in surgical resection of brain gliomas has showed that the technique can contribute to decreasing postoperative motor deficit.[76] However, the reconstruction of CST remains challenging in the region of the centrum semiovale due to its crossing with the superior longitudinal fasciculus, corpus callosum and short u-fibers. As a result, most tractography algorithms based on the single tensor model fail to represent pyramidal fibers that arise from the lateral portion of the primary motor cortex corresponding to the face and tongue area. The evaluation of the anatomical accuracy and clinical relevance of deterministic, probabilistic and global tractography algorithms in the reconstruction of the PT in low-grade and high-grade gliomas by 19 international tractography groups has demonstrated a large variability among tractography methods.[77] Recent studies comparing the accuracy and precision of deterministic and probabilistic algorithms on preoperative DTI data acquired on brain tumor patients show that, despite the poor sensitivity of the single tensor model in the reconstruction of lateral motor pathways, probabilistic algorithms have fewer false negative findings than deterministic tracking methods.[67] In addition, sensitivity performance can be further improved by combining Q-ball imaging with probabilistic tracking algorithms on HARDI data.[78] Similarly, multitensor tractography using a b value of 1000 s/mm^2 shows promise for depicting fiber tracts from the face and tongue regions which were not previously detected by conventional single tensor tractography methods.[79,80]

Language pathways

Tractography of the arcuate fasiculus (AF) and the inferior fronto-occipital fasciculus (IFOF), two major pathways involved in language function, is an area of growing interest in neurosurgery clinics.[81] The AF is involved in the dorsal stream of language processing, and has been described as composed of a direct and an indirect pathway. The direct long segment of the bundle connects Broca's and Wernicke's regions. The indirect pathway runs laterally, and is composed of an anterior segment connecting Broca's area to Geschwind's region located in the inferior parietal lobule, and a posterior segment connecting Wernicke's to Geschwind's region.[82] The IFOF is a ventral stream of language processing that connects the ventral occipital lobe to the orbitofrontal cortex by means of the temporal lobe.[81]

Tractography reconstruction of language pathways can be used to plan brain tumor resection[83,84] and to predict postoperative language recovery.[85] In addition, tractography reconstructions have been combined with intraoperative direct electrical stimulation (DES) to help in preserving language fascicles during surgery.[86,87] Studies in patients presenting with a tumor in the white matter of the left dominant hemisphere along the course of the language pathways show that probabilistic tractography can outperform streamline deterministic algorithms in the estimation of the extent of the AF.[88] Similarly, a global search approach dedicated to anatomical connectivity analysis has demonstrated better reliability than deterministic streamline and TEND algorithms in the reconstruction of language pathways in neurosurgical cases, with results similar to time-consuming global or probabilistic approaches.[89] Other approaches include high definition fiber tracking methods that show promising results for the depiction of the different branches of the AF in the frontal and temporal regions, as well as the identification of the terminal arborization of the arcuate tract.[90] The combined use of HARDI and compressed sensing, an advanced signal processing technique, in a series of eight patients presenting with a glioma in the temporal lobe resulted in improved tractography reconstruction of language pathways in areas of disturbed diffusion signal, while overcoming the issue of long acquisition time.[91]

However, the technical difficulties in tracing the path of the AF in diffusion data are enhanced by the individual variability in the location of language function in the brain. Electrical stimulation mapping have demonstrated that language sites can be located far beyond the traditional boundaries of Broca's area as illustrated on Figure 4.6.[92] The connectivity of the white matter pathways that support the language function in humans still remains controversial.[93] In that context, the association of

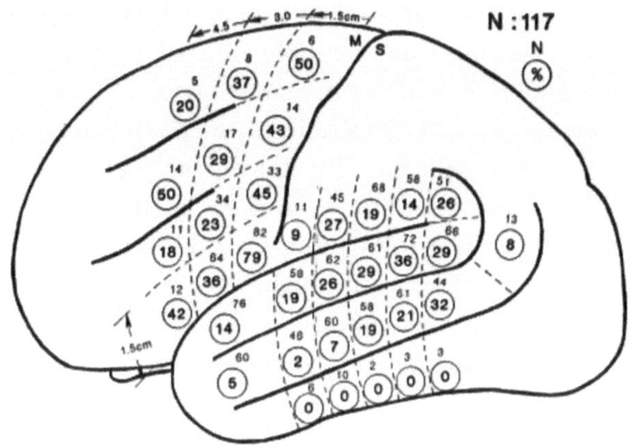

Figure 4.6 Variability in language localization in the left dominant hemisphere in 117 patients. Upper number represents the number of patients with a site in a given zone. Lower numbers in circle represent the percentage of patients with sites of evoked naming errors. From Ojemann et al.[92], reproduced with permission from Journal of Neurosurgery Publishing Group.

tractography maps and intraoperative electrical stimulation findings offers a unique opportunity to revisit the anatomo-functional connectivity of language.[94]

Visual pathways

The optic radiations (ORs) carry visual information to the visual cortex. The tracts emanate from the lateral geniculate body (LGN), run around the roof of the occipital horn of the lateral ventricle and terminate in the primary visual cortex in the calcarine sulcus. The OR are composed of three bundles: the posterior bundle, the central bundle, and the anterior bundle. The posterior bundle runs straight from the LGN to the occipital cortex; the central bundle makes an anterior curve and courses along the lateral wall of the atrium and occipital horn of the lateral ventricle. The anterior bundle or Meyer's loop detours entirely around the tip of the temporal horn, passes posteriorly around the lateral ventricles and converges on the lower lip of the calcarine fissure. The Meyer's loop transmits visual information from the contralateral superior field of the eyes, and knowledge of the location of this anatomical structure is essential to avoid quadrantanopia during resection of lesions located in the temporal lobe. However, the anterior extent of the Meyer's loop presents high intersubject variability which makes the prediction of its precise location in an individual patient difficult.[95] Several studies have demonstrated the use of DTI tractography to depict the OR and the Meyer's loop in healthy subjects.[96–99] As a result, DTI tractography has been introduced as a potential predictor of the visual field deficit caused by anterior temporal lobectomy.[100–103] Due to the complexity of the anatomical structure, several exploratory approaches have been proposed for tracing the OR, including increasing the number of diffusion sensitizing gradient directions,[104] using a DWI STEAM MRI sequence with interpolated 0.9 mm isotropic voxels,[105] and seeding from multiple fiducial volumes.[106,107] However, as of today, there is no consensus as to the best tractography approach and tract tracing parameters to reconstruct the OR. Probabilistic methods have demonstrated better results than deterministic approaches in healthy subjects and epilepsy patients, mainly because they allow for more possible pathway orientations at each voxel.[108,109] Still, the curved course of the OR and sharp turn of the Meyer's loop as well as their proximity to other white matter tracts, which include the uncinate fasciculus, the inferior longitudinal fasciculus, and the inferior occipito-temporal fasciculus, remain a challenge for most tractography techniques.

The role of tractography in the surgical treatment of brain tumors

Brain tumors are space-occupying lesions that can affect adjacent white matter pathways involved in motor, vision, and language function. The history of modern brain tumor surgery goes back over a century, and surgical techniques have been continuously optimized by a community of highly trained personnel. The goal of surgical treatment is to maximize the extent of resection while preserving eloquent cortex and subcortical white matter to avoid postoperative neurological deficits. Thus, the knowledge of the location and integrity of critical pathways is crucial to the neurosurgeon. In that context, diffusion tractography has been introduced as a new tool to complement

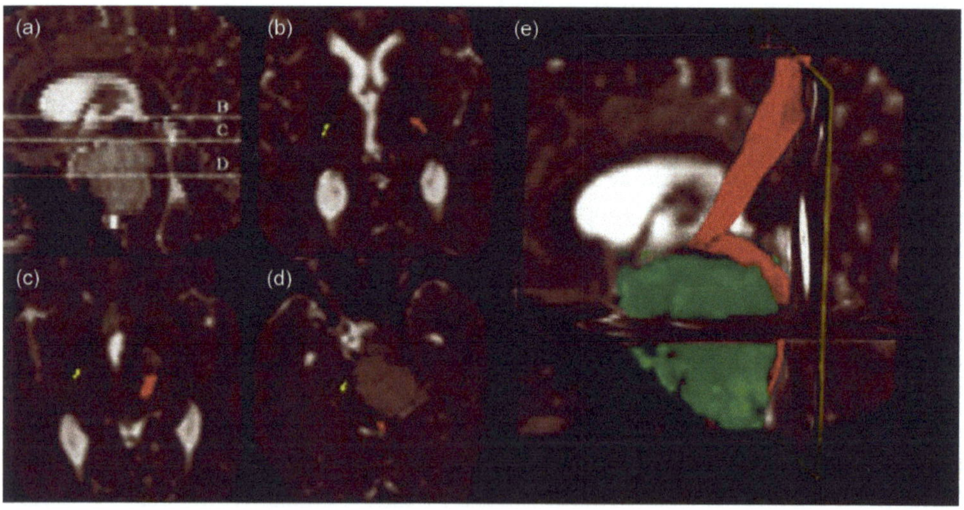

Figure 4.7 Deformation of the corticospinal tract in a patient with a meningioma. (a) Midsagittal T2-weighted image showing the extent of the tumor. (b–d) T2-weighted axial images with superimposed left (yellow) and right (red) corticospinal tract. The section level is indicated on (a). (e) Three-dimensional view of a surface-rendered model of the tumor (green) and tractography reconstruction of the CST (red). From Stieltjes et al.[54], reproduced with permission from Elsevier.

the apparatus of brain mapping techniques. Figure 4.7 shows the first published three-dimensional 3D view of the tractography reconstruction of the CST in a patient with a meningioma.[54]

Today the technology is the only noninvasive method to map the organization of white matter tracts during surgical planning.[110] The possibility of generating individualized reconstructions of the trajectory and integrity of white matter pathways in the vicinity of a tumor represents a major technological advance in brain mapping.[55,111] In particular, tractography maps can help depict the pathological effects of brain tumor growth on white matter pathways, which include displacement, disruption, infiltration, and destruction of fibers.[112,113] In addition, as tractography is noninvasive, tract reconstructions can be repeated multiple times from a single DWI acquisition, without any risk to the patient. Interactive tractography applications have been developed to assist clinicians in the exploration and interpretation of DTI data. These applications allow clinical researchers to display specific fibers using box or ellipsoid-shaped ROIs that can be interactively positioned in a set of precomputed pathways.[114] Similarly, dynamic seed points have been proposed to generate tractography reconstructions "on-the-fly," to help neurosurgeons define the location and trajectory of tracts around a tumor (Figures 4.8 and 4.9).[115] Other interactive applications include tools that show the tracts within a certain distance of a tumor boundary,[115] as well as hulls around fiber pathways with a size that can be varied to represent a percentile of enclosed paths.[116]

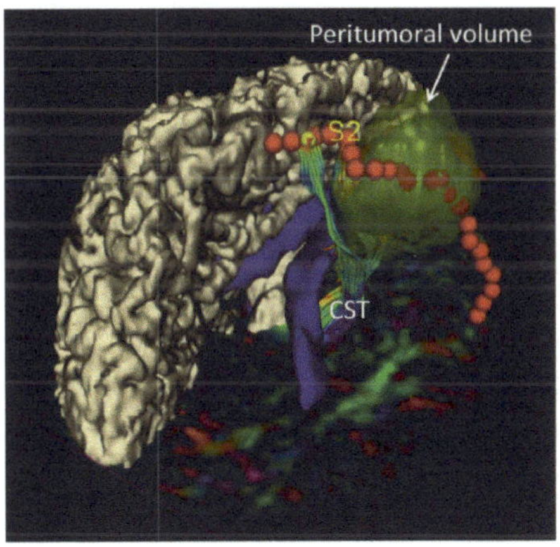

Figure 4.8 Tractography "on the fly" in a recurrent/residual WHO grade II oligodendroglioma. The figure shows 3D surface models of the tumor recurrence (dark green) and peritumoral region (light green) encroaching the motor cortex area (red fiducials) overlaid on a DEC image. A portion of the corticospinal tract has been generated from a fiducial S2 (yellow) positioned in the motor cortex area. The seed can be moved in the peritumoral region to explore the spatial relationship between the tract and the tumor. 3D models of the lateral ventricles (blue) and cortical surface of the left hemisphere (white) have been added for anatomical reference.

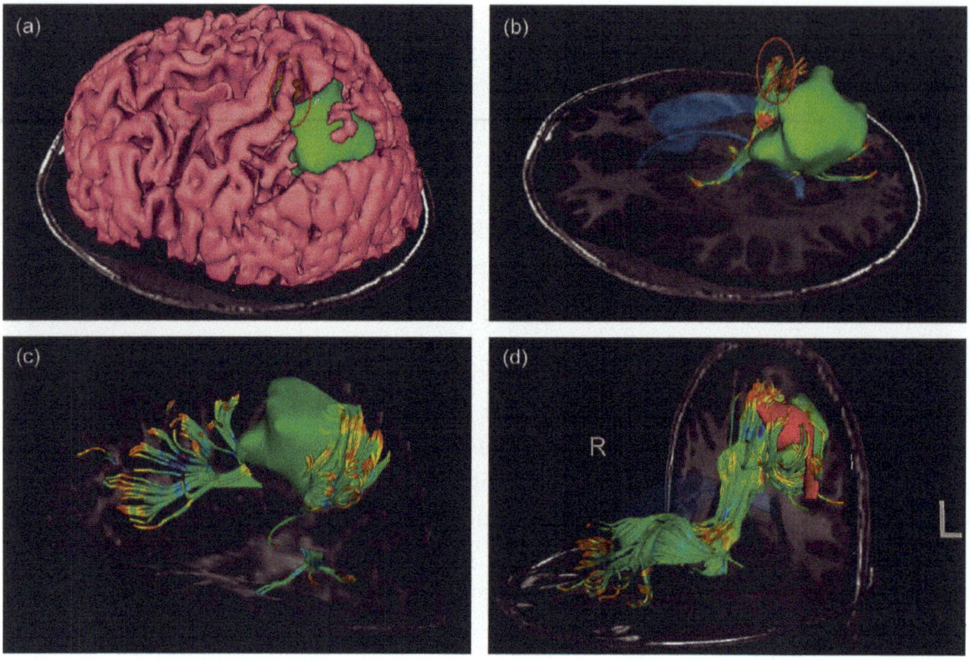

Figure 4.9 Surgical planning in WHO grade II astrocytoma. The frontoparietal tumor (green) is displayed with and without cortical surface rendering (pink) (a and b). The tractography reconstructions show the corticospinal tract (red ellipse) defined by manual seeding (b), parietal tracts generated from posterior dynamic seeding (c), and fiber projections generated by seeding from right-hand fMRI activation area (magenta) (d). From Golby et al.[115], reproduced with permission from Kraus Back and Neck Institute.

Tractography as a tool to enhance brain mapping

Tractography can be used to complement the information available from other brain-mapping techniques. In particular, the anatomical information provided by tractography reconstructions can be combined with the functional information provided by functional MRI (fMRI) and DES data, to provide neurosurgeons three-dimensional visualization of subcortical structures connected to eloquent cortical areas.

fMRI is a noninvasive functional mapping technique that measures the changes in the cerebral blood flow using blood oxygenation level dependent (BOLD) contrast when a subject is performing a task. The technique has gained increased interest in the neurosurgical community to identify the location of cortical motor and language areas, and their relationship with a tumor.[117–120] The use of fMRI activation maps as seeding ROIs for tractography has enabled the visualization of tracts connected to specific eloquent areas (Figure 4.9).[75,121] Such a combined approach can demonstrate the displacement of white matter bundles,[122] and provide useful information when tumorous, necrotic, or edematous tissues make the identifications of anatomical landmarks challenging. Still, both tractography and fMRI have known limitations, and the absence of ground truth knowledge of the location of the so-called eloquent tracts that they help identify make the validation of both techniques difficult. The combined use of tractography reconstructions and fMRI activation maps can allow the evaluation of one technique against the other, thereby providing neurosurgeons useful information based on their level of agreement.

Preoperative tractography reconstructions can be integrated with intraoperative DES data acquired during awake mapping in surgical resection of brain tumors located in eloquent areas. As the clinical gold standard for functional brain mapping, DES allows the identification of eloquent cortical sites and subcortical pathways located within the margins of the tumor boundaries.[123–128] Seeding tracts from DES sites can provide identification of eloquent white matter pathways such as the CST and the AF, and help neurosurgeons understand their involvement with a lesion.[86,87,129,130] In addition, as DES is an invasive technique, tractography can be used as a guide to the locations of eloquent tracts during surgical planning. Finally, DES information can provide ground truth to evaluate the reliability of tractography reconstructions,[131] and to quantify the sensitivity of fiber tracking algorithms.[67]

Combining preoperative fMRI activation maps and tractography reconstructions with intraoperative DES has the potential to provide neurosurgeons richer information on the location of eloquent peritumoral pathways.[132] Clinical studies are needed to evaluate the impact that using functional information to guide tractography can have on the accuracy of peritumoral fiber reconstructions.

Tractography as a tool to enhance neuronavigation

Preoperative DTI tractography reconstructions can be integrated into a neuronavigation workstation to provide a three-dimensional 3D visualization of white matter tracts of interest and their relationship to critical anatomical structures during surgery.[133,134] The visualization of the preoperative trajectory of the tracts has proven to be helpful to neurosurgeons in carrying out interventions in essential cortical areas when compared with the use of standard navigation systems without tractography data.[76] The combined use of fiber tracking and subcortical stimulation offers clinical anatomical–functional information that can aid in the preservation of PT.[135] Wrapping surfaces around individual streamlines has been proposed to simplify the visualization of the border of white matter bundles, and integrate safety margins to compensate for image distortion and registration error.[136] Furthermore, intraoperative diffusion MRI acquisitions have opened up the possibility of assessing the displacement of white matter structures during surgery using two-dimensional maps of the principal eigenvector[137] as well as tractography reconstructions of critical pathways such as the PT.[84,138,139] The co-registration of pre- and intraoperative DTI data has proven to be useful in depicting shifting of major white matter pathways due to surgically induced brain deformation.[84,140]

Current limitations of tractography

DTI-based tractography is gaining increased interest in clinical settings, mainly due to the implementation of diffusion sequences on most MRI scanners and the availability of commercial and open-source tractography software. While tractography reconstructions can provide clinically relevant information to assist during presurgical mapping and intraoperative navigation, the technology is based on complex mathematical models that rely on many assumptions. Therefore, the utility and usability of tractography tools for neurosurgical decision-making are within the constraints imposed by the resolution of diffusion MRI data and the unsolved challenges of the technology.

A complex image acquisition and postprocessing pipeline

From a practical point of view, tractography is at the end of a complex image acquisition and postprocessing pipeline that remains hardware-, software-, and operator-dependent. Factors influencing the accuracy and reproducibility of tractography reconstructions include the characteristics of the MRI scanner, the parameters of the acquisition sequence and the quality of the DWI data. Preprocessing of DWI data has been shown to help mitigate artifacts in diffusion-weighted acquisitions which can result in errors that propagate through the entire tractography pipeline. In particular, the correction of EPI image distortions should be included in any diffusion MRI study, for both DTI and HARDI schemes.[141] In addition, factors such as the model of diffusion, the fiber tracking algorithm and tract selection parameters play a critical role in the accuracy of tractography results.[77,142–145]

A macroscopic picture of microscopic structures

Tractography reconstructions provide an indirect macroscopic representation of underlying microscopic structures. A single DWI voxel, typically in the range of 1.5–5 mm in size, includes tens of thousands of axons. This poses critical challenges in kissing fibers regions where white matter pathways such as the corpus callosum and the cingulum run adjacent to each other, or in crossing fibers regions where pathways intersect, such as the superior longitudinal fasciculus and the CST in the centrum semiovale. In such regions, the diffusion tensor model which can only resolve a single fiber direction within an imaging voxel is not adapted to predict the orientation of the underlying fibers. In recent years many technical developments have attempted to overcome the limitations of DTI for clinical use.[146] High definition fiber tracking (HDFT) which combines HARDI and DSI techniques has shown improvement in the depiction of the trajectory of fibers through complex crossing areas in both healthy subjects and neurosurgical cases.[90] However, the scanning acquisition time of 45 minutes is still a limitation to the transfer of the technology to clinical routine. Other approaches combining HARDI acquisitions within a clinically feasible time frame of 13 minutes and probabilistic Q-ball tractography demonstrated improved results when compared to standard DTI methods for the delineation of connectivity in regions of complex architecture.[78]

Limitations in pathological regions

In addition to the difficulties posed by complex anatomical regions with multiple fiber populations, tractography algorithms can be confounded by disrupted anatomy caused by tumoral tissues and edema. The heterogeneity of brain tumors within and across tumor types, combined with the heterogeneity of their pathological effects on white matter fibers makes the setting of cerebral neoplasms highly complex. Tumors can disrupt, displace and infiltrate white matter. In addition, the presence and nature of peritumoral edema is different across tumor types. Peritumoral edema is considered pure vasogenic edema in meningiomas and metastatic tumors, whereas the periphery of infiltrative gliomas includes vasogenic and neoplastic edema.[32] Both vasogenic edema and tumor infiltrated edema create a change in the free water content of tissue that results in lowered FA values. As a consequence, while the structure of the underlying fibers might remain unaffected, tractography reconstructions can become inaccurate and inconsistent in edematous regions. This poses critical issues for neurosurgical planning of glioma resection as differentiating between destroyed white matter tracts from intact white matter is crucial. Studies evaluating the effects of different FA thresholds on the reconstruction of fibers in the vicinity of gliomas demonstrated that an FA threshold in the range of 0.15–0.2 allowed for the reconstruction of tracts that may be infiltrated by tumor cells.[147] New tractography methods using two-tensor unscented Kalman filtering has shown promising results in the depiction of the arcuate fasciculus in the setting of peritumoral edema.[148] Other approaches using generalized q-sampling imaging (GQI) demonstrated better performance than DTI-based methods in the visualization of fiber tracts located in peritumoral edematous regions.[149] Similarly, a prospective study on small group of patients demonstrated the feasibility of combining HARDI techniques and compressed sensing to trace fiber through peritumoral edema with time constraints compatible with the clinical routine.[91] Finally, the introduction of restriction sprectrum imaging (RSI) has shown promising results for distinguishing separable pools of water within tissue based on their intrinsic diffusion characteristics.[150]

Limitations inherent to brain shift

A practical limitation on the use of tractography results for neurosurgical decision-making comes from the problem of brain shift. Most commercial neuronavigation workstations use DTI tractography maps generated from preoperative diffusion MRI data. However, brain deformations due to gravity and loss of CSF occurring after the craniotomy and dural opening result in a misalignment between the preoperative map and the intraoperative position of the brain. An average displacement of white matter tracts of 2.7 mm, with an intraoperative shift varying from an inward shift of 8 mm to an outward shift of 15 mm, has been reported during gliomas resection.[140] As a consequence, the use of neuronavigation systems require that neurosurgeons track changes visually by looking at anatomical landmarks such as vessels, cortical gyri and sulci, and compensate for the effect of brain shift mentally. Intraoperative MRI acquisitions can be used to generate tractography reconstruction of the displaced white matter tracts and help understand their spatial relationship to the surgical resection cavity.[84,88,139] However, intraoperative DTI tractography remains challenging due to severe distortion of diffusion-weighted images at the interface between air and tissue in the resection cavity. In addition, the number of diffusion sensitizing gradients is usually smaller than in preoperative imaging due to the time constraints imposed by the intraoperative setting. Finally, the cost and complexity of intraoperative MRI systems remains a limiting factor to its widespread availability. Other methods including the integration of ultrasound imaging with neuronavigation systems and the overlay of intraoperative 3D ultrasound images of the tumor onto preoperative imaging can help surgeons assess and compensate for brain shift.[151,152] Today, brain shift remains an active area of investigation, and automatic ultrasound-based updates of preoperative MRI data have been proposed to compensate for the brain deformations.[153] Nonrigid registration of preoperative tractography reconstruction to intraoperative space also shows promising results for solving the problem of brain shift.[103,154]

Consequences of tractography errors

The limitations of tractography techniques presented in the previous section result in tractography errors that can be described as false-negative and false-positive tracts.

False-negative tracts are anatomical pathways that tractography algorithms fail to reconstruct. Examples of false-negative tracts are the missing lateral corticobulbar fibers that originate in the face, lip and tongue area in the primary motor cortex. Missing tracts can pose critical issues in neurosurgical settings as the absence or interruption of a bundle in a tractography map does not necessarily imply that the tract is anatomically destroyed. As a consequence, false-negative tracts can lead the neurosurgeons to underestimate the size of critical white matter structures and can result in postoperative neurological deficits.[155] When a tract close to a tumor appears as cut off in a DTI tractography reconstruction, it is difficult for the surgeon to know if the tumor has infiltrated the white matter or if the tract is missing because the fiber tracking algorithm cannot find it. Therefore, negative tractography findings should not be used to preclude the presence of eloquent tracts.[156]

False-positive tracts are fibers that do not represent clinical reality. Examples of false-positive tracts are corticospinal fibers that cross to the other hemisphere.[66] Spurious fibers have a similar appearance as the fibers of the expected tract, except that they are anatomically erroneous. In neurosurgical settings, false-positive reconstructions can obstruct the visualization of the tracts of interest or lead to an incomplete tumor resection.

The interpretation of tractography results requires a solid neuroanatomical knowledge, and both false-negative and false-positive tracts add to the difficulties of assessing the anatomical accuracy of tractography reconstructions. In particular, in high-grade gliomas presenting with necrosis and tumor infiltrated edema, it can be difficult to tell if a tract is missing due to the pathology or due to a tractography error. In that context, tractography should not be used as a standalone brain-mapping tool for neurosurgical decision support. Intraoperative electrical cortical and subcortical stimulation during brain surgery can provide ground truth information on the location of the termination of white matter pathways and help neurosurgeons assess the validity of tractography reconstructions during an intervention.[67]

As the complexity of diffusion MRI data has fostered many exploratory activities in new computational approaches for tracing white matter pathways, there is a need to validate tractography findings in order to establish the clinical utility and usability of the techniques. International community-based efforts have started working on the standardized evaluation of the performances of tractography methods on clinical diffusion MRI data acquired on brain tumor patients.[77,157] These efforts bring together practicing neurosurgeons and diffusion MRI researchers to implement a set of standards and guidelines to ascertain quality features of diffusion MRI tractography for neurosurgical guidance.

CONCLUSION

Diffusion MRI provides a unique window into the organization of the brain white matter. Tractography reconstructions allow the visualization of the trajectory of white matter pathways noninvasively. In recent years, considerable advances have been made in the acquisition and postprocessing of diffusion-weighted MR images. With the availability of diffusion sequences on most clinical MR scanners and the development of various fiber-tracking software packages, tractography is becoming part of the brain-mapping apparatus of numerous surgical centers. While tractography reconstructions can provide clinically relevant information on peritumoral white matter anatomy for the planning of brain tumor resection, the technique should not be used without awareness and understanding of its current limitations. Validation of tractography derived information will help establish the clinical utility and usability of the technique.

Diffusion MRI is an active area of research that produces an unprecedented wealth of information on white matter anatomy. Future refinement of diffusion MRI tools for mapping white matter pathways will continue to advance tissue characterization, surgical planning, and follow-up of brain tumor patients.

REFERENCES

1. Le Bihan D, Breton E, Lallemand D, Grenier P, Cabanis E, Laval-Jeantet M. MR imaging of intravoxel incoherent motions: application to diffusion and perfusion in neurologic disorders. *Radiology.* 1986;161:401–407.
2. Beaulieu C. The basis of anisotropic water diffusion in the nervous system—a technical review. *NMR Biomed.* 2002;15:435–455.
3. Stejskal EO, Tanner JE. Spin diffusion measurements: spin echoes in the presence of a time-dependent field gradient. *J Chem Phys.* 1965;42:288.
4. Jones DK. The effect of gradient sampling schemes on measures derived from diffusion tensor MRI: a Monte Carlo study. *Magn Reson Med.* 2004;51:807–815.
5. Berlot R, Metzler-Baddeley C, Jones DK, O'Sullivan MJ. CSF contamination contributes to apparent microstructural alterations in mild cognitive impairment. *Neuroimage.* 2014;92:27–35.
6. Jeurissen B, Leemans A, Tournier J-D, Jones DK, Sijbers J. Investigating the prevalence of complex fiber configurations in white matter tissue with diffusion magnetic resonance imaging. *Hum Brain Mapp.* 2013;34:2747–2766.
7. Manjón JV, Coupé P, Concha L, Buades A, Collins DL, Robles M. Diffusion weighted image denoising using overcomplete local PCA. *PLoS One.* 2013;8(9):e73021.
8. Warach S, Gaa J, Siewert B, Wielopolski P, Edelman RR. Acute human stroke studied by whole brain echo planar diffusion-weighted magnetic resonance imaging. *Ann Neurol.* 1995;37:231–241.
9. Schaefer PW, Grant PE, Gonzalez RG. Diffusion-weighted MR imaging of the brain. *Radiology.* 2000;217:331–345.
10. Maier SE, Sun Y, Mulkern RV. Diffusion imaging of brain tumors. *NMR Biomed.* 2010;23:849–864.
11. Padhani AR, Liu G, Mu-koh D, Chenevert TL, Thoeny HC, Ross BD, et al. Diffusion-weighted magnetic resonance imaging as a cancer biomarker: consensus and recommendations. Neoplasia. 2009; 11: 102–125.
12. Ji JX, Zhang X. Parallel and sparse MR imaging: methods and instruments: Part 2. *Quant Imaging Med Surg.* 2014;4:68–70.
13. Jezzard P, Barnett AS, Pierpaoli C. Characterization of and correction for eddy current artifacts in echo planar diffusion imaging. *Magn Reson Med.* 1998;39:801–812.
14. Jenkinson M. Fast, automated, N-dimensional phase-unwrapping algorithm. *Magn Reson Med.* 2003;49:193–197.
15. Funai AK, Fessler JA, Yeo DTB, Olafsson VT, Noll DC. Regularized field map estimation in MRI. *IEEE Trans Med Imaging.* 2008;27:1484–1494.
16. Kybic J, Thévenaz P, Nirkko A, Unser M. Unwarping of unidirectionally distorted EPI images. *IEEE Trans Med Imaging.* 2000;19:80–93.
17. Merhof D, Soza G, Stadlbauer A, Greiner G, Nimsky C. Correction of susceptibility artifacts in diffusion tensor data using non-linear registration. *Med Image Anal.* 2007;11:588–603.

18. Huang H, Ceritoglu C, Li X, Qiu A, Miller MI, van Zijl PCM, et al. Correction of B0 susceptibility induced distortion in diffusion-weighted images using large-deformation diffeomorphic metric mapping. *Magn Reson Imaging*. 2008;26:1294–1302.
19. Daga P, Pendse T, Modat M, White M, Mancini L, Cardoso MJ, et al. Susceptibility artefact correction using dynamic graph cuts: application to neurosurgery. *Med Image Anal*. 2014;18:1132–1142.
20. Le Bihan D, Poupon C, Amadon A, Lethimonnier F. Artifacts and pitfalls in diffusion MRI. *J Magn Reson Imaging*. 2006;24:478–488.
21. Pierpaoli C. Artifacts in diffusion MRI. In: Derek K. Jones, ed. *Diffusion MRI: Theory, Methods, and Applications*. New York, NY: Oxford University Press 2010; 303–318.
22. Turner R, Le Bihan D, Scott Chesnicks A. Echo-planar imaging of diffusion and perfusion. *Magn Reson Med*. 1991;19:247–253.
23. Gudbjartsson H, Maier SE, Mulkern RV, Mórocz IA, Patz S, Jolesz FA. Line scan diffusion imaging. *Magn Reson Med*. 1996;36:509–519.
24. Maier SE, Gudbjartsson H, Patz S, Hsu L, Lovblad KO, Edelman RR, et al. Line scan diffusion imaging: characterization in healthy subjects and stroke patients. *Am J Roentgenol*. 1998;171:85–93.
25. Pruessmann KP, Weiger M, Scheidegger MB, Boesiger P. SENSE: sensitivity encoding for fast MRI. *Magn Reson Med*. 1999;42:952–962.
26. Oguz I, Farzinfar M, Matsui J, Budin F, Liu Z, Gerig G, et al. DTIPrep: quality control of diffusion-weighted images. *Front Neuroinform*. 2014;8:4.
27. Basser PJ, Mattiello J, LeBihan D. Estimation of the effective self-diffusion tensor from the NMR spin echo. *J Magn Reson*. 1994;103:247–254.
28. Pierpaoli C, Basser PJ. Toward a quantitative assessment of diffusion anisotropy. *Magn Reson Med*. 1996;36:893–906.
29. Pajevic S, Pierpaoli C. Color schemes to represent the orientation of anisotropic tissues from diffusion tensor data: application to white matter fiber tract mapping in the human brain. *Magn Reson Med*. 2000;43:921.
30. Cortez-Conradis D, Favila R, Isaac-Olive K, Martinez-Lopez M, Rios C, Roldan-Valadez E. Diagnostic performance of regional DTI-derived tensor metrics in glioblastoma multiforme: simultaneous evaluation of p, q, L, Cl, Cp, Cs, RA, RD, AD, mean diffusivity and fractional anisotropy. *Eur Radiol*. 2013;23:1112–1121.
31. Lee EJ, Ahn KJ, Lee EK, Lee YS, Kim DB. Potential role of advanced MRI techniques for the peritumoural region in differentiating glioblastoma multiforme and solitary metastatic lesions. *Clin Radiol*. 2013;68:689–697.
32. Min Z, Niu C, Rana N, Ji H, Zhang M. Differentiation of pure vasogenic edema and tumor-infiltrated edema in patients with peritumoral edema by analyzing the relationship of axial and radial diffusivities on 3.0T MRI. *Clin Neurol Neurosurg*. 2013;115:1366–1370.
33. Lemercier P, Paz Maya S, Patrie JT, Flors L, Leiva-Salinas C. Gradient of apparent diffusion coefficient values in peritumoral edema helps in differentiation of glioblastoma from solitary metastatic lesions. *Am J Roentgenol*. 2014;203:163–169.
34. Hoefnagels FWA, De Witt Hamer P, Sanz-Arigita E, Idema S, Kuijer JPA, Pouwels PJW, et al. Differentiation of edema and glioma infiltration: proposal of a DTI-based probability map. *J Neurooncol*. 2014;120:187–198.
35. Jensen JH, Helpern JA, Ramani A, Lu H, Kaczynski K. Diffusional kurtosis imaging: the quantification of non-Gaussian water diffusion by means of magnetic resonance imaging. *Magn Reson Med*. 2005;53:1432–1440.
36. Van Cauter S, Veraart J, Sijbers J, Peeters RR, Himmelreich U, De Keyzer F, et al. Gliomas: diffusion kurtosis MR imaging in grading. *Radiology*. 2012;263:492–501.
37. Tuch DS, Reese TG, Wiegell MR, Makris N, Belliveau JW, Wedeen VJ. High angular resolution diffusion imaging reveals intravoxel white matter fiber heterogeneity. *Magn Reson Med*. 2002;48:577–582.
38. Alexander AL, Hasan KM, Lazar M, Tsuruda JS, Parker DL. Analysis of partial volume effects in diffusion-tensor MRI. *Magn Reson Med*. 2001;45:770–780.

39. Tournier J-D, Calamante F, Connelly A. Determination of the appropriate b value and number of gradient directions for high-angular-resolution diffusion-weighted imaging. *NMR Biomed.* 2013;26:1775–1786.
40. Assemlal H-E, Tschumperlé D, Brun L, Siddiqi K. Recent advances in diffusion MRI modeling: angular and radial reconstruction. *Med Image Anal.* 2011;15:369–396.
41. Descoteaux M, Angelino E, Fitzgibbons S, Deriche R. Regularized, fast, and robust analytical Q-ball imaging. *Magn Reson Med.* 2007;58:497–510.
42. Tournier J-D, Calamante F, Gadian DG, Connelly A. Direct estimation of the fiber orientation density function from diffusion-weighted MRI data using spherical deconvolution. *Neuroimage.* 2004;23:1176–1185.
43. Tournier J-D, Calamante F, Connelly A. Robust determination of the fibre orientation distribution in diffusion MRI: non-negativity constrained super-resolved spherical deconvolution. *Neuroimage.* 2007;35:1459–1472.
44. Assaf Y, Basser PJ. Composite hindered and restricted model of diffusion (CHARMED) MR imaging of the human brain. *Neuroimage.* 2005;27:48–58.
45. Tuch DS. Q-ball imaging. *Magn Reson Med.* 2004;52:1358–1372.
46. Jansons KM, Alexander DC. Persistent angular structure: new insights from diffusion MRI data. Dummy version. *Inf Process Med Imaging.* 2003;18:672–683.
47. Ozarslan E, Shepherd TM, Vemuri BC, Blackband SJ, Mareci TH. Resolution of complex tissue microarchitecture using the diffusion orientation transform (DOT). *Neuroimage.* 2006;31:1086–1103.
48. Sotiropoulos SN, Jbabdi S, Xu J, Andersson JL, Moeller S, Auerbach EJ, et al. Advances in diffusion MRI acquisition and processing in the Human Connectome Project. *Neuroimage.* 2013;80:125–143.
49. Wedeen VJ, Wang RP, Schmahmann JD, Benner T, Tseng WYI, Dai G, et al. Diffusion spectrum magnetic resonance imaging (DSI) tractography of crossing fibers. *Neuroimage.* 2008;41:1267–1277.
50. Menzel MI, Tan ET, Khare K, Sperl JI, King KF, Tao X, et al. Accelerated diffusion spectrum imaging in the human brain using compressed sensing. *Magn Reson Med.* 2011;66:1226–1233.
51. Setsompop K, Kimmlingen R, Eberlein E, Witzel T, Cohen-Adad J, McNab JA, et al. Pushing the limits of in vivo diffusion MRI for the Human Connectome Project. *Neuroimage.* 2013;80:220–233.
52. Mori S, Crain BJ, Chacko VP, van Zijl PC. Three-dimensional tracking of axonal projections in the brain by magnetic resonance imaging. *Ann Neurol.* 1999;45:265–269.
53. Basser PJ, Pajevic S, Pierpaoli C, Duda J, Aldroubi A. In vivo fiber tractography using DT-MRI data. *Magn Reson Med.* 2000;44:625–632.
54. Stieltjes B, Kaufmann WE, van Zijl PC, Fredericksen K, Pearlson GD, Solaiyappan M, et al. Diffusion tensor imaging and axonal tracking in the human brainstem. *Neuroimage.* 2001;14:723–735.
55. Mori S, Frederiksen K, van Zijl PCM, Stieltjes B, Kraut MA, Solaiyappan M, et al. Brain white matter anatomy of tumor patients evaluated with diffusion tensor imaging. *Ann Neurol.* 2002;51:377–380.
56. Lazar M, Weinstein DM, Tsuruda JS, Hasan KM, Arfanakis K, Meyerand ME, et al. White matter tractography using diffusion tensor deflection. *Hum Brain Mapp.* 2003;18:306–321.
57. Catani M, Howard RJ, Pajevic S, Jones DK. Virtual in vivo interactive dissection of white matter fasciculi in the human brain. *Neuroimage.* 2002;17:77–94.
58. Catani M, Thiebaut de Schotten M. A diffusion tensor imaging tractography atlas for virtual in vivo dissections. *Cortex.* 2008;44:1105–1132.
59. Behrens TEJ, Johansen-Berg H, Woolrich MW, Smith SM, Wheeler-Kingshott CAM, Boulby PA, et al. Non-invasive mapping of connections between human thalamus and cortex using diffusion imaging. *Nat Neurosci.* 2003;6:750–757.
60. Parker GJM, Alexander DC. Probabilistic Monte Carlo based mapping of cerebral connections utilising whole-brain crossing fibre information. *Inf Process Med Imaging.* 2003;18:684–695.
61. Jones DK, Pierpaoli C. Confidence mapping in diffusion tensor magnetic resonance imaging tractography using a bootstrap approach. *Magn Reson Med.* 2005;53:1143–1149.
62. Friman O, Farnebäck G, Westin C-F. A Bayesian approach for stochastic white matter tractography. *IEEE Trans Med Imaging.* 2006;25:965–978.

63. Behrens TEJ, Berg HJ, Jbabdi S, Rushworth MFS, Woolrich MW. Probabilistic diffusion tractography with multiple fibre orientations: what can we gain? *Neuroimage.* 2007;34:144–155.
64. Jones DK. Tractography gone wild: probabilistic fibre tracking using the wild bootstrap with diffusion tensor MRI. *IEEE Trans Med Imaging.* 2008;27:1268–1274.
65. Eickhoff SB, Jbabdi S, Caspers S, Laird AR, Fox PT, Zilles K, et al. Anatomical and functional connectivity of cytoarchitectonic areas within the human parietal operculum. *J Neurosci.* 2010;30:6409–6421.
66. Jones DK. Challenges and limitations of quantifying brain connectivity in vivo with diffusion MRI. Imaging in Medicine. 2010;2:341–355.
67. Mandelli ML, Berger MS, Bucci M, Berman JI, Amirbekian B, Henry RG. Quantifying accuracy and precision of diffusion MR tractography of the corticospinal tract in brain tumors. *J Neurosurg.* 2014;121:349–358.
68. Mangin JF, Fillard P, Cointepas Y, Le Bihan D, Frouin V, Poupon C. Toward global tractography. *Neuroimage.* 2013;80:290–296.
69. Mangin J-F, Poupon C, Cointepas Y, Rivière D, Papadopoulos-Orfanos D, Clark CA, et al. A framework based on spin glass models for the inference of anatomical connectivity from diffusion-weighted MR data: a technical review. *NMR Biomed.* 2002;15:481–492.
70. Fillard P, Poupon C, Mangin J-F. A novel global tractography algorithm based on an adaptive spin glass model. *Med Image Comput Comput Assist Interv.* 2009;12:927–934.
71. Parker GJM, Wheeler-Kingshott CAM, Barker GJ. Estimating distributed anatomical connectivity using fast marching methods and diffusion tensor imaging. *IEEE Trans Med Imaging.* 2002;21:505–512.
72. Jbabdi S, Bellec P, Toro R, Daunizeau J, Pélégrini-Issac M, Benali H. Accurate anisotropic fast marching for diffusion-based geodesic tractography. *Int J Biomed Imaging.* 2008;320195.
73. Reisert M, Mader I, Anastasopoulos C. Global fiber reconstruction becomes practical. *Neuroimage.* 2011;54:955–962.
74. Lemkaddem A, Skioldebrand D, Dal Palu A, Thiran J-P, Daducci A. Global tractography with embedded anatomical priors for quantitative connectivity analysis. *Front Neurol.* 2014;5232.
75. Holodny AI, Ollenschleger MD, Liu W-C, Schulder M, Kalnin AJ. Identification of the corticospinal tracts achieved using blood-oxygen-level-dependent and diffusion functional mr imaging in patients with brain tumors. *Am J Neuroradiol.* 2001;22:83–88.
76. Wu J-S, Zhou L-F, Tang W-J, Mao Y, Hu J, Song Y-Y, et al. Clinical evaluation and follow-up outcome of diffusion tensor imaging-based functional neuronavigation: a prospective, controlled study in patients with gliomas involving pyramidal tracts. *Neurosurgery.* 2007; 61:935–948.
77. Pujol S, Golby A, Wells W, Pierpaoli C, Chauvin L, Mamata H, et al. The DTI Challenge initiative on the standardized evaluation of DTI tractography for neurosurgical planning. *Proceedings of the 100th Annual Meeting of the Radiological Society of North America.* November 30–December 5, Chicago, IL; 2014.
78. Bucci M, Mandelli ML, Berman JI, Amirbekian B, Nguyen C, Berger MS, et al. Quantifying diffusion MRI tractography of the corticospinal tract in brain tumors with deterministic and probabilistic methods. *NeuroImage Clin.* 2013;3:361–368.
79. Yamada K, Sakai K, Hoogenraad FGC, Holthuizen R, Akazawa K, Ito H, et al. Multitensor tractography enables better depiction of motor pathways: initial clinical experience using diffusion-weighted MR imaging with standard b-value. *Am J Neuroradiol.* 2007;28:1668–1673.
80. Qazi AA, Radmanesh A, O'Donnell L, Kindlmann G, Peled S, Whalen S, et al. Resolving crossings in the corticospinal tract by two-tensor streamline tractography: method and clinical assessment using fMRI. *Neuroimage.* 2009;47(Suppl 2):T98–T106.
81. Muthusami P, James J, Thomas B, Kapilamoorthy TR, Kesavadas C. Diffusion tensor imaging and tractography of the human language pathways: moving into the clinical realm. *J Magn Reson Imaging.* 2013;40:1041–1053.
82. Catani M, Jones DK, Ffytche DH. Perisylvian language networks of the human brain. *Ann Neurol.* 2005;57:8–16.

83. Henning Stieglitz L, Seidel K, Wiest R, Beck J, Raabe A. Localization of primary language areas by arcuate fascicle fiber tracking. *Neurosurgery.* 2012;70:56–64.
84. Zhao Y, Chen X, Wang F, et al. Integration of diffusion tensor-based arcuate fasciculus fibre navigation and intraoperative MRI into glioma surgery. *J Clin Neurosci.* 2012;19:255–261.
85. Kinoshita M, Nakada M, Okita H, Hamada J-I, Hayashi Y. Predictive value of fractional anisotropy of the arcuate fasciculus for the functional recovery of language after brain tumor resection: a preliminary study. *Clin Neurol Neurosurg.* 2014;117:45–50.
86. Bello L, Gambini A, Castellano A, Carrabba G, Acerbi F, Fava E, et al. Motor and language DTI Fiber Tracking combined with intraoperative subcortical mapping for surgical removal of gliomas. *Neuroimage.* 2008;39:369–382.
87. Vassal F, Schneider F, Sontheimer A, Lemaire J-J, Nuti C. Intraoperative visualisation of language fascicles by diffusion tensor imaging-based tractography in glioma surgery. *Acta Neurochir (Wien).* 2013;155:437–448.
88. Li Z, Peck KK, Brennan NP, Jenabi M, Hsu M, Zhang Z, et al. Diffusion tensor tractography of the arcuate fasciculus in patients with brain tumors: comparison between deterministic and probabilistic models. *J Biomed Sci Eng.* 2013;6:192–200.
89. Richter M, Zolal A, Ganslandt O, Buchfelder M, Nimsky C, Merhof D. Evaluation of diffusion-tensor imaging-based global search and tractography for tumor surgery close to the language system. *PLoS One.* 2013;8:e50132.
90. Fernandez-Miranda JC, Pathak S, Engh J, Jarbo K, Verstynen T, Yeh F-C, et al. High-definition fiber tractography of the human brain: neuroanatomical validation and neurosurgical applications. *Neurosurgery.* 2012;71:430–453.
91. Kuhnt D, Bauer MHA, Egger J, Richter M, Kapur T, Sommer J, et al. Fiber tractography based on diffusion tensor imaging compared with high-angular-resolution diffusion imaging with compressed sensing: initial experience. *Neurosurgery.* 2013;72(Suppl 1):165–175.
92. Ojemann G, Ojemann J, Lettich E, Berger M. Cortical language localization in left, dominant hemisphere. An electrical stimulation mapping investigation in 117 patients. *J Neurosurg.* 1989;71:316–326.
93. Dick AS, Tremblay P. Beyond the arcuate fasciculus: consensus and controversy in the connectional anatomy of language. *Brain.* 2012;135:3529–3550.
94. Duffau H. The anatomo-functional connectivity of language revisited. New insights provided by electrostimulation and tractography. *Neuropsychologia.* 2008;46:927–934.
95. Ebeling U, Reulen HJ. Neurosurgical topography of the optic radiation in the temporal lobe. *Acta Neurochir (Wien).* 1988;92:29–36.
96. Yamamoto T, Yamada K, Nishimura T, Kinoshita S. Tractography to depict three layers of visual field trajectories to the calcarine gyri. *Am J Ophthalmol.* 2005;140:781–785.
97. Nilsson D, Starck G, Ljungberg M, Ribbelin S, Jönsson L, Malmgren K, et al. Intersubject variability in the anterior extent of the optic radiation assessed by tractography. *Epilepsy Res.* 2007;77:11–16.
98. Sherbondy AJ, Dougherty RF, Napel S, Wandell BA. Identifying the human optic radiation using diffusion imaging and fiber tractography. *J Vis.* 2008;8:12.1–11.
99. Wang Y-XJ, Zhu X-L, Deng M, Siu DYW, Leung JCS, Chan Q, et al. The use of diffusion tensor tractography to measure the distance between the anterior tip of the Meyer loop and the temporal pole in a cohort from Southern China. *J Neurosurg.* 2010;113:1144–1151.
100. Yogarajah M, Focke NK, Bonelli S, Cercignani M, Acheson J, Parker GJM, et al. Defining Meyers looptemporal lobe resections, visual field deficits and diffusion tensor tractography. *Brain.* 2009;132:1656–1668.
101. Chen X, Weigel D, Ganslandt O, Buchfelder M, Nimsky C. Prediction of visual field deficits by diffusion tensor imaging in temporal lobe epilepsy surgery. *Neuroimage.* 2009;45:286–297.
102. Winston GP, Mancini L, Stretton J, Ashmore J, Symms MR, Duncan JS, et al. Diffusion tensor imaging tractography of the optic radiation for epilepsy surgical planning: a comparison of two methods. *Epilepsy Res.* 2011;97:124–132.

103. Daga P, Winston G, Modat M, White M, Mancini L, Cardoso MJ, et al. Accurate localization of optic radiation during neurosurgery in an interventional MRI suite. *IEEE Trans Med Imaging*. 2012;31:882–891.
104. Yamamoto A, Miki Y. Diffusion tensor fiber tractography of the optic radiation: analysis with 6-, 12-, 40-, and 81-directional motion-probing gradients, a preliminary study. *Am J Neuroradiol*. 2007;28(1):92–96.
105. Hofer S, Karaus A, Frahm J. Reconstruction and dissection of the entire human visual pathway using diffusion tensor MRI. *Front Neuroanat*. 2010;4:15.
106. Tao X, Wang Z, Gong W, Jiang Q, Shi Z. A new study on diffusion tensor imaging of the whole visual pathway fiber bundle and clinical application. *Chin Med J (Engl)*. 2009;122:178–182.
107. Wu W, Rigolo L, O'Donnell LJ, Norton I, Shriver S, Golby AJ. Visual pathway study using in vivo diffusion tensor imaging tractography to complement classic anatomy. *Neurosurgery*. 2012;70:145–156.
108. Anastasopoulos C, Reisert M, Kiselev VG, Nguyen-Thanh T, Schulze-Bonhage A, Zentner J, et al. Local and global fiber tractography in patients with epilepsy. *Am J Neuroradiol*. 2014;35:291–296.
109. Lilja Y, Ljungberg M, Starck G, Malmgren K, Rydenhag B, Nilsson DT. Visualizing Meyer's loop: a comparison of deterministic and probabilistic tractography. *Epilepsy Res*. 2014;108:481–490.
110. Berman J. Diffusion MR tractography as a tool for surgical planning. *Magn Reson Imaging Clin N Am*. 2009;17:205–214.
111. Clark CA, Barrick TR, Murphy MM, Bell BA. White matter fiber tracking in patients with space-occupying lesions of the brain: a new technique for neurosurgical planning? *Neuroimage*. 2003;20:1601–1608.
112. Witwer BP, Moftakhar R, Hasan KM, Deshmukh P, Haughton V, Field A, et al. Diffusion-tensor imaging of white matter tracts in patients with cerebral neoplasm. *J Neurosurg*. 2002;97:568–575.
113. Jellison BJ, Field AS, Medow J, Lazar M, Salamat MS, Alexander AL. Diffusion tensor imaging of cerebral white matter: a pictorial review of physics, fiber tract anatomy, and tumor imaging patterns. *Am J Neuroradiol*. 2004;25:356–369.
114. Sherbondy A, Akers D, Mackenzie R, Dougherty R, Wandell B. Exploring connectivity of the brain's white matter with dynamic queries. *IEEE Trans Vis Comput Graph*. 2005;11:419–430.
115. Golby AJ, Kindlmann G, Norton I, Yarmarkovich A, Pieper S, Kikinis R. Interactive diffusion tensor tractography visualization for neurosurgical planning. *Neurosurgery*. 2011;68:496–505.
116. Enders F, Sauber N, Merhof D, Hastreiter P, Nimsky C, Stamminger M. Visualization of white matter tracts with wrapped streamlines. In: *VIS* 05. *IEEE Visualization*, 2005. IEEE; 2005:51–58.
117. Nimsky C, Ganslandt O, Kober H, Moller M, Ulmer S, Tomandl B, et al. Integration of functional magnetic resonance imaging supported by magnetoencephalography in functional neuronavigation. *Neurosurgery*. 1999;44:1249–1255.
118. Jannin P, Morandi X, Fleig OJ, Le Rumeur E, Toulouse P, Gibaud B, et al. Integration of sulcal and functional information for multimodal neuronavigation. *J Neurosurg*. 2002;96:713–723.
119. Krishnan R, Raabe A, Hattingen E, Szelényi A, Yahya H, Hermann E, et al. Functional magnetic resonance imaging-integrated neuronavigation: correlation between lesion-to-motor cortex distance and outcome. *Neurosurgery*. 2004;55:904–914.
120. Bizzi A, Blasi V, Falini A, Ferroli P, Cadioli M, Danesi U, et al. Presurgical functional MR imaging of language and motor functions: validation with intraoperative electrocortical mapping. *Radiology*. 2008;248:579–589.
121. Hendler T, Pianka P, Sigal M, Kafri M, Ben-Bashat D, Constantini S, et al. Delineating gray and white matter involvement in brain lesions: three-dimensional alignment of functional magnetic resonance and diffusion-tensor imaging. *J Neurosurg*. 2003;99:1018–1027.
122. Schonberg T, Pianka P, Hendler T, Pasternak O, Assaf Y. Characterization of displaced white matter by brain tumors using combined DTI and fMRI. *Neuroimage*. 2006;30:1100–1111.
123. Skirboll S, Ojemann G, Berger M, Lettich E, Winn H. Functional cortex and subcortical white matter located within gliomas. *Neurosurgery*. 1996;38:678–684.

124. Duffau H, Capelle L, Sichez J, Faillot T, Abdennour L, Law Koune JD, et al. Intra-operative direct electrical stimulations of the central nervous system: the Salpêtrière experience with 60 patients. *Acta Neurochir (Wien).* 1999;141:1157–1167.
125. Duffau H, Capelle L, Sichez N, Denvil D, Lopes M, Sichez J-P, et al. Intraoperative mapping of the subcortical language pathways using direct stimulations. An anatomo-functional study. *Brain.* 2002;125:199–214.
126. Duffau H, Capelle L, Denvil D, Sichez N, Gatignol P, Taillandier L, et al. Usefulness of intraoperative electrical subcortical mapping during surgery for low-grade gliomas located within eloquent brain regions: functional results in a consecutive series of 103 patients. *J Neurosurg.* 2003;98:764–778.
127. Keles GE, Lundin DA, Lamborn KR, Chang EF, Ojemann G, Berger MS. Intraoperative subcortical stimulation mapping for hemispherical perirolandic gliomas located within or adjacent to the descending motor pathways: evaluation of morbidity and assessment of functional outcome in 294 patients. *J Neurosurg.* 2004;100:369–375.
128. Saito T, Tamura M, Muragaki Y, Maruyama T, Kubota Y, Fukuchi S, et al. Intraoperative cortico-cortical evoked potentials for the evaluation of language function during brain tumor resection: initial experience with 13 cases. *J Neurosurg.* 2014;121:827–838.
129. Berman JI, Berger MS, Mukherjee P, Henry RG. Diffusion-tensor imaging-guided tracking of fibers of the pyramidal tract combined with intraoperative cortical stimulation mapping in patients with gliomas. *J Neurosurg.* 2004;101:66–72.
130. Kamada K, Todo T, Masutani Y, Aoki S, Ino K, Takano T, et al. Combined use of tractography-integrated functional neuronavigation and direct fiber stimulation. *J Neurosurg.* 2005;102:664–672.
131. Leclercq D, Duffau H, Delmaire C, Capelle L, Gatignol P, Ducros M, et al. Comparison of diffusion tensor imaging tractography of language tracts and intraoperative subcortical stimulations. *J Neurosurg.* 2010;112:503–511.
132. González-Darder JM, González-López P, Talamantes F, Quilis V, Cortés V, García-March G, et al. Multimodal navigation in the functional microsurgical resection of intrinsic brain tumors located in eloquent motor areas: role of tractography. *Neurosurg Focus.* 2010;28:E5.
133. Coenen VA, Krings T, Mayfrank L, Polin RS, Reinges MH, Thron A, et al. Three-dimensional visualization of the pyramidal tract in a neuronavigation system during brain tumor surgery: first experiences and technical note. *Neurosurgery.* 2001;49:86–92.
134. Nimsky C, Ganslandt O, Fahlbusch R. Implementation of fiber tract navigation. *Neurosurgery.* 2006;58:ONS-292–ONS-303.
135. Mikuni N, Okada T, Enatsu R, Miki Y, Hanakawa T, Urayama S, et al. Clinical impact of integrated functional neuronavigation and subcortical electrical stimulation to preserve motor function during resection of brain tumors. *J Neurosurg.* 2007;106:593–598.
136. Merhof D, Meister M, Bingol E, Nimsky C, Greiner G. Isosurface-based generation of hulls encompassing neuronal pathways. *Stereotact Funct Neurosurg.* 2009;87:50–60.
137. Mamata Y, Mamata H, Nabavi A, Kacher DF, Pergolizzi RS, Schwartz RB, et al. Intraoperative diffusion imaging on a 0.5 Tesla interventional scanner. *J Magn Reson Imaging.* 2001;13:115–119.
138. Nimsky C, Ganslandt O, Hastreiter P, Wang R, Benner T, Sorensen AG, et al. Preoperative and intraoperative diffusion tensor imaging-based fiber tracking in glioma surgery. *Neurosurgery.* 2005;56:130–137.
139. Nimsky C, Ganslandt O, Merhof D, Sorensen AG, Fahlbusch R. Intraoperative visualization of the pyramidal tract by diffusion-tensor-imaging-based fiber tracking. *Neuroimage.* 2006;30:1219–1229.
140. Nimsky C, Ganslandt O, Hastreiter P, Wang R, Benner T, Sorensen AG, et al. Intraoperative diffusion-tensor MR imaging: shifting of white matter tracts during neurosurgical procedures—Initial experience. *Radiology.* 2005;234:218–225.
141. Irfanoglu M, Walker L, Sarlls J. Effects of image distortions originating from susceptibility variations and concomitant fields on diffusion MRI tractography results. *Neuroimage.* 2012;61:275–288.
142. Reich DS, Smith SA, Jones CK, Zackowski KM, van Zijl PC, Calabresi PA, et al. Quantitative characterization of the corticospinal tract at 3T. *Am J Neuroradiol.* 2006; 27:2168–2178.

143. Tensaouti F, Lahlou I, Clarisse P, Lotterie JA, Berry I. Quantitative and reproducibility study of four tractography algorithms used in clinical routine. *J Magn Reson Imaging*. 2011;34:165–172.
144. Fillard P, Descoteaux M, Goh A, Gouttard S, Jeurissen B, Malcolm J, et al. Quantitative evaluation of 10 tractography algorithms on a realistic diffusion MR phantom. *Neuroimage*. 2011;56:220–234.
145. Feigl GC, Hiergeist W, Fellner C, Schebesch KMM, Doenitz C, Finkenzeller T, et al. Magnetic resonance imaging diffusion tensor tractography: evaluation of anatomic accuracy of different fiber tracking software packages. *World Neurosurg*. 2014;81:144–150.
146. Nimsky C. Fiber tracking—we should move beyond diffusion tensor imaging. *World Neurosurg*. 2014;82:35–36.
147. Stadlbauer A, Nimsky C, Buslei R, Salomonowitz E, Hammen T, Buchfelder M, et al. Diffusion tensor imaging and optimized fiber tracking in glioma patients: histopathologic evaluation of tumor-invaded white matter structures. *Neuroimage*. 2007;34:949–956.
148. Chen Z, Tie Y, O'Donnell L, Buchfelder M, Nimsky C. Resolving the challenges of peritumoral edema in tracing arcuate fasciculus for surgical planning using two-tensor unscented Kalman filter tractography. *Neuro Oncol*. 2014;16(5):v140.
149. Zhang H, Wang Y, Lu T, Qiu B, Tang Y, Ou S, et al. Differences between generalized q-sampling imaging and diffusion tensor imaging in the preoperative visualization of the nerve fiber tracts within peritumoral edema in brain. *Neurosurgery*. 2013;73:1044–1053.
150. White NS, Leergaard TB, D'Arceuil H, Bjaalie JG, Dale AM. Probing tissue microstructure with restriction spectrum imaging: histological and theoretical validation. *Hum Brain Mapp*. 2013;34:327–346.
151. Berntsen EM, Gulati S, Solheim O, Kvistad KA, Torp SH, Selbekk T, et al. Functional magnetic resonance imaging and diffusion tensor tractography incorporated into an intraoperative 3-dimensional ultrasound-based neuronavigation system: impact on therapeutic strategies, extent of resection, and clinical outcome. *Neurosurgery*. 2010;67:251–264.
152. Coenen V, Krings T, Weidemann J, Hans F-J, Reinacher P, Gilsbach JM, et al. Sequential visualization of brain and fiber tract deformation during intracranial surgery with three-dimensional ultrasound: an approach to evaluate the effect of brain shift. *Neurosurgery*. 2005;56:133–141.
153. Rasmussen I, Lindseth F, Rygh OM, Berntsen EM, Selbekk T, Xu J, et al. Functional neuronavigation combined with intra-operative 3D ultrasound: initial experiences during surgical resections close to eloquent brain areas and future directions in automatic brain shift compensation of preoperative data. *Acta Neurochir (Wien)*. 2007;149:365–378.
154. Archip N, Clatz O, Whalen S, Kacher D, Fedorov A, Kot A, et al. Non-rigid alignment of preoperative MRI, fMRI, and DT-MRI with intra-operative MRI for enhanced visualization and navigation in image-guided neurosurgery. *Neuroimage*. 2007;35:609–624.
155. Kinoshita M, Yamada K, Hashimoto N, Kato A, Izumoto S, Baba T, et al. Fiber-tracking does not accurately estimate size of fiber bundle in pathological condition: initial neurosurgical experience using neuronavigation and subcortical white matter stimulation. *Neuroimage*. 2005;25:424–429.
156. Duffau H. Diffusion tensor imaging is a research and educational tool, but not yet a clinical tool. *World Neurosurg*. 2013;82(1–2):e43–e45.
157. DTI Tractography Challenge Working Group, MICCAI 2014 DTI Challenge. 2014. Available from: <http://dti-challenge.org>.

CHAPTER 5

Background on Stereotaxis

David W. Roberts
Geisel School of Medicine at Dartmouth, Section of Neurosurgery, Dartmouth-Hitchcock Medical Center, Lebanon, NH, USA

INTRODUCTION

The origins of modern surgical stereotaxy can be traced to the rapidly developing investigation of the central nervous system during the nineteenth century. Understanding of the neuroanatomic substrate of clinical neurology, exemplified by the work of such neurologists as Broca, Jackson, and Ferrier among others, was paralleled by studies of nonhuman brain utilizing newly adapted technologies for electrical stimulation or ablation of neural tissue. The need to overcome the methodological limitations of these laboratory studies led to technological strategies and tools that, in turn, found clinical application when the time was opportune.

Any review of the development of stereotactic technique focuses on the well-recognized, pioneering work of Sir Victor Horsley and Robert Clarke in conceiving and implementing a mechanical device for accurate localization of instruments in their animal work, although the larger context of the work is often underappreciated. This chapter will begin with these early forays into neurological investigation, trace the adaptation of this work into the clinical arena, and follow the further developments and applications of the same fundamental principles in the setting of more recent neuroimaging and computational advances.

Technologically, a major conceptual leap a little over a hundred years ago enabled remarkable investigational and clinical achievement, and that same concept is inherent in many of the remarkable interventions in use and under development today.

LABORATORY WORK

From the broadest perspective, the seeds of stereotaxy were planted when the theories of Dr. Franz Joseph Gall (1758–1828), attempting to localize functions to specific areas of the brain, threatened the tenets of the preceding century's conceptualization of the brain as a holistic organ. Although Gall's association with phrenology has diminished appreciation for his considerable contributions to neuroanatomy, as Macdonald Critchley has eloquently pointed out in *The Divine Banquet of the Brain,*[1] his focus of attention on anatomic features played an instrumental role in promotion of both an orientation toward localization and a subsequent wave of investigation of

A. Golby (Ed): Image-Guided Neurosurgery
DOI: http://dx.doi.org/10.1016/B978-0-12-800870-6.00005-4

cortical topography and function. That the phreno-physiometric caliper of phrenologists F. Bridges and N. Morgan may have even foreshadowed later stereotactic frames (as well as the encephalometer of Zernov) is intriguingly suggested by Critchley.[1]

In the latter nineteenth century, investigations of the cerebral cortex, the work of Eduard Hitzig and Gustav Theodore Fritsch published in 1870, were seminal. Demonstrating that electrical stimulation of specific areas of the dog brain produced muscular contractions on the opposite side of the body refuted the theories of holistic brain function advocated by Pierre Flourens. Subsequent investigators explored all of the cortex seeking organizational insights, and it was as part of this movement that Horsley became involved. Marshall and Magoun's detailed history of this fascinating development describes Hitzig's subsequent work in nonhuman primates, David Ferrier's further confirmation of these observations and attempt to apply them to the human brain, and the thread of this work to the Horsley—Clarke frame.[2]

A colleague of Ferrier's at the National Hospital at Queen Square, Horsley engaged in similar work. His famous surgical case of the excision of an epileptogenic scar from the frontal cortex, under the encouragement of another colleague, Hughlings Jackson, applied this understanding of cortical localization. In the laboratory, elucidation of cortical localization with particular respect to the motor system became a sustained and coherent effort. He and Charles Beevor mapped the central cortex of the orangutan brain using a millimeter grid and electrical stimulation. By the early 1890s, Horsley had gained wide recognition for his continued work on the localization of function in the mammalian brain using electrical stimulation. Shortly after the turn of the century, Horsley's investigative attention become focused on the cerebellum, work undertaken with another colleague, Robert Clarke. As with all investigation of brain function being done at that time, the results and conclusions derived from either ablative or stimulatory techniques were confounded by imprecise localization or collateral damage. As they specifically sought to understand the efferent connectivity of the cerebellum, these methodologic limitations were fully recognized. In their 1908 paper in *Brain* (see Figure 5.1) they write:

> *On making a general survey of the subject before us, and considering the most promising methods of research, we were confronted with the following preliminary difficulty: The nuclei of the cerebellum in monkeys, dogs, and cats are small, deeply situated, and not very accessible for excitation experiments, while it is evident that to get results of any value by Marchi's degeneration method, lesions must be precisely limited to the nuclei, or, if possible, to parts of them, and that such lesions must not only be accurately localized but also produced without noteworthy injury to other structures, for we had already observed the difficulties and confusion which such complications have introduced into the discussion of the subject. An essential preliminary, therefore, to further progress was to find some method which would satisfy these conditions, viz., a means of producing lesions of the cerebellar nuclei which should be accurate in position, limited to any desired degree in extent, and involving as little injury as possible to other structures....*

THE STRUCTURE AND FUNCTIONS OF THE CEREBELLUM EXAMINED BY A NEW METHOD.

By SIR VICTOR HORSLEY, F.R.S., F.R.C.S.,

AND

R. H. CLARKE, M.A., M.B.

(*From the Laboratory of Pathological Chemistry, University College, London.*)

PART I.—METHODS.

I.—INTRODUCTION.
II.—RECTILINEAR TOPOGRAPHY.
III.—STEREOTAXIC INSTRUMENT.
IV.—ELECTROLYSIS.
V.—EXCITATION.

I.—INTRODUCTION.

THE methods and experiments described in the following pages are the direct outcome of an investigation into the anatomical relations of the cortex of the cerebellum to its nuclei and peduncles, and to the rest of the brain and spinal cord. An account of that research was published in *Brain* in the spring of 1905.

When we began that work (1903) the view had been gaining ground that there was no direct path from the cortex of the cerebellum to the peduncles or to the spinal cord, and had been advanced by distinguished observers, especially Ferrier and Turner, Risien Russell and Thomas, who expressed themselves more or less definitely in favour of this opinion, and supported it with observations furnished by their own experiments. But although the evidence adduced established a strong probability we did not consider that it amounted to proof, as the conclusions were founded on lesions involving both cortex and nuclei, or complicated with injuries to other parts. Nor were all the conclusions of the authors absolutely definite. Marchi originally described a direct descending path in the spinal cord derived from the cerebellum. Ramon y Cajal spoke of this tract in a rather ambiguous way, leaving the reader in some doubt whether he recognized the tract himself or was merely quoting Marchi by calling it the *via descendente*. Ramon y Cajal also described some fibres passing from the cerebellar cortex to the superior peduncle.

Figure 5.1 The first page of Horsley and Clarke's 1908 paper, published in *Brain*, in which they describe their stereotaxic instrument.[3]

While the scientific direction and motivation may have come from Horsley, there is no question that the technological development of the Horsley–Clarke stereotactic instrument was the work of Clarke. Faced with the challenge enumerated above and fortuitously with enforced time off recovering from pneumonia, Clarke devised an instrument whose localization was based not on cortical features but rectilinear

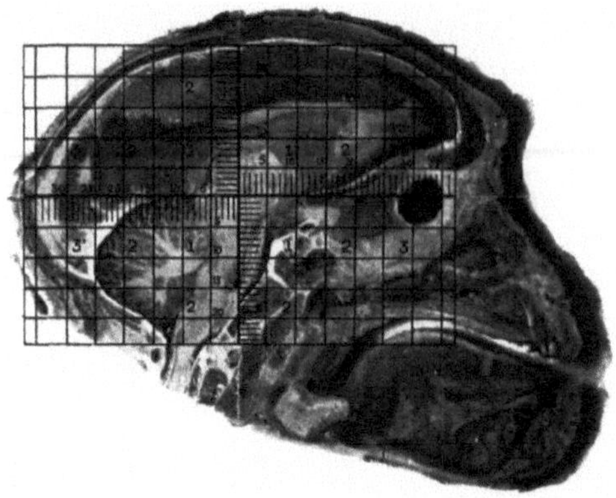

Figure 5.2 A sagittal section of a (frozen) Rhesus head, with an overlying glass plate with millimeter markings, from Horsley and Clarke's 1908 paper. Anatomic structure visualized in such a section could be surgically localized in a living Rhesus, using the structure's spatial coordinates and the stereotaxic instrument.

coordinates by which structures identified on a previously prepared brain slice (see Figure 5.2) could be located on a mechanical apparatus, or frame, attached to the animal's cranium. The apparatus itself was machined at Palmer and Company by James Swift (see Figures 5.3 and 5.4). Again from the 1908 paper:

> *We find, however, that a practicable and, on the whole, satisfactory solution of this problem may be attained by dividing the cranium into eight segments, by three section planes at right angles to each other, e.g., sagittal, horizontal, and frontal. As a result of these sections, each segment presents the three internal surfaces of a cube, and every point in it can be identified by rectilinear measurements from those surfaces or section planes, i.e., from their internal boundaries. By this means the irregular curved surface, which corresponds to the three outer sides of the cube, is not involved in any way and needs no further consideration. In short, instead of employing the usual method of endeavoring to project the detailed structure of the interior of the encephalon on to the surface of the head, we measure the position of the deep parts of the brain by their relation to three section planes. The advantage of applying this principle, which, so far as we know, is new in the topography of the brain, to the localization of structural detail for the identification and record of lesions, and for the mechanical direction of an insulated needle for excitation or electrolysis, is obvious, and its utility will become more evident as we proceed to consider the details of its practical application.*

Complementing this accurate and precise localization device was a needle or probe by which electrolytic lessoning could be performed in a minimally invasive and reproducible fashion. This technical development is reported in the same publication, and while clearly an important part of their stereotactic methodology, it is independent of the fundamental localization principles that are the primary concern of this review. Interestingly, historians have cited the antecedent work of either Carl Ludwig or C. Dittmar in his laboratory, using a two-coordinate system, as reported by Tigerstedt in his *Handbuch* of 1912.[2,4,5]

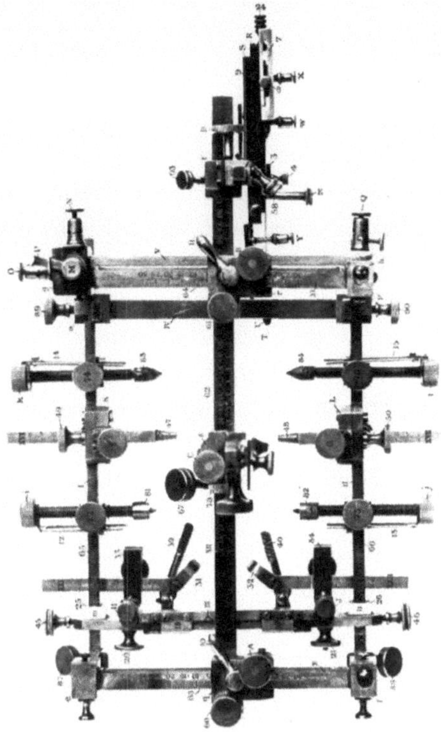

Figure 5.3 Clarke's stereotaxic instrument, by which a needle could be directed in three planes.

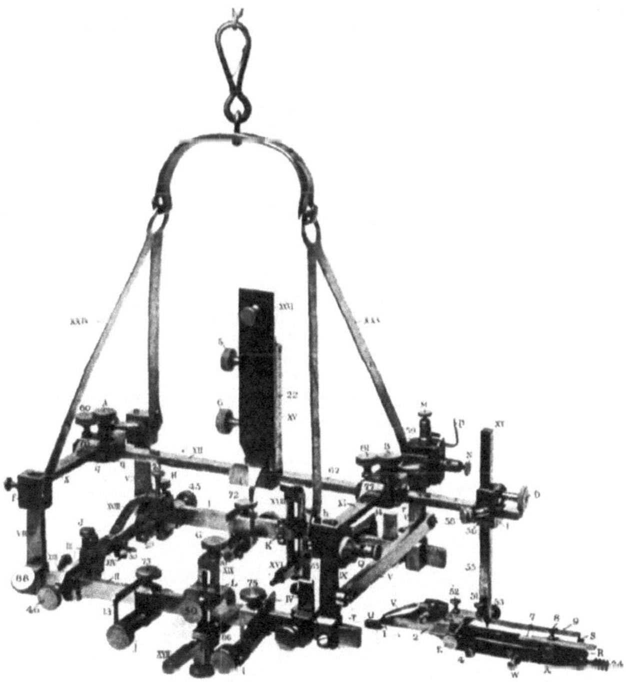

Figure 5.4 Clarke's stereotaxic instrument (another perspective).

Horsley's prominent stature and accomplishments in medicine far outshone those of Clarke, and the latter's contribution to stereotaxy has been underappreciated. Throughout his subsequent career, Horsley referred to the stereotactic frame described in 1906 and 1908 as Clarke's instrument, as did the young, visiting American neurosurgeon who worked with that instrument, Ernest Sachs.

The investigative work using the frame proposed by Horsley and Clarke in their preliminary publications was never brought to conclusion. Sachs, spending several years in London under Horsley, did go on to carry out and publish rigorous investigations of the thalamus and optic apparatus, using the stereotactic frame.[6] His detailed, 92-page 1909 paper in *Brain* titled "On the structure and functional relations of the optic thalamus" reports the results of acute stimulation and lesioning studies in monkey and cat, including investigation of the pre- and post-central cortex, the anterior, medial, lateral, and central thalamus, Forel's field, and red nucleus. A testimony to Horsley's generosity was his removal of his name from that important work, telling the young Sachs that if he were co-author, others would think him and not Sachs the principal investigator and author (personal communication from Ernest Sachs, Jr). Before leaving London, Sachs had a second Horsley–Clarke frame produced by Palmer and Company and brought it with him on his return to the United States and Washington University in St. Louis, MO.

Further investigative work with the original frame was also carried out by Aubrey Mussen, a visiting physician from Montreal, who around the same time as Sachs carried out studies investigating stimulation of the hypoglossal nucleus in cat and monkey.[7] Further stereotactic work was performed by Kinnier Wilson, who in collaboration with Horsley evaluated lesioning and its effects in 25 monkeys. F. J. F. Barrington, another investigator using the frame, studied the effects of lesioning on micturition in cats.[8]

Clarke, whose close friendship with Horsley early on waned after their early collaboration, went on to refine and improve his original apparatus, adding the capability of delivering insertion needles not just rectilinearly but along angled trajectories as well. A further refined third instrument was later produced for Mussen, who was then working at Johns Hopkins Hospital and who would use the apparatus for investigation of the cerebellum and red nucleus in cats and monkeys.[9] Not coincidentally, a detailed description of his later instrument as well as a stereotaxic atlas of the monkey brain produced together with E. E. Henderson was published as a special volume of Johns Hopkins Hospital Reports in 1920.[10]

As a laboratory tool for neuroanatomic and physiologic investigation, the Horsley–Clarke instrument became a major part of the fundamental investigative infrastructure for over a decade at Northwestern's Institute of Neurology. The institute's director, S.W. Ranson, had had an exact copy of the Horsley–Clarke instrument made by a machinist at the medical school, subsequent versions and simplifications of which followed at both Northwestern and the University of Chicago. Krieg relates well the enthusiasm and productivity of this center of activity that included the work of Ranson, Magoun, Fisher, Ingram, Kabat, Teague, Pitts, Clark, Gerard, Sugar, Masserman, Haertig, and others. Using stimulation, recording, and lesioning, these two centers in Chicago extensively investigated the mesencephalon, corticofugal pathways, and hypothalamus.[11]

FIRST HUMAN STEREOTAXIS

It was not until the late 1940s that the generally accepted first report of stereotaxic technique in humans was published. The advantages of stereotaxic technique over conventional craniotomy for creation of small, accurate subcortical lesions were readily apparent to Ernest Spiegel who had been using a modification of Horsley and Clarke's stereotaxic apparatus in his laboratory for more than 15 years.[12,13] The historic 1947 report of Spiegel and Wycis in *Science*—"Stereotaxic Apparatus for Operations on the Human Brain"—is a paragon of brevity, taking up barely more than a single page, including two figures and two references.[14] The rationale for the adaptation of the previous laboratory methodology is explicit: "Exposure of subcortical areas usually necessitates rather extensive operations. It seemed desirable, therefore, to adapt the stereotaxic technic for use on the human brain. This technique, employed thus far for animal experimentation only (1), permits one to insert a wire or a cannula accurately into a desired subcortical area with minimal injury to the cerebral cortex or the white matter."[14]

Their "stereoencephalotome," or stereotactic frame, consisted of a base ring that was firmly secured to the patient's head using a plaster cast. Attached to the base ring was a superstructure on which a needle carrier could be moved along millimeter-scaled sagittal and frontal planes as well as rotated within those planes for angled trajectories (see Figures 5.5 and 5.6). The targeting process utilized radiographs with and without air in the ventricular system, from which the coordinates of the lesion and entry point were determined. After trepanation of the skull, a wire or cannula was advanced through the dura via the needle holder, and thermocoagulation, aspiration, or instillation of fluid performed. The clinical application alluded to in this paper is psychosurgery, and studying the patients with H. Freed, lesions were made within the medial thalamus, "...in order to reduce the emotional reactivity by a procedure much less drastic than frontal lobotomy."[14] Foreshadowing much of the field, he and Wycis go on to note their investigations of the technique across a broad spectrum of conditions including pain, movement disorders, trigeminal neuralgia, cysts, and tumor.

The advantage of adapting the stereotactic approach advanced by Horsley and Clarke over conventional craniotomy was apparent to Spiegel, "...when he proposed to Wycis to replace prefrontal lobotomy, with its extensive gross tissue damage and many undesirable side effects, by circumscribed lesions of the dorsomedial

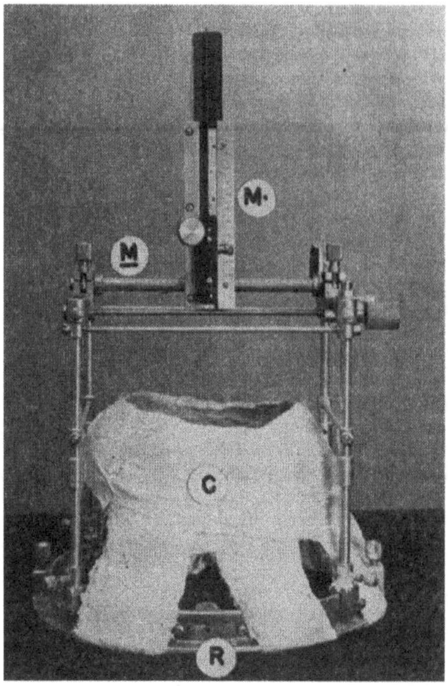

Figure 5.5 Side view of Spiegel and Wycis's stereotaxic apparatus. *From Reference [14], reprinted with permission from AAAS.*

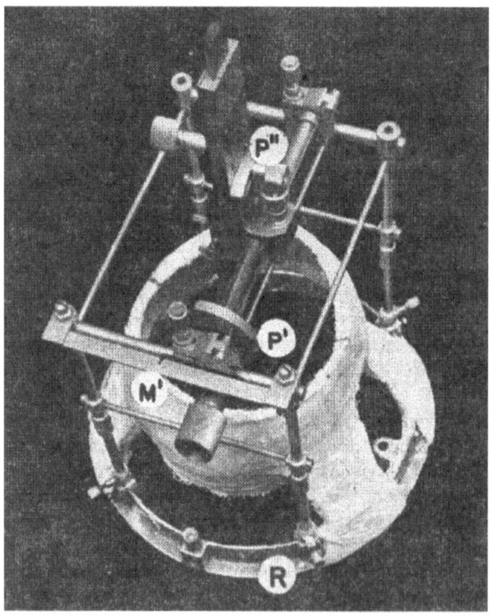

Figure 5.6 Spiegel and Wycis's stereotaxic apparatus, seen from above. *From Reference [14], reprinted with permission from AAAS.*

nucleus... ."[15] The long delay in bringing stereotactic methodology from the laboratory was attributed by Spiegel, in his later writing, to the significantly greater variation in brain morphology in humans compared to cats or nonhuman primates. Whereas greater uniformity of size and shape in those species enabled targeting brain structures in reference to external landmarks on the head, human variability required the greater accuracy enabled by referencing to nearby periventricular structures within the brain, visualization of which was made possible by later imaging techniques of roentgenological ventriculography using air, Pantopaque or Conray.[15] "The cardinal difference between Horsley and Clarke's procedure in experimental animals and the use of guided instruments in the human subcortex is the *necessity,* in patients*, to use intracerebral reference points or lines*. These are chiefly structures around the third ventricle (the pineal gland, anterior and posterior commissure, intercommissural line)... ."[15] Spiegel and Wycis went on to refine their instrument through at least four subsequent models.

The idea of using Horsley and Clarke's laboratory method for clinical surgery had been appreciated as early as 1918, when Mussen had had a similar apparatus built in London with specifications designed for the human head. The instrument being similarly dependent on a relatively constant relationship between external landmarks and potential intracerebral targets, there is no evidence that the apparatus was ever used clinically.[9]

Generally underrecognized today, like Mussen's frame, but in this instance actually clinically implemented is the apparatus and methodology of Kirschner, who reported in 1933 the use of a guidance system for lesioning of the gasserian ganglion, a structure less variable in its relationship to external features.[16] Not a coordinate-based methodology relying on internal landmarks, it was not recognized in the stereotactic tradition.

Lastly, in any discussion of stereotactic precedence, the work of Zernov must be acknowledged. He described in 1889 an encephalometer by which surface areas of the human brain could be localized. Its applications were apparently at least in part anthropomorphic, focusing on group differences in cranial indices. Like Horsley's earliest investigation of cortical function relative to surface features, the methodology predated the concept of spatial localization and co-registration based upon a coordinate geometry, and for this reason the general consensus has not accorded this device priority as the first stereotactic system.[17–19]

Following Spiegel and Wycis's report, with its fundamental modification of Horsley and Clarke's technique using intracerebral reference points rather than skull features, there was enormous activity in the development of stereotactic instrumentation for human use and in subsequent clinical application. Lars Leksell returned from visiting Philadelphia and designed a stereotactic frame system for both radiofrequency and Cobalt 60 gamma radiation lesioning. Distinct in its targeting principle, the arc-centered system placed the target at the center of a rotating arc attached to which was a moveable electrode carrier, enabling any number of trajectories to the preset target. Conceptually, a nonconstrained trajectory enabled ease of use for RF lesioning as well as the utilization of multiple highly collimated radiation beams for radiosurgery. A clever solution to parallax—the spiral diagram—eliminated the need for long focal beams with the localizing pneumoencephalography.[20,21] Although this frame has evolved through multiple models, the underlying arc-centered principle remains unchanged.

In Freiburg, Reichert together with colleagues including Hassler, Wolff, and Mundinger, developed a stereotactic frame that also uses a probe carrier on an arc attached to a base ring.[22,23] Although a trajectory in this system is thus defined by spherical coordinates, its original implementation utilized a phantom base on which the Cartesian coordinates derived from intraoperative radiography were used to place a target. The arc-based superstructure would then be placed upon the phantom and the trajectory mechanically adjusted and defined by touching its probe to the phantom target; the apparatus would then be transferred to the patient's base platform. Borrowing from one of Clarke's later instruments, such a work flow eliminated the need to actually calculate spherical coordinates; with the later advent of computational resources in the operating room, direct calculation of settings eliminated the need of the phantom.[24] A number of other arc-based systems also evolved, one of the more prominent of which was that of Schaltenbrand in Würzburg, an elaborate system utilizing five-meter teleradiography to minimize parallax and movement of the head, rather than the frame, to the targeted trajectory.[25] The Todd–Wells system, an arc-based system which became popular in the United States, similarly adjusted the head position rather than the frame.[26] The Brown–Roberts–Wells, of which more will be discussed later, shared this heritage of arc-based trajectory definition.[27]

Others in Spiegel's footsteps approached the requirements of co-registration and probe or electrode guidance differently. Also working in the late 1940s, Talairach developed a stereotactic instrument with grids consisting of side-plates with finely spaced openings. Using anterior–posterior and lateral teleradiography, specific holes in the grids could be selected for electrode placement, with depth determined on the orthogonal film. Further planning of such electrode placement could be performed with the aid of superimposed angiography or pneumoencephalography, performed with the frame in place. Such a system is particularly well-suited for efficient placement of a large number of electrodes, as in the stereotactic investigation of medically intractable seizure epilepsy.[28]

Another type of stereotactic instrumentation was more minimalist in approach, attaching to the skull at or near the trajectory entry site. Bailey, who had earlier reported a modified Horsley–Clarke apparatus for nonhuman investigation,[29] presented a burr hole adaptation in 1951.[30] Austin and Lee reported a subsequent ball-and-socket adaptation in 1958,[31] and Rand,[32] Ward,[33] Walker,[34] and Ray[35] each described their similar devices over the next decade. Targeting with these devices was generally achieved by iteratively aligning a probe with a desired trajectory, rather than by reference to an external coordinate system. Although larger than smaller burr hole devices, Cooper's chemopallidectomy and chemothalamectomy "guide" functioned similarly.[36] Interestingly, some of the most current iterations of image-guidance biopsy and electrode placement systems, including those for use with intraoperative MRI, are essentially adaptations of this type of apparatus now coupled with computational resources that enable co-registration to real-time images.

The large number of individual stereotactic frames developed in the years immediately following the report of Spiegel and Wycis precludes individual description of all. Thirty-five years later, Spiegel looked back at "over three dozen models of stereoencephalotomes" that had been developed since his and Wycis's report.[15] Gildenberg notes that more than 40 stereotactic apparatuses were designed in the late 1940s and the 1950s, citing activity around the world led by, in addition to those already mentioned, Narabayashi, Guiot, Gillingham, Laitinen, Toivakka, Siegfried, C. and G. Bertrand, Velasco-Suarez, Escobedo, Obrador, Kandel, Bechtereva, among others.[19] Gildenberg goes on to cite the first meeting of the International Society for Research in Stereoencephalotomy in Philadelphia in 1966 and Spiegel's estimate that more than 37,000 stereotactic procedures had been performed by 1969.[19] The major clinical indications for all these procedures were movement disorders (where open surgery had already demonstrated a beneficial effect but the same disadvantage alluded to by Spiegel and Wycis with respect to psychosurgery application), psychiatric conditions, and intractable pain. With the discovery of L-dopa and its wide utilization in Parkinson's disease and with psychosurgical applications falling rapidly into societal disfavor, the following decade would be very different.

DEVELOPMENT OF COMPUTATIONAL STEREOTAXY AND IMAGE GUIDANCE

The revolutionary advances in neuroimaging in the 1970s, and in particular computed tomography (CT) imaging, had a tremendous influence in generating new interest, excitement and growth in stereotactic neurosurgery. First, CT scanning of the head brought a diagnostic tool with unprecedented sensitivity and spatial resolution for intracranial tumor. Where functional disorders had previously represented the primary indication for stereotactic surgery, tumor for which safe and efficient biopsy was needed created a new, high-volume indication. All of the technology's advantages over open craniotomy for the treatment of psychiatric or movement disorders were appreciated in tumor biopsy. Second, CT scanning was readily adaptable into stereotactic methodology. Whereas special knowledge, familiarity and interest in functional neurophysiology, stereotactic atlases, and esoteric techniques to account for such challenges as radiologic parallax limited the number of stereotactic neurosurgeons to a small subset, CT scanning and the availability of computational resources to facilitate co-registration and targeting opened the field at a time when there was a need.

Adapting stereotactic frames and methodology to CT imaging was a major activity of the late 1970s and early 1980s. Some changes in instrumentation were required, to some extent for imaging compatibility, but especially for co-registration. Accurate, digital tomographic planes afforded new opportunities to localize imaged structures with respect to the coordinate space of a stereotactic frame. Whether direct or indirect targeting was required, the generation of stereotactic coordinates could be accomplished quickly as imaged structure within the brain could be related to the imaged structure of the frame. Relatively simple attachments to a frame and minor constraints on head positioning within a gantry provided immediate solutions to the registration requirement.[37–39] One stereotactic system, the Brown–Roberts–Wells, took advantage of available computational resources and was designed specifically for such tomographic scanning.[27] Translation of these co-registration strategies to other new neuroimaging modalities like MRI was relatively straightforward.

Computational resources, however, changed the field in other, equally profound ways. Sheldon and Jacques had earlier described a tulip-shaped retractor system that integrated with a stereotactic frame for guidance during open craniotomy for tumor,[40] but it was Kelly and his team who recognized early on the capability provided by computational resources readily accessible to the operating surgeon. As neuroradiological advances led the technologies providing multiple sources of preoperative information about an individual patient, computers in the operating room environment now had the ability to co-register these various imaging techniques into one multimodality data set. Co-registering this preoperative set of one or multiple CT or MRI scans with the surgical field could then be accomplished using a stereotactic frame attached to the patient. In Kelly's implementation, this frame-based system included a larger than usual arc to which an operating microscope, rather than an electrode carrier, was attached. Together, this system enabled co-registration of the surgeon's field of view through the operating microscope with appropriately segmented and reformatted preoperative radiologic studies.[41–43]

In all of these implementations, computers facilitated the registration of preoperative imaging with a stereotactic frame. Digitization of the operating room, however, enabled more. Stereotactic frames functioned by serving in three essential roles: they defined a coordinate space in the operative field, they facilitated co-registration between imaging and that surgical field, and they made that co-registration is useful to the surgeon by guidance of some kind of effector, usually in the form of an electrode, a biopsy needle, or a highly collimated radiation beam. The computer could serve in all of these roles and in the process render the intermediary of a stereotactic frame unnecessary.

Definition of a coordinate space, or specification of a unique address system for a given space, can be accomplished with a carefully machined, mechanical apparatus, such as a stereotactic frame, but computers spawned a host of nonmechanical digitizers. The original Dartmouth system adapted a two-dimensional sonic digitizer used in the shoe industry to cut leather to function in three-dimensional space: a broad bandwidth acoustic click emitted by a spark gap held at a fiducial point on the patient's scalp was detected by an array of microphones suspended above the surgical field, and from the respective times of flight, the location of that spark gap relative to the microphone array could be geometrically determined.[44] In this manner, a large portion of operating room space became stereotactic.

The second function of co-registration was now accomplished using a set of a small number of what would be called fiducial points, identifiable on both the preoperative CT scan (small glass beads were taped over each point) and in the operating room using the sonic digitizer. The computer could now derive the spatial transformation relating the three-dimensional coordinate space of the imaging study to the three-dimensional coordinate space of the operating room.[44]

Lastly, the information of interest made possible by that co-registration could be provided to the surgeon in several ways. Tracking a handheld spark gap positioned on the not-yet prepped scalp, the spark gap's location could be depicted on the appropriate corresponding CT slice. The primary device tracked by the system was the operating microscope. Attached to the microscope was a bracket holding three additional spark gaps, and the same microphone array could now track the position and orientation of the microscope's optical axis. Its focal point could be displayed on a corresponding CT or MRI image plane but the primary display mode of the system was a heads-up display accomplished by the attachment of a small cathode ray tube to a beam-splitter on the microscope. In this manner, the surgeon could see the outline of a segmented object of interest, such as a tumor, superposed on the surgical field in the correct position, orientation, and scale.[44] (A similar heads-up display was implemented in Kelly's system the same year.)

In eliminating the stereotactic frame, the computer-based system implemented what was variously called "frameless stereotaxy," "neuronavigation," or "image-guidance." Conceptually such a system employed transformational stereotaxy to rapidly move in either direction between coordinate spaces: whereas traditional stereotaxy independently selected a target in atlas or image space and set the apparatus to reach a dependent corresponding target in the surgical field, matrix transformations enabled the surgeon to make an object of interest in either imaging space or the surgical field the independent location and determine its corresponding location in the other. Further, whereas traditional stereotaxy necessarily focused on one target or at most a small number, computational power in a computer calculated the correspondence between every point in the respective coordinate spaces. Lastly, freed from the encumbrances of a mechanical frame, such a system was fully compatible with conventional craniotomies, be they for tumor, epilepsy, or other indication. Not requiring a calvarium to attach a stereotactic frame to, its application was no longer constrained to the intracranial compartment.[45] Not least, the technology required little stereotactic expertise to operate. The net effect of all of this was to open stereotactic technique to a wide field of surgeries and surgeons.

Independently, Watanabe in Tokyo developed a frameless system based upon an articulated arm whose known arm lengths and joint angulations made it a digitizer, as well as paired-point registration and a graphics display.[46] Barnett and colleagues in Cleveland[47,48] and Bucholz and colleagues[49,50] in St. Louis early on pioneered key developments; Heilbrun in Salt Lake City championed the concept and hosted an early symposium for enthusiasts. Kelly's team adapted an electromagnetic digitizer.[51] Multidisciplinary teams including Maciunas, Adler, Drake, Guthrie, McDermott, Zamorano, Kalfas, Alexander, Black, Olivier, Schlondorf, Reinhardt, and a great many others joined the effort. Together all helped troubleshoot and refine the nascent next stage of stereotaxy and shepherd the technology through industry to wider dissemination.

Conceptualization of stereotaxy into component tasks of coordinate space definition, co-registration, and effector was and remains useful in consideration and exploration of subsequent refinements and developments. Any number of digitizing technologies could be substituted for the sonic digitizer, and were. Watanabe's own development of a frameless system in Tokyo used an articulated arm both to locate fiducial points for registration and to locate points of interest during surgery.[46] Articulated arms were subsequently utilized in several early commercial systems. Alternatively, optical camera systems tracking either light-emitting diodes or small reflector spheres proved robust and accurate, though constrained by a line-of-sight requirement, and these rapidly proliferated and remain popular today. Optical systems with more sophisticated algorithms, as employed in machine vision and not dependent on tracked LEDs or reflectance spheres, have great potential but are still largely in development. Electromagnetic-field-based digitizing systems, popular in early artificial- and augmented-reality systems used in simulators and in games, do not have line-of-sight requirements (they are, however, vulnerable to perturbation by ferromagnetic material), enabling direct tracking within the body and have been implemented in open[51] as well as endovascular procedures.

Co-registration of coordinate spaces can be accomplished using paired points[44,52] (independent of the digitizing technology being used), and that strategy is the most commonly employed in commercial surgical navigation systems today. It is also possible to compute the transformation matching one surface contour to another, an approach first popularized for imaging studies by Pelizzari and Chen.[53–55] Camera (or fluoroscopic) systems can match image perspectives or projections, a utility applicable to machine-vision approaches, radiosurgery, or registration quality control.[56] With image datasets such as ultrasound, CT, or MRI, matching of three-dimensional information can be similarly employed.[57–59] Combinations of all of these, either at one time or iteratively, can and have been utilized. Accuracy, speed, efficiency, and user-friendliness are essential attributes; fully automated schemes are especially appealing.

The limitation of using preoperatively acquired spatial information during a procedure when over time co-registration is degraded by tissue displacement, deformation, or removal has been recognized since the earliest days of image guidance[60–63] and recognition of the inherent risks this represents has generated strategies to minimize its effects; awareness alone is invaluable. One solution, of course, has been to simply re-image the patient during surgery, using the new dataset as an updated map. Intraoperative CT, MRI, and ultrasound systems have all been commercially developed for this.[64–66] Alternatively, since the surgical field is subject to known physical laws, including those of gravity and loss of buoyancy, a number of groups have generated predictive models to update imaging, with varying degrees of success.[67–69] Hybrid strategies utilizing more easily acquired, sparse but limited imaging data (such as from intraoperative ultrasound, or optical information from the operating microscope's field of view) to inform models that can then deform less sparse imaging data sets such as MRI have been developed and offer efficiency and potential cost-effectiveness.[70–72]

The third component of the effector classically was an inserted probe, and with image-guidance systems, graphic displays and heads-up displays are widespread. As digitized space and co-registration represent part of an infrastructure, however, a diverse set of technologies for diagnostic and therapeutic interventions becomes possible. As early as Leksell's first system, direction of highly collimated radiation beams was envisioned and implemented.[73] Tracking of a handheld articulated arm or a wired (or wireless) pointer was universal from the earliest days of frameless systems. Co-registered laser instrumentation for tissue ablation was incorporated by Kelly.[41] Most enticingly, co-registered robotic instrumentation is increasingly being integrated into the surgical field today.[74–79] (See Chapter 19 for a detailed discussion of robotics in neurosurgery.)

All of these technologies are topics in themselves, but the most fundamental stereotactic principles enable and facilitate their integration into surgical intervention. These principles provide a conceptual framework in which new implementations and applications can be developed. From a small and often esoteric discipline first in the laboratory and then in a handful of highly subspecialized operating rooms, stereotaxy has now entered the mainstream of surgical practice.

REFERENCES

1. Critchley M. *The Divine Banquet of the Brain*. New York: Raven Press; 1979.
2. Marshall L, Magoun H. The Horsley-Clarke stereotaxic instrument: the beginning. *Carrier*. 1990;27:15.
3. Horsley V, Clarke RH. The structure and functions of the cerebellum examined by a new method. *Brain*. 1908;31:45, 47, 54, 64, 67, 50.
4. Gildenberg P. Stereotactic surgery: present and past. In: Heilbrun MP, ed. *Stereotactic Neurosurgery*. Baltimore: Williams & Wilkins; 1988:1–15.
5. Dittmar C. Ueber die Lage des sogenannten Gefaesszentrums in der Medulla oblongata. *Bersaechs Ges Wiss Leipzig Math Phys)*. 1873;25:449–469.
6. Sachs E. On the structure and functional relations of the optic thalamus. *Brain*. 1909;32:95–186.
7. Mussen AT. Note on the movements of the tongue from stimulation of the twelfth nucleus, root, and nerve. *Brain*. 1909;32:206–208.
8. Marshall LH, Magoun HW. The Horsley-Clarke stereotaxic instrument: the first three instruments. *Kopf Carrier*. 1991;28:1–5.
9. Picard C, Olivier A, Bertrand G. The first human stereotaxic apparatus. The contribution of Aubrey Mussen to the field of stereotaxis. *J Neurosurg*. 1983;59:673–676.
10. Clarke RH. Investigation of the central nervous system. I. Methods and instruments. Johns Hopkins Hospital Reports 1920; Special volume.
11. Krieg WJS. *Stereotaxy*. Evanston, IL: Brain Books; 1975.
12. Spiegel EA, Wycis HT. *Stereoencephalotomy (Thalamotomy and Related Procedures). Part I: Methods and Stereotaxic Atlas of the Human Brain*. New York: Grune & Stratton; 1952.
13. Spiegel EA, Miller HR. Stand for aseptic operations with Horsley-Clarke's stereotaxic instrument. *J Lab Clin Med*. 1941;26:1657–1658.
14. Spiegel EA, Wycis HT, Marks M, Lee AJ. Stereotaxic apparatus for operations on the human brain. *Science*. 1947;106:349–350.
15. Spiegel EA. *Guided Brain Operations: Methodological and Clinical Developments in Stereotactic Surgery [and] Contributions to the Physiology of Subcortical Structures*. Basel, Switzerland: S.Karger; 1982.
16. Kirschner M. Die Punktionstechnik und Elektrokoagulation des Ganglion Gasseri. Uber "gezielte" Operationen. *Arch Klin Chir*. 1933;176:581–620.
17. Zernov D. Encephalometer: device for estimation of parts of the brain in human. *Proc Soc Physicomed Moscow Univ*. 1889;2:70–80.
18. Altukhov N. Encephalometric investigations of hte brain relative to the sex, age, and skull indexes. Moscow, Russia; 1891.
19. Gildenberg P. The history of stereoactic and functional neurosurgery. In: Gildenberg P, Tasker R, eds. *Textbook of Stereotactic and Functional Neurosurgery*. New York: McGraw-Hill; 1998:5–19.
20. Leksell L. A stereotaxic apparatus for intracerebral surgery. *Acta Chir Scand*. 1949;99:229–233.
21. Leksell L. *Stereotaxis and Radiosurgery*. Springfield, IL: Charles C Thomas; 1971.
22. Riechertt T, Wolff M. Die Entwicklung und klinische Bedeutung der gezielten Hirnoperationen. *Medsche Klin*. 1951;46:609–611.
23. Riechert T. *Stereotatic Brain Operations. Methods, Clinical Aspects, Indications*. Bern, Switzerland: Hans Huber; 1980.
24. Mundinger F, Birg W, Klar M. Computer assisted stereotactic brain operations. *Appl Neurophysiol*. 1978;41:169–182.
25. Riechert T, Spuler H. Instrumentation of stereotaxy. In: Schaltenbrand G, Walker AE, eds. *Stereotaxy of the Human Brain: Anatomical, Physiological and Clinical Applications*. 2nd ed. Stuttgart: Georg Thieme Verlag; 1982:350–363.
26. Todd E. *Todd-Wells Manual of Stereotaxic Procedures*. Randollph, MA: Codman & Shurtleff; 1967.
27. Brown RA, Roberts TS, Osborn AG. Stereotaxic frame and computer software for CT-directed neurosurgical localization. *Invest Radiol*. 1980;15:308–312.
28. Talairach J, Szikla G. *Atlas of Stereotaxic Anatomy of the Telencephalon*. Paris, France: Masson & C; 1967.

29. Bailey P, Davis EW. A modifiation of the Horsley-Clarke apparatus. *J Neuropathol Exp Neurol.* 1943;2:99–101.
30. Bailey P, Stein SU. *A stereotaxic instrument for use on the human brain. Studies in Medicine.* Springfield, IL: C C Thomas; 1951:40–49.
31. Austin B, Lee A. A plastic ball and socket type of stereotaxic detector. *J Neurosurg.* 1958;15:264–268.
32. Rand RW. A stereotaxic instrument for pallidothalamectomy. *J Neurosurg.* 1961;18:258–260.
33. Ward A. Symposium in Parkinson's disease. *J Neurosurg.* 1966;24:466.
34. Walker E. Symposium on Parkinson's disease. *J Neurosurg.* 1966;24:466.
35. Ray CD. A new simplified stereotaxic instrument. *Confin Neurol.* 1967;29.
36. Cooper IS. Surgical treatment of Parkinsonism. *Br Med J.* 1961;1:1248–1249.
37. Bergstrom M, Greitz T. Stereotaxic computed tomography. *AJR Am J Roentgenol.* 1976;127:167–170.
38. Lunsford LD, Leksell L, Jernberg B. Probe holder for stereotactic surgery in the CT scanner. A technical note. *Acta Neurochir.* 1983;69:297–304.
39. Boethius J, Bergstrom M, Greitz T. Stereotaxic computerized tomography with a GE 8800 scanner. *J Neurosurg.* 1980;52:794–800.
40. Shelden CH, McCann G, Jacques S, et al. Development of a computerized microstereotaxic method for localization and removal of minute CNS lesions under direct 3-D vision. Technical report. *J Neurosurg.* 1980;52:21–27.
41. Kelly PJ, Alker Jr. GJ. A method for stereotactic laser microsurgery in the treatment of deep-seated CNS neoplasms. *Appl Neurophysiol.* 1980;43:210–215.
42. Kelly PJ, Alker Jr. GJ, Goerss S. Computer-assisted stereotactic microsurgery for the treatment of intracranial neoplasms. *Neurosurgery.* 1982;10:324–331.
43. Kall BA, Kelly PJ, Goerss SJ. The computer as a stereotactic surgical instrument. *Neurol Res.* 1986;8:201–208.
44. Roberts DW, Strohbehn JW, Hatch JF, Murray W, Kettenberger H. A frameless stereotaxic integration of computerized tomographic imaging and the operating microscope. *J Neurosurg.* 1986;65:545–549.
45. Brodwater BK, Roberts DW, Nakajima T, Friets EM, Strohbehn JW. Extracranial application of the frameless stereotactic operating microscope: experience with lumbar spine. *Neurosurgery.* 1993;32:209–213:discussion 13.
46. Watanabe E, Watanabe T, Manaka S, Mayanagi Y, Takakura K. Three-dimensional digitizer (neuronavigator): new equipment for computed tomography-guided stereotaxic surgery. *Surg Neurol.* 1987;27:543–547.
47. Barnett GH, Kormos DW, Steiner CP, Weisenberger J. Intraoperative localization using an armless, frameless stereotactic wand. Technical note. *J Neurosurg.* 1993;78:510–514.
48. Barnett GH, Kormos DW, Steiner CP, Weisenberger J. Use of a frameless, armless stereotactic wand for brain tumor localization with two-dimensional and three-dimensional neuroimaging. *Neurosurgery.* 1993;33:674–678.
49. Smith KR, Frank KJ, Bucholz RD. The NeuroStation—a highly accurate, minimally invasive solution to frameless stereotactic neurosurgery. *Comput Med Imaging Graph.* 1994;18:247–256.
50. Bucholz RD, Ho HW, Rubin JP. Variables affecting the accuracy of stereotactic localization using computerized tomography. *J Neurosurg.* 1993;79:667–673.
51. Goerss SJ, Kelly PJ, Kall B, Stiving S. A stereotactic magnetic field digitizer. *Stereotact Funct Neurosurg.* 1994;63:89–92.
52. Heilbrun MP, Koehler S, MacDonald P, Siemionow V, Peters W. Preliminary experience using an optimized three-point transformation algorithm for spatial registration of coordinate systems: a method of noninvasive localization using frame-based stereotactic guidance systems. *J Neurosurg.* 1994;81:676–682.
53. Friets EM, Strohbehn JW, Roberts DW. Curvature-based nonfiducial registration for the Frameless Stereotactic Operating Microscope. *IEEE Trans Biomed Eng.* 1995;42:867–878.

54. Henderson JM, Smith KR, Bucholz RD. An accurate and ergonomic method of registration for image-guided neurosurgery. *Comput Med Imaging Graph*. 1994;18:273–277.
55. Pelizzari CA, Chen GT, Spelbring DR, Weichselbaum RR, Chen CT. Accurate three-dimensional registration of CT, PET, and/or MR images of the brain. *J Comput Assist Tomogr*. 1989;13:20–26.
56. Heilbrun MP, McDonald P, Wiker C, Koehler S, Peters W. Stereotactic localization and guidance using a machine vision technique. *Stereotact Funct Neurosurg*. 1992;58:94–98.
57. Ji S, Wu Z, Hartov A, Roberts DW, Paulsen KD. Mutual-information-based image to patient re-registration using intraoperative ultrasound in image-guided neurosurgery. *Med Phys*. 2008;35:4612–4624.
58. Ji S, Roberts DW, Hartov A, Paulsen KD. Combining multiple true 3D ultrasound image volumes through re-registration and rasterization. *Medical Image Computing and Computer-Assisted Intervention: MICCAI International Conference on Medical Image Computing and Computer-Assisted Intervention* 2009; London: 12:795–802.
59. Ji S, Roberts DW, Hartov A, Paulsen KD. Intraoperative patient registration using volumetric true 3D ultrasound without fiducials. *Med Phys*. 2012;39:7540–7552.
60. Kelly P. Stereotactic excision of brain tumors. In: Thomas DGT, ed. *Stereotactic and Image Directed Surgery of Brain Tumours*. Edinburgh: Churchill Livingstone; 1993:89–109.
61. Roberts DW, Hartov A, Kennedy FE, Miga MI, Paulsen KD. Intraoperative brain shift and deformation: a quantitative analysis of cortical displacement in 28 cases. *Neurosurgery*. 1998;43:749–758: discussion 58–60.
62. Hill DL, Maurer Jr. CR, Maciunas RJ, Barwise JA, Fitzpatrick JM, Wang MY. Measurement of intraoperative brain surface deformation under a craniotomy. *Neurosurgery*. 1998;43:514–526: discussion 27–8.
63. Dorward NL, Alberti O, Velani B, et al. Postimaging brain distortion: magnitude, correlates, and impact on neuronavigation. *J Neurosurg*. 1998;88:656–662.
64. Black PM, Moriarty T, Alexander 3rd E, et al. Development and implementation of intraoperative magnetic resonance imaging and its neurosurgical applications. *Neurosurgery*. 1997;41:831–842: discussion 42–5.
65. Schulder M, Jacobs A, Carmel PW. Intraoperative MRI and adjuvant radiosurgery. *Stereotact Funct Neurosurg*. 2001;76:151–158.
66. Nimsky C, Ganslandt O, von Keller B, Fahlbusch R. Preliminary experience in glioma surgery with intraoperative high-field MRI. *Acta Neurochir Suppl*. 2003;88:21–29.
67. Paulsen KD, Miga MI, Kennedy FE, Hoopes PJ, Hartov A, Roberts DW. A computational model for tracking subsurface tissue deformation during stereotactic neurosurgery. *IEEE Trans Biomed Eng*. 1999;46:213–225.
68. Miga MI, Paulsen KD, Lemery JM, et al. Model-updated image guidance: initial clinical experiences with gravity-induced brain deformation. *IEEE Trans Med Imaging*. 1999;18:866–874.
69. Dumpuri P, Thompson RC, Dawant BM, Cao A, Miga MI. An atlas-based method to compensate for brain shift: preliminary results. *Med Image Anal*. 2007;11:128–145.
70. Roberts DW, Miga MI, Hartov A, et al. Intraoperatively updated neuroimaging using brain modeling and sparse data. *Neurosurgery*. 1999;45:1199–1206:discussion 206–7.
71. Lunn KE, Paulsen KD, Liu F, Kennedy FE, Hartov A, Roberts DW. Data-guided brain deformation modeling: evaluation of a 3-D adjoint inversion method in porcine studies. *IEEE Trans Biomed Eng*. 2006;53:1893–1900.
72. Fan X, Ji S, Hartov A, Roberts DW, Paulsen KD. Stereovision to MR image registration for cortical surface displacement mapping to enhance image-guided neurosurgery. *Med Phys*. 2014;41:102302.
73. Leksell L. The stereotaxic method and radiosurgery of the brain. *Acta Chir Scand*. 1951; 102:316–319.
74. Young RF. Application of robotics to stereotactic neurosurgery. *Neurol Res*. 1987;9:123–128.

75. Koyama H, Uchida T, Funakubo H, Takakura K, Fankhauser H. Development of a new microsurgical robot for stereotactic neurosurgery. *Stereotact Funct Neurosurg*. 1990;54–55:462–467.
76. Benabid AL, Hoffmann D, Lavallee S, et al. Is there any future for robots in neurosurgery? *Adv Tech Stand Neurosurg*. 1991;18:3–45.
77. Drake JM, Joy M, Goldenberg A, Kreindler D. Computer- and robot-assisted resection of thalamic astrocytomas in children. *Neurosurgery*. 1991;29:27–33.
78. Haase J. Neurosurgical tools and techniques—modern image-guided surgery. *Neurol Med Chir*. 1998;38(Suppl):303–307.
79. Sutherland GR, Latour I, Greer AD, Fielding T, Feil G, Newhook P. An image-guided magnetic resonance-compatible surgical robot. *Neurosurgery*. 2008;62:286–292:discussion 92–3.

CHAPTER 6

Role of Computers and Image Processing in Image-Guided Brain Tumor Surgery

Nabgha Farhat, Tina Kapur, and Ron Kikinis
Surgical Planning Lab, Department of Radiology, Brigham and Women's Hospital, Harvard Medical School, Boston, MA, USA

INTRODUCTION

Computers and image processing pervade every stage of image-guided brain tumor surgery, from preoperative planning to intraoperative guidance to postoperative follow-up. An understanding of the role of computers and computing in neurosurgery requires familiarity with a few core concepts, including image segmentation, tractography, registration, and navigation. Segmentation involves identifying meaningful structures and regions within an image, such as normal anatomical tissue or pathology. Further subdivision of segmented tissues may also be called parcellation. Tractography refers to the modeling of white matter tracts from diffusion magnetic resonance (MR) images. Registration or image alignment describes the process of identifying a transformation that aligns one data set with another, fusing complementary information sources for planning and intraprocedural guidance. Navigation allows the movement of surgical instruments during the procedure to be shown relative to diagnostic images.

These functionalities are integrated in image-guided neurosurgery in order to provide the surgeon with intuitive and interactive visualizations that inform the planning, progress, and follow-up of the procedure (Figure 6.1). We begin this chapter with a quick overview of the role that these computing and navigation techniques play in neurosurgery, and then discuss each of the techniques in detail. The implementation of these techniques in 3D Slicer (a free software application for medical image analysis) will also be presented, to provide the reader with concrete descriptions and images of how these processes are carried out.

Medical image computing and navigation in the neurosurgery workflow

Preoperative phase

In brain tumor resection, the preoperative planning phase includes multimodal imaging and extraction of relevant information from those images. Patient- and procedure-specific decisions are made in choosing the imaging modalities, based on the objective of

A. Golby (Ed): Image-Guided Neurosurgery
DOI: http://dx.doi.org/10.1016/B978-0-12-800870-6.00006-6

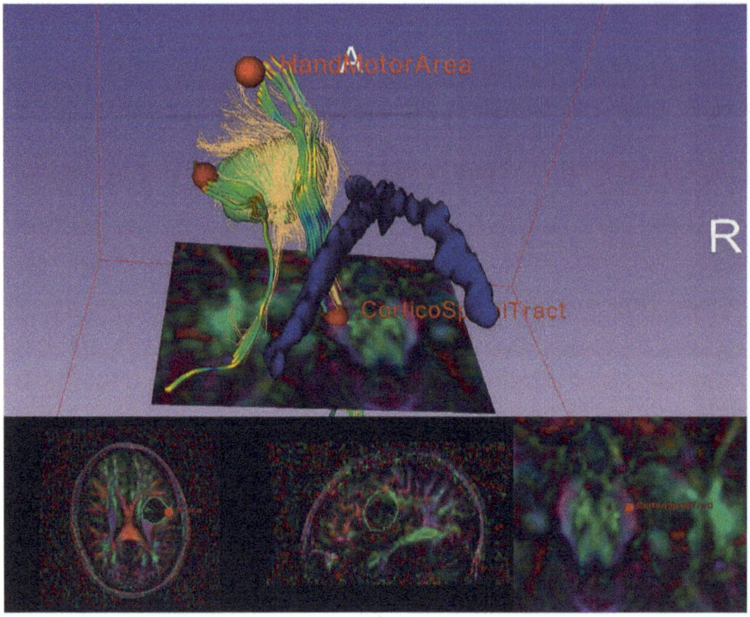

Figure 6.1 Computational techniques in neurosurgery. Image segmentation, tractography, registration, and navigation are used in neurosurgery to provide visualizations that inform the surgeon's decisions. In this example, the tumor to be resected (green) is shown along with neighboring white matter tracts and models of the ventricles (dark blue). Probes (shown as red spheres) have been positioned in Broca's area and in the corticospinal tract area in the mid-brain. Streamline tractography originating in those probes reveal the corticospinal tract all the way to the hand motor region as well as the arcuate fasciculus connecting the Broca and Wernicke areas. Integrated displays such as these aid in preoperative planning, intraoperative guidance, and postoperative assessment.

the intervention. Through a combination of modalities that provide anatomical information, such as computed tomography (CT), magnetic resonance imaging (MRI), and diffusion tensor imaging (DTI), and modalities that provide functional information, such as functional MRI (fMRI), positron emission tomography (PET), or transcranial magnetic stimulation (TMS), the care-providing team creates a plan that minimizes functional damage while maximizing tumor removal. To this end, image processing is critical in translating the raw imaging data to useful information that informs the surgeon's decisions. From modalities that provide anatomical details, a technique called segmentation is frequently used to identify important structures contained in the image (such as the lesion or tumor to be resected, ventricles, or eloquent cortices). Tractography is used to model white matter tracts from diffusion MRI, providing information about fiber bundles that may be at risk of damage during the procedure. Registration enables the fusion of these multimodal images, aligning them in one space so the procedure is planned with both anatomical and functional information considered.

Intraoperative phase

As the neurosurgical intervention takes place, the surgeon uses and updates information from the planning phase to inform the decisions made during the procedure. Planning data, including segmentations, tractography, and any other image processing, should be accessible during the course of the procedure (rather than being mentally recollected by the physician). To this end, image-to-image registration between pre- and intraoperative images is critical to enable the useful translation of planning data. A relationship between these images and the physical space of the patient (image-to-patient registration) is also necessary to ensure that the surgeon is making physical movements that are in agreement with the planned surgical approach. Navigation systems are based on image-to-patient registration and can track the location of the surgeon's tools and display them on pre- or intraoperatively acquired images, creating a direct and useful application of imaging during the procedure.

Postoperative phase

Assessment of the outcome of neurosurgical interventions is critical to monitor complications, plan follow-up treatment, and ensure minimal tumor recurrence. The image processing techniques described previously for surgical planning, including segmentation and tractography, become important once again in the postoperative setting to measure tumor volume and white matter reorganization after surgery.

This chapter will discuss segmentation, tractography, registration, and navigation both broadly and within this described context of image-guided brain tumor resection. The implementation of these functionalities in 3D Slicer will also be presented, giving the reader concrete tools, images, and resources to apply these computation techniques.

3D slicer

3D Slicer (Slicer)[1] is a free, open source software application for medical image analysis that is actively used in neurosurgical planning, guidance, and follow-up. Started as a master's thesis in 1995, it is developed today mostly by professional engineers in close collaboration with algorithm developers and application domain scientists. Initially envisioned as a neurosurgical guidance, visualization, and analysis system, Slicer has evolved into an integrated platform over the last decade.[2]

Slicer is a multiplatform application that is distributed under a BSD-style license agreement, which is a liberal license for computer software that permits free distribution of derivative software for academic and commercial use. Hence, image analysis tools developed within 3D Slicer can be adopted directly by the industry collaborators and ensure wide distribution.

User and developer community mailing lists have been active for the last 10 years, and currently include over a thousand subscribers, demonstrating sources of Slicer

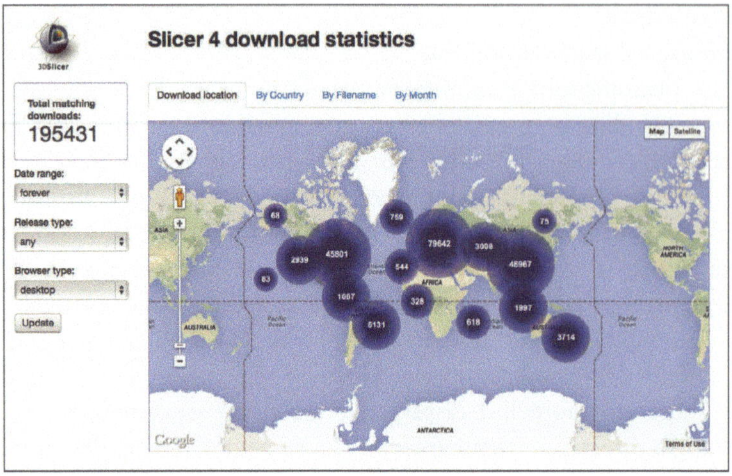

Figure 6.2 3D Slicer download statistics. Although 3D Slicer began as a Master's thesis in Boston, it is now used and developed worldwide. This map shows the cumulative number of Slicer downloads worldwide since its release.

expertise that go well beyond the community of the core Slicer developers. Its development has grown into a community effort, as numerous groups and individual users around the world are continuously improving it (Figure 6.2).

Slicer is designed with the intention of carrying out interactive, subject-specific work, which means that surgeons can afford to make case-by-case judgments as they use its various tools. This flexibility is critical to the complex neurosurgery environment. 3D Slicer provides functionality—algorithm implementations and device interfaces—that support segmentation, tractography, registration, and navigation for neurosurgery.

SEGMENTATION

Segmentation refers to a process in which each voxel of a medical image is labeled by tissue type, and like voxels are grouped together to identify various structures. These segmented structures can then be displayed as three-dimensional (3D) models. Without the use of segmentation, the task of using two-dimensional (2D) image slices to visualize spatial relationships between different anatomical features is a difficult procedure carried out exclusively in the surgeon's mind.[3] Segmentation provides the surgeon with 3D information that enables a more intuitive understanding of these spatial relationships. Segmentations generated from different images can also be displayed together, allowing for the integration of information from more than one source. When segmented tissue is further subdivided, especially prior to structural connectivity analysis through DTI, this process may also be referred to as parcellation.

Manual segmentation describes the process by which a user delineates structure boundaries by inspection, without any algorithmic support. Automated segmentation makes use of image features to separate structures automatically, with no (or minimal) user input required. Often, an interactive segmentation protocol that involves both manual and automated segmentation is used in practice. Automated segmentation algorithms can be based on a number of different features of the image volume to be segmented. Some algorithms, for example, rely on voxel intensity alone, setting threshold values to segment regions with specific grayscale intensities, which can be especially useful for CT images. Other algorithms employ user-identified seed points for region growing, extracting all voxels connected to the initial seed with the same intensity value. Still others apply graph theory to segmentation, by creating a graph partitioning problem in which each voxel of the image is a node and nearby nodes are connected with edges. Of these edges, a subset is then chosen for segmentation. Interested readers are referred to[4] for a more detailed review of various algorithms and their applications.

Some fully automatic algorithms for tumor segmentation exist, but are not practical at this point as they require over one hour of processing time on a high power computer.[5] Although these algorithms are not practical today, they hold promise for the future.

Application of segmentation to brain tumor resection

Irrespective of what type of segmentation algorithm is used, the identification of structures via segmentation enables the surgeon to make a preoperative assessment of the spatial orientation of a lesion with respect to anatomical landmarks. Lesion location relative to areas such as the precentral, postcentral, left superior temporal, and left inferior front gyri enable the surgeon to anticipate the challenges associated with reaching that lesion.[6] Linear distances and volumes can also be measured to features such as blood vessels, nerves, the brain stem, and the thalamus to improve procedure planning.[7] In cerebrovascular neurosurgery, segmentation from MR angiograms can inform decisions about what type of intervention to pursue (surgery, radiosurgery, etc.).[8] Intraoperatively, segmentation also simply makes the lesion more visible on imaging which is important during surgery when the surgeon's attention is divided between the surgical field and the imaging.

Postoperatively, segmentation enables accurate understanding of changes in various brain structures, including the extent of tumor resection. Residual tumor after resection is a strong predictor of tumor recurrence, especially in the case of gliomas.[9] A comprehensive postoperative assessment may therefore enable the care-providing team to make decisions about future treatment options and plans.

Segmentation is limited, however, by the extent to which the physician can identify boundaries of the structure to be segmented. Abnormal tissue appearance in MRI may not fully indicate the extent of tumor infiltration in, for example, diffuse low-grade gliomas.[10] Edema associated with the tumor is also difficult to segment, because it often appears with a gradient boundary rather than a clear edge. Despite this limitation, segmentation can certainly be used to provide the surgeon with spatial orientation, inform preoperative planning for the approach to the lesion, and maximize lesion resection.

Segmentation in 3D slicer

A variety of segmentation tools are implemented within 3D Slicer, enabling users to segment images manually, automatically, or through a combination of different techniques. This flexibility allows the user to select methods that are most effective for the specific imaging modality and structure to be segmented.

The *Editor* module of 3D Slicer is used to create and manipulate label maps. A label map is a 3D scalar volume in which each voxel contains a number indicating the tissue type at that location. These label maps are associated with a Color Node that maps the numbers at each voxel into colors and text strings. In the case of tumor segmentation, for example, each voxel that is part of the tumor would contain the same number in the label map, while nontumor voxels would contain different numbers. This label map could then be used to generate a 3D model of the tumor by modeling only those voxels that contain the label value for tumor tissue.

The effects featured in the *Editor* module range in complexity and take advantage of different characteristics of the source volume. Depending on the type of structure to be segmented, different effects or combinations of effects might be applied. The *Paintbrush* effect, for example, simply paints over the image with a brush of adjustable size. The *Threshold* effect applies a label value to all voxels of the image volume that fall within a user-specified range of intensity values. Using similar threshold principles, the *Level Tracing* effect defines a closed outline in which all the voxels comprising the outline have the same intensity as the clicked voxel. Semi-automatic segmentation tools are also integrated in the *Editor* module, including the *GrowCut* effect, which is a competitive region-growing algorithm that uses an input sample segmentation to create a full segmentation of the desired structure[11] (Figure 6.3).

Once an initial segmentation has been achieved, other *Editor* effects are used for post-processing. The *Erode* and *Dilate* effects shrink or grow the existing segmentation, which can be used to refine its edges. Different connectivity methods can also be applied, such as the *Save Island* effect to remove all segmentation that is not directly connected to the clicked structure. Similarly, the *Identify Island* assigns a different label value to every spatially disjoint structure in the segmentation. An informatics capability has also been integrated into the *Editor* to allow it to split, edit, and merge multiple labels.

In carrying out segmentation, these effects might be combined to obtain the best results. The user may choose, for instance, to use the *GrowCut* effect to segment a lesion

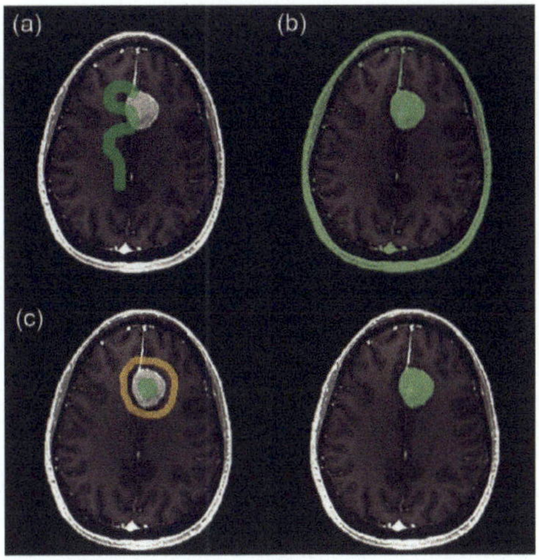

Figure 6.3 ***Examples of* Editor *tools.*** A variety of tools are implemented in the *Editor* module to aid in image segmentation. (a) The *Paintbrush* tool: the user has assigned the green label to a curved line of voxels; (b) The *Threshold* tool: the user has assigned the green label to all voxels above a certain brightness value; (c) User input for GrowCut effect (left) and output segmentation (right). The user assigns the green label to seed the region to be segmented and orange to seed the region to be excluded from the segmentation. The GrowCut effect is then implemented to automatically complete the segmentation.

and then manually refine the edges with the *Paintbrush* effect. Interested readers are referred to the Slicer User Manual[12] for a more complete description of *Editor* effects, as well as other modules (outside of the *Editor*) that can be used for segmentation.

After a label map containing segmentations for all relevant structures has been created, 3D surface models can be generated in 3D Slicer (Figure 6.4). The *Make Model* effect in the *Editor* module can be used for a quick translation of the label volume to 3D models. In this case, each label value is modeled individually by applying the effect to every value in isolation. For increased control of modeling parameters, and to generate multiple models at once, the user can apply the *Model Maker* module instead. With these tools, the surgeon can generate a 3D reconstruction that is most useful for planning purposes.

Segmented structures can then be used as a starting point for further analysis. For example, the *Label Statistics* module provides a simple tool to measure the volume, pixel count, or image intensity of a segmented region in an image volume. By applying this module to both pre- and postoperative images, a quantified comparison of these metrics is possible. Output values can include volume (in cubic millimeters or centimeters), number of pixels designated by that label value, and the minimum, maximum, and standard deviation of intensity values that are part of that label value. Thus, the clinician can analyze changes in tumor volume and enhancement as well as overall brain anatomy, which might inform planning for future therapy.

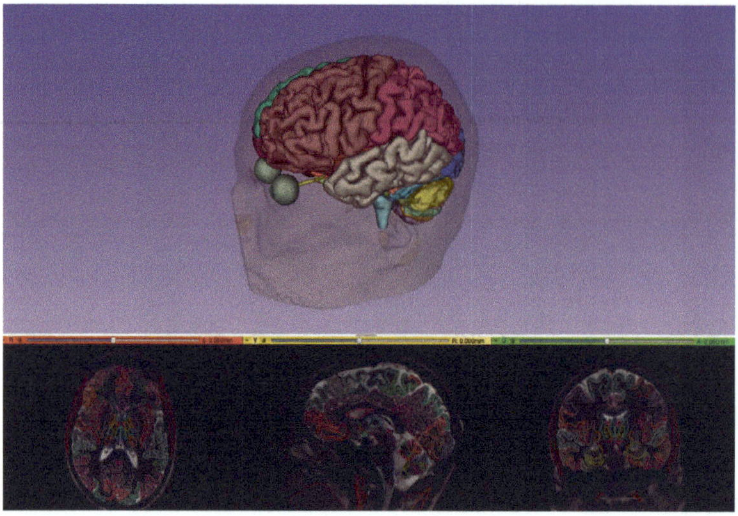

Figure 6.4 Models made from image segmentation. 3D surface models (top) were generated after segmentation of various brain structures in 3D Slicer. Label maps delineating these structures are also shown in axial, sagittal, and coronal slice views (bottom).

TRACTOGRAPHY

Diffusion MRI provides information about the directionality of water diffusion. By applying different pulsed magnetic field gradients, the tendency of water to diffuse in one particular direction (anisotropy) or in all directions equally (isotropy) is measured. At each voxel in the acquired image, parameters describe the overall direction and rate of water diffusion. This information is then used to estimate the existence and trajectories of white matter tracts, using a technique called diffusion tensor imaging (DTI), under the assumption that a heavy preference for diffusion in a particular direction indicates the existence of myelinated axons.[13] This modeling of fiber bundles is known as tractography.

Several FDA approved products have the ability to perform tractography. This includes iPlan® FiberTracking software (Brainlab AG, Feldkirchen, Germany) and StealthDTI™ software (Medtronic, Minneapolis, MN, USA).

Application of tractography to brain tumor resection

Preoperatively, tractography enables the surgeon to model large white matter tracts in the vicinity of the surgical site and plan an approach to the lesion that minimizes damage to these regions. Modeling of the pyramidal tract, corpus callosum, and optic radiation has been shown to improve neurosurgical planning as well as postoperative assessment in cases of cerebral tumors.[14] For patients with tumors in functionally eloquent areas, visualization of white matter tracts is especially useful for maximizing tumor resection while minimizing neurological deficit.[15,16] Interactive exploration of white matter connectivity is possible in some systems and allows the surgeon to get a better understanding of the organization of the white matter anatomy in a region of interest.[17]

Navigation and instrument tracking tools are also making intraoperative tractography possible, a development that is discussed further in the "Navigation in 3D Slicer" section of this chapter. These tools enable the surgeon to probe white matter architecture during the intervention itself, by using the location of a tracked tool touching the tissue in question to "seed" tracts. Postoperatively, tractography may be used as a follow-up tool to assess changes in white matter structure after surgery.[18]

Tractography may be limited, however, by the location of the tracts to be modeled and by pathology. Deflection of subcortical white matter is difficult to model with DTI tractography, and changes caused by edema or necrosis may alter diffusion calculations. Certain pathologies that result in reduced anisotropy, such as multiple sclerosis, also make accurate tractography difficult. Nevertheless, tractography has the potential to provide quantitative and qualitative information that enables the surgeon to make decisions based on an understanding of the spatial organization of white matter bundles.[19] (Please see Chapter 4 for more detailed discussion of diffusion MRI and imaging of white matter anatomy.)

Tractography in 3D slicer

There are a number of 3D Slicer modules dedicated to fiber tractography. Users can import diffusion images, carry out interactive tractography on loaded images, and analyze scalar diffusion measurements. The *DTI Import* and *DTI Export* modules provide flexibility by allowing the user to load and save diffusion images, including NifTi (a widely used format proposed by the Neuroimaging Informatics Technology Initiative [NifTi] Data Format Working Group).[20]

Once images are loaded in 3D Slicer, the *Tractography Interactive Seeding* module can be used to visualize fiber bundles based on dynamic fiducial points or model vertices. Fiducial point seeding displays all white matter tracts that pass through a user-specified seed point. The size of the seeded region, measured in millimeter radius around the chosen point, is adjustable. The location of the fiducial point can be interactively moved, and as movements are made, the module updates displayed tracts so that only those emanating from the current position of the fiducial point are displayed. In preoperative planning, the surgeon may choose to move the point along the boundary of a lesion, for example, to visualize the tracts passing nearby.

When models of important structures have been made (by segmentation, as described above), white matter bundles can also be traced from the vertices of those models (Figure 6.5). In this case, an isocontour shell that is a user-chosen distance inward or outward from the model is used for tractography seeding. To reduce latency

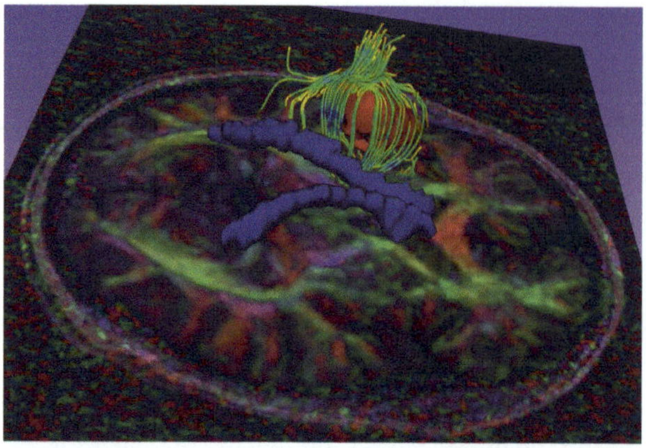

Figure 6.5 Tractography interactive seeding. White matter tracts that surround a segmented tumor are modeled and displayed alongside the tumor (red) and ventricles (dark blue). An axial slice view of the diffusion MRI is also visualized for anatomical context.

of the modeling, the user defines a maximum number of seed points to be used for tractography within this shell. As the distance from the model is changed, the module updates the displayed tracts so that only those that fall within the specified distance from the model are visible. The user may choose to model fiber tracts near the lesion, ventricles, functional areas defined by fMRI or other modality, or other important structures using this module. As a planning tool, this enables the surgeon to choose a path to the lesion that minimizes functional damage.

Alternatively, the *Tractography Label Map Seeding* and *Fiber Bundle Label Select* modules allow the user to interactively visualize white matter tracts based on label maps. All the tracts emanating from a particular label map can be visualized, and the map can be manipulated in the *Editor* module to change the seeding region. If a full brain mask is used for seeding, full brain tractography is performed. Label maps with multiple label values delineating different structures can be resolved on the *Fiber Bundle Label Select* module, so that only the tracts emanating from specific label values are displayed while other label values are excluded. This gives the surgeon flexibility with regards to which white matter tracts to display, facilitating the visualization of only those tracts that are relevant to surgical planning.

A developing module in 3D Slicer, called *UKF Tractography*, implements an unscented Kalman filter in white matter bundle modeling.[21] This module allows the user to apply a two-tensor algorithm (as opposed to the one-tensor algorithm applied in all the previously described modules), which better represents crossing and branching fibers.[22] It also enables representation of a larger number of fiber tracts overall (Figure 6.6). With these improvements, tractography is becoming increasingly applicable to surgical planning as it gains accuracy and robustness.

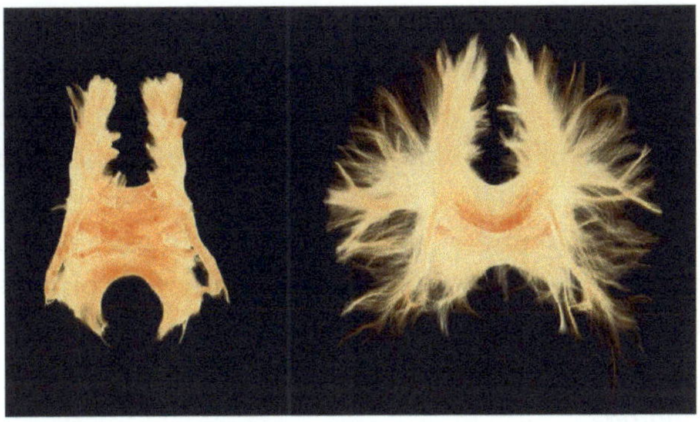

Figure 6.6 UKF tractography. Fibers of the corpus callosum are modeled using a single tensor algorithm (left) and two-tensor algorithm (right). Note the improved resolution and larger number of overall tracts made possible through two-tensor tractography.

REGISTRATION

The goal of registration is to create a spatial alignment between two different images of the same underlying tissue or patient. One of the two images (called the reference or source image) is fixed in space while the other image (called the target or sensed image) is transformed such that it aligns with the first. The process of calculating and applying this spatial transform defines the process of registration.

The transform that maps the target image to the reference image may take a number of different forms. One of the simplest is a rigid transform, in which three translations and three rotations are possible, giving a total of six degrees of freedom. Transforms that additionally allow three directions of scaling and three directions of skews, for a total of 12 degrees of freedom, are called affine. Nonrigid transforms allow for further deformation and even greater degrees of freedom. Often, organ deformation is very complex and goes beyond the scope of affine transformations alone. However, affine transformations can be useful for overcoming errors, such as geometric distortion in MRI, introduced by the scanning procedure.[23]

Registration algorithms may also be classified based on the type of information that is used from the images to create the transformation. Some algorithms make use of geometrical features in the images (such as points, lines, and surfaces) and rely on these to create a transform that maps the features of one image to the features of the other image. Other registration algorithms are based on image intensity values, creating a transform that optimizes some type of voxel similarity measure. These similarity measures might be level sets that contain all voxels with the same intensity value, or with intensity values that fall within a given range. Still other algorithms use boundaries, or surfaces, that tend to be easily distinguishable in medical images (such as the skin surface). Interested readers are referred to[23] for a more exhaustive discussion of registration techniques and algorithms.

Application of registration to brain tumor resection

Registration enables effective preoperative planning via the alignment of multimodal images that provide different types of information. In this case, the subject is the same but the method and time of acquisition varies. Complementary structural and functional images are aligned in order to provide the surgeon with comprehensive information that informs the planning of the intervention. In cases in which the tumor is adjacent to functional areas, accurate registration is critical to plan a surgical approach that minimizes functional damage.

Intraoperative imaging

In some settings, the intraoperative configuration of the surgical site can be assessed with updated imaging. Intraoperative imaging ranges from relatively simple and non workflow disrupting ultrasound, to resource intense and workflow disrupting MRI. (For a detailed discussion of intraoperative imaging, please see Chapter 7.) As the intervention takes place, intraoperative MR imaging is carried out to identify any residual tumor tissue, which enables more effective tumor removal.[24] These intraoperative MR images are often lower quality than preoperative scans and currently do not provide the same extent of functional information. In order to overcome this, functional images are aligned to high-quality preoperative MR images, and these preoperative images are nonrigidly registered with intraoperative images.[25] This registration makes planning information, such as the location of white matter bundles relative to tumor tissue, more readily applicable to the intraprocedural setting. The surgeon can therefore make an assessment about the progress of the surgery and decide how to proceed during the course of the intervention.

Registration between images and the physical space of the patient bed also becomes critical during the intervention, so that the movement of surgical tools during the procedure can be understood in the context of acquired images. This image-to-physical-space registration problem is further discussed in the "Navigation" section of this chapter.

Registration in 3D slicer

As with segmentation, the registration capabilities of 3D Slicer range in the amount of user input and interaction required. From strictly manual transformations to fully automatic registration, the range of modules enables the user to solve different types of registration problems (multimodal, across different time points, etc.). The coordinate system infrastructure provides the foundation upon which registration is accomplished. While the data are kept in their original voxel space, all objects in Slicer share the same Cartesian anatomical coordinate system called RAS (right anterior superior).

To perform a manual rigid registration, the *Transforms* module can be used in order to create and manipulate a transformation matrix. Three directions of translation and three directions of rotation can be controlled with sliding bars, and the resulting transform can be applied to the target image to roughly register it with the reference image. Interactive rigid registration is often useful as an initialization step for more sophisticated (and less robust) higher-order algorithms.

Automated, intensity-based registration can be carried out with the *BRAINSfit* module, which uses mutual information to rapidly and robustly register image volumes (Figure 6.7). The user can choose to perform rigid, affine, or nonrigid registration, depending on image sources and similarity. A B-spline transformation (which is a piecewise polynomial transformation) can be carried out with a grid of controlled size, allowing for 27 or greater degrees of freedom in the transformation. The module can be adjusted to output either a new image volume containing the registered image, or a transformation matrix that can then be further manipulated manually in the *Transformations* module.

Pairs of manually selected fiducial points can also be used to perform image alignment in the *Fiducial Registration* module, which uses the iterative closest point (ICP) algorithm as a way to register point clouds. The user designates a set of points as "fixed landmarks" on the reference image and another set of anatomically corresponding "moving landmarks" on the target image. After the user chooses between a simple translation and rigid registration, the module aligns the two sets of points and

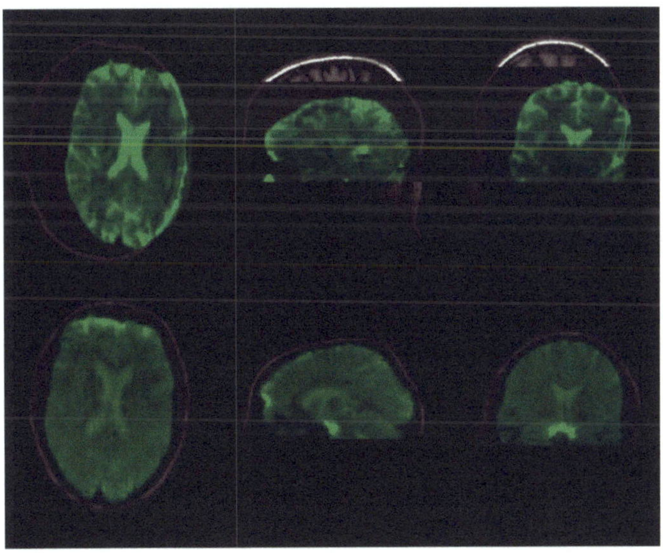

Figure 6.7 Registration of structural and functional imaging. T1-weighted MRI and fMRI are shown prior to registration (top row) and after registration (bottom row). Alignment of the two images was achieved via the BRAINSfit module.

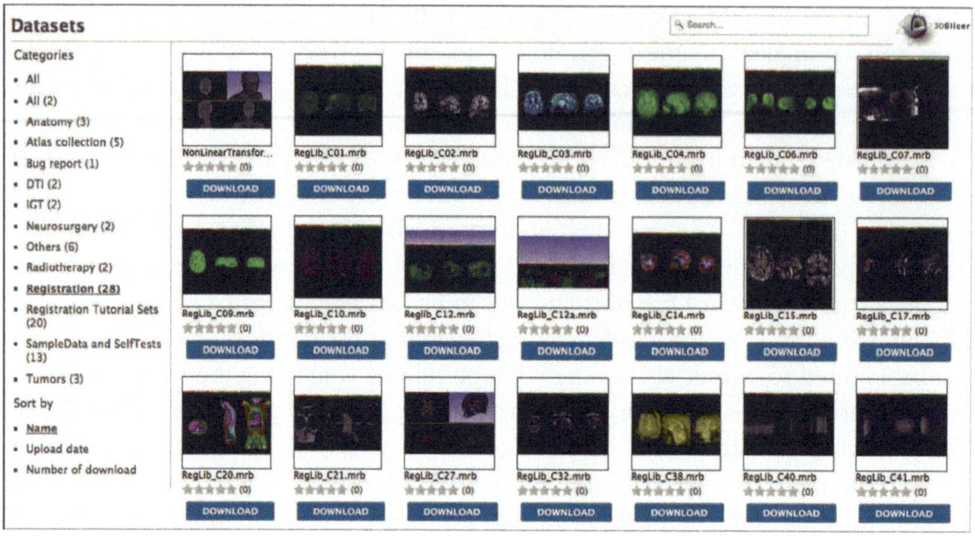

Figure 6.8 Image Registration Data Repository. The Image Registration Data Repository provides users with a number of sample cases and example workflows for registration. This resource highlights different registration problems (inter-modality, inter-patient, etc.) and solutions.

gives a transformation matrix as an output. This particular type of registration is most useful when the user can clearly identify corresponding landmarks on both the reference and target images.

The *ACPC Transform* module enables orientation of the brain volume with respect to the line connecting the anterior and posterior commissures as well as the mid-sagittal plane. A set of two seed points is used to delineate the line connecting the anterior and posterior commissures, which is then aligned with the anterior–posterior line of the image volume. A second set of points is used to delineate the mid-sagittal plane (the plane that separates the two brain hemispheres), which is aligned with the anterior–superior plane of the image volume. The surgeon may choose to apply the ACPC transform in order to normalize the orientation of all acquired images.

Slicer provides a library of examples, with both a sample data repository and matched solution workflows[26] (Figure 6.8). Interested readers are also referred to the Slicer Usual Manual[12] for a detailed description of the different registration implementations in 3D Slicer.

These registration implementations allow the surgeon to align preoperative images with each other to enable accurate planning of the intervention. They are also used to align preoperative and intraoperative images, making a translation of planning information to the procedure more feasible.

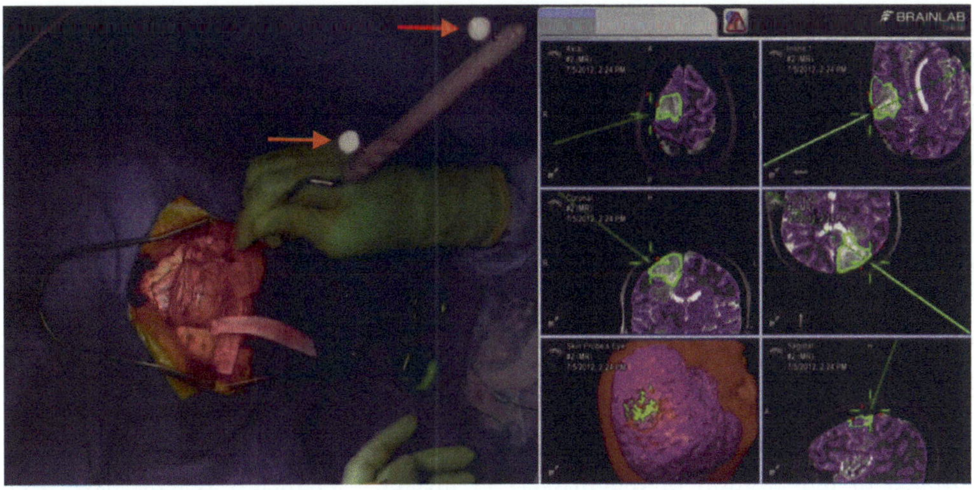

Figure 6.9 Passive optical tracking. Through the BrainLab Cranial Navigation Application (Brainlab AG, Feldkirchen, Germany), reflective spheres (red arrows) are used as markers to achieve passive optical tracking during brain tumor resection. The surgeon's tracked tool is visualized on images acquired during preoperative planning (the tool is displayed as a green line). Segmentation of the tumor and white matter is also visualized on these images, giving the surgeon information about the position of the tool with respect to both lesion and functional areas.

NAVIGATION

The use of a navigation system during neurosurgery creates a direct relationship between pre- or intraoperative images and the physical space of the patient. Ideally, a navigation system should track the position of the surgeon's tools and intuitively display their position on image volumes. The surgeon can then make movements in physical space that improve lesion resection or minimize functional damage, as informed by the previously described image processing (e.g., tumor segmentation and tractography of nearby fiber bundles).

Several different techniques have evolved in order to achieve device tracking. Two common types of optical tracking systems, both having a large field of view and being highly accurate, are videometric and infrared systems. Videometric optical tracking systems use marker patterns placed on the patient that are recognized by calibrated cameras. Infrared (IR) systems use charge-coupled device (CCD) cameras to recognize two different types of markers: active IR systems use light-emitting diode (LED) markers while passive IR systems use retroreflective spheres (Figure 6.9). Passive IR systems are wireless, while active IR tracking systems are wired due to the need for powering the LED markers. These optical systems in general, though they are highly accurate, are limited because a direct line of sight is required between the markers and cameras.[27]

Electromagnetic tracking systems (EMTS) have evolved to eliminate this need, but presently compromise accuracy in order to do so. EMTS rely on small solenoids that can be localized in a pulsed magnetic field of known geometry. Because the solenoids themselves can be quite small, and because a direct line of sight is not required to detect them, EMTS are highly desirable in the complex operating room setting. However, the pulsed magnetic field is susceptible to distortion (particularly by metal tools which are ubiquitous in surgery), making the detection much less accurate than the direct visualization of markers that is present in optical tracking systems.[28]

Irrespective of what type of tracking is used, the need to relate the tracked objects to imaging creates an image-to-physical space registration problem. In practice, this problem is often solved with a rigid registration. One of the earliest approaches was to attach a rigid stereotactic frame to the patient, which contained imaging markers that were visible in different modalities. Coordinates of the physical markers could then be mapped to the position of the markers in images.[29] Frameless systems have been devised to be less cumbersome, and might include markers rigidly attached to the skull,[30] markers attached to the skin,[31] or bite blocks attached to the teeth,[32] among others. Surface registration techniques have also emerged and are presently broadly available on several commercial platforms. To use these, the exposed bone or skin surface is scanned with lasers or a tracked probe to enable registration with the skin surface in medical images.[33] Once this registration is complete, the tracked devices can then be visualized on the imaging space.

Navigation in 3D slicer

The *OpenIGTLink Interface* module provides a platform for network communication with external software and hardware that uses the OpenIGTLink protocol.[34] This protocol provides an easy-to-implement standard mechanism for communication among computers and devices in operating rooms. The protocol is extensible and allows for the exchange of information including images, tracking information, device control, and device monitoring. Within the *OpenIGTLink Interface* module in Slicer, a number of functions are possible to create an integrated navigation system. The data import and export tools enable exchange of position, linear transform, and image data between any OpenIGTLink-compliant software and 3D Slicer. The module can manage multiple connections at the same time, allowing for simultaneous connection to a number of different trackers or devices. For visualization that is useful and adaptable, the module can also control the image volume reslicing plane based on device tracking. Thus, as the surgeon moves a tool, a relevant resliced plane of the image is displayed.

A variety of clinical applications of this module are being tested. One application is a neurosurgical manipulator system, in which a master–slave configuration allows the surgeon to operate a master manipulator that dictates the movements of a slave

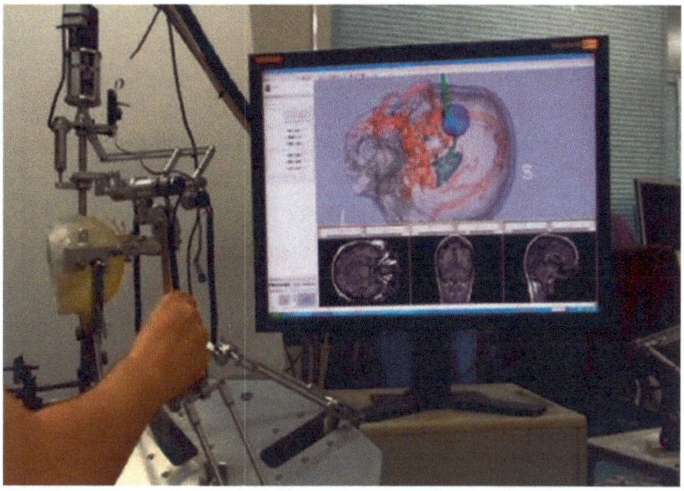

Figure 6.10 Neurosurgical manipulator system. An operator is shown controlling the master manipulator to dictate the movement of a slave manipulator in simulated image-guided neurosurgery. The position of the slave is tracked and displayed in real time along with segmented models of the tumor and surrounding vasculature. *Image courtesy of Dr. Jumpei Arata, Kyushu University, Japan.*

manipulator (Figure 6.10). The position of the slave is measured with the Optotrak optical tracking system (Northern Digital Inc., Ontario, Canada) as well as encoders implemented within the slave itself. OpenIGTLink enables communication between the slave and master manipulators for real-time control of the device, as well as communication between the tracking interface and navigation software, for visualization of the surgical tool with the radiological images. Initial trials show that tracking data is successfully obtained every 100 ms from the optical tracking system, allowing the surgeon to interactively control the slave manipulator from the master manipulator, using 3D Slicer as a console.[34]

The OpenIGTLink protocol can also be applied to the translation of tractography as an intraoperative tool rather than a planning tool alone. A platform for real-time intraoperative tractography was developed, allowing the surgeon to visualize white matter tracts emanating from the tip of surgical tools used during the procedure. The VectorVision Cranial (VVCranial) surgical navigation system is used to perform optical surgical tool tracking (Brainlab AG, Feldkirchen, Germany). The tracking information is received via the *OpenIGTLink Interface* module in 3D Slicer, and the position information of the tracked tool tip is used as a seed point for tractography. The user may designate a radius around the tool tip for which tractography is performed, or select an extended location away from the tool tip to display white matter tracts at a given depth away from the surgical tool. This enables the surgeon to use tractography information in real-time, immediately prior to the tool reaching the tissue itself.[35]

CONCLUSION

In this chapter, we describe the role of computation and image processing in brain tumor resection via the description of some core processes: image segmentation, tractography, registration, and navigation. Medical image computing is of increasing importance due to the increase in image modalities, complexity, and sheer amount of data produced. Over the last decades, the technology has matured from something that was done in a few academic places to a core technology for a variety of clinical applications. This trend continues. 3D Slicer is a software platform designed to ease the translation of technology from academic prototypes to tools for clinical research.

REFERENCES

1. 3D Slicer: A multi-platform, free and open source software package for visualization and medical image computing. 3D Slicer official website. <www.slicer.org>; 2015; Accessed 10.01.15.
2. Fedorov A, Beichel R, Kalpathy-Cramer J, et al. 3D Slicer as an image computing platform for the quantitative imaging network. *Magn Reson Imaging*. 2012;30(9):1323–1341.
3. Kaus MR, Warfield SK, Nabavi A, Black PM, Jolesz FA, Kikinis R. Automated segmentation of MR images of brain tumors. *Radiology*. 2001;218:586–591.
4. Pham DL, Xu C, Prince JL. Current methods in medical image segmentation. *Annu Rev Biomed Eng*. 2000;2:315–337.
5. Prastawa M, Bullitt E, Ho S, Gerig G. A brain tumor segmentation framework based on outlier detection. *Med Image Anal*. 2004;8(3):275–283.
6. Hu X, Tan KK, Levin DN, et al. Three-dimensional magnetic resonance images of the brain: application to neurosurgical planning. *J Neurosurg*. 1990;72:433–440.
7. Kikinis R, Gleason PL, Moriarty TM, et al. Computer-assisted interative three-dimensional planning for neurosurgical procedures. *J Neurosurg*. 1995;38(4):640–651.
8. Nakajima S, Atsumi H, Bhalerao AB, et al. Computer-assisted surgical planning for cerebrovascular neurosurgery. *J Neurosurg*. 1997;41(2):403–410.
9. Ekinci G, Akpinar IN, Baltacioglu F, et al. Early-postoperative magnetic resonance imaging in glial tumors: prediction of tumor regrowth and recurrence. *Eur J Radiol*. 2003;45(2):99–107.
10. Pallud J, Varlet P, Devaux B, et al. Diffuse low-grade oligodendrogliomas extend beyond MRI-defined abnormalities. *Neurology*. 2010;74(21):1724–1731.
11. Egger J, Kapur T, Fedorov A, et al. GBM volumetry using the 3D Slicer medical image computing platform. *Sci Rep*. 2013;3:1364.
12. 3D Slicer Documentation/4.4. Slicer Wiki. Last modified 25 December 2014. Available at: <http://www.slicer.org/slicerWiki/index.php/Documentation/Release>.
13. Beaulieu C. The basis of anisotropic water diffusion in the nervous system: a technical review. *NMR Biomed*. 2002;15:435–455.
14. Yu CS, Li KC, Xuan Y, Ji XM, Qin W. Diffusion tensor tratography in patients with cerebral tumors: a helpful technique for neurosurgical planning and postoperative assessment. *Eur J Radiol*. 2005;56:197–204.
15. Chen X, Weigel D, Ganslandt O, Fahlbusch R, Buchfelder M, Nimsky C. Diffusion tensor-based fiber tracking and intraoperative neuronavigation for the resection of a brainstem cavernous angioma. *Surg Neurol*. 2007;68(3):285–291.
16. Romano A, Ferrante M, Cipriani V, et al. Role of magnetic resonance tractography in the preoperative planning and intraoperative assessment of patients with intra-axial brain tumours. *Radiol Med*. 2007;112(6):906–920.
17. Golby AJ, Kindlmann G, Norton I, Yarmarkovich A, Pieper S, Kikinis R. Interactive diffusion tensor tractography visualization for neurosurgical planning. *Neurosurgery*. 2011;68(2):496–505.

18. Lazar M, Alexander AL, Thottakara PJ, Badie B, Field AS. White matter reorganization after surgical resection of brain tumors and vascular malformations. *Am J Neuroradiol.* 2006;27:1258–1271.
19. Bammer R, Acar B, Moseley ME. In vivo MR tractography using diffusion imaging. *Eur J Radiol.* 2003;45:223–234.
20. NIfTI-1 Data Format. Neuroimaging Informatics Technology Initiative. 2015. Available at: <http://nifti.nimh.nih.gov/nifti-1/>.
21. Malcolm JG, Shenton ME, Rathi Y. Neural tractography using an unscented Kalman filter. *Inf Process Med Imaging.* 2009;21:126–138.
22. Malcolm JG, Shenton ME, Rathi Y. Filtered multi-tensor tractography. *IEEE Trans Med Imaging.* 2010;29(9):1664–1675.
23. Hill DLG, Batchelor PG, Holden M, Hawkes DJ. Topical review: medical image registration. *Phys Med Biol.* 2001;46:R1–R45.
24. Wirtz CR, Knauth MA, Subert MMB, Sartor K, Kunze S, Tronnier VM. Clinical evaluation and follow-up results for intraoperative magnetic resonance imaging in neurosurgery. *Neurosurgery.* 2000; 46(5):1112–1120.
25. Risholm P, Golby AJ, Wells WM. Multi-modal image registration for preoperative planning and image guided neurosurgical procedures. *Neurosurg Clin N Am.* 2011;22(2):197–206.
26. Datasets. Slicer Datastore. 2015. Available from: <http://slicer.kitware.com/midas3/slicerdatastore>.
27. Cleary K, Peters TM. Image-guided interventions: technology review and clinical applications. *Annu Rev Biomed Eng.* 2010;12:119–142.
28. Glossop ND. Advantages of optical compared with electromagnetic tracking. *J Bone Joint Surg.* 2009;91(Suppl 1):23–28.
29. Peters TM, Clark JA, Olivier A, et al. Integrated stereotaxic imaging with CT, MR imaging, and digital subtraction angiography. *Radiology.* 1986;161(3):821–826.
30. Maurer CR, Fitzpatrick JM, Wang MY, Galloway RL, Maciunas RJ, Allen GS. Registration of head volume images using implantable fiducial markers. *IEEE Trans Med Imaging.* 1997;16:447–462.
31. Roberts DW, Strohbehn JW, Hatch JF, Murray W, Kettenberger H. A frameless stereotaxic integration of computerized tomographic imaging and the operating microscope. *J Neurosurg.* 1986;65: 545–549.
32. Fenlon MR, Jusczyzck AS, Edwards PJ, King AP. Locking acrylic resin dental stent for image guided surgery. *J Prosthetic Dentistry.* 2000;83:482–485.
33. Grimson WEL, Ettinger GJ, White SJ, Lozano-Perez T, Wells WM, Kikinis R. An automatic registration method for frameless stereotaxy, image guided surgery, and enhanced reality visualization. *IEEE Trans Med Imaging.* 1996;15:129–140.
34. Tokuda J, Fischer GS, Papademetris X, et al. OpenIGTLink: an open network protocol for image-guided therapy environment. *Int J Med Robot.* 2009;5(4):423–434.
35. Elhawary H, Liu H, Patel P, et al. Intraoperative real-time querying of white matter tracts during frameless stereotactic neuronavigation. *Neurosurgery.* 2011;68(2):506–516.

CHAPTER 7

Intraoperative Imaging

Christopher Nimsky and Barbara Carl
Department of Neurosurgery, University Marburg, Marburg, Germany

INTRODUCTION: INTRAOPERATIVE IMAGING MODALITIES

Soon after the discovery of X-rays in 1895 by Röntgen, the idea of applying X-rays in the operating room emerged. Conventional fluoroscopy and angiography were the first imaging methods used intraoperatively. Ultrasound and computed tomography (CT) followed later, although their initial image quality was less than satisfactory for neurosurgical procedures. Compared to CT, magnetic resonance imaging (MRI) provides for an excellent soft tissue contrast necessary for cranial imaging.

The first attempts were made in the late 1970s and the 1980s to use ultrasound and CT during neurosurgical operations to determine the extent of resection. At that time, though, image quality was quite limited, so that neither method gained wide use in neurosurgery. Since then, MRI has become the method of choice for preoperative diagnosis of brain tumors and epilepsy. However, the closed-bore design and the strong fringe fields of the first MR scanners prevented their use in the operating room. Then, in the mid-1990s, the concept of intraoperative imaging experienced a renaissance with the development of open MR systems. For the first time, these systems made it possible to use MR for intraoperative imaging.

Methods such as standard X-ray fluoroscopy are used as quick and reliable navigation tools depicting where a surgical instrument is located in the patient in relation to bony landmarks; these approaches have been used commonly in transsphenoidal and spine surgery. Ultrasound is another method for direct localization, with the advantages of real-time imaging capability. In this respect it may be used as a method to localize a lesion after craniotomy, such as a cavernoma or metastasis not visible directly at the brain surface. All tomographic methods that can display a sectional view of the anatomy, including ultrasound, CT, and of course MRI, are major tools that allow determining the extent of a resection.

Intraoperative imaging might also cover a discussion about endoscopy techniques, but these are not in the scope of this chapter. Another emerging field in the last few years as an alternative or adjunct to the use of classical imaging machines is the application of fluorescence methods—that is, application of a fluorescent dye, a fluorophore that is able to resolve vasculature or tumor extent in the surgical field by applying special filters to the operating microscope.

A. Golby (Ed): Image-Guided Neurosurgery
DOI: http://dx.doi.org/10.1016/B978-0-12-800870-6.00007-8

Since the benefits of improved extent of resection in surgery for gliomas have become increasingly accepted in recent years,[1] intraoperative methods to optimize the extent of resection are receiving more attention from the neurosurgical community. The combination of preserving function while maximizing the extent of resection seems to be the optimal intraoperative treatment strategy.

A recent opinion statement by M. Berger[2] summarizes: "Survival of patients with both low- and high-grade gliomas is enhanced with maximal tumor resection. The pursuit of more aggressive extent of resection must be balanced with preservation of functional pathways. Several innovations in neurosurgical oncology have expanded our understanding of individualized patient neuroanatomy, physiology, and function. Emerging imaging technologies as well as intraoperative techniques have expanded our ability to resect maximal amounts of tumor while preserving essential function. Stimulation mapping of language and motor pathways is well-established for the safe resection of intrinsic brain lesions. Additional techniques including neuronavigation, fluorescence-guided microsurgery using 5-aminolevulinic acid, intraoperative magnetic resonance imaging, and high-frequency ultrasonography can all be used to improve extent of resection in glioma patients."

However, there are also recent, quite skeptical interpretations of the role of technology in resection of gliomas. From the Cochrane database: "There is low to very low quality evidence (according to GRADE criteria) that image guided surgery using iMRI, 5-ALA or DTI-neuronavigation increases the proportion of patients with high grade glioma that have a complete tumor resection on post-operative MRI. There is a theoretical concern that maximizing the extent of resection may lead to more frequent adverse events but this was poorly reported in the included studies. Effects of image guided surgery on survival and QoL are unclear. Further research, including studies of ultrasound guided surgery, is needed."[3]

To give an overview on intraoperative imaging to support tumor resection is the major focus of this chapter. Additionally, intraoperative imaging has a major role in spine, stereotactic, as well as vascular neurosurgery, which will be addressed briefly in the subsections on the different imaging modalities.

NAVIGATION AND IMAGING

Typical real-time intraoperative imaging modalities like fluoroscopy and ultrasound are tools for immediate localization. However, CT and MRI are typically used purely for diagnostic imaging in radiological units; such preoperative image data can be displayed in the operating room. Image-guided surgery transformed the techniques and principles of classical frame-based stereotaxy, in which a coordinate system is applied to the head, to the modern era in the sense of frameless stereotaxy, which is clinically known as neuronavigation, or just navigation.

Navigation allows the visualization of the essentials of pre- and intraoperative imaging in the surgical field, so that the image data provide an immediate intraoperative feedback. The most important aspect is to prevent increased neurological deficits despite increased resections that might result from the attempt to remove initially overlooked tumor remnants that are detected by intraoperative imaging. Intraoperative imaging should be accompanied by a close integration of navigation.[4–6]

In standard navigation the physical space of the surgical field is registered to the 3D image space, which is based on anatomical data from MRI or CT. In contrast to pointer-based navigation, microscope-based navigation provides a more intuitive data visualization directly in the surgical field. Pointer-based systems only delineate the position of an instrument, e.g., typically the tip of a pointer, in the image space, so that during surgery when navigation information is needed the surgical workflow is interrupted by necessitating that the surgeon is looking away from the surgical field to a navigation screen. Microscope-based navigation has the advantage of heads-up displays superimposing additional information on the surgical field by color contours or semi-transparent 3D objects, while in parallel still the position of an instrument, such as the autofocus position of the microscope, may be displayed on the navigation screen.

Navigation accuracy is influenced by a variety of factors; among them being the so-called application accuracy, factors relating to an unwanted movement of the registration coordinate system (positional shift), and intraoperative events like brain deformation, which is known as brain shift. Brain shift compensation by intraoperative image updating will be discussed in the subsequent subsections. For all three imaging modalities providing 3D imaging data (ultrasound, CT and MRI), several setups have been developed to compensate for the effects of brain shift.

The overall application accuracy is influenced by the quality of imaging, by the technical accuracy of the system itself, and by the quality of patient registration, which defines the process of registering image space and real/surgical space.[7]

Standard anatomical navigation is based on anatomical information only, which has become a routine tool in neurosurgical departments. This technique was developed further by the integration of additional information obtained by other modalities, resulting in the so-called multimodal navigation. An initial step in establishing multimodal navigation was the development of functional navigation, in which preoperative data from magnetoencephalography (MEG)[8–10] and functional magnetic resonance imaging (fMRI)[5,11] define the site of eloquent cortical brain areas such as the motor and speech areas. This method of functional navigation allowed more thorough resections of tumors in high-risk zones with low morbidity. Integration of diffusion tensor imaging (DTI) data delineating the course of major white matter tracts extended this concept to subcortical areas,[12,13] while the co-registration of positron emission tomography (PET) data and information from MR spectroscopy (MRS) added metabolic information leading to true multimodal navigation.[11–18]

INTRAOPERATIVE X-RAY FLUOROSCOPY AND INTRAOPERATIVE ANGIOGRAPHY

The application of X-rays was the first imaging modality introduced in routine operating procedures. In 1980 Rey[19] concludes: "Intraoperative control in neurosurgery requires mainly fluoroscopy, with the possibility of a single exposure for checking purposes." This "checking purpose" would nowadays perhaps be named navigation or quality control.

Fluoroscopy is still used routinely today as a reliable and quick means for intraoperative orientation, e.g. in transsphenoidal procedures to identify the trajectory to the sella turcica, in stereotactic surgeries,[20] as well as in spine procedures for level identification and assisting pedicle screw trajectory adjustment.[21] Catheter and electrode placements may be easily verified during surgery, e.g., placement of atrial shunt catheters, road mapping in deep brain stimulation (DBS) procedures, spinal stimulation, and epilepsy surgery.

Further developments of fluoroscopic imaging, such as in C-arm technology, led to volumetric imaging by acquiring multiple fluoroscopic images about an isocentric point in space, providing axial plane tomographic images that may be reconstructed into an accurate 3D volume (Iso-C-arm technology).[22] The authors conclude: "The usefulness of the Iso-C lies in its convenience and diversity. It provides quick, seamless, and accurate data acquisition for intraoperative imaging with or without navigation. It can be used during traditional open spinal and cranial base approaches or in conjunction with minimally invasive approaches such as thoracoscopy, vertebroplasty, biopsy, and minimally invasive pedicle screw placement. This new technology is readily adaptable to any operating room."[22]

X-ray fluoroscopy is also the basis of angiography, which is applied in vascular procedures as well. Technological progress like in the standard c-arm systems has led to the development of biplanar systems, with flat panel detectors, where one imaging arm can function as a tool for 3D imaging, e.g. DynaCT.[23] High-contrast structures, such as bone, calcified lesions, and metallic materials can be visualized by DynaCT, as well as by multidetector CT. Although contrast differentiation of soft tissue such as cerebral cortex, muscle, and hematoma on DynaCT are inferior to multidetector CT scans, the images have been used as intraoperative CT (iCT)-like images. However, DynaCT images acquired from flat panel detector-based systems were found to be superior to those images acquired from image intensifier-based systems.[23]

Modern robotic mounting devices further improved these systems recently, which have found broad application in neuroradiological units, but also in so-called hybrid-OR concepts where angiography units are placed in an operating room environment.[24]

INTRAOPERATIVE FLUORESCENCE TECHNIQUES AND OTHERS

There are two major applications of fluorescence techniques in neurosurgical procedures.[25] One is the intravenous application of a fluorescent biomarker to visualize vasculature, which is helpful in aneurysm, angioma, and bypass surgery. Indocyanine green (ICG) video angiography yields high resolution, real-time images of arterial, capillary, and venous flow of cerebral vasculature. Although it gives adequate information about the clipped neck, parent/branching artery and perforator involvement, it has some limitations such as viewing the neck residuals located behind the aneurysm, thick walled atherosclerotic vessels, and thrombosed aneurysms. In arteriovenous malformation (AVM) surgery, it is useful in detecting the residual nidus in diffuse-type AVM, but cannot be relied in deep-seated AVMs. The technique also provides exact information about the anastomosis site in EC-IC bypass, thus, avoiding early bypass graft failure. ICG video angiography is a simple, reliable, and quick method to pick up subtle findings in cerebrovascular procedures. In selected aneurysm cases, other technologies such as endoscopy or intraoperative digital substraction angiography may be helpful, and in deep-seated AVMs, navigation may be required as an adjunct to confirm intraoperative findings.[26] A recent paper demonstrated the application of ICG also in combination with endoscopy.[27] ICG video angiography is a quick and safe method of intraoperatively mapping the angioarchitecture of superficial AVMs, but less helpful for deep-seated lesions. This modality alone does not improve identification of residual disease or clinical outcomes. Surgeon experience with extensive study of preoperative vascular imaging is paramount to achieving acceptable clinical outcomes. Formal angiography remains the gold standard for evaluation of AVM obliteration.[28]

On the other hand there is fluorescence guidance aimed at maximizing the extent of high-grade glioma resection. Different fluorophores, i.e. fluorescent biomarkers, including 5-aminolevulinic acid (5-ALA),[29] have been examined with the use of several imaging techniques.[30] 5-ALA is the most examined fluorophore applied to glioma surgery; however, there are potential alternatives like hypericin which might have advantages also in respect to photodynamic therapy.[31,32] It is still unclear whether fluorescent molecules have specific affinity for neoplastic cells or whether they just reflect a leaky local vasculature and inflammatory change.[33] These techniques seem to be safe and effective for increasing gross total resection,[29] but the development of more tumor-specific fluorophores is needed to resolve problems with subjective interpretation of fluorescent signal at tumor margins. Techniques such as quantum dots and polymer or iron oxide-based nanoparticles have shown promise as potential future tools.[30]

Confocal imaging techniques,[34] as well as other local imaging techniques, such as optical coherence tomography[35,36] could provide a level of imaging that goes beyond the classical operating microscope to a more histological level offering interesting alternatives in the future.

INTRAOPERATIVE ULTRASOUND

In 1978 Reid[37] reported on the application of intraoperative ultrasound during surgery of a cystic cervical astrocytoma. Intraoperative ultrasound is a truly real-time imaging modality. This ability to depict real-time anatomic data during a surgical procedure is a valuable surgical adjunct and can guide surgical decision making. Intraoperative ultrasound is a rapid and effective way to localize and characterize diseases of the brain, osseous spine, and adjacent spinal cord, providing accurate localization of anatomy and reducing the risk of injury to the spinal cord or surrounding brain parenchyma.[38]

Despite more than 30 years of use in neurosurgical operating theaters and its quite frequent use by many neurosurgeons, a major breakthrough is missing. One obstacle to the greatest impact for the ultrasound in neurosurgery is that much expertise is needed for interpretation of ultrasound images.[39] The limitations of intraoperative ultrasound for determining the extent of resection are the difficult image interpretation, needing a long learning curve, as well as limitations in tissue differentiation.

Even the combination of intraoperative ultrasound with navigation systems, which allows an easier ultrasound image interpretation by the parallel display of corresponding MRI or CT slices for easier orientation and identification of structures,[40] as well as techniques like native 3D ultrasound,[41] have not yet solved this challenge completely. Establishing grading systems,[42] as well as emerging techniques to apply ultrasound contrast media[43] might offer new possibilities for intraoperative ultrasound to determine the extent of a resection reliably. Also, recent technological advancements in intraoperative ultrasound technology including higher frequencies and reduced probe dimensions might extend the application of ultrasound technology in neurosurgery.[44]

There are several case reports and case series on the application of intraoperative ultrasound in glioma[45–47] and pituitary tumor surgery.[48–51] There are no large prospective studies on the effect of intraoperative ultrasound on glioma resection. In a larger retrospective study on 192 glioblastoma patients it has been shown by Saether et al.[52] that survival has improved in this patient group over the time period in which intraoperative ultrasound and neuronavigation were introduced and established in their neurosurgical department.

As will be described for intraoperative MRI in the following sections, there are also attempts to combine intraoperative ultrasound with functional navigation, including fMRI and DTI.[53] Unfortunately, only a few groups have managed to establish such a working setup. A few studies comparing different intraoperative imaging modalities have demonstrated that ultrasound is a helpful adjunct as a real-time imaging modality; however, the sensitivity of intraoperative MRI was higher.[54,55]

Of course intraoperative ultrasound as well as duplex sonography have a major role in vascular neurosurgery.[56,57] Here the advantages of intraoperative ultrasound become clear: it is easily available, no excessive preparation is necessary when suddenly needed during surgery, and it is a real-time modality.

INTRAOPERATIVE CT

The first intraoperative application of CT technology was published in 1979 by Shalit et al.[58] The entire surgical procedure for a recurrent glioma was carried out on the table of a CT scanner. In his overview on X-ray control in the operating theatre Rey[19] states: "As a conclusion, let us have a glance into the future. Shalit from Tel Aviv has started using CT scan control during tumor surgery, and has assessed the quality of removal in two cases of gliomas." Lunsford[59] and Okudera[60,61] published similar early attempts in the next years and adapted CT imaging technology to the operating room environment. However, no large case series were published.

The next step in technical development was a fully mobile CT scanner that was used in the operating room, as well as in the intensive care unit by several groups.[62–65] The main investigations were stereotactic procedures,[64] tumor resections, and also the first spinal applications.[62] This setting, as realized in the first intraoperative MRI systems that were investigated in the same time frame, combined intraoperative imaging with navigation systems.[63,64]

With the imminent development of low- and high-field MRI systems for intraoperative use, the focus shifted away from iCT, which had its strengths in depicting bony structures, but which provided only restricted soft-tissue contrast. Other limitations were fixed slice position, challenges in OR workflow, and the requirement for radiation protection for the whole neurosurgical and anesthesiology OR staff.

About 10 years after the fully mobile CT scanner was investigated, fixed CT systems on rails were integrated in operating room settings, often in close combination and integration with navigation systems.[66]

Modern multislice CT scanning offers a very reasonable technology for intraoperative imaging, especially in conjunction with a navigation system for intraoperative updating of the data set. Unlike intraoperative MRI, the system can be installed in a preexisting operating environment without the need for special surgical instruments. It increases the safety of the patient, the surgeon, and the anesthesiologist without the need to significantly change existing protocols and workflows. Imaging and updating of the navigation system can be performed at any time during surgery. Multidisciplinary use increases utilization of the system, thus improving the cost-efficiency relationship.[66] Hosodoa et al. showed prolonged survival in a series of low-grade gliomas due to usage of iCT, while extent of resection correlated to longer survival.[67]

The extent of tumor resection in transsphenoidal surgery could be reliably assessed by iCT in a study by Lee at al.[68] Compared with postoperative MR imaging, iCT provided good guidance for residual tumor resection, and it not only held significant promise for maximizing the radicality of tumor resection but also increased the percentage of complete removals without any major complications. Similar benefit was found in a recent study investigating the application of DynaCT in skull base surgery.[69]

iCT has a classical application in stereotactic procedures, like the application of the mobile CT-like O-arm for DBS procedures.[70–72]

In vascular neurosurgery iCT systems are used to visualize the angioarchitecture, the extent of vascular anomaly resection or clipping,[73] to give information about the perfusion,[74] and to delineate the extent of hematoma evacuation. iCT images can be obtained and merged with those obtained using other modalities such as MRI if desired. Depending on the iCT system, there may be limitations due to lower resolution compared to standard multislice systems.[56]

Spine surgery with its increasing technically complex procedures is a main field for iCT technology[75,76] because of its ability to visualize bony structures in great detail, and to confirm implant placement.[77–82] Compared to intraoperative fluoroscopy in spine surgery for pedicle screw placement, iCT provided better results.[21] In a study by Costa et al.[83] the overall median accuracy of screw placement using the assistance of a navigation system was 93.5%; when preoperative CT data set images were used the rate of accuracy was 91.8%; whereas with the iCT acquisition the accuracy was 95.2%. The iCT acquisition and subsequent automerging of the surgical anatomy with the 3D model also led to a significant reduction in surgical time. With the use of a system allowing iCT scan acquisition, the overall need for re-operation for misplaced screws was almost completely avoided. Additionally, use of a spinal navigation system allows a significant reduction in exposure to ionizing radiation for the surgeons and nurses.[83,84] Bandela et al. confirmed this by comparing the use of iCT-based navigation systems with fluoroscopic-based methods.[85]

INTRODUCTION TO INTRAOPERATIVE MRI

Intraoperative imaging has gained increasing interest in the last two decades. If intraoperative imaging depicts that a resection is incomplete, tumor residues that were initially missed can be removed during the same operation. In contrast to the subjective estimation by the neurosurgeon, intraoperative imaging allows an objective evaluation of the intraoperative situation, thus acting as quality control during surgery.[5,86–92]

Many brain tumors, particularly low-grade gliomas, have the look and feel of normal brain. Without iMRI, achieving gross total resection without being unduly aggressive is virtually impossible. With iMRI, however, the neurosurgeon can evaluate the brain at any time during resection, and thus can both avoid eloquent structures and achieve a more complete resection.[93]

Due to limited imaging quality, the first attempts in the 1980s to apply ultrasound and CT during neurosurgical procedures were frustrating. Since then, MRI has become the method of choice for the preoperative diagnosis of brain tumors and epilepsy. However, the closed-bore design and the strong fringe fields of the first MR scanners prevented their use in the operating room.

In the mid-1990s with the development of open MR systems the concept of intraoperative imaging experienced a renaissance.[6,86,94] The first designs were based on low-field magnets with magnetic field strengths up to 0.5T.[95] The use of MR scanners in the operating environment proved safe and reliable, as well as applicable to neurosurgical procedures, even if these procedures had to be adapted to the MR environment to a certain extent. Several reports on intraoperative low-field MRI for large numbers of patients have been published.[96–101]

In contrast to a specially developed MR scanner dedicated to operating room use as pioneered by P. Black, F. Jolesz, and General Electric Medical Systems at the Brigham and Women's hospital in Boston,[86] at the Universities in Erlangen and Heidelberg together with Siemens Medical, a low-field MR scanner (0.2T MR Magnetom Open) was adapted for surgical use.[6,94] In addition to intraoperative imaging, an integral part of the Erlangen concept was the option of applying neuronavigation simultaneously, with a preference for microscope-based neuronavigation, where the extent and localization of a tumor are superimposed on the microscope field of view through contours. Aside from standard neuronavigation based on anatomical information only, which has become a routine tool in many neurosurgical departments, preoperative functional data from MEG or fMRI defining patient-specific localization of eloquent brain areas, such as the motor and speech areas, can be integrated, resulting in so-called functional neuronavigation.[10,11,102–104] Between March 1996 and July 2001 intraoperative low-field MRI was performed in the Erlangen setting in 330 patients.[101] Among them were 240 craniotomies, 59 transsphenoidal approaches, and 31 burr hole procedures. The simultaneous use of intraoperative MRI and functional neuronavigation allowed preservation of neurological function despite extended resections.

Soon after the implementation of intraoperative MRI technology it became obvious that the most important indications for intraoperative imaging included: gliomas,[98–100,105,106] hormonally inactive pituitary tumors,[107–110] and pharmacoresistant epilepsy.[111–114] Intraoperative MRI also enables compensation for brain shift by an update of the navigation system with intraoperative image data.[115–118]

However, the diagnostic quality of the intraoperative low-field MR systems could not compete with the image quality of routine neuroradiological diagnosis generally performed with high-field magnets. Advances in scanner design, including those due to active magnetic shielding, have made it possible to adapt modern high-field scanners to the surgical environment. Two major different high-field concepts were initially realized.[88,90] Basically, as with intraoperative low-field magnet concepts, there are two possibilities: either taking a standard diagnostic scanner and adapting it to the operating environment, as done in Minneapolis with a Philips scanner,[88,119] or designing a high-field scanner specifically dedicated for the requirements of an operating room, as implemented in Calgary with a standard magnet from the commercial line being ceiling-mounted and moved into the appropriate imaging position during surgery.[90,120] The Erlangen approach to realize intraoperative high-field MR scanning combined with microscope-based neuronavigation resembled the Minneapolis setup, necessitating some kind of intraoperative patient transport for intraoperative imaging. The active magnetic shielding of modern high-field magnets results in the 5G zone being relatively close to the scanner, so that the adaptation of a rotating operating table enables combining intraoperative high-field MRI with microscope-based neuronavigation.[121] The main operating position where navigation can be applied is located in the fringe-field of the scanner; there the use of all standard neurosurgical equipment is possible. In this way, the concept of intraoperative MRI with integration of microscope-based neuronavigation in the low magnetic fringe field[4] could be applied to high-field magnets. The operating room was appropriately reconstructed between August 2001 and March 2002, with the first patients operated on using intraoperative high-field MRI and integrated microscope-based neuronavigation at the end of April 2002. Over 2500 patients have been operated on with this configuration since its inception.

Meanwhile, developments in the implementation of intraoperative MRI have expanded at the extremes of magnetic field strength. On one end there is the development of a mobile iMRI system with magnetic field strengths between 0.12T and 0.15T.[122–124] At the other extreme, there are many centers moving from intraoperative 1.5T to 3T[125–127] as the magnetic field strength of 3T has become standard in routine neuroradiological diagnostics.

Latest developments in technology relate to the integration of robotics in 3T intraoperative imaging.[128–130] (See Chapter 19 for a full discussion of robotics in image-guided neurosurgery.)

One of the most prominent arguments against intraoperative MRI are the high initial costs for the installation of such systems. There are only few detailed published economic analyses: one on low-field settings by Makary which did not demonstrate an adequate justification for installation of such low-field systems mainly due to a lack of clinical benefit.[131] In contrast Hall et al. have shown a cost benefit in their analysis of a high-field system,[87] which without doubt has clear benefits in image quality and imaging spectrum compared to low-field systems. A major cost-saving approach is the dual use of intraoperative systems, which is nowadays often applied especially for the high-field strength systems, which allows the parallel use of a scanner for both intraoperative use and for diagnostic scanning. Such an approach could enhance the cost-effectiveness of iMRI systems. Alternatively, some sites try to find low-cost solutions, such as moving the patient out of the operating room to a standard diagnostic scanner while the head is still open or just closed temporarily, which however clearly interrupts the surgical workflow, so that these approaches are not real intraoperative imaging but more some kind of interdissection imaging.[132]

Intraoperative MRI in glioma surgery

Despite quite skeptical comments still advocating a conceptual limitation of intraoperative MRI for glioma surgery[133] being published, there is increasing evidence that glioma surgery benefits from intraoperative MRI.

The review by Kubben et al.[91] gives a comprehensive overview[92] to address the added value of iMRI-guided resection of glioblastoma multiforme compared with conventional neuronavigation-guided resection, with respect to extent of tumor resection (EOTR), quality of life, and survival. Twelve nonrandomized cohort studies matched all selection criteria and were used for qualitative synthesis.[100,105,106,134–142] Most of the studies included descriptive statistics of patient populations of mixed pathology, and iMRI systems of varying field strengths between 0.15T and 1.5T. Most studies provided information on EOTR, but did not always mention how iMRI affected the surgical strategy. Only a few studies included information on quality of life or survival for subpopulations with glioblastoma multiforme or high-grade glioma. Several limitations and sources of bias were apparent, which affected the conclusions drawn and might have led to overestimation of the added value of iMRI-guided surgery for resection of glioblastoma multiforme. Based on the available literature, there is at best level 2 evidence that iMRI-guided surgery is more effective than conventional neuronavigation-guided surgery in increasing EOTR, enhancing quality of life, or prolonging survival after resection of glioblastoma multiforme.

In 2011, Senft et al.[143] reported a randomized trial analyzing the effect of intraoperative low-field MRI on the extent of resection in surgery for contrast-enhancing gliomas. The rate of complete tumor resections was significantly higher in surgeries done with intraoperative MRI than it was in those without, and the rates of postoperative neurological deficits did not differ between the groups. The investigation was the first randomized trial investigating the application of intraoperative MRI in patients with glioma, which is notable because participants recruited to departments with intraoperative MRI technology are often unwilling to risk missing out on a potentially useful surgical addition and decline randomization. This challenge explains why no randomized studies of intraoperative MRI have successfully been completed in the past 15 years before, despite the technique being available at different sites with various setups ranging from ultra-low field (0.12T)[122] to high-field (3T) MRI scanners. Retrospective studies of glioma surgery with intraoperative MRI showed a significant reduction of remaining tumor volume after intraoperative imaging,[144] but such studies can provide, at best, only level 2 evidence of superiority.[91] The expectation was that an increased extent of resection would translate into prolonged survival, as was shown in a multicenter phase 3 trial of surgery for glioblastoma comparing conventional surgery with fluorescence-guided surgery.[29] Retrospective analysis in surgery for glioblastoma with intraoperative high-field MRI guidance showed a significant increase in median overall survival when greater than 98% extent of resection was achieved.[145] Other promising data from a single-center study showed that age, Karnofsky performance score, and extent of resection were predictive of survival in a consecutive series of 500 patients with glioblastoma.[146] Furthermore, resections that were as small as 78% of the intended size corresponded with a survival benefit. These data support the notion of achieving the maximum possible resection combined with preservation of neurological function.

A paper by Mehdorn et al. from the same year also depicted a clear benefit in survival for patients being operated on with the guidance of high-field iMRI.[147] A recent study by Mohammadi et al.[148] supports these findings, as stated in the comment by Haydon & Dacey[149] it can be concluded: "Although mainly descriptive, these data contribute to the growing literature regarding the impact of iMRI on glioma management. Prospective data addressing the impact of iMRI on patient survival remain wanting. Such studies will perhaps appear in the near future."

Technical refinements in the iMRI setting allowing integration of electrophysiological monitoring,[150] as well as attempts to perform awake craniotomny in an iMRI environment[151–155] increase the potential beneficial application in glioma surgery.

Investigations comparing intraoperative fluorescence techniques with intraoperative MRI by Coburger et al.[156] showed that in high-grade gliomas, imaging results of 5-ALA and iMRI were significantly different at the border zone; 5-ALA had a higher sensitivity and a lower specificity for tumor detection than Gd-DTPA–enhanced iMRI. For detection of infiltrating tumor at the border of the resection cavity, 5-ALA was superior to Gd-DTPA–enhanced iMRI concerning both sensitivity and specificity. Thus, the use of 5-ALA in addition to iMRI might be beneficial to maximize extent of resection. This synergistic effect was also seen by others[157,158]; however, there are also reports concluding that iMRI was superior to 5-ALA supported surgery.[159] Berger[160] comments: "With our increasing knowledge that extent of resection significantly impacts not only progression-free survival but also overall survival, we as neurosurgeons are searching for the next biggest impact in patient outcome with our surgical techniques. Resections beyond the enhancement depicted by MRI and now beyond the 5-ALA–positive tissue can also impact outcome."

Intraoperative MRI in pituitary tumor surgery

Intraoperative MRI is a major tool to investigate the extent of resection in pituitary tumor surgery, mainly in transsphenoidal resections of pituitary adenoma. All of various intraoperative MR settings have been used for transsphenoidal surgery, showing more or less satisfying results. Applications showing beneficial effects are documented for ultra-low-field iMRI,[161,162] low-field iMRI,[110,163,164] and high-field iMRI.[165–170] There is not much doubt that the high-field settings have greater ability to resolve detailed anatomy especially with regard to evaluation of the extent of removal for intra- and parasellar tumor parts. Intraoperative MRI at low-field magnetic field strengths had its limitations in the evaluation of craniopharyngioma surgery since resections defined as complete in intraoperative imaging did not exclude recurrences.[171] Investigations at 1.5T clearly demonstrated improved image quality.[172] However, long-term analysis data are still missing. (See Chapter 10 for a detailed discussion of image guided pituitary surgery.)

Further indications for intraoperative MRI

Only a few attempts to apply intraoperative MRI in spine surgery have been published to date. Woodard et al. published a small patient series from the original double donut configuration. Due to the real-time capabilities of the MR system used, with the patient being operated in the scanner itself, level localization was easily possible, as well as in the majority of cases the extent of spinal decompression could be documented.[173] Recently, from Sutherland's group a paper was published demonstrating how intraoperative MRI was applied in transoral approaches for upper cervical spine pathologies.[174]

Besides glioma surgery and pituitary tumor surgery, intraoperative MRI is applied in a variety of other lesions, such as to evaluate the removal of cavernomas,[56,175] and there are several reports on applying iMRI in pediatric neurosurgery.[176–179]

Surgery inside an MR scanner provides an inherent coordinate system easily allowing stereotactic procedures.[124,180,181] Compared to standard frameless procedures, applying high-field iMRI results in an increased stereotactic accuracy.[182] Also, epilepsy surgery was investigated by iMRI, which could depict the extent of tailored resections reliably.[112–114,183] (For detailed discussions of image guidance in functional neurosurgery and epilepsy surgery, please see Chapters 13 and 11, respectively.)

Intraoperative MRI and brain shift

Combining iMRI and microscope-based navigation enables intraoperative compensation for the effects of brain shift by an immediate updating the image information. Updated navigation with intraoperative image data allows a reliable identification of tumor remnants. Microscope-integrated heads-up displays visualize the segmented tumor remnant in the surgical field facilitating the precise localization and orientation in the resection cavity.[115,116,184–186]

However, these updates typically contain standard anatomical data only, so that functional data that were integrated in the preoperative navigation planning are lost.

Intraoperative high-field MRI can augment standard anatomical imaging with intraoperative fMRI, as well as intraoperative DTI; that is, intraoperative MRI beyond standard imaging is possible.

Intraoperative DTI

Pre- and intraoperative DTI for tractography of major white matter tracts in glioma surgery can be performed using a 1.5 Tesla MR scanner.[12,13,185,187] Intraoperative fiber tract visualization using a software solution running on the MR scanner platform needed less than one minute, so that the whole evaluation could be performed during surgery. The interactive 3D display with co-registered b0 diffusion images gave a quick and intuitive overview of the position of major white matter tracts. Thus, fiber tracking is not only a method for preoperative neurosurgical visualization but also for further intraoperative planning. Only in one patient of the initial investigation series was a neurological aggravation (2.7%) observed; the deficit was not felt to be related to a misinterpretation of fiber tracking. The measured extent of shifting of the major white matter tracts in glioma surgery corresponded well to previous data on brain shift of the so-called deep tumor margin, which was reported to be up to more than one centimeter.[116,188] Furthermore, the individually unpredictable direction and great interindividual variability of white matter tract shifting confirmed previous findings.[115,116,189] The absolute amount of shifting correlated with the tumor volume: i.e., in larger tumors greater deformations were likely to occur. However, the direction of white matter tract shifting, whether in the outward or inward direction with respect to the craniotomy opening, seemed to be unpredictable. Even the opening of the ventricular system was not a reliable parameter to predict inward shifting due to the loss of cerebrospinal fluid.

The knowledge of the actual position of major white matter tracts during glioma resection helps to prevent too extensive resections that could potentially damage major white matter tracts and subsequently result in postoperative neurological deficits. When data from fiber tracking are integrated into a navigational setup, preferably with the simultaneous application of fMRI, serving as seed regions for DTI fiber tracking algorithms, it is essential that the effects of brain shift, which clearly affect the spatial position of major white matter tracts, are compensated for. In contrast to mathematical models,[190–194] which still have great restrictions simulating the brain shift behavior for deep brain structures, intraoperative DTI is a reliable possibility to obtain actual data for fiber tracking representing the intraoperative situation after substantial portions of a glioma are removed and further guidance is needed. Besides integration of DTI data for the course of the corticospinal tract, reconstructions of the arcuate fasciculus[195–197] as well as the optic radiation[198,199] find their way in the intraoperative clinical neurosurgical routine.

The implementation of a DTI tracking algorithm in the navigation software allows intraoperative updating of the navigation system with the intraoperative DTI data in less than 5–10 minutes, thus compensating for the effects of brain shift not only for standard 3D anatomical data but also for the position of major white matter tracts.[13]

Such intraoperative DTI updated data delineating the course of major white matter tracts is a prerequisite for a real electrophysiological validation of the white matter tract data. Reports on comparisons between subcortical electrical stimulations and preoperative DTI data showed some inconsistencies, which were probably due to the effects of brain shift.[200,201] Maesawa et al. have demonstrated that intraoperative tractography could locate the corticospinal tract more accurately than preoperative tractography.[202] A very recent paper found that tumor-to-corticospinal tract distance measurements taken from preoperative DTI-based tractography can reliably be used for interpretation of subcortical motor evoked potential response during resection surgery for nonenhancing tumors.[203]

Maximal safety may require combining electrophysiological brain mapping with functional navigation that integrates fMRI/MEG data and DTI-based fiber tracking acquired before or during surgery. Intraoperative electrocortical mapping can identify cortical eloquent brain areas; subcortical electrical stimulation helps to identify major white matter tracts during surgery. Recent studies emphasize that functional navigation and subcortical stimulation are complementary methods that may facilitate the preservation of pyramidal tracts.

Future research will have to be in the field of quantification and reduction of spatial inaccuracies of the raw DTI data,[204] as well as to improve sequence design, tracking parameters, and algorithms. Besides progress in sequence development with reduced image distortion, de-noising increased number of diffusion directions, and higher resolution of the raw data, further progress will also relate to a more accurate reconstruction of neural connectivity patterns. Correct identification of areas of fiber crossings is not possible by standard DTI because of its inability to resolve more than a single axon direction within each imaging voxel. Techniques that can resolve multiple axon directions within a single voxel may solve the problem of white matter fiber crossings, as well as white matter insertions into the cortex.

Further challenges relate to the effects of edema surrounding a tumor where fiber tracking is performed. Effects of edema, the resection cavity, and tumor remnants may directly impede the correct tracking so that either existing fibers are not visualized at all or even erroneous tracking may result. There are various technical attempts to approach the limitations of DTI-based tractography; however, an agreed standard, or ideal solution, has not yet been defined.[205–207] It will be important to compare the different approaches especially in respect to their reliability and also clinical applicability.[208]

A recent study investigated differences in the metabolism of fiber tract alterations between gliomas grade II to IV by correlation of fiber density values with metabolite concentrations measured by fiber density mapping and MR spectroscopic imaging. Structural integrity of fiber tracts was assessed as the fiber density ipsilateral-to-contralateral ratio (FD-ICR). Metabolite concentrations for choline-containing compounds and *N*-acetyl-aspartate were computed and correlated to FD-ICR values after co-registration with anatomic MR imaging. In tumor areas, choline-containing compound concentrations of altered fiber tracts were significantly different between low- and high-grade glioma and showed different courses for the correlations of FD-ICR and choline-containing compounds. In high-grade glioma, increasing fiber destruction was associated with a massive progression in cell membrane proliferation. Peritumor fiber structures showed significantly decreased *N*-acetyl-aspartate concentrations for all patients, but only patients with glioblastoma multiforme had significantly decreased fiber density compared with the contralateral side. Glioma grades II and III had significantly higher peritumor FD-ICR than glioblastoma multiforme. A multiparametric MR imaging strategy that provides information about both structural integrity and metabolism of the tumor is required for detailed assessment of glioma-related fiber tract alterations.[209] (Please see Chapter 4 for a detailed discussion of diffusion imaging and tractography.)

Intraoperative fMRI

By 1998 Gering et al. had demonstrated that fMRI is possible in a low-field iMRI setting.[210] In a 1.5T intraoperative magnet shielded electrical peripheral nerve stimulation of the median and tibial nerve was used to implement intraoperative fMRI as a passive stimulation paradigm.[211] For electrical stimulation an electromagnetically shielded coaxial lead was developed, while shielding was achieved by connecting the conductor's shielding-mesh to the MRI cage. The impulse generator was located outside the radiofrequency-shielded cabin and the conductor was threaded through a waveguide array into the actual operating theatre. After induction of anesthesia and patient positioning, the stimulation electrodes were attached and the motor threshold was defined. After initial anatomical and functional MR scans, two further data sets were acquired during and at the end of the surgical procedure. The block-design stimulation paradigm alternated four rest and four activation periods. For functional imaging, slices parallel to the anterior-posterior commissural plane were acquired as T2*-weighted echo planar imaging sequences. The fMRI data were analyzed during acquisition by an online statistical evaluation package installed on the MR scanner console. In addition, phase reversal of somatosensory evoked potentials was used for verification of intraoperative fMRI. In four anesthetized patients with lesions in the vicinity of the central region a total of eleven fMRI measurements were successfully acquired and analyzed online. Activation was found in the somatosensory cortex, which could be confirmed by intraoperative phase reversal for each measurement. No neurological deteriorations or complications due to the stimulation technique were observed. Thus, intraoperative fMRI is technically feasible, allowing a real-time identification of eloquent brain areas despite brain shift.[211] Similar results could also be reproduced applying a sophisticated setup with a low-field (0.3T) scanner.[212]

In a recent publication, a new technique that combines awake surgery and i-fMRI, named "awake" i-fMRI (ai-fMRI) was proposed. Ai-fMRI was applied to the real-time localization of sensor motor areas during awake craniotomy in seven patients. The results showed that ai-fMRI could successfully detect activations in the bilateral primary sensor motor areas and supplementary motor areas for all patients, indicating the feasibility of this technique in eloquent area localization. The reliability of ai-fMRI was further validated using intraoperative stimulation mapping in two of the seven patients. Comparisons between the pre-fMRI-derived localization result and the ai-fMRI derived result showed that the former was subject to a heavy brain shift and led to incorrect localization, while the latter approach solved that problem.[213]

Nevertheless, the clinical necessity for intraoperative fMRI is debatable, since the position of cortical eloquent brain areas can be identified and marked after dural opening when applying preoperative data-based functional navigation. The shifting of the cortical eloquent areas during a procedure can be observed by the surgeon without much guesswork, so that a time-consuming fMRI update procedure might only be of interest in very selected cases, for example in which the updated fMRI information is used as seed regions for fiber tracking algorithms to reconstruct the arcuate fasciculus [196,214] applying intraoperative DTI to update reconstructed language fibers.

Intraoperative MR spectroscopy and other techniques

Intraoperative MR spectroscopy (MRS) is still limited by the effect of the brain–air interface, so there remain some distinct challenges for updating MRS information during surgery. Applying chemical shift imaging (CSI) using a 1.5T system, we were not able to get reliable data due to the close air interface in the vicinity of the resection cavity. Measurements on the healthy hemisphere, on the other hand, were possible; however, these were of no clinical relevance. Perhaps modifying the operative field prior to imaging could allow for intraoperative MRS measurements. In the literature there are some reports of single voxel measurements that were applied with the attempt to differentiate between tumor tissue and surgically induced changes at the resection border.[215,216] There are also only rare reports on the intraoperative application of perfusion measurements[217] to delineate tumor remnants, as well as the application of techniques like arterial spin labeling in visualization of vasculature.

OUTLOOK

By 1980 Rey[19] raised the question as to whether the neurosurgeons in the future will operate on their own with the aid of the neuroradiologist, or if the neurosurgical procedure will be performed in the neuroradiological department with the aid of the neurosurgeon.

Maybe the Advanced Multimodality Image Guided Operating (AMIGO) Suite in Boston[218] is the answer as it demonstrates all current features in a sort of translational test bed. Multidisciplinary teams can treat patients with the benefit of intraoperative imaging using multiple modalities. Real-time anatomical imaging modalities such as X-ray and ultrasound are combined with cross-sectional digital imaging systems such as CT, MRI, and PET. Additionally, molecular image-guided therapy, using multiple molecular probes such as PET, optical imaging, and targeted mass spectrometry, is being evaluated to increase the sensitivity and specificity of cancer detection. The application of these technologies is expected to improve the ability to define tumor margins in order to more completely excise or thermally ablate tumors. In addition to multimodality imaging there are numerous devices geared towards image-guided therapy, including several navigational systems, robotic devices, and therapy delivery systems that help physicians to localize and treat tumors and other targeted abnormalities. The AMIGO suite represents and encourages multidisciplinary cooperation and collaboration among teams of surgeons, interventional radiologists, imaging physicists, computer scientists, biomedical engineers, nurses, and technologists to reach the common goal of delivering the safest and the most effective state-of-the-art therapy to patients in a technologically advanced but patient-friendly environment.[219]

Multimodal navigation enables the removal of a tumor close to eloquent brain areas with low postoperative deficits, while additional intraoperative imaging ensures that the maximum extent of the resection can be achieved. Intraoperative imaging allows updating of the preoperative image data compensating for the effects of brain shift. All kinds of intraoperative imaging technologies beyond standard anatomical imaging add further safety for complex tumor resections.

REFERENCES

1. Sanai N, Berger MS. Glioma extent of resection and its impact on patient outcome. *Neurosurgery.* 2008;62(4):753–764:discussion 264–266.
2. Hervey-Jumper SL, Berger MS. Role of surgical resection in low- and high-grade gliomas. *Curr Treat Options Neurol.* 2014;16(4):284.
3. Barone DG, Lawrie TA, Hart MG. Image guided surgery for the resection of brain tumours. *Cochrane Database Syst Rev.* 2014;1:CD009685.
4. Nimsky C, Ganslandt O, Kober H, Buchfelder M, Fahlbusch R. Intraoperative magnetic resonance imaging combined with neuronavigation: a new concept. *Neurosurgery.* 2001;48(5):1082–1089:discussion 89–91.

5. Nimsky C, Ganslandt O, Von Keller B, Romstock J, Fahlbusch R. Intraoperative high-field-strength MR imaging: implementation and experience in 200 patients. *Radiology.* 2004;233(1):67–78.
6. Steinmeier R, Fahlbusch R, Ganslandt O, et al. Intraoperative magnetic resonance imaging with the magnetom open scanner: concepts, neurosurgical indications, and procedures. A preliminary report. *Neurosurgery.* 1998;43(4):739–748.
7. Steinmeier R, Rachinger J, Kaus M, Ganslandt O, Huk W, Fahlbusch R. Factors influencing the application accuracy of neuronavigation systems. *Stereotact Funct Neurosurg.* 2000;75(4):188–202.
8. Ganslandt O, Buchfelder M, Hastreiter P, Grummich P, Fahlbusch R, Nimsky C. Magnetic source imaging supports clinical decision making in glioma patients. *Clin Neurol Neurosurg.* 2004;107(1): 20–26.
9. Ganslandt O, Steinmeier R, Kober H, et al. Magnetic source imaging combined with image-guided frameless stereotaxy: a new method in surgery around the motor strip. *Neurosurgery.* 1997;41(3):621–628.
10. Ganslandt O, Fahlbusch R, Nimsky C, et al. Functional neuronavigation with magnetoencephalography: outcome in 50 patients with lesions around the motor cortex. *J Neurosurg.* 1999;91(July):73–79.
11. Nimsky C, Ganslandt O, Kober H, et al. Integration of functional magnetic resonance imaging supported by magnetoencephalography in functional neuronavigation. *Neurosurgery.* 1999;44(6): 1249–1255:discussion 55–56.
12. Nimsky C, Ganslandt O, Fahlbusch R. 1.5 T: intraoperative imaging beyond standard anatomic imaging. *Neurosurg Clin N Am.* 2005;16(1):185–200:vii.
13. Nimsky C, Ganslandt O, Fahlbusch R. Implementation of fiber tract navigation. *Neurosurgery.* 2006;58(4 Suppl 2):ONS-292–ONS-304.
14. Nimsky C, Ganslandt O, Hastreiter P, et al. Preoperative and intraoperative diffusion tensor imaging-based fiber tracking in glioma surgery. *Neurosurgery.* 2005;56(1):130–137:discussion 38.
15. Stadlbauer A, Ganslandt O, Buslei R, et al. Gliomas: histopathologic evaluation of changes in directionality and magnitude of water diffusion at diffusion-tensor MR imaging. *Radiology.* 2006;240(3): 803–810.
16. Stadlbauer A, Moser E, Gruber S, Nimsky C, Fahlbusch R, Ganslandt O. Integration of biochemical images of a tumor into frameless stereotaxy achieved using a magnetic resonance imaging/magnetic resonance spectroscopy hybrid data set. *J Neurosurg.* 2004;101(2):287–294.
17. Stadlbauer A, Nimsky C, Buslei R, et al. Proton magnetic resonance spectroscopic imaging in the border zone of gliomas: correlation of metabolic and histological changes at low tumor infiltration—initial results. *Invest Radiol.* 2007;42(4):218–223.
18. Stadlbauer A, Prante O, Nimsky C, et al. Metabolic imaging of cerebral gliomas: spatial correlation of changes in O-(2-18F-fluoroethyl)-L-tyrosine PET and proton magnetic resonance spectroscopic imaging. *J Nucl Med.* 2008;49(5):721–729.
19. Rey A. X-ray control in the operating theatre. *Acta Neurochir (Wien).* 1980;55(1–2):3–13.
20. Weise L, Eibach S, Seifert V, Setzer M. Intraoperative 3D fluoroscopy in stereotactic surgery. *Acta Neurochir (Wien).* 2012;154(5):815–821.
21. Waschke A, Walter J, Duenisch P, Reichart R, Kalff R, Ewald C. CT-navigation versus fluoroscopy-guided placement of pedicle screws at the thoracolumbar spine: single center experience of 4,500 screws. *Eur Spine J.* 2013;22(3):654–660.
22. Hott JS, Deshmukh VR, Klopfenstein JD, et al. Intraoperative Iso-C C-arm navigation in craniospinal surgery: the first 60 cases. *Neurosurgery.* 2004;54(5):1131–1136:discussion 36–37.
23. Irie K, Murayama Y, Saguchi T, et al. Dynact soft-tissue visualization using an angiographic C-arm system: initial clinical experience in the operating room. *Neurosurgery.* 2008;62(3 Suppl 1):266–272: discussion 72.
24. Starke RM, Dumont AS. Intraoperative imaging and assessment of cerebral blood flow in cerebrovascular surgery: hybrid operating rooms, intraoperative angiography and MRI, Doppler ultrasound, cerebral blood flow probes, endoscopic assistance, indocyanine green videography, and laser speckle contrast imaging. *World Neurosurg.* 2014;82(6):e693–e696.
25. Roberts DW. Applications of fluorescent technology in neurosurgery. *Neurosurg Focus.* 2014;36(2):E2.
26. Balamurugan S, Agrawal A, Kato Y, Sano H. Intra operative indocyanine green video-angiography in cerebrovascular surgery: an overview with review of literature. *Asian J Neurosurg.* 2011;6(2):88–93

27. Mielke D, Malinova V, Rohde V. Comparison of intraoperative microscopic and endoscopic ICG-angiography in aneurysm surgery. *Neurosurgery.* 2014;10(Suppl 3):418–425.
28. Zaidi HA, Abla AA, Nakaji P, Chowdhry SA, Albuquerque FC, Spetzler RF. Indocyanine green angiography in the surgical management of cerebral arteriovenous malformations: lessons learned in 130 consecutive cases. *Neurosurgery.* 2014;10(Suppl 2):246–251.
29. Stummer W, Pichlmeier U, Meinel T, et al. Fluorescence-guided surgery with 5-aminolevulinic acid for resection of malignant glioma: a randomised controlled multicentre phase III trial. *Lancet Oncol.* 2006;7(5):392–401.
30. Li Y, Rey-Dios R, Roberts DW, Valdes PA, Cohen-Gadol AA. Intraoperative fluorescence-guided resection of high-grade gliomas: a comparison of the present techniques and evolution of future strategies. *World Neurosurg.* 2014;82(1–2):175–185.
31. Ritz R, Daniels R, Noell S, et al. Hypericin for visualization of high grade gliomas: first clinical experience. *Eur J Surg Oncol.* 2012;38(4):352–360.
32. Ritz R, Scheidle C, Noell S, et al. In vitro comparison of hypericin and 5-aminolevulinic acid-derived protoporphyrin IX for photodynamic inactivation of medulloblastoma cells. *PLoS One.* 2012;7(12):e51974.
33. Bi WL, Laws Jr. ER. Searching for the light: fluorescence guidance in glioma resection. *World Neurosurg.* 2014;82(1–2):54–55.
34. Martirosyan NL, Georges J, Eschbacher JM, et al. Potential application of a handheld confocal endomicroscope imaging system using a variety of fluorophores in experimental gliomas and normal brain. *Neurosurg Focus.* 2014;36(2):E16.
35. Kantelhardt SR, Finke M, Schweikard A, Giese A. Evaluation of a completely robotized neurosurgical operating microscope. *Neurosurgery.* 2013;72(Suppl 1):19–26.
36. Assayag O, Grieve K, Devaux B, et al. Imaging of non-tumorous and tumorous human brain tissues with full-field optical coherence tomography. *Neuroimage Clin.* 2013;2:549–557.
37. Reid MH. Ultrasonic visualization of a cervical cord cystic astrocytoma. *Am J Roentgenol.* 1978;131(5):907–908.
38. Sosna J, Barth MM, Kruskal JB, Kane RA. Intraoperative sonography for neurosurgery. *J Ultrasound Med.* 2005;24(12):1671–1682.
39. Moiyadi AV. Objective assessment of intraoperative ultrasound in brain tumors. *Acta Neurochir (Wien).* 2014;156(4):703–704.
40. Muns A, Meixensberger J, Arnold S, et al. Integration of a 3D ultrasound probe into neuronavigation. *Acta Neurochir (Wien).* 2011;153(7):1529–1533.
41. Unsgaard G, Rygh OM, Selbekk T, et al. Intra-operative 3D ultrasound in neurosurgery. *Acta Neurochir (Wien).* 2006;148(3):235–253:discussion 53.
42. Mair R, Heald J, Poeata I, Ivanov M. A practical grading system of ultrasonographic visibility for intracerebral lesions. *Acta Neurochir (Wien).* 2013;155(12):2293–2298.
43. Prada F, Perin A, Martegani A, et al. Intraoperative contrast enhanced ultra-sound (iCEUS) for brain surgery. *Neurosurgery.* 2014;74(5):542–552.
44. Serra C, Stauffer A, Actor B, et al. Intraoperative high frequency ultrasound in intracerebral high-grade tumors. *Ultraschall Med.* 2012;33(7):E306–E312.
45. Hammoud MA, Ligon BL, elSouki R, Shi WM, Schomer DF, Sawaya R. Use of intraoperative ultrasound for localizing tumors and determining the extent of resection: a comparative study with magnetic resonance imaging. *J Neurosurg.* 1996;84(5):737–741.
46. Le Roux PD, Berger MS, Wang K, Mack LA, Ojemann GA. Low grade gliomas: comparison of intraoperative ultrasound characteristics with preoperative imaging studies. *J Neurooncol.* 1992;13(2):189–198.
47. Woydt M, Krone A, Becker G, Schmidt K, Roggendorf W, Roosen K. Correlation of intra-operative ultrasound with histopathologic findings after tumour resection in supratentorial gliomas. A method to improve gross total tumour resection. *Acta Neurochir (Wien).* 1996;138(12):1391–1398.
48. Ram Z, Bruck B, Hadani M. Ultrasound in pituitary tumor surgery. *Pituitary.* 1999;2(2):133–138.
49. Arita K, Kurisu K, Tominaga A, et al. Trans-sellar color Doppler ultrasonography during transsphenoidal surgery. *Neurosurgery.* 1998;42(1):81–85:discussion 86.

50. Watson JC, Shawker TH, Nieman LK, DeVroom HL, Doppman JL, Oldfield EH. Localization of pituitary adenomas by using intraoperative ultrasound in patients with Cushing's disease and no demonstrable pituitary tumor on magnetic resonance imaging. *J Neurosurg*. 1998;89(6): 927–932.
51. Atkinson JL, Kasperbauer JL, James EM, Lane JI, Nippoldt TB. Transcranial-transdural real-time ultrasonography during transsphenoidal resection of a large pituitary tumor. Case report. *J Neurosurg*. 2000;93(1):129–131.
52. Saether CA, Torsteinsen M, Torp SH, Sundstrom S, Unsgard G, Solheim O. Did survival improve after the implementation of intraoperative neuronavigation and 3D ultrasound in glioblastoma surgery? A retrospective analysis of 192 primary operations. *J Neurol Surg A Cent Eur Neurosurg*. 2012;73(2):73–78.
53. Berntsen EM, Gulati S, Solheim O, et al. Functional magnetic resonance imaging and diffusion tensor tractography incorporated into an intraoperative 3-dimensional ultrasound-based neuronavigation system: impact on therapeutic strategies, extent of resection, and clinical outcome. *Neurosurgery*. 2010;67(2):251–264.
54. Gerganov VM, Samii A, Giordano M, Samii M, Fahlbusch R. Two-dimensional high-end ultrasound imaging compared to intraoperative MRI during resection of low-grade gliomas. *J Clin Neurosci*. 2011;18(5):669–673.
55. Gerganov VM, Samii A, Akbarian A, Stieglitz L, Samii M, Fahlbusch R. Reliability of intraoperative high-resolution 2D ultrasound as an alternative to high-field strength MR imaging for tumor resection control: a prospective comparative study. *J Neurosurg*. 2009;111(3):512–519.
56. Goren O, Monteith SJ, Hadani M, Bakon M, Harnof S. Modern intraoperative imaging modalities for the vascular neurosurgeon treating intracerebral hemorrhage. *Neurosurg Focus*. 2013;34(5):E2.
57. Heiroth HJ, Etminan N, Steiger HJ, Hanggi D. Intraoperative Doppler and Duplex sonography in cerebral aneurysm surgery. *Br J Neurosurg*. 2011;25(5):586–590.
58. Shalit MN, Israeli Y, Matz S, Cohen ML. Intra-operative computerized axial tomography. *Surg Neurol*. 1979;11(5):382–384.
59. Lunsford LD, Parrish R, Albright L. Intraoperative imaging with a therapeutic computed tomographic scanner. *Neurosurgery*. 1984;15(4):559–561.
60. Okudera H, Kyoshima K, Kobayashi S, Sugita K. Intraoperative CT scan findings during resection of glial tumours. *Neurol Res*. 1994;16(4):265–267.
61. Okudera H, Kobayashi S, Kyoshima K, Gibo H, Takemae T, Sugita K. Development of the operating computerized tomographic scanner system for neurosurgery. *Acta Neurochir (Wien)*. 1991;111 (1–2):61–63.
62. Hum B, Feigenbaum F, Cleary K, Henderson FC. Intraoperative computed tomography for complex craniocervical operations and spinal tumor resections. *Neurosurgery*. 2000;47(2):374–380:discussion 80–81.
63. Matula C, Rossler K, Reddy M, Schindler E, Koos WT. Intraoperative computed tomography guided neuronavigation: concepts, efficiency, and work flow. *Comput Aided Surg*. 1998;3(4):174–182.
64. Grunert P, Muller-Forell W, Darabi K, et al. Basic principles and clinical applications of neuronavigation and intraoperative computed tomography. *Comput Aided Surg*. 1998;3(4):166–173.
65. Butler WE, Piaggio CM, Constantinou C, et al. A mobile computed tomographic scanner with intraoperative and intensive care unit applications. *Neurosurgery*. 1998;42(6):1304–1310:discussion 10–11.
66. Uhl E, Zausinger S, Morhard D, et al. Intraoperative computed tomography with integrated navigation system in a multidisciplinary operating suite. *Neurosurgery*. 2009;64(5 Suppl 2):231–239:discussion 39–40.
67. Hosoda T, Takeuchi H, Hashimoto N, et al. Usefulness of intraoperative computed tomography in surgery for low-grade gliomas: a comparative study between two series without and with intraoperative computed tomography. *Neurol Med Chir (Tokyo)*. 2011;51(7):490–495.
68. Lee CC, Lee ST, Chang CN, et al. Volumetric measurement for comparison of the accuracy between intraoperative CT and postoperative MR imaging in pituitary adenoma surgery. *Am J Neuroradiol*. 2011;32(8):1539–1544.

69. Mori R, Joki T, Matsuwaki Y, Karagiozov K, Murayama Y, Abe T. Initial experience of real-time intraoperative C-arm computed-tomography-guided navigation surgery for pituitary tumors. *World Neurosurg*. 2013;79(2):319–326.
70. Holloway K, Docef A. A quantitative assessment of the accuracy and reliability of O-arm images for deep brain stimulation surgery. *Neurosurgery*. 2013;72(1 Suppl Operative):47–57.
71. Burchiel KJ, McCartney S, Lee A, Raslan AM. Accuracy of deep brain stimulation electrode placement using intraoperative computed tomography without microelectrode recording. *J Neurosurg*. 2013;119(2):301–306.
72. Smith AP, Bakay RA. Frameless deep brain stimulation using intraoperative O-arm technology. Clinical article. *J Neurosurg*. 2011;115(2):301–309.
73. Schnell O, Morhard D, Holtmannspotter M, Reiser M, Tonn JC, Schichor C. Near-infrared indocyanine green videoangiography (ICGVA) and intraoperative computed tomography (iCT): are they complementary or competitive imaging techniques in aneurysm surgery? *Acta Neurochir (Wien)*. 2012;154(10):1861–1868.
74. Schichor C, Rachinger W, Morhard D, et al. Intraoperative computed tomography angiography with computed tomography perfusion imaging in vascular neurosurgery: feasibility of a new concept. *J Neurosurg*. 2010;112(4):722–728.
75. Scheufler KM, Franke J, Eckardt A, Dohmen H. Accuracy of image-guided pedicle screw placement using intraoperative computed tomography-based navigation with automated referencing. Part II: thoracolumbar spine. *Neurosurgery*. 2011;69(6):1307–1316.
76. Scheufler KM, Franke J, Eckardt A, Dohmen H. Accuracy of image-guided pedicle screw placement using intraoperative computed tomography-based navigation with automated referencing, part I: cervicothoracic spine. *Neurosurgery*. 2011;69(4):782–795:discussion 95.
77. Drazin D, Kim TT, Polly Jr. DW, Johnson JP. Introduction: intraoperative spinal imaging and navigation. *Neurosurg Focus*. 2014;36(3):Introduction.
78. Hsieh JC, Drazin D, Firempong AO, Pashman R, Johnson JP, Kim TT. Accuracy of intraoperative computed tomography image-guided surgery in placing pedicle and pelvic screws for primary versus revision spine surgery. *Neurosurg Focus*. 2014;36(3):E2.
79. Kim TT, Drazin D, Shweikeh F, Pashman R, Johnson JP. Clinical and radiographic outcomes of minimally invasive percutaneous pedicle screw placement with intraoperative CT (O-arm) image guidance navigation. *Neurosurg Focus*. 2014;36(3):E1.
80. Rivkin MA, Yocom SS. Thoracolumbar instrumentation with CT-guided navigation (O-arm) in 270 consecutive patients: accuracy rates and lessons learned. *Neurosurg Focus*. 2014;36(3):E7.
81. Zausinger S, Scheder B, Uhl E, Heigl T, Morhard D, Tonn JC. Intraoperative computed tomography with integrated navigation system in spinal stabilizations. *Spine (Phila Pa 1976)*. 2009;34(26): 2919–2926.
82. Dinesh SK, Tiruchelvarayan R, Ng I. A prospective study on the use of intraoperative computed tomography (iCT) for image-guided placement of thoracic pedicle screws. *Br J Neurosurg*. 2012;26 (6):838–844.
83. Costa F, Cardia A, Ortolina A, Fabio G, Zerbi A, Fornari M. Spinal navigation: standard preoperative versus intraoperative computed tomography data set acquisition for computer-guidance system: radiological and clinical study in 100 consecutive patients. *Spine (Phila Pa 1976)*. 2011;36(24): 2094–2098.
84. Costa F, Tomei M, Sassi M, et al. Evaluation of the rate of decompression in anterior cervical corpectomy using an intra-operative computerized tomography scan (O-Arm system). *Eur Spine J*. 2012;21(2):359–363.
85. Bandela JR, Jacob RP, Arreola M, Griglock TM, Bova F, Yang M. Use of CT-based intraoperative spinal navigation: management of radiation exposure to operator, staff, and patients. *World Neurosurg*. 2013;79(2):390–394.
86. Black PM, Moriarty T, Alexander III E, et al. Development and implementation of intraoperative magnetic resonance imaging and its neurosurgical applications. *Neurosurgery*. 1997;41(4):831–845.
87. Hall WA, Kowalik K, Liu H, Truwit CL, Kucharezyk J. Costs and benefits of intraoperative MR-guided brain tumor resection. *Acta Neurochir Suppl*. 2003;85:137–142.

88. Hall WA, Liu H, Martin AJ, Pozza CH, Maxwell RE, Truwit CL. Safety, efficacy, and functionality of high-field strength interventional magnetic resonance imaging for neurosurgery. *Neurosurgery*. 2000;46(3):632–642.
89. Nimsky C, Ganslandt O, Fahlbusch R. Comparing 0.2 Tesla with 1.5 Tesla intraoperative magnetic resonance imaging analysis of setup, workflow, and efficiency. *Acad Radiol*. 2005;12(9):1065–1079.
90. Sutherland GR, Kaibara T, Louw D, Hoult DI, Tomanek B, Saunders J. A mobile high-field magnetic resonance system for neurosurgery. *J Neurosurg*. 1999;91(5):804–813.
91. Kubben PL, ter Meulen KJ, Schijns OE, ter Laak-Poort MP, van Overbeeke JJ, van Santbrink H. Intraoperative MRI-guided resection of glioblastoma multiforme: a systematic review. *Lancet Oncol*. 2011;12(11):1062–1070.
92. Nimsky C. Intraoperative MRI in glioma surgery: proof of benefit? *Lancet Oncol*. 2011;12(11): 982–983.
93. Bradley WG. Achieving gross total resection of brain tumors: intraoperative MR imaging can make a big difference. *Am J Neuroradiol*. 2002;23(3):348–349.
94. Tronnier VM, Wirtz CR, Knauth M, et al. Intraoperative diagnostic and interventional magnetic resonance imaging in neurosurgery. *Neurosurgery*. 1997;40(5):891–902.
95. Mislow JM, Golby AJ, Black PM. Origins of intraoperative MRI. *Neurosurg Clin N Am*. 2009;20 (2):137–146.
96. Schwartz RB, Hsu L, Wong TZ, et al. Intraoperative MR imaging guidance for intracranial neurosurgery: experience with the first 200 cases. *Radiology*. 1999;211(2):477–488.
97. Seifert V, Zimmermann M, Trantakis C, et al. Open MRI-guided neurosurgery. *Acta Neurochir (Wien)*. 1999;141(5):455–464.
98. Black PM, Alexander III E, Martin C, et al. Craniotomy for tumor treatment in an intraoperative magnetic resonance imaging unit. *Neurosurgery*. 1999;45(3):423–433.
99. Schneider JP, Schulz T, Schmidt F, et al. Gross-total surgery of supratentorial low-grade gliomas under intraoperative MR guidance. *Am J Neuroradiol*. 2001;22(1):89–98.
100. Wirtz CR, Knauth M, Staubert A, et al. Clinical evaluation and follow-up results for intraoperative magnetic resonance imaging in neurosurgery. *Neurosurgery*. 2000;46(5):1112–1122.
101. Nimsky C, Ganslandt O, Tomandl B, Buchfelder M, Fahlbusch R. Low-field magnetic resonance imaging for intraoperative use in neurosurgery: a 5 year experience. *Eur Radiol*. 2002;12(11):2690–2703.
102. Kober H, Moller M, Nimsky C, Vieth J, Fahlbusch R, Ganslandt O. New approach to localize speech relevant brain areas and hemispheric dominance using spatially filtered magnetoencephalography. *Hum Brain Mapp*. 2001;14(4):236–250.
103. Kober H, Nimsky C, Moller M, Hastreiter P, Fahlbusch R, Ganslandt O. Correlation of sensorimotor activation with functional magnetic resonance imaging and magnetoencephalography in presurgical functional imaging: a spatial analysis. *Neuroimage*. 2001;14(5):1214–1228.
104. Kober H, Nimsky C, Vieth J, Fahlbusch R, Ganslandt O. Co-registration of function and anatomy in frameless stereotaxy by contour fitting. *Stereotact Funct Neurosurg*. 2002;79(3–4):272–283.
105. Bohinski RJ, Kokkino AK, Warnick RE, et al. Glioma resection in a shared-resource magnetic resonance operating room after optimal image-guided frameless stereotactic resection. *Neurosurgery*. 2001;48(4):731–744.
106. Knauth M, Wirtz CR, Tronnier VM, Aras N, Kunze S, Sartor K. Intraoperative MR imaging increases the extent of tumor resection in patients with high-grade gliomas. *Am J Neuroradiol*. 1999;20(9):1642–1646.
107. Martin CH, Schwartz R, Jolesz F, Black PM. Transsphenoidal resection of pituitary adenomas in an intraoperative MRI unit. *Pituitary*. 1999;2:155–162.
108. Pergolizzi Jr. RS, Nabavi A, Schwartz RB, et al. Intra-operative MR guidance during transsphenoidal pituitary resection: preliminary results. *J Magn Reson Imaging*. 2001;13(1):136–141.
109. Bohinski RJ, Warnick RE, Gaskill-Shipley MF, et al. Intraoperative magnetic resonance imaging to determine the extent of resection of pituitary macroadenomas during transsphenoidal microsurgery. *Neurosurgery*. 2001;49(5):1133–1144.
110. Fahlbusch R, Ganslandt O, Buchfelder M, Schott W, Nimsky C. Intraoperative magnetic resonance imaging during transsphenoidal surgery. *J Neurosurg*. 2001;95(3):381–390.

111. Schwartz TH, Marks D, Pak J, et al. Standardization of amygdalohippocampectomy with intraoperative magnetic resonance imaging: preliminary experience. *Epilepsia*. 2002;43(4):430–436.
112. Kaibara T, Myles ST, Lee MA, Sutherland GR. Optimizing epilepsy surgery with intraoperative MR imaging. *Epilepsia*. 2002;43(4):425–429.
113. Buchfelder M, Fahlbusch R, Ganslandt O, Stefan H, Nimsky C. Use of intraoperative magnetic resonance imaging in tailored temporal lobe surgeries for epilepsy. *Epilepsia*. 2002;43(8):864–873.
114. Buchfelder M, Ganslandt O, Fahlbusch R, Nimsky C. Intraoperative magnetic resonance imaging in epilepsy surgery. *J Magn Reson Imaging*. 2000;12:547–555.
115. Nabavi A, Black PM, Gering DT, et al. Serial intraoperative magnetic resonance imaging of brain shift. *Neurosurgery*. 2001;48(4):787–798.
116. Nimsky C, Ganslandt O, Cerny S, Hastreiter P, Greiner G, Fahlbusch R. Quantification of, visualization of, and compensation for brain shift using intraoperative magnetic resonance imaging. *Neurosurgery*. 2000;47(5):1070–1079:discussion 79–80.
117. Nimsky C, Ganslandt O, Hastreiter P, Fahlbusch R. Intraoperative compensation for brain shift. *Surg Neurol*. 2001;56(6):357–364:discussion 64–65.
118. Wirtz CR, Bonsanto MM, Knauth M, et al. Intraoperative magnetic resonance imaging to update interactive navigation in neurosurgery: method and preliminary experience. *Comput Aided Surg*. 1997;2:172–179.
119. Hall WA, Martin AJ, Liu H, Nussbaum ES, Maxwell RE, Truwit CL. Brain biopsy using high-field strength interventional magnetic resonance imaging. *Neurosurgery*. 1999;44(4):807–814.
120. Kaibara T, Saunders JK, Sutherland GR. Advances in mobile intraoperative magnetic resonance imaging. *Neurosurgery*. 2000;47(1):131–138.
121. Nimsky C, Ganslandt O, Keller VB, Fahlbusch R. Preliminary experience in glioma surgery with intraoperative high-field MRI. *Acta Neurochir Suppl*. 2003;88:21–29.
122. Livne O, Harhel R, Hadani M, Spiegelmann R, Feldman Z, Cohen ZR. Intraoperative magnetic resonance imaging for resection of intra-axial brain lesions. A decade of experience using low-field MRI, Polestar N-10, 20, 30 systems. *World Neurosurg*. 2014;82(5):770–776.
123. Schulder M, Liang D, Carmel PW. Cranial surgery navigation aided by a compact intraoperative magnetic resonance imager. *J Neurosurg*. 2001;94(6):936–945.
124. Schulder M, Spiro D. Intraoperative MRI for stereotactic biopsy. *Acta Neurochir Suppl*. 2011;109:81–87.
125. Kim PD, Truwit CL, Hall WA. Three-Tesla high-field applications. *Neurosurg Clin N Am*. 2009;20(2):173–178.
126. Lang MJ, Kelly JJ, Sutherland GR. A moveable 3-Tesla intraoperative magnetic resonance imaging system. *Neurosurgery*. 2011;68(1 Suppl Operative):168–179.
127. Truwit CL, Hall WA. Intraoperative magnetic resonance imaging-guided neurosurgery at 3-T. *Neurosurgery*. 2006;58(4 Suppl 2):ONS-338–ONS-345:discussion ONS-45–ONS-46.
128. Lacey C, Sutherland G. Advancing neurosurgery through translational research. *Neurosurgery*. 2013;72(Suppl 1):176–181.
129. Sutherland GR, Lama S, Gan LS, Wolfsberger S, Zareinia K. Merging machines with microsurgery: clinical experience with neuroArm. *J Neurosurg*. 2013;118(3):521–529.
130. Sutherland GR, Wolfsberger S, Lama S, Zarei-nia K. The evolution of neuroArm. *Neurosurgery*. 2013;72(Suppl 1):27–32.
131. Makary M, Chiocca EA, Erminy N, et al. Clinical and economic outcomes of low-field intraoperative MRI-guided tumor resection neurosurgery. *J Magn Reson Imaging*. 2011;34(5):1022–1030.
132. Ramina R, Coelho Neto M, Giacomelli A, et al. Optimizing costs of intraoperative magnetic resonance imaging. A series of 29 glioma cases. *Acta Neurochir (Wien)*. 2010;152(1):27–33.
133. Duffau H. The conceptual limitation to rely on intraoperative MRI in glioma surgery. *World Neurosurg*. 2014;82(5):601–603.
134. Nimsky C, Ganslandt O, Buchfelder M, Fahlbusch R. Glioma surgery evaluated by intraoperative low-field magnetic resonance imaging. *Acta Neurochir Suppl*. 2003;85:55–63.
135. Hirschberg H, Samset E, Hol PK, Tillung T, Lote K. Impact of intraoperative MRI on the surgical results for high-grade gliomas. *Minim Invasive Neurosurg*. 2005;48(2):77–84.

136. Schneider JP, Trantakis C, Rubach M, et al. Intraoperative MRI to guide the resection of primary supratentorial glioblastoma multiforme—a quantitative radiological analysis. *Neuroradiology*. 2005;47(7):489–500.
137. Busse H, Schmitgen A, Trantakis C, Schober R, Kahn T, Moche M. Advanced approach for intraoperative MRI guidance and potential benefit for neurosurgical applications. *J Magn Reson Imaging*. 2006;24(1):140–151.
138. Muragaki Y, Iseki H, Maruyama T, et al. Usefulness of intraoperative magnetic resonance imaging for glioma surgery. *Acta Neurochir Suppl*. 2006;98:67–75.
139. Nimsky C, Ganslandt O, Buchfelder M, Fahlbusch R. Intraoperative visualization for resection of gliomas: the role of functional neuronavigation and intraoperative 1.5 T MRI. *Neurol Res*. 2006;28(5):482–487.
140. Hatiboglu MA, Weinberg JS, Suki D, et al. Impact of intraoperative high-field magnetic resonance imaging guidance on glioma surgery: a prospective volumetric analysis. *Neurosurgery*. 2009;64(6):1073–1081:discussion 81.
141. Lenaburg HJ, Inkabi KE, Vitaz TW. The use of intraoperative MRI for the treatment of glioblastoma multiforme. *Technol Cancer Res Treat*. 2009;8(2):159–162.
142. Senft C, Franz K, Blasel S, et al. Influence of iMRI-guidance on the extent of resection and survival of patients with glioblastoma multiforme. *Technol Cancer Res Treat*. 2010;9(4):339–346.
143. Senft C, Bink A, Franz K, Vatter H, Gasser T, Seifert V. Intraoperative MRI guidance and extent of resection in glioma surgery: a randomised, controlled trial. *Lancet Oncol*. 2011;12(11):997–1003.
144. Kuhnt D, Ganslandt O, Schlaffer SM, Buchfelder M, Nimsky C. Quantification of glioma removal by intraoperative high-field magnetic resonance imaging: an update. *Neurosurgery*. 2011;69(4):852–862: discussion 62–63.
145. Kuhnt D, Becker A, Ganslandt O, Bauer M, Buchfelder M, Nimsky C. Correlation of the extent of tumor volume resection and patient survival in surgery of glioblastoma multiforme with high-field intraoperative MRI guidance. *Neuro Oncol*. 2011;13(12):1339–1348.
146. Sanai N, Polley MY, McDermott MW, Parsa AT, Berger MS. An extent of resection threshold for newly diagnosed glioblastomas. *J Neurosurg*. 2011;115(1):3–8.
147. Mehdorn HM, Schwartz F, Dawirs S, Hedderich J, Dorner L, Nabavi A. High-field iMRI in glioblastoma surgery: improvement of resection radicality and survival for the patient? *Acta Neurochir Suppl*. 2011;109:103–106.
148. Mohammadi AM, Sullivan TB, Barnett GH, et al. Use of high-field intraoperative magnetic resonance imaging to enhance the extent of resection of enhancing and nonenhancing gliomas. *Neurosurgery*. 2014;74(4):339–350.
149. Haydon D, Dacey RJ. Use of high-field intraoperative magnetic resonance imaging to enhance the extent of resection of enhancing and nonenhancing gliomas (comment). *Neurosurgery*. 2014;74(4):349.
150. Senft C, Forster MT, Bink A, et al. Optimizing the extent of resection in eloquently located gliomas by combining intraoperative MRI guidance with intraoperative neurophysiological monitoring. *J Neurooncol*. 2012;109(1):81–90.
151. Tuominen J, Yrjana S, Ukkonen A, Koivukangas J. Awake craniotomy may further improve neurological outcome of intraoperative MRI-guided brain tumor surgery. *Acta Neurochir (Wien)*. 2013;155(10):1805–1812.
152. Lu J, Wu J, Yao C, et al. Awake language mapping and 3-Tesla intraoperative MRI-guided volumetric resection for gliomas in language areas. *J Clin Neurosci*. 2013;20(9):1280–1287.
153. Peruzzi P, Puente E, Bergese S, Chiocca EA. Intraoperative MRI (ioMRI) in the setting of awake craniotomies for supratentorial glioma resection. *Acta Neurochir Suppl*. 2011;109:43–48.
154. Parney IF, Goerss SJ, McGee K, Huston 3rd J, Perkins WJ, Meyer FB. Awake craniotomy, electrophysiologic mapping, and tumor resection with high-field intraoperative MRI. *World Neurosurg*. 2010;73(5):547–551.
155. Goebel S, Nabavi A, Schubert S, Mehdorn HM. Patient perception of combined awake brain tumor surgery and intraoperative 1.5-T magnetic resonance imaging: the Kiel experience. *Neurosurgery*. 2010;67(3):594–600:discussion 00.

156. Coburger J, Engelke J, Scheuerle A, et al. Tumor detection with 5-aminolevulinic acid fluorescence and Gd-DTPA-enhanced intraoperative MRI at the border of contrast-enhancing lesions: a prospective study based on histopathological assessment. *Neurosurg Focus*. 2014;36(2):E3.
157. Tsugu A, Ishizaka H, Mizokami Y, et al. Impact of the combination of 5-aminolevulinic acid-induced fluorescence with intraoperative magnetic resonance imaging-guided surgery for glioma. *World Neurosurg*. 2011;76(1–2):120–127.
158. Eyupoglu IY, Hore N, Savaskan NE, et al. Improving the extent of malignant glioma resection by dual intraoperative visualization approach. *PLoS One*. 2012;7(9):e44885.
159. Roder C, Bisdas S, Ebner FH, et al. Maximizing the extent of resection and survival benefit of patients in glioblastoma surgery: high-field iMRI versus conventional and 5-ALA-assisted surgery. *Eur J Surg Oncol*. 2014;40(3):297–304.
160. Berger MS. Use of 5-aminolevulinic acid helps see the way beyond MRI. *Neurosurg Focus*. 2014;36(2):E4.
161. Hlavica M, Bellut D, Lemm D, Schmid C, Bernays RL. Impact of ultra-low-field intraoperative magnetic resonance imaging on extent of resection and frequency of tumor recurrence in 104 surgically treated nonfunctioning pituitary adenomas. *World Neurosurg*. 2013;79(1):99–109.
162. Berkmann S, Fandino J, Muller B, Remonda L, Landolt H. Intraoperative MRI and endocrinological outcome of transsphenoidal surgery for non-functioning pituitary adenoma. *Acta Neurochir (Wien)*. 2012;154(4):639–647.
163. Vitaz TW, Inkabi KE, Carrubba CJ. Intraoperative MRI for transphenoidal procedures: short-term outcome for 100 consecutive cases. *Clin Neurol Neurosurg*. 2011;113(9):731–735.
164. Pergolizzi Jr. RS, Nabavi A, Schwartz RB, et al. Intra-operative MR guidance during trans-sphenoidal pituitary resection: preliminary results. *J Magn Reson Imaging*. 2001;13(1):136–141.
165. Coburger J, Konig R, Seitz K, Bazner U, Wirtz CR, Hlavac M. Determining the utility of intraoperative magnetic resonance imaging for transsphenoidal surgery: a retrospective study. *J Neurosurg*. 2014;120(2):346–356.
166. Szerlip NJ, Zhang YC, Placantonakis DG, et al. Transsphenoidal resection of sellar tumors using high-field intraoperative magnetic resonance imaging. *Skull Base*. 2011;21(4):223–232.
167. Boellis A, Espagnet MC, Romano A, et al. Dynamic intraoperative MRI in transsphenoidal resection of pituitary macroadenomas: a quantitative analysis. *J Magn Reson Imaging*. 2014;40(3):668–673.
168. Netuka D, Masopust V, Belsan T, Kramar F, Benes V. One year experience with 3.0 T intraoperative MRI in pituitary surgery. *Acta Neurochir Suppl*. 2011;109:157–159.
169. Nimsky C, von Keller B, Ganslandt O, Fahlbusch R. Intraoperative high-field magnetic resonance imaging in transsphenoidal surgery of hormonally inactive pituitary macroadenomas. *Neurosurgery*. 2006;59(1):105–114:discussion 05–14.
170. Fahlbusch R, Keller B, Ganslandt O, Kreutzer J, Nimsky C. Transsphenoidal surgery in acromegaly investigated by intraoperative high-field magnetic resonance imaging. *Eur J Endocrinol*. 2005;153(2):239–248.
171. Nimsky C, Ganslandt O, Hofmann B, Fahlbusch R. Limited benefit of intraoperative low-field magnetic resonance imaging in craniopharyngioma surgery. *Neurosurgery*. 2003;53(1):72–80:discussion 80–81.
172. Hofmann BM, Nimsky C, Fahlbusch R. Benefit of 1.5-T intraoperative MR imaging in the surgical treatment of craniopharyngiomas. *Acta Neurochir (Wien)*. 2011;153(7):1377–1390:discussion 90.
173. Woodard EJ, Leon SP, Moriarty TM, Quinones A, Zamani AA, Jolesz FA. Initial experience with intraoperative magnetic resonance imaging in spine surgery. *Spine (Phila Pa 1976)*. 2001;26(4):410–417.
174. Dhaliwal PP, Hurlbert RJ, Sutherland GS. Intraoperative magnetic resonance imaging and neuronavigation for transoral approaches to upper cervical pathology. *World Neurosurg*. 2012;78(1–2):164–169.
175. Gralla J, Ganslandt O, Kober H, Buchfelder M, Fahlbusch R, Nimsky C. Image-guided removal of supratentorial cavernomas in critical brain areas: application of neuronavigation and intraoperative magnetic resonance imaging. *Minim Invasive Neurosurg*. 2003;46(2):72–77.

176. Vitaz TW, Hushek S, Shields CB, Moriarty T. Intraoperative MRI for pediatric tumor management. *Acta Neurochir Suppl.* 2003;85:73–78.
177. Nimsky C, Ganslandt O, Gralla J, Buchfelder M, Fahlbusch R. Intraoperative low-field magnetic resonance imaging in pediatric neurosurgery. *Pediatr Neurosurg.* 2003;38(2):83–89.
178. Abernethy LJ, Avula S, Hughes GM, Wright EJ, Mallucci CL. Intra-operative 3-T MRI for paediatric brain tumours: challenges and perspectives. *Pediatr Radiol.* 2012;42(2):147–157.
179. Avula S, Mallucci CL, Pizer B, Garlick D, Crooks D, Abernethy LJ. Intraoperative 3-Tesla MRI in the management of paediatric cranial tumours—initial experience. *Pediatr Radiol.* 2012;42(2):158–167.
180. Nabavi A, Gering DT, Kacher DF, et al. Surgical navigation in the open MRI. *Acta Neurochir Suppl.* 2003;85:121–125.
181. Hall WA, Liu H, Truwit CL. Navigus trajectory guide. *Neurosurgery.* 2000;46(2):502–504.
182. Tanaka S, Puffer RC, Hoover JM, et al. Increased frameless stereotactic accuracy with high-field intraoperative magnetic resonance imaging. *Neurosurgery.* 2012;71(2 Suppl Operative):ons321–ons327: discussion ons27–ons28.
183. Kelly JJ, Hader WJ, Myles ST, Sutherland GR. Epilepsy surgery with intraoperative MRI at 1.5 T. *Neurosurg Clin N Am.* 2005;16(1):173–183.
184. Ferrant M, Nabavi A, Macq B, et al. Serial registration of intraoperative MR images of the brain. *Med Image Anal.* 2002;6(4):337–359.
185. Hastreiter P, Rezk-Salama C, Soza G, et al. Strategies for brain shift evaluation. *Med Image Anal.* 2004;8(4):447–464.
186. Nimsky C, von Keller B, Schlaffer S, et al. Updating navigation with intraoperative image data. *Top Magn Reson Imaging.* 2009;19(4):197–204.
187. Nimsky C, Ganslandt O, Hastreiter P, et al. Intraoperative diffusion tensor imaging: shifting of white matter tracts during neurosurgical procedures—initial experience. *Radiology.* 2005;234(1):218–225.
188. Dorward NL, Alberti O, Velani B, et al. Postimaging brain distortion: magnitude, correlates, and impact on neuronavigation. *J Neurosurg.* 1998;88(4):656–662.
189. Keles GE, Lamborn KR, Berger MS. Coregistration accuracy and detection of brain shift using intraoperative sononavigation during resection of hemispheric tumors. *Neurosurgery.* 2003;53(3):556–564.
190. Chen I, Coffey AM, Ding S, et al. Intraoperative brain shift compensation: accounting for dural septa. *IEEE Trans Biomed Eng.* 2011;58(3):499–508.
191. Soza G, Grosso R, Labsik U, et al. Fast and adaptive finite element approach for modeling brain shift. *Comput Aided Surg.* 2003;8(5):241–246.
192. Miga MI, Roberts DW, Kennedy FE, et al. Modeling of retraction and resection for intraoperative updating of images. *Neurosurgery.* 2001;49(1):75–84:discussion 84–85.
193. Roberts DW, Miga MI, Hartov A, et al. Intraoperatively updated neuroimaging using brain modeling and sparse data. *Neurosurgery.* 1999;45(5):1199–1206:discussion 206–207.
194. Vigneron LM, Noels L, Warfield SK, Verly JG, Robe PA. Serial FEM/XFEM-based update of preoperative brain images using intraoperative MRI. *Int J Biomed Imaging.* 2012;2012:872783.
195. Zhao Y, Chen X, Wang F, et al. Integration of diffusion tensor-based arcuate fasciculus fibre navigation and intraoperative MRI into glioma surgery. *J Clin Neurosci.* 2012;19(2):255–261.
196. Kuhnt D, Bauer MH, Becker A, et al. Intraoperative visualization of fiber tracking based reconstruction of language pathways in glioma surgery. *Neurosurgery.* 2012;70(4):911–919:discussion 19–20.
197. Richter M, Zolal A, Ganslandt O, Buchfelder M, Nimsky C, Merhof D. Evaluation of diffusion-tensor imaging-based global search and tractography for tumor surgery close to the language system. *PLoS One.* 2013;8(1):e50132.
198. Sun GC, Chen XL, Zhao Y, et al. Intraoperative high-field magnetic resonance imaging combined with fiber tract neuronavigation-guided resection of cerebral lesions involving optic radiation. *Neurosurgery.* 2011;69(5):1070–1084:discussion 84.
199. Chen X, Weigel D, Ganslandt O, Buchfelder M, Nimsky C. Prediction of visual field deficits by diffusion tensor imaging in temporal lobe epilepsy surgery. *Neuroimage.* 2009;45(2):286–297.
200. Kamada K, Todo T, Masutani Y, et al. Combined use of tractography-integrated functional neuronavigation and direct fiber stimulation. *J Neurosurg.* 2005;102(4):664–672.

201. Kinoshita M, Yamada K, Hashimoto N, et al. Fiber-tracking does not accurately estimate size of fiber bundle in pathological condition: initial neurosurgical experience using neuronavigation and subcortical white matter stimulation. *Neuroimage*. 2005;25(2):424–429.
202. Maesawa S, Fujii M, Nakahara N, Watanabe T, Wakabayashi T, Yoshida J. Intraoperative tractography and motor evoked potential (MEP) monitoring in surgery for gliomas around the corticospinal tract. *World Neurosurg*. 2010;74(1):153–161.
203. Shahar T, Rozovski U, Marko NF, et al. Preoperative imaging to predict intraoperative changes in tumor-to-corticospinal tract distance: an analysis of 45 cases using high-field intraoperative magnetic resonance imaging. *Neurosurgery*. 2014;75(1):23–30.
204. Merhof D, Soza G, Stadlbauer A, Greiner G, Nimsky C. Correction of susceptibility artifacts in diffusion tensor data using non-linear registration. *Med Image Anal*. 2007;11(6):588–603.
205. Farquharson S, Tournier JD, Calamante F, et al. White matter fiber tractography: why we need to move beyond DTI. *J Neurosurg*. 2013;118(6):1367–1377.
206. Fernandez-Miranda JC. Editorial: beyond diffusion tensor imaging. *J Neurosurg*. 2013;118(6):1363–1365:discussion 65–66.
207. Tournier JD, Mori S, Leemans A. Diffusion tensor imaging and beyond. *Magn Reson Med*. 2011;65(6):1532–1556.
208. Nimsky C. Fiber tracking—we should move beyond diffusion tensor imaging. *World Neurosurg*. 2014;82(1–2):35–36.
209. Stadlbauer A, Hammen T, Buchfelder M, et al. Differences in metabolism of fiber tract alterations in gliomas: a combined fiber density mapping and magnetic resonance spectroscopic imaging study. *Neurosurgery*. 2012;71(2):454–463.
210. Gering DT, Weber DM. Intraoperative, real-time, functional MRI. *J Magn Reson Imaging*. 1998;8(1):254–257.
211. Gasser T, Ganslandt O, Sandalcioglu E, Stolke D, Fahlbusch R, Nimsky C. Intraoperative functional MRI: implementation and preliminary experience. *Neuroimage*. 2005;26(3):685–693.
212. Gasser T, Szelenyi A, Senft C, et al. Intraoperative MRI and functional mapping. *Acta Neurochir Suppl*. 2011;109:61–65.
213. Lu JF, Zhang H, Wu JS, et al. "Awake" intraoperative functional MRI (ai-fMRI) for mapping the eloquent cortex: is it possible in awake craniotomy? *Neuroimage Clin*. 2012;2:132–142.
214. Kuhnt D, Bauer MH, Egger J, et al. Fiber tractography based on diffusion tensor imaging compared with high-angular-resolution diffusion imaging with compressed sensing: initial experience. *Neurosurgery*. 2013;72(Suppl 1):165–175.
215. Pamir MN, Ozduman K, Yildiz E, Sav A, Dincer A. Intraoperative magnetic resonance spectroscopy for identification of residual tumor during low-grade glioma surgery: clinical article. *J Neurosurg*. 2013;118(6):1191–1198.
216. Roder C, Skardelly M, Ramina KF, et al. Spectroscopy imaging in intraoperative MR suite: tissue characterization and optimization of tumor resection. *Int J Comput Assist Radiol Surg*. 2014;9(4):551–559.
217. Roder C, Bender B, Ritz R, et al. Intraoperative visualization of residual tumor: the role of perfusion-weighted imaging in a high-field intraoperative magnetic resonance scanner. *Neurosurgery*. 2013;72(2 Suppl Operative):ONS151–ONS158.
218. Jolesz FA. Intraoperative imaging in neurosurgery: where will the future take us? *Acta Neurochir Suppl*. 2011;109:21–25.
219. National Center for Image Guided Therapy (NCIGT.org). Advanced Multimodality Image Guided Operating (AMIGO) Suite. Available from: <http://www.ncigt.org/pages/AMIGO>; 2014.

SECTION II

Applications

CHAPTER 8

Image-Guided Brain Biopsy

Peter C. Warnke and Ashley Ralston
Section of Neurosurgery, The University of Chicago, Chicago, IL, USA

BRAIN BIOPSY V. ADVANCED IMAGING: WHY DO WE STILL NEED BRAIN BIOPSIES IN THE AGE OF PET, MRS AND MOLECULAR IMAGING?

The golden age of neuroimaging came with a plethora of promises to clinicians including the one that brain biopsies, in particular of neoplastic lesions, will soon be obsolete and will be replaced by noninvasive characterization of tissue using a combination of MR spectroscopy, quantitative characterization of blood volume, apparent diffusion coefficient maps, and other sophisticated imaging.[1,2] The combination of these and metabolic imaging as well as receptor binding characteristics established by positron emission tomography (PET) truly should replace biopsies, but never has, for a reason.[3]

With the tremendous advances in molecular imaging we can now diagnose early stages of Alzheimer's disease with a reasonable sensitivity and specificity employing specific ligands to beta-amyloid linked to positron-emitting isotopes. So why do we still biopsy tumors and other lesions in the brain and will we continue to do so in the foreseeable future?

The first question has to be answered along the lines of the diagnostic gold standard of biopsies and the respective diagnostic yields when compared to imaging diagnoses. The standard has been set by the large series of stereotactic biopsies, in particular serial biopsies up to an impressive series comprising 5000 cases.[4] The diagnostic yield in a large variety of tumors basically covering the whole WHO spectrum is 95.4% with a correct immediate intraoperative diagnosis in 90.3% of cases. The series antecedes even the advent of modern molecular techniques in neuropathology, which has increased the diagnostic yield even further by looking at loss of heterozygosity 1p,19q, IDH-1 mutations PTEN, p53, EGFR (epidermal growth factor receptor) mutations, and numerous others.[5]

The second part of the question has a speculative angle but, still continuing the molecular theme, an extrapolation into the future is relatively safe. In tumors alone the sequential editions of the WHO classification of brain tumors have shown an exponential growth of molecular markers in use now to obtain a specific diagnosis and actually reduced the merely morphological component drastically. Furthermore, the correlation of multiple molecular markers with therapeutic efficacy and every

A. Golby (Ed): Image-Guided Neurosurgery
DOI: http://dx.doi.org/10.1016/B978-0-12-800870-6.00008-X

aspect of prognosis make it mandatory for the treating physician to have as much molecular information as possible before making a treatment decision.[6] This will eventually lead to personalized treatment. So if the diagnosis comprising cellular type, WHO grade and the molecular profile is needed, then tissue is essential and mandatory.[7] Hence, biopsy is here to stay and likely the need for it will increase. As our molecular understanding of the disease process is likewise increasing with accelerating speed in nonneoplastic diseases of the brain, including demyelinating diseases, autoimmune processes, vasculitis, to name a few, even in the realm of nontumor lesions the need for a biopsy to achieve a specific and "molecular" diagnosis, if anything, will become more urgent.[8,9]

Historically, radiologic diagnosis employing morphological imaging has led to misdiagnosis in 20–30% of cases when compared with histologic diagnosis.[10] In a study of 174 supratentorial gliomas, lesions were separated into high grade and low grade based on magnetic resonance imaging (MRI) characteristics.[11] Those with necrosis, edema, irregular contrast enhancement, and involvement of white matter were classified as high grade, whereas those with diffuse hypointensity on T1, hyperintensity on T2, and minimal mass effect or edema were characterized as low grade. Higher diagnostic accuracy was seen with tumors diagnosed as high grade compared with a low grade diagnosis based on imaging alone. Of the 174 biopsies performed, 95% of the patients received a conclusive histological diagnosis, which is on par with most recent studies regarding stereotactic biopsies. Only 10% of those diagnosed as high grade based on imaging had a different histologic diagnosis which changed treatment course, whereas 30% of those presumed to be low grade based on imaging were discordant. Of the presumed low-grade tumors which were misdiagnosed, 10 were pathologically higher grade, anaplastic astrocytomas, and 10 were not tumor. Therefore, this data suggests that even simple tissue diagnosis is necessary to prevent undertreatment of high-grade gliomas and mistreatment of lesions which are not tumor, including inflammatory or infectious diagnoses. Certain lesion locations, such as those in the corpus callosum (CC), may increase the accuracy of imaging diagnosis, as 10 out of 10 lesions located in the CC in this group were high-grade gliomas but other studies have found no correlation of location and accuracy of imaging diagnosis.[12]

Variable terminology has been used to define concordance and discordance of histologic diagnosis when compared to preoperative diagnosis based on imaging. Some studies use discordance to mean either the grade or histology was incorrectly diagnosed, and some require both to be incorrect. In a review of 3166 stereotactic procedures, 506 (16%) cases showed some discordance between MRI and histologic diagnosis.[11]

This is clearly unacceptable for clinical decision-making, especially when aggressive treatment with concomitant side effects is contemplated. And thus for tumors, but also for non-neoplastic lesions, tissue remains essential to obtain a definitive diagnosis. Adding MR spectroscopy, PET, or other advanced molecular imaging

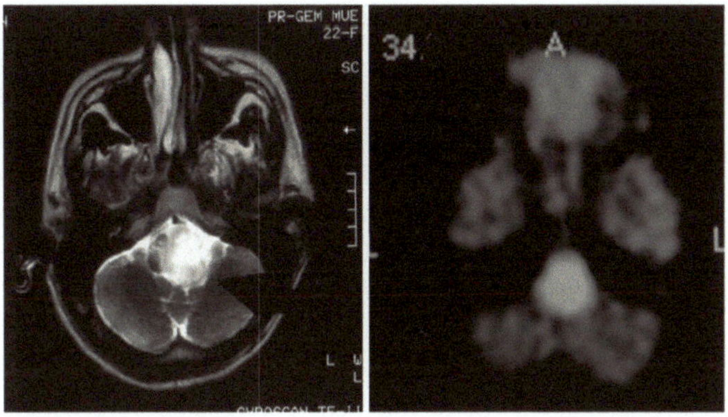

Figure 8.1 Pontine glioma with T2-weighted MRI (left) and FDG-PET (right). The high glucose utilization on the FDG-PET was suspicious for a malignant glioma but a serial biopsy proved this to be a pilocytic astrocytoma WHO I and the patient (9 years old) survived 7.3 years.

techniques has not significantly affected the rate of misdiagnosis as the sensitivity has increased, in particular with MRS, but not the specificity.[13,14]

In some instances functional imaging can even be misleading, indicating a high-grade lesion characterized by increased 18F-deoxyglucose utilization when the biopsy shows a low-grade lesion WHO I (Figure 8.1, Pontine glioma). This should not be confused with functional/physiologic imaging utilized to guide a biopsy rather than with the goal of providing a diagnosis. In this scenario functional imaging can be very helpful and increase the diagnostic yield while simultaneously reducing the potential sampling error.[15–19]

As mentioned above, molecular characterization of tumors is increasing and may become the gold standard as the field of neuro-oncology advances. Most recently proteomics has been introduced as an addition to genetic analysis of tumors and the technology has been scaled down and analysis has been accelerated using mass spectroscopy so that this technology could even be used intraoperatively.[20,21] Again this is only possible with tissue available for analysis.

MULTIMODAL IMAGING FOR BRAIN BIOPSY: CT, MR, PET, MRS, SPECT

Brain biopsies are performed daily and currently neurosurgeons generally employ contrast-enhanced MRI for localization and target selection.[12,22,23] Head computed tomography (CT) with contrast was the standard imaging modality when this procedure was first performed; however, with the superior anatomical resolution of MRI and current MRI capabilities available at most hospitals, CT scans are reserved for those patients who cannot undergo MRI. MRI-compatible stereotactic frames are standard now, as are MRI-compatible fiducials, both for frame-based as well as frameless stereotactic approaches. A remaining issue pertaining to all MRI-based procedures is the degree of image distortion. As the magnetic field of the scanner is not completely homogeneous, this leads to distortions which are more pronounced in the periphery of the field of view. The distortions are minute with lower field strengths but increase with higher field strengths of 3 Tesla and above, which have become the modern standard due to better image quality. Even 7 T and 9.4 T MRI are now employed clinically and render superb image quality.

Whereas image distortion is only marginally clinically relevant for larger or centrally located lesions, image distortion is a factor in smaller, more peripheral targets with a diameter of less than 10 mm.[24] An elegant solution to MR distortion and its impact on targeting accuracy is the fusion with a stereotactic CT which reformats the MRI to the superb linear geometry of the CT scan without losing the spatial and anatomical resolution. As a standard feature now available with all commercial systems for stereotactic targeting regardless of frame-based or frameless image fusion has basically eliminated the problem of MRI distortion with rigid fiducials.[25,26]

While CT and MRI form the anatomical and geometric basis for biopsies, other functional modalities have greatly added to the armamentarium for stereotactic biopsies. Those are PET and here predominantly fluorodeoxy-glucose utilization FDG-PET and amino acid PET, MR spectroscopy, mostly proton spectroscopy, imaging of regional cerebral blood volume, and to a lesser degree single photon emission tomography with 201 Thallium reflecting Na/K-ATPase activity.[16–18,27,28] Combining these techniques allows an elegant targeting of the biologically relevant tumor parts (Figure 8.2).

These image modalities have primarily been employed to target the biologically most active area of a tumor, which usually is also the most malignant portion determining the prognosis of the patient. Likewise identification of the areas with the highest degree of malignancy in a given tumor is crucial to find the appropriate therapy. This has been investigated in numerous, though mostly retrospective, single institution investigations.

A study of 20 patients with brainstem gliomas who underwent planning and biopsies via trajectories with both contrast MRI and FDG-PET scans assessed the benefit of adding PET to intraoperative planning. Contrast enhancement on MRI did not correlate with areas of high FDG uptake or areas of increased tumor grade from biopsy samples. However, areas with high FDG/methionine-PET uptake yielded a diagnosis in 20/20 patients. MR-guided biopsies failed to achieve a diagnose in 2 of 8 patients and 4 other patients had a higher-grade tumor when using the PET-guided trajectory fused to the stereotactic MRI as compared with the contrast MRI plan alone.[29]

FDG-PET has been shown in smaller studies to accurately differentiate between low and high-grade glioma; however, this technique is unable to differentiate between diverse histopathologic diagnoses. When comparing FDG uptake at conventional (20 minutes)

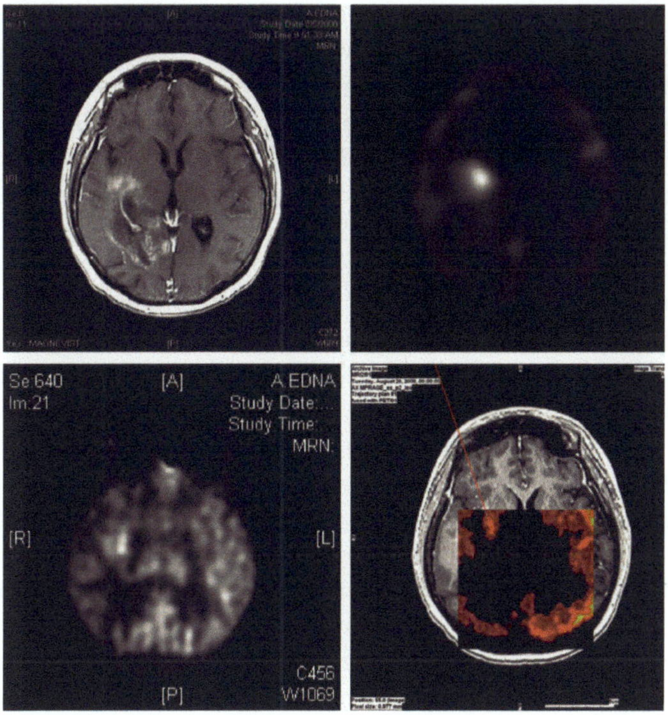

Figure 8.2 MRI (top left), TI-SPECT (top right), Arterial spin labeling blood flow image (lower left), and stereotactic trajectory targeting the SPECT zone of high NA/K ATPase fused to the stereotactic MRI.

and delayed (30 minutes) intervals, there was a significant difference between low and high grade. In general, the low-grade tumors were hypometabolic and the high-grade tumors were hypermetabolic. Nonetheless, FDG uptake was unable to differentiate between astrocytoma, oligodendroglioma, or oligoastrocytoma.[30]

Another option involves the addition of magnetic resonance spectroscopy (MRS), which, when properly performed, can increase diagnostic accuracy. This technique is operator dependent and requires facilities capable of performing such imaging techniques reliably and reproducibly.[31,32]

The major disadvantage of MRS over PET is the fact that no voxel-based fusion for larger regions of interest for MRS is possible without a third morphological MRI (slab) to then interpolate the voxels. Using chemical shift imaging fusion with the anatomical MRI is possible but still carries the burden of interpolation.

A comparison of 30 patients who underwent biopsy guided by MRS and 39 with only morphological MRI achieved a diagnosis in all 30 patients with MRS guidance and in 35 patients (90%) with MRI guidance.[3] Though this study shows a trend toward improved accuracy in MRS, the diagnostic success in the MRI-guided biopsies is lower than the current average of most studies. Sampling bias may have influenced these results, as most of the patients had only one biopsy specimen sampled during surgery. Multiple studies have shown increased rates of diagnosis with increasing number of biopsy specimens.

Similar findings have been shown with thallium (Tl) single photon emission computed tomography (SPECT). A study adding Tl SPECT to MRI planning for brain biopsies investigated whether the increased Tl uptake correlated with increased histologic grade. Though there was a trend toward increased grade in areas of higher Tl uptake, this was not statistically significant.[33]

Technetium 99 m (Tc-GHA SPECT) has been shown to have lower diagnostic accuracy when compared with contrast-enhanced MRI. Twenty-nine patients underwent SPECT and then biopsy, with 6 classified as low grade and 14 as high grade based on SPECT. Histopathologic diagnosis correctly diagnosed low-grade lesions in 4 of the 6; however, this again did not differentiate between astrocytic and oligodendroglial pathologies. Of the high-grade lesions, variable histology included anaplastic astrocytoma, oligodendroglioma, glioblastoma, and one was found to be metastatic adenocarcinoma.[34]

Although MRS or Tl SPECT integration into MRI surgical planning may increase the diagnostic yield of brain biopsies, this calls for a referral of these respective patients to a center where all the aforementioned technologies are available and validated. Possibility of various pathologies confirms the need for tissue sampling, as histological diagnosis remains elusive of current imaging technology. Combining the above information with the lack of definitive imaging diagnosis when utilizing MRI, PET, or SPECT leads to the conclusion that lesions should undergo brain biopsy prior to initiating treatment.

OPEN, FRAMELESS, OR FRAME-BASED STEREOTACTIC BIOPSIES

Biopsies are performed with stereotactic guidance either frame-based or frameless and referenced to the image modalities mentioned in the previous section for targeting. So-called open biopsies or excisional biopsies which carry the additional risk of a craniotomy should be avoided as modern imaging including tractography and fMRI should allow the neurosurgeon *a priori* to make a decision whether a tumor is resectable or not, if a resection is contemplated. For non-neoplastic or questionable lesions, an open approach is obsolete as there is no advantage over a stereotactic biopsy but only increased morbidity and the perceived advantage of "more tissue" can be overcome by performing serial biopsies throughout the whole lesion, which renders a complete histological and molecular profile as compared to a surface specimen from an open biopsy.[12,35]

For intraventricular lesions or tumors extending into the ventricle, endoscopic biopsies have been advocated under vision but the diagnostic yield compared to image-guided stereotactic biopsies has been disappointing while the incidence of hemorrhages has been quite high, albeit mostly clinically asymptomatic.[36]

Stereotactic systems with an excellent mechanical stability and using rigid fiducials for imaging with an accuracy between 0.3 and 0.7 mm are traditionally used and have been validated in more than 10,000 cases. Either arc-centered systems (Leksell Elekta, CRW, Integra) or polar coordinate systems (Riechert, Inomed, MHT, Freiburg, Germany) have been used in large series and shown excellent results. Frameless systems relying on skin fiducials have been used as well for over 10 years and have shown reliable accuracy and similar diagnostic yields. The term "frameless" in this context is somewhat of a semantic trick given that the patient's head still needs to be immobilized and usually gets fixed with Mayfield pins. This reduces the "frameless" part to three pins instead of the classic four pins used for stereotactic frames. It obviates the need to place the patient in the frame prior to the start of the procedure, and allows the separation of the image acquisition from the biopsy procedure. With the advent of intraoperative CT and MRI and image fusion of the rather lengthy MRI imaging with an intraoperative CT which takes three minutes with modern helical scanners (our preferred approach), this advantage has become rather relative. The argument of higher versatility of frameless systems again is rather relative given that the aiming bow of a frame, in particular the Riechert frame, can be removed with one twist leaving the whole operative field free—for example, if after a biopsy the surgeon decides to proceed to an open resection in the same session. Given the high congruency of intraoperative diagnoses with the results from permanent sections we have used this approach not infrequently.[4]

One important distinction favoring the use of frame-based systems is the ability to perform serial biopsies through the vicinity of a lesion, the maximum diameter of the lesion, and back out into the surrounding tissue again.

This approach—favored by highly experienced groups—has the advantage that it renders much more information about a given tumor than just a histological diagnosis. The growth pattern, the invasiveness, and the actual tumor border can be determined based on real cellular architecture and not only on imaging which can be deceiving.[37,38] Furthermore, by taking a biopsy literally every millimeter, a molecular and cellular profile of a tumor can be constructed which has far-reaching clinical implications, such as planning of radiation fields.[39,40] The latter becomes even more important when highly conformal therapies like proton beam therapy or intensity-modulated radiotherapy or stereotactic radiosurgery are used subsequently. Incorporating biopsy findings and trajectories into radiosurgery planning thus has become an integral part of interstitial radiosurgery, for example (Figure 8.3).

Stereotactic biopsies using intraoperative MR imaging has been advocated as safer and potentially resulting in a higher diagnostic yield. Comparison with data from large classic stereotactic series have not substantiated this claim, nor has it shown reduced morbidity.[41–43]

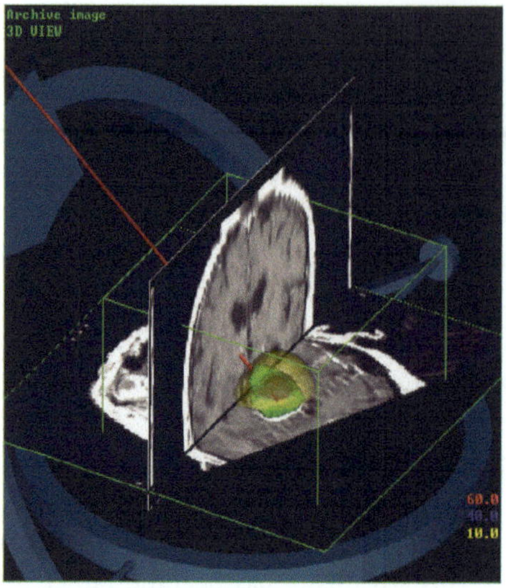

Figure 8.3 Three-dimensional reconstruction of the tumor volume with superimposed biopsy trajectory and isodose configuration for a low-grade glioma treated with 125 interstitial radiosurgery as part of the biopsy procedure.

BIOPSY INSTRUMENTS: SELDINGER NEEDLE, BACKLUND SPIRAL, MICROFORCEPS

Many different instruments are available for frame-based or frameless stereotactic biopsy. Two main options exist: side-cutting needles or microforceps (Figure 8.4). The Seldinger needle, Nashold needle, and Backlund spiral are some examples of side-cutting instruments. These produce a sample size of $2 \times 8-10$ mm. Advocates of this instrument cite the preserved cytoarchitecture as an advantage over microforceps, which generally deliver a smaller biopsy sample of $1-2\ \text{mm}^3$. However, the side-cutting needles tend to have higher incidence of symptomatic hemorrhage, which may be related to the increased sample size or the tearing nature as the needle turns within the tumor tissue. Most importantly serial biopsies through a lesion are not possible with the side-cutting needles or only at the expense of multiple passes increasing the risk of hemorrhage further. Microforceps are introduced through a biopsy cannula which remains in the same trajectory and is not moved during a serial biopsy as only the microforceps is advanced per mm using a calibrated depth gauge.

In a study of 150 patients all of whom had biopsies performed with a side-cutting Sedan needle, a hemorrhage rate of 4.7% was encountered (3.3% asymptomatic and 1.4% symptomatic).[44] Another study, which mainly employed 1 mm cupped forceps, achieved a diagnosis in 96% of 500 patients. Using the cupped forceps in this study was associated with symptomatic hemorrhage in 2/500 patients (0.4%).[45]

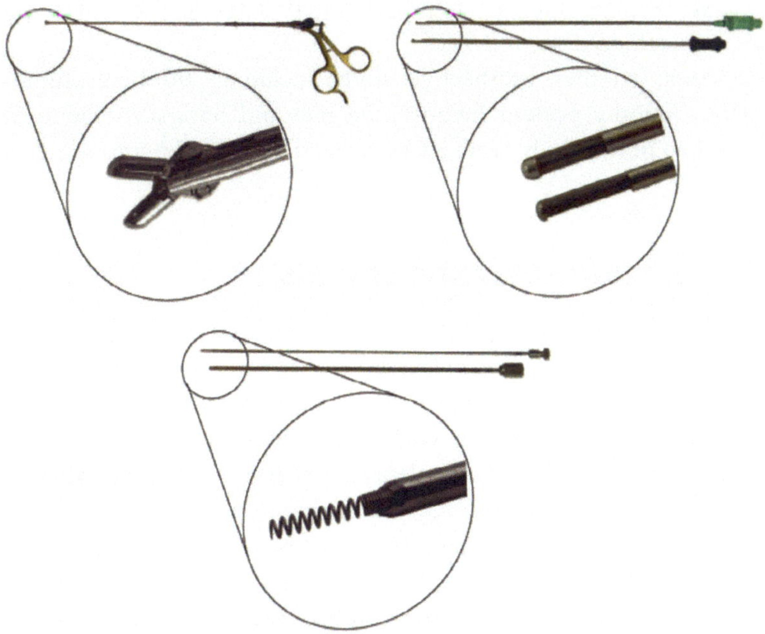

Figure 8.4 Different biopsy instruments. From top to bottom: Microforceps, Seldinger needle, Backlund spiral.

The ability to compare the hemorrhage rate between these two instruments is limited as the studies were performed in different hospitals, by different surgeons, with many unmatched variables.

Prior retrospective studies have shown symptomatic hemorrhage rates of 2.5–4.8%.[46–48] Bernstein, Sawin, and Kulkarni all showed a statistically significant increase in symptomatic hemorrhage with malignant gliomas, inconsistent with the study by Kreth.[19,50] However, there was an increase in silent hemorrhages consistent with other studies. Bernstein's biopsies utilized microforceps in one-third and a side-cutting needle in two-thirds of patients. Though numbers were not given, the study reports no difference in hemorrhage rates between the two instruments.

Along with the above retrospective studies, two prospective studies specified the use of different instruments. Over a two-year period, 326 patients underwent frame-based stereotactic biopsy, and 98% achieved a diagnosis. Planning was achieved with contrast head CT guidance and all biopsies were taken with cupped microforceps producing 1 mm^3 tissue samples. Patients with malignant gliomas or those located in eloquent areas were not at risk for increased morbidity; however, malignant gliomas did have an increased rate of silent bleeds found on routine postoperative head CT

($p = 0.02$). Overall, the use of cupped microforceps in this study produced a symptomatic hemorrhage rate of 0.9%.[50]

In comparison, another prospective study including 90 brain tumor biopsies showed a 10% clinically evident hemorrhage rate and 55% silent hemorrhage rate. The biopsies taken in this study were all performed with a side-cutting needle.[49]

NEOPLASTIC V. NON-NEOPLASTIC LESIONS

Brain biopsies have remained the gold standard for diagnosis of several nonfocal, nononcological disease processes. Even in presumed tumor cases, interestingly, a single series showed 28 of 158 biopsies (18%) did not have a neoplastic diagnosis; these had other etiologies including demyelination, as well as infectious and inflammatory reactions.[51] Though an uncommon diagnosis, acute demyelination in tumefactive MS can appear as a ring-enhancing lesion on CT or MRI. These should undergo stereotactic biopsy for diagnosis while minimizing risk, compared with open biopsy or attempted resection. Consideration of acute demyelination should be given to younger patients with no history of cancer and suspicion of MS based on history and physical exam.[52]

Multifocal lesions represent another subset of patients where stereotactic biopsy may be useful. In a single center review, 158 of 2081 patients who underwent stereotactic biopsy had multifocal lesions. Many of these multifocal lesions remain gliomas after diagnosis, with 57% low grade, 10% AA, and 11% GBM. Lymphomas are also commonly multifocal in nature, with a 10% incidence in this cohort. Some of the more rare entities included vascular angiopathy, tuberculoma or brain abscess.[53] In addition, stereotactic biopsy has been successfully employed for the diagnosis of intracerebral Whipple's disease in patients presenting with progressive dementia and multifocal T2 hyperintense lesions on brain MRI.[54]

One study of 64 patients attempted to diagnose diffuse disease causing encephalopathy, compared with either focal or multifocal disease.[55] This was diagnostic in 34 patients, abnormal but without definitive diagnosis in 21 patients and nondiagnostic in 9 patients. In 20 patients (31%) it changed management. Those patients with focal or multifocal disease had a higher probability of achieving a diagnosis. Most nondiagnostic samples came from those patients with diffuse disease and showed evidence of nonspecific gliosis and inflammation.

Vasculitis can be diffuse or multifocal and often can require tissue sampling of meninges, as well as brain tissue. Therefore, diagnostic procedures for the evaluation of vasculitis often involve open biopsies.

Stereotactic biopsy may also be helpful in the diagnosis of brain abscesses, followed by subsequent needle drainage in order to debulk the abscess enough to allow adequate antibiotic treatment.

Though less common in the era of HAART therapy (highly active anti-retroviral therapy), CNS involvement by HIV virus, and predilection for opportunistic infections leads to specific consideration for brain biopsy in patients affected by AIDS. Many patients presenting with low CD4 counts and multiple intracranial lesions are empirically treated for toxoplasmosis. Only when they fail treatment is further work-up undertaken. Several additional etiologies are common in AIDS patients, including primary CNS lymphoma (PCNSL), progressive multifocal leukoencephalopathy (PML), and other less common tumors. In a large series of AIDS patients, 243 underwent stereotactic biopsy for diagnosis of concerning brain lesions which did not respond to conventional treatment for toxoplasmosis. Separate from a diagnosis of toxoplasmosis in 38 patients (15%), other diagnoses included PCNSL (33%), PML (29%), encephalitis (12%), abscess (3.6%), and other tumors (3%). It is important to note these biopsies were performed on patients already treated for toxoplasmosis, and those who responded were not included in the review. As such, the distribution of etiologies is likely biased against toxoplasmosis in this study.[56]

Another smaller cohort showed the difference in distribution, with 31% toxoplasmosis, 29% PML, and 18% non-Hodgkin's lymphoma. In 89% a definitive diagnosis was made while 11% were nondiagnostic.[57]

Some studies have shown that empiric toxoplasmosis treatment may lead to excessive toxicities (liver, bone marrow) and may affect tissue diagnosis if a biopsy is indicated after treatment failure. Cryptococcal treatment, as well as dexamethasone, may affect the inflammatory response in the brain and results of the biopsy sample. Some authors, therefore, suggest early biopsy in order to confirm diagnosis and employ appropriate treatments with the aim of improving survival.[58]

NONFOCAL DISEASE: LEUKENCEPHALOPATHIES AND NEURODEGENERATIVE DISEASES

Diagnosis of varying neurodegenerative processes is often completed at autopsy. Alzheimer's is often clinically misdiagnosed as frontotemporal dementia (FTD) or corticobasalar dementia (CBD). Several imaging techniques have been studied in an attempt to accurately diagnose these processes prior to death. In general, these diseases lead to decreased brain volume, which can be assessed by measuring volumes of the ventricles, gray matter, and certain anatomic structures including the hippocampus. Combining parietal gray matter volume on MRI with corpus callosal white matter volume on DTI increased diagnostic accuracy; however, this method only reached 89% accuracy. As new treatments are emerging, correct diagnosis is of increasing importance, for enrollment in trials as well as initiating appropriate treatment. In selected cases, stereotactic brain biopsy offers a minimally invasive method of accurately diagnosing these patients in order to enroll in trials and guide proper treatment.[59]

Evaluation for JC virus, commonly associated with progressive multifocal leukoencephalopathy (PML) in immunocompromised patients, can be performed via DNA polymerase chain reaction (PCR) of cerebrospinal fluid. CSF evaluation shows sensitivity of 93% and a specificity of 99%; however, if CSF evaluation does not reveal a diagnosis, stereotactic brain biopsy is recommended for prompt diagnosis and treatment.

RESULTS: DIAGNOSTIC YIELD

Historically, diagnostic yield of biopsies has been variable, as low as 30% for nonfocal disease and as high as 97–99% for focal lesions larger than 15 mm.[60,61] Stereotactic technique (frame versus frameless) affects this less than the number of specimens and the availability of a dedicated neuropathologist.

To construct the composite image of a lesion via serial biopsies takes an experienced neuropathologist but also allows analysis of tumor heterogeneity and growth pattern with a high degree of reliability. To later correlate histological and molecular findings with image characteristics we routinely implant a titanium marker which is the size of a voxel at the lowest point of the biopsy trajectory (0.8 × 0.8 mm) through the biopsy cannula which guides the biopsy forceps. It is pushed to target position by inserting the inner stylet (MHT Freiburg) (Figure 8.5).

This also allows further sequential postoperative imaging with MRI and CT to relate treatment effect to the originally biopsied tumor. Last, but not least, it eliminates the debate whether the lesion has truly been reached as the postoperative MRI documents the biopsy site beyond any doubt. Sampling error, which can be massively reduced via serial biopsies as well, should thus be a rare exception to the rule as a maximum number of biopsies is taken every millimeter through the maximum diameter of the tumor.

A large, retrospective review compared intraoperative pathology findings with the final diagnoses. Of 5000 biopsies performed on 4589 patients, 95.4% resulted in a diagnosis. Intraoperative diagnosis was concordant with final diagnosis in 90% of biopsies, with complete correlation of histology and grade in 81% and partial correlation (either histology or grade) in 9%.[4] Anaplastic astrocytomas and oligodendrogliomas had greater discordance, whereas grade II astrocytomas and glioblastomas had higher concordance when comparing intraoperative diagnosis with the final pathologic diagnosis.

Decreased diagnostic yield was seen in infratentorial lesions, with those in the midbrain only correctly diagnosed in 73.5% and cerebellum only 72.6%.

Though the above study is the largest, several smaller studies have also been completed. One group of 326 patients had similar results, though with a lower correct

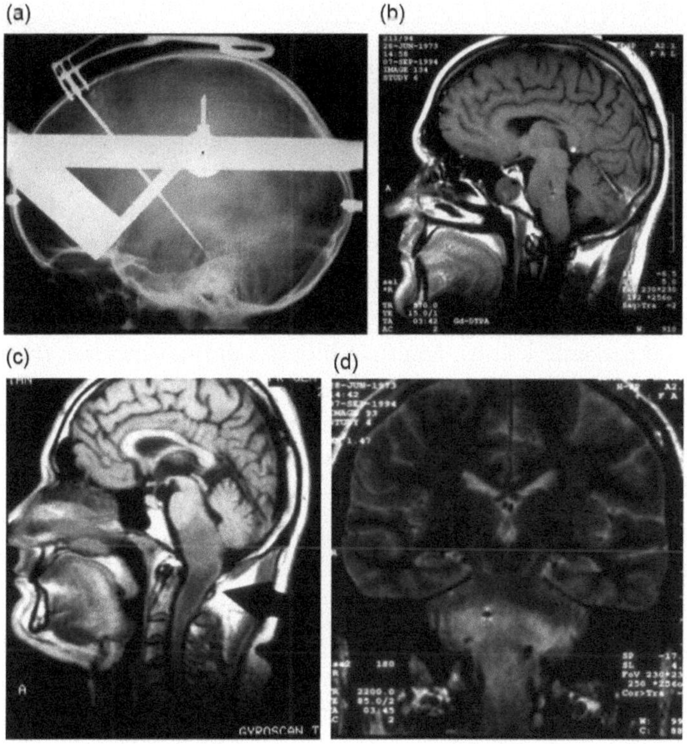

Figure 8.5 Biopsy trajectory for brainstem tumors. (a) A precoronal approach is chosen to biopsy through the maximum diameter of this pontine lesion. The biopsy forceps is opened. (b) A titanium marker is left at the most caudal part of the biopsy trajectory. (c) Diffuse pontomedullary lesion which is best approached by a transfrontal approach. (d) Titanium marker implanted in a paramedian pontine lesion.

final diagnosis rate (84%). Certain pathologies had a lower yield of stereotactic biopsy, including oligodendroglioma and anaplastic oligodendroglioma. An appreciation for the clinical history, as well as location of the tumor, leads to increased rates of diagnosis.[50]

A smaller cohort of 20 patients suspected to have low-grade gliomas based on imaging underwent stereotactic biopsies. Half of the lesions were confirmed to be low-grade astrocytomas; however, 9 biopsies revealed grade III astrocytoma and one patient was diagnosed with encephalitis. Therefore, treatment plans were affected in 50% of the patients. Importance must be given to obtaining histologic diagnosis in order to prevent undertreating higher-grade tumors or inappropriately treating non-neoplastic lesions.[22]

In comparison with a historic control group, diagnostic yield was increased with the availability of intraoperative pathologic diagnosis via frozen section analysis. Diagnostic yield was increased from 89% to 98% in otherwise similar cohorts.[62]

COMPLICATIONS

Brain biopsies are considered minimally invasive when compared with open craniotomies for management of focal and some nonfocal lesions. As minimally invasive as possible, they still require passing an instrument through brain tissue in order to reach selected targets. As instruments pass, there is a hemorrhage risk which several studies have shown to be between 1 and 4%. Hemorrhage risk is increased in patients with poorly controlled hypertension and hyperglycemia. Patients with blood glucose greater than 200 mg/dL had a significantly increased risk of postoperative deficits.[63] Certain locations have shown trends toward increased hemorrhage risk, including basal ganglia and thalamic lesions. Multiple biopsies did not increase the risk for morbidity, when taken through a single trajectory. However, increasing the number of needle passes was associated with an increased risk of deficit in basal ganglia and thalamic biopsies. A trend toward significance was seen in patients with increased international normalized ratio (INR) (>1.4) or platelets less than 150,000. A review of several studies, totaling 7624 patients, averaged an overall hemorrhage rate of 2.4% and 0.8% mortality.[64] Other complications include infections and seizures, which occurred in less than 1% in all studies.

In the above-mentioned retrospective review of 5000 biopsies, similar morbidity (2.7%) and mortality (0.7%) rates were seen as compared with other studies. Overall, the main complications of brain biopsies remain hemorrhagic, with 2.7% morbidity and 1.3% of patients left with residual permanent deficits. Of importance, this study utilized serial biopsies (ranging from 2–10 samples) with microforceps, and did not encounter increased risks of morbidity or mortality.[4]

A prospective study by Kreth was in concurrence with these large retrospective reviews. Number of samples taken was not shown to increase hemorrhage risk; however, the use of more than two tracts for biopsy sampling has shown a trend toward increased hemorrhage risk. Two tracts were used in 22/326 patients with a range of 4–25 tissue samples per patient. Tumor location, histology, and mass effect were not significant factors in determining risk of hemorrhage by univariate or multivariate analysis. Karnofsky performance score, age, and number of biopsy samples were also not found to be significant factors.[50]

Interestingly, patients who underwent brain biopsy with a preoperative diagnosis of AIDS had higher hemorrhage rates, with symptomatic hemorrhage in 4 of 32 patients (12.5%) and 3 mortalities. Due to the increased hemorrhage rate and need for diagnosis, a preoperative coagulopathy protocol was instituted in order to reduce the hemorrhage and mortality. This involved transfusing FFP and platelets as well as administering DDAVP in order to promote clot formation and stability. Since instituting the protocol, 9 complications (4%) and 5 deaths (2%) occurred in 218 patients.[56]

Rare complications include secondary tumor formation along the biopsy trajectory. Seeding of tumor cells is very rare and, in few case reports, has been associated with biopsy performed using a spiral needle (Backlund) without a cannula.[65]

BIOPSIES IN ELOQUENT, FUNCTIONAL AREAS: BRAINSTEM, PINEAL REGION, SUPRASELLAR REGION

Despite numerous reports showing the contrary, it has remained conventional wisdom that the pineal region, the brainstem and the suprasellar region pose an increased risk for stereotactic biopsies.

With high resolution MRI and MR angiography fused to a CT scan and with frame-based precision, these regions can be reached safely and a high diagnostic yield can be achieved.[66,7,67,68] Approaches to pathologies in these regions have to be individually tailored around the vascular architecture which needs to be visualized accordingly.

One review of 30 stereotactic brainstem biopsies had a diagnostic yield of 93%, with two temporary neurologic deficits. Two main approaches were utilized—transfrontal and transcerebellar. For midline lesions the transfrontal approach proved safer.[69]

In the previously mentioned review by Tilgner et al., 32% of the biopsies were taken from eloquent areas including basal ganglia, thalamus, cerebellum, and brainstem. These showed slightly lower rates of diagnosis, though this was not found to be statistically significant. There was also no increase in hemorrhage risk based on location alone.[4]

We prefer the transfrontal route to the brainstem as it follows a trajectory parallel to the major fiber tracts, hence reducing the risk of interrupting functional pathways.

Routinely we perform brainstem biopsies as serial biopsies covering the whole diameter of the lesion (Figure 8.5). The approaches to the pineal region depend on the location/dislocation of the internal cerebral veins. Either a frontal precoronal trajectory can be chosen or, if the veins interfere, a temporoparietal trajectory approaching the tumor below the level of the internal cerebral veins.

FUTURE DIRECTIONS: MOLECULAR NEUROPATHOLOGY

As illustrated previously in this chapter, tissue pathology has the highest sensitivity and specificity for diagnosis when compared with all imaging methods—there are no current imaging methods which can render a specific biological diagnosis. Separate from accurate diagnosis, as oncology moves forward, many tumor specific therapies are being developed along the molecular fingerprint of an individual tumor making molecular profiling of tumors mandatory and thus mandating a tissue biopsy. Molecular neuropathology is becoming increasingly important in determining prognosis, outcomes, and treatment responsiveness and will become the gold standard in the foreseeable future. With single cell PCR now available and increasingly sensitive molecular techniques employed, stereotactic biopsies will certainly provide sufficient tissue and remain the most elegant and least invasive technique to provide material.

CONCLUSION

Stereotactic brain biopsies are an extremely safe and reliable method to provide a histological diagnosis in numerous CNS diseases. This is not limited to neoplastic processes. Probably more important in the future when molecular diagnosis and molecular fingerprinting of individual disease is fully implemented, stereotactic biopsies will be used even more to establish the molecular basis of personalized therapy when it comes to CNS disease. In conjunction with functional and even molecular imaging, targeting will be refined as it should be when an organ as eloquent as the brain is concerned.

REFERENCES

1. Broggi M, Broggi G. Stereotactic biopsy: an established procedure, but still modern? *World Neurosurg*. 2014;S1878–8750(14):00896–1.
2. Thompson TP, Lunsford LD, Kondziolka D. Distinguishing recurrent tumor and radiation necrosis with positron emission tomography versus stereotactic biopsy. *Stereotact Funct Neurosurg*. 1999;73(1–4):9–14.
3. Chernov MF, Muragaki Y, Ochiai T, et al. Spectroscopy-supported frame-based image-guided stereotactic biopsy of parenchymal brain lesions: comparative evaluation of diagnostic yield and diagnostic accuracy. *Clin Neurol Neurosurg*. 2009;111(6):527–535.
4. Tilgner J, Herr M, Ostertag C, Volk B. Validation of intraoperative diagnoses using smear preparations from stereotactic brain biopsies: intraoperative versus final diagnosis—influence of clinical factors. *Neurosurgery*. 2005;56(2):257–265.
5. Eigenbrod S, Trabold R, Brucker D, et al. Molecular stereotactic biopsy technique improves diagnostic accuracy and enables personalized treatment strategies in glioma patients. *Acta Neurochir*. 2014;156(8):1427–1440.
6. Thon N, Eigenbrod S, Grasbon-Frodl EM, et al. Novel molecular stereotactic biopsy procedures reveal intratumoral homogeneity of loss of heterozygosity of 1p/19q and TP53 mutations in World Health Organization grade II gliomas. *J Neuropathol Exp Neurol*. 2009;68(11):1219–1228.
7. Kickingereder P, Willeit P, Simon T, Ruge MI. Diagnostic value and safety of stereotactic biopsy for brainstem tumors: a systematic review and meta-analysis of 1480 cases. *Neurosurgery*. 2013;72(6):873–881.
8. Arun T, Tomassini V, Sbardella E, et al. Targeting ASIC1 in primary progressive multiple sclerosis: evidence of neuroprotection with amiloride. *Brain*. 2013;136(Pt 1):106–115.
9. Shyam Babu C, Satishchandra P, Mahadevan A, et al. Usefulness of stereotactic biopsy and neuroimaging in management of HIV-1 Clade C associated focal brain lesions with special focus on cerebral toxoplasmosis. *Clin Neurol Neurosurg*. 2013;115(7):995–1002.
10. Arbit E, Galicich JH. Importance of image-guided stereotactic biopsy to confirm diagnosis in an oncological setting. *Ann Surg Oncol*. 1994;1(5):368–372.

11. Callovini GM. Is it appropriate to redefine the indication for stereotactic brain biopsy in the MRI era? Correlation with final histological diagnosis in supratentorial gliomas. *Minim Invasive Neurosurg.* 2008;51(2):109–113.
12. Kuhn SA, Romeike B, Walter J, Kalff R, Reichart R. Multiplanar MRI-CT fusion neuronavigation-guided serial stereotactic biopsy of human brain tumors: proof of a strong correlation between tumor imaging and histopathology by a new technical approach. *J Cancer Res Clin Oncol.* 2009;135(9):1293–1302.
13. Setzer M, Herminghaus S, Marquardt G, et al. Diagnostic impact of proton MR-spectroscopy versus image-guided stereotactic biopsy. *Acta Neurochir.* 2007;149(4):379–386.
14. Burtscher IM, Skagerberg G, Geijer B, Englund E, Stahlberg F, Holtas S. Proton MR spectroscopy and preoperative diagnostic accuracy: an evaluation of intracranial mass lesions characterized by stereotactic biopsy findings. *Am J Neuroradiol.* 2000;21(1):84–93.
15. Pirotte B, Goldman S, David P, et al. Stereotactic brain biopsy guided by positron emission tomography (PET) with [F-18]fluorodeoxyglucose and [C-11]methionine. *Acta Neurochir Suppl.* 1997;68:133–138.
16. Levivier M, Goldman S, Pirotte B, et al. Diagnostic yield of stereotactic brain biopsy guided by positron emission tomography with [18F]fluorodeoxyglucose. *J Neurosurg.* 1995;82(3):445–452.
17. Bader JB, Samnick S, Moringlane JR, et al. Evaluation of l-3-[123I]iodo-alpha-methyltyrosine SPET and [18F]fluorodeoxyglucose PET in the detection and grading of recurrences in patients pretreated for gliomas at follow-up: a comparative study with stereotactic biopsy. *Eur J Nucl Med.* 1999;26(2):144–151.
18. Reithmeier T, Cordeiro J, Mix M, Trippel M, Rottenburger C, Nikkhah G. Impact of automated hotspot detection for (18)FET PET-guided stereotactic biopsy. *Acta Neurochir Suppl.* 2013;117:93–99.
19. Goldman S, Levivier M, Pirotte B, et al. Regional glucose metabolism and histopathology of gliomas. A study based on positron emission tomography-guided stereotactic biopsy. *Cancer.* 1996;78(5):1098–1106.
20. Santagata S, Eberlin LS, Norton I, et al. Intraoperative mass spectrometry mapping of an onco-metabolite to guide brain tumor surgery. *Proc Natl Acad Sci U S A.* 2014;111(30):11121–11126.
21. Agar NY, Golby AJ, Ligon KL, et al. Development of stereotactic mass spectrometry for brain tumor surgery. *Neurosurgery.* 2011;68(2):280–289.
22. Kondziolka D, Lunsford LD. The role of stereotactic biopsy in the management of gliomas. *J Neurooncol.* 1999;42(3):205–213.
23. McGirt MJ, Bulsara KR, Cummings TJ, et al. Prognostic value of magnetic resonance imaging-guided stereotactic biopsy in the evaluation of recurrent malignant astrocytoma compared with a lesion due to radiation effect. *J Neurosurg.* 2003;98(1):14–20.
24. Waters JD, Gonda DD, Reddy H, Kasper EM, Warnke PC, Chen CC. Diagnostic yield of stereotactic needle-biopsies of sub-cubic centimeter intracranial lesions. *Surg Neurol Int.* 2013;4(Suppl 3):S176–181.
25. Kim HY, Lee SI, Jin SJ, Jin SC, Kim JS, Jeon KD. Reliability of stereotactic coordinates of 1.5-tesla and 3-tesla MRI in radiosurgery and functional neurosurgery. *J Korean Neurosurg Soc.* 2014;55(3):136–141.
26. Choudhri AF, Chin EM, Klimo P, Boop FA. Spatial distortion due to field inhomogeneity in 3.0 tesla intraoperative MRI. *Neuroradiol J.* 2014;27(4):387–392.
27. Massager N, David P, Goldman S, et al. Combined magnetic resonance imaging- and positron emission tomography-guided stereotactic biopsy in brainstem mass lesions: diagnostic yield in a series of 30 patients. *J Neurosurg.* 2000;93(6):951–957.
28. Goldman S, Levivier M, Pirotte B, et al. Regional methionine and glucose uptake in high-grade gliomas: a comparative study on PET-guided stereotactic biopsy. *J Nucl Med.* 1997;38(9):1459–1462.
29. Pirotte BJ, Lubansu A, Massager N, Wikler D, Goldman S, Levivier M. Results of positron emission tomography guidance and reassessment of the utility of and indications for stereotactic biopsy in children with infiltrative brainstem tumors. *J Neurosurg.* 2007;107(5 Suppl):392–399.

30. Mertens K, Acou M, Van Hauwe J, et al. Validation of 18F-FDG PET at conventional and delayed intervals for the discrimination of high-grade from low-grade gliomas: a stereotactic PET and MRI study. *Clin Nucl Med.* 2013;38(7):495–500.
31. Frati A, Pichierri A, Bastianello S, et al. Frameless stereotactic cerebral biopsy: our experience in 296 cases. *Stereotact Funct Neurosurg.* 2011;89(4):234–245.
32. Son BC, Kim MC, Choi BG, et al. Proton magnetic resonance chemical shift imaging (1H CSI)-directed stereotactic biopsy. *Acta Neurochir.* 2001;143(1):45–49.
33. Hemm S, Rigau V, Chevalier J, et al. Stereotactic coregistration of 201Tl SPECT and MRI applied to brain tumor biopsies. *J Nucl Med.* 2005;46(7):1151–1157.
34. Santra A, Kumar R, Sharma P, Bal C, Julka PK, Malhotra A. Detection of recurrence in glioma: a comparative prospective study between Tc-99m GHA SPECT and F-18 FDG PET/CT. *Clin Nucl Med.* 2011;36(8):650–655.
35. Kourbeti IS, Vakis AF, Ziakas P, et al. Infections in patients undergoing craniotomy: risk factors associated with post-craniotomy meningitis. *J Neurosurg.* 2014:1–7.
36. Balossier A, Blond S, Touzet G, et al. Endoscopic versus stereotactic procedure for pineal tumour biopsies: Comparative review of the literature and learning from a 25-year experience. *Neurochirurgie.* 2014; pii: S0028–3770(14):00079–4.
37. Jenkinson MD, du Plessis DG, Smith TS, Joyce KA, Warnke PC, Walker C. Histological growth patterns and genotype in oligodendroglial tumours: correlation with MRI features. *Brain.* 2006;129 (Pt 7):1884–1891.
38. Kelly PJ, Daumas-Duport C, Scheithauer BW, Kall BA, Kispert DB. Stereotactic histologic correlations of computed tomography- and magnetic resonance imaging-defined abnormalities in patients with glial neoplasms. *Mayo Clin Proc.* 1987;62(6):450–459.
39. Walker C, du Plessis DG, Fildes D, et al. Correlation of molecular genetics with molecular and morphological imaging in gliomas with an oligodendroglial component. *Clin Cancer Res.* 2004;10 (21):7182–7191.
40. Walker C, du Plessis DG, Joyce KA, et al. Molecular pathology and clinical characteristics of oligodendroglial neoplasms. *Ann Neurol.* 2005;57(6):855–865.
41. Burkhardt JK, Neidert MC, Woernle CM, Bozinov O, Bernays RL. Intraoperative low-field MR-guided frameless stereotactic biopsy for intracerebral lesions. *Acta Neurochir.* 2013;155(4):721–726.
42. Schulder M, Spiro D. Intraoperative MRI for stereotactic biopsy. *Acta Neurochir Suppl.* 2011;109:81–87.
43. Quinn J, Spiro D, Schulder M. Stereotactic brain biopsy with a low-field intraoperative magnetic resonance imager. *Neurosurgery.* 2011;68(1 Suppl Operative):217–224.
44. Shakal AA, Mokbel EA. Hemorrhage after stereotactic biopsy from intra-axial brain lesions: incidence and avoidance. *J Neurol Surg Part A, Central Eur Neurosurg.* 2014;75(3):177–182.
45. Apuzzo ML, Chandrasoma PT, Cohen D, Zee CS, Zelman V. Computed imaging stereotaxy: experience and perspective related to 500 procedures applied to brain masses. *Neurosurgery.* 1987;20 (6):930–937.
46. Kongkham PN, Knifed E, Tamber MS, Bernstein M. Complications in 622 cases of frame-based stereotactic biopsy, a decreasing procedure. *Can J Neurol Sci.* 2008;35(1):79–84.
47. Sawin PD, Hitchon PW, Follett KA, Torner JC. Computed imaging-assisted stereotactic brain biopsy: a risk analysis of 225 consecutive cases. *Surg Neurol.* 1998;49(6):640–649.
48. Hall WA, Liu H, Martin AJ, Truwit CL. Comparison of stereotactic brain biopsy to interventional magnetic-resonance-imaging-guided brain biopsy. *Stereotact Funct Neurosurg.* 1999;73(1–4):148–153.
49. Kulkarni AV, Guha A, Lozano A, Bernstein M. Incidence of silent hemorrhage and delayed deterioration after stereotactic brain biopsy. *J Neurosurg.* 1998;89(1):31–35.
50. Kreth FW, Muacevic A, Medele R, Bise K, Meyer T, Reulen HJ. The risk of haemorrhage after image guided stereotactic biopsy of intra-axial brain tumours—a prospective study. *Acta Neurochir.* 2001;143(6):539–545.
51. Whiting DM, Barnett GH, Estes ML, et al. Stereotactic biopsy of non-neoplastic lesions in adults. *Cleve Clin J Med.* 1992;59(1):48–55.

52. Maarouf M, Kuchta J, Miletic H, et al. Acute demyelination: diagnostic difficulties and the need for brain biopsy. *Acta Neurochir.* 2003;145(11):961–969.
53. Meshkini A, Shahzadi S, Alikhah H, Naghavi-Behzad M. Role of stereotactic biopsy in histological diagnosis of multiple brain lesions. *Asian J Neurosurg.* 2013;8(2):69–73.
54. Mendel E, Khoo LT, Go JL, Hinton D, Zee CS, Apuzzo ML. Intracerebral Whipple's disease diagnosed by stereotactic biopsy: a case report and review of the literature. *Neurosurgery.* 1999;44 (1):203–209.
55. Wong SH, Jenkinson MD, Faragher B, Thomas S, Crooks D, Solomon T. Brain biopsy in the management of neurology patients. *Eur Neurol.* 2010;64(1):42–45.
56. Gildenberg PL, Gathe Jr. JC, Kim JH. Stereotactic biopsy of cerebral lesions in AIDS. *Clin Infect Dis.* 2000;30(3):491–499.
57. Alesch F, Armbruster C, Budka H. Diagnostic value of stereotactic biopsy of cerebral lesions in patients with AIDS. *Acta Neurochir.* 1995;134(3–4):214–219.
58. Nicolato A, Gerosa M, Piovan E, et al. Computerized tomography and magnetic resonance guided stereotactic brain biopsy in nonimmunocompromised and AIDS patients. *Surg Neurol.* 1997;48 (3):267–276.
59. McMillan CT, Avants BB, Cook P, Ungar L, Trojanowski JQ, Grossman M. The power of neuroimaging biomarkers for screening frontotemporal dementia. *Hum Brain Mapp.* 2014;35(9):4827–4840.
60. Heper AO, Erden E, Savas A, et al. An analysis of stereotactic biopsy of brain tumors and nonneoplastic lesions: a prospective clinicopathologic study. *Surg Neurol.* 2005;64(Suppl 2):S82–88.
61. Gralla J, Nimsky C, Buchfelder M, Fahlbusch R, Ganslandt O. Frameless stereotactic brain biopsy procedures using the Stealth Station: indications, accuracy and results. *Zentralbl Neurochir.* 2003;64 (4):166–170.
62. Dammers R, Schouten JW, Haitsma IK, Vincent AJ, Kros JM, Dirven CM. Towards improving the safety and diagnostic yield of stereotactic biopsy in a single centre. *Acta Neurochir.* 2010;152 (11):1915–1921.
63. McGirt MJ, Woodworth GF, Coon AL, et al. Independent predictors of morbidity after image-guided stereotactic brain biopsy: a risk assessment of 270 cases. *J Neurosurg.* 2005;102(5):897–901.
64. Chen CC, Hsu PW, Erich Wu TW, et al. Stereotactic brain biopsy: Single center retrospective analysis of complications. *Clin Neurol Neurosurg.* 2009;111(10):835–839.
65. Marx T, Rainov NG, Heidecke V, Burkert W. Secondary tumor formation after stereotactic biopsy for intracerebral metastatic disease. *Surg Neurol.* 2001;55(1):41–45.
66. Regis J, Bouillot P, Rouby-Volot F, Figarella-Branger D, Dufour H, Peragut JC. Pineal region tumors and the role of stereotactic biopsy: review of the mortality, morbidity, and diagnostic rates in 370 cases. *Neurosurgery.* 1996;39(5):907–912:discussion 912–904.
67. Yap L, Crooks D, Warnke P. Low grade astrocytoma of the pituitary stalk. *Acta Neurochir.* 2007;149 (3):307–311:discussion 311–302.
68. Dellaretti M, Reyns N, Touzet G, et al. Stereotactic biopsy for brainstem tumors: comparison of transcerebellar with transfrontal approach. *Stereotact Funct Neurosurg.* 2012;90(2):79–83.
69. Goncalves-Ferreira AJ, Herculano-Carvalho M, Pimentel J. Stereotactic biopsies of focal brainstem lesions. *Surg Neurol.* 2003;60(4):311–320.

CHAPTER 9

Multimodal Image-Guided Brain Tumor Resection

Xiaolei Chen
Department of Neurosurgery, Chinese PLA General Hospital, Beijing, China

INTRODUCTION

Over the past few decades, diagnostic imaging of brain tumors has made great progress. The currently used imaging modalities include computerized tomography (CT), magnetic resonance imaging (MRI), positron emission tomography (PET), magnetoencephalography (MEG), single-photon-emission computed tomography (SPECT), and others. Each modality has its specific advantages and disadvantages, implying the combination of different techniques to optimize the sensitivity and specificity of detection and, if possible, histological confirmation of brain tumors.

However, it is still challenging for the neurosurgeon to translate these images to the surgical field. A critical aspect of brain tumor resection is to adequately differentiate between tumor and normal tissue and, consequently, to determine the tumor margin that is key for the oncologic prognosis of the patient. Another aspect, which is equally critical, is to depict and protect the surrounding eloquent neurological structures (functional brain cortex and major white matter fiber tracts), which are usually "invisible" in the surgical field.

Bringing imaging technologies into the operating room has the potential of traversing the gap between neuroradiology and neurosurgery, resulting in intraoperative image-guided neurosurgery.

Image-guided neurosurgery, also known as neuronavigation, is currently considered one of the most important achievements in modern neurosurgery. It allows the visualization of the results of pre- and possibly intraoperative imaging in the surgical field, so that the image data can provide information for depicting tumor margin and nearby eloquent structures. The main goal is to achieve maximal brain tumor resection, without causing new neurological deficits.

Over the past decades, a significant amount of progress has occurred in the field of image-guided neurosurgery for neuro-oncology. For brain tumor resection, many reports have focused on the multimodal functional brain imaging and navigation. The aim of this chapter is to outline the current modalities of image-guided brain tumor resection. We focus on the clinical application of this technology for accurately defining the tumor margin; localization of surrounding eloquent structures; and correction

A. Golby (Ed): Image-Guided Neurosurgery
DOI: http://dx.doi.org/10.1016/B978-0-12-800870-6.00009-1

brain shift. Additionally, multimodal image-guided neurosurgery for patients with preoperative neurological deficits and comparison between preoperative functional mapping and intraoperative direct cortical stimulation (DCS)-based brain mapping are also discussed.

TYPES OF IMAGE GUIDANCE IN BRAIN TUMOR RESECTION

Frameless anatomical navigation for brain tumor resection

Frame-based computer-assisted anatomical navigation was first described by Kelly in 1982.[1] In 1989, Roberts reported a frameless microscopic-based navigation system.[2] In standard anatomical navigation, also known as frameless stereotaxy, the real space of the surgical field is registered to the three-dimensional (3D) image space, which is based on anatomical data only. The ensuing possibility to objectively determine the contour and margin of the tumor during surgery is highly appreciated. Thus, image-guided, or navigated brain tumor resection has gained increasing interest in the last three decades. So far, standard frameless anatomical navigation, based on anatomical information only, has become a routine tool in many neurosurgical departments. However, it is also essential to depict the surrounding eloquent neurologic structures, so that neurological function can be protected. Unfortunately, with anatomical navigation only, the surrounding eloquent structures are usually "invisible" in the surgical field. To solve this problem, anatomical navigation was enhanced by the integration of further information from other modalities, resulting in so-called multimodal navigation.

Multimodal image-guided brain tumor resection

It is critical to balance maximal tumor resection with the goal of minimal injury of neurologic function. With intraoperative imaging, maximizing the extent of resection (EoR) is now possible. Recent studies have demonstrated a survival advantage for patients with brain tumors with a resection extent of 98% or greater, particularly in younger patients with good Karnofsky performance scores (KPSs).[3–6] In addition, multimodal navigation can be very helpful to minimize postoperative neurological deficits.[7–9] With the advancements in surgical techniques and perioperative imaging technology, multimodal image-guided brain tumor resection is now technically feasible. With this technology, it is now possible to maximally remove malignant intrinsic gliomas, even when functionally critical areas are involved, with minimally increased morbidity. Images and processed information from various sources, representing functional brain cortex, major white matter tracts, different metabolic status, etc., can be integrated and clinically used for navigation.

For resection of tumors involving functional brain cortex, functional imaging may be combined with frameless neuronavigation to allow the intraoperative localization of both the lesion and eloquent brain areas such as the motor area, the visual cortex, and language-related areas.[10]

In a retrospective study, Ganslandt et al. analyzed how the decision for glioma resection was influenced by MEG.[11] When functional data were used in combination with frameless stereotactic devices, the postoperative overall morbidity was 6.2%. These data can also be interpreted as guiding more careful patient selection with the help of advanced preoperative brain mapping. Hence, preoperative identification of eloquent brain areas has an impact in the risk evaluation in glioma surgery, as well as in guiding intraoperative decision-making via functional navigation to reduce the risk for postoperative neurological deficits. Clinically, MEG-based navigation is less frequently used, mainly because of its limited availability and high cost of investment in the hardware. In recent years, blood oxygenation level dependent (BOLD)-based functional MRI (fMRI) has been widely adopted for noninvasive functional brain mapping. Images are acquired while the patient performs tasks that involve the area of brain that is likely to be at risk during the planned surgery. Postprocessing then creates statistical maps which show brain areas where blood flow changes were correlated with task performance. These data are then coregistered with conventional high-resolution anatomic MRI and appear on the neuronavigation images. From the late 1990s, some small series of patients have been reported, demonstrating the clinical impact of this technique for the resection of brain tumors involving eloquent cerebral cortex.[12–14] However, to date, there is still no prospective multicenter trial supporting the use of fMRI in brain tumor surgery.

fMRI or MEG can help to localize functional brain cortex; however, during tumor resection, damage to deeper structures, such as major white matter tracts, can also cause neurological deficits. Diffusion tensor imaging (DTI) can be used to display the course of major white matter tracts, such as the pyramidal tract. Knowledge of the course of these tracts in relationship to a tumor helps to reduce the incidence of new postoperative neurological deficits.[15,16] The registration of diffusion imaging data with the navigational data[17–19] facilitates the intraoperative preservation of these eloquent structures.

Besides functional and structural data, metabolic imaging data are also available for integration into a multimodal navigation setup. PET and magnetic resonance spectroscopy (MRS) imaging may provide metabolic information on the diffuse tumor margin. Integration of metabolic maps into the neuronavigation datasets enables a spatial correlation of the tumor metabolic status and histo-pathological findings.[20,21]

It is often difficult to delineate the heterogeneous structure of gliomas based only on conventional anatomical MRI images. Present state-of-the-art tissue imaging using T2-weighted MRI and contrast-enhanced MRI as a technique for visualizing regions where the blood–brain barrier is damaged may not fully delineate the extent of tumors and can result in ambiguous or misleading results.[22,23] Establishing the position and the extent of the border zone between tumor and normal brain tissue is one of the major problems in therapy planning.

PET has been used for metabolic imaging-based delineation of the tumor and image-guided resection.[24] Current work is investigating the clinical value of glioma-specific tracers for surgical guidance for both biopsy[25,26] and resection.[27] These PET images, in Digital Imaging and Communications in Medicine (DICOM) format, can be integrated in a standard neuronavigation system for intraoperative multimodal image-guided brain tumor resection. However, the use of this technique is hampered by the limited access of PET tracers, particularly those with strong differential affinity for brain tumors.

Proton magnetic resonance spectroscopic imaging (1H-MRSI) is a relatively low-cost noninvasive tool for investigating the spatial distribution of metabolic changes in brain lesions, which is more broadly available than PET. For MRS imaging, calculation of metabolic maps by integrating the peak area of a metabolite of interest for each voxel is a common method to allow visualization of these changes.[28,29] Several studies have reported increased levels of choline-containing compounds (Cho) and a reduction in signal intensity of the *N*-acetyl-aspartate (NAA) and creatine (Cr) in brain tumors.[30–32] Choline-containing compounds are composed of choline, phosphocholine, and glycerophosphocholine. The signal of these metabolites is often elevated in the presence of tumor tissue which is thought to be due to increased membrane synthesis in rapidly dividing tumor cells.[33,34] NAA is regarded as a neuronal marker mainly contained within neurons.[35] The range of Cho increase and NAA decrease is compatible with the range of tumor infiltration.[22,36] In principle, metabolic maps of NAA and Cho allow the differentiation of necrosis, solid tumor, and varying degrees of tumor infiltration as well as tissue edema. Resection and biopsy of brain tumors guided by delineation and segmentation based on a MRI/MRS hybrid dataset was first reported by Stadlbauer et al. in 2004.[21]

Optical image-guided brain tumor resection

The use of intraoperative fluorescence has proven to be a useful tool in multiple neurosurgical applications. In recent years, fluorescent modules have been successfully integrated with the operating microscope and have become widely used.

For brain tumor resection, the main fluorophore for which an integrated fluorescent module is available is 5-aminolevulinic acid (5-ALA). Successful neurosurgical application of 5-ALA induced fluorescence for high-grade glioma resection was demonstrated by a European randomized controlled trial conducted in 2006.[37] Since then, this technology has been adopted in the neurosurgical international community as an adjunct for high-grade glioma surgery. In the United States, current regulatory barriers have limited the use of this fluorescent agent.[38] The results of an ongoing multicenter trial will determine if its use will be approved in routine practice.

5-ALA elicits synthesis and accumulation of fluorescing porphyrins, predominantly protoporphyrin IX (PPIX), in malignant glioma tissue and is being used for their intraoperative identification.[37–41] This fluorescence can be made visible macroscopically with the aid of standard surgical microscopes equipped with optical filters[38,42,43] and is highly predictive of malignant glioma tissue.[38,40,42,44,45] The lack of visible fluorescence in adjacent tissue, however, is not highly predictive of normal tissue, because biopsies taken from these areas have frequently been infiltrated by tumor cells, corresponding to a low negative predictive value. Nonvisual methods such as high-resolution spectrometry[46,47] allow PPIX to be detected with greater sensitivity in these regions.

In recent years, a new fluorophore, fluorescein sodium, which has been widely used in medical applications for over five decades, has been successfully integrated in brain tumor resection procedure. It has primarily been used in the field of ophthalmology. In the field of neurosurgery, it is currently injected intrathecally to facilitate detection of cerebrospinal fluid (CSF) leaks during endoscopic transnasal surgery.[48] Historically, fluorescein has been used in neurosurgery with multiple purposes, including surgery of arteriovenous malformations, aneurysm surgery, bypass operations, and tumor surgery.

Even though fluorescein fluorescence appears to have a promising potential in neurosurgery, practical intraoperative fluorescent imaging technology has not been available in the operative field. Recently, a new microscope-integrated fluorescent module was developed to allow the surgeon to observe fluorescein fluorescence using low doses of this drug while being able to continue microsurgery under operating room microscope in an almost natural light environment.[49]

Fluorescein fluorescence been observed through a microscope-integrated module appears to be a promising and potentially versatile practical tool in the operating room without interfering with microsurgical workflow.[49] The ability to see fluorescence in real time through the operator's oculars along with the anatomy of surrounding brain structures allows the surgeon to better appreciate the anatomy in the surgical field. The use of fluorescein fluorescence appears to be a valuable tool in oncologic neurosurgery. This fluorescence technology allows the surgeon to localize and demarcate lesions located close to the surface, even when inspection under white light does not show a significant abnormality.

The ability to see the nonfluorescent tissues in nearly natural colors allows the surgeon to proceed with tumor resection continuously under fluorescent mode guidance. Another advantage of this technology over the currently used 5-ALA is its ease of use. 5-ALA needs to be administered orally at least 3 hours prior to anesthesia induction, which can be inconvenient for operative schedules, and the patient needs to be in a darkened room for 24 hours after surgery.[38] But fluorescein sodium only needs to be administered 2 hours before exposing the tumor with a relatively low dose (1 mg/kg). Another big advantage is that 5-ALA, so far, is still not officially allowed to be routinely used clinically in several countries, like China and the United States, while fluorescein sodium doesn't have this issue. Although fluorescein fluorescence has potential advantages compared to 5-ALA fluorescence, its accuracy and specificity are not established. Because, theoretically, fluorescein fluorescence can be detected in any place where the blood–brain barrier is damaged, which occurs frequently around the tumor resection cavity, the specificity and accuracy of this technique may be compromised.

Multimodal image guidance supported by intraoperative imaging

Intraoperative imaging has been used to detect tumor remnants in gliomas that could undergo further resection in 20–60% of cases.[9,50–52] Intraoperative high-field MRI in combination and close integration with microscope-based navigation serving as a common interface for the presentation of multimodal data in the surgical field seems to be one of the most promising surgical setups to avoid unwanted tumor remnants while preserving neurological function. Navigation alone, while useful, is degraded in accuracy as the resection proceeding due to brain shift. With updated intraoperative imaging, navigation data can be updated, so that brain shift can be compensated for, and initially missed tumor remnants can be localized reliably.

Gross total resection usually implies removal of the preoperatively defined contrast-enhancing tumor tissue for high-grade gliomas and of T2/FLAIR signal change for low-grade gliomas. Postoperative MRI documentation represents the current gold standard in surgical outcome assessment but, ideally, resection control should be performed using intraoperative imaging or, as an alternative, with early postoperative MRI (preferably within 48 hours following surgery). Assessment of tumor volume prior to and following surgery is instrumental for evaluation of surgical success and multimodal therapy outcome, but requires definition of accurate and validated volumetric methods to measure the EoR.

Although methods to accurately measure pre- and postoperative tumor volume exist, no established objective protocols to evaluate the extent of tumor resection have been defined. Many academic neurosurgical centers are using established preoperative and postoperative MRI with T2/FLAIR sequences to make 3D volumetric measurements in their clinical routine. Creation of computer-assisted tumor volume algorithms, which could enable comparison of surgical techniques and clinical trials, will lead to further progress in objective evaluation of tumor resection. Furthermore, the introduction of intraoperative imaging techniques such as intraoperative MRI, neurophysiologically and biologically active fluorescence markers, and their various combinations, now permits direct visualization of tumor tissue and enables intraoperative assessment.[9,17,37,44,53–55]

Multimodal functional data, such as fMRI or DTI data can also be acquired intraoperatively and directly used for intraoperative navigation updating, which in the clinical routine might be a time-consuming effort, especially when in the case of, e.g., visualization of speech connecting fiber tracts, some sophisticated time-consuming nonstandard tracking algorithms have to be applied.

Alternatively nonlinear registration techniques or sophisticated techniques from pattern recognition analysis may allow a matching of preoperative MR datasets containing functional information with intraoperative MR image volumes.[56] This might also be an approach in cases where intraoperative MRI is not available, but other imaging modalities can provide intraoperative 3D information about the brain configuration, so that high-resolution multimodality data can be registered nonlinearly onto the "low-quality" intraoperative data. Such an alternative to intraoperative MRI might be intraoperative ultrasound, especially intraoperative 3D ultrasound.[57–59] Whether the image quality to evaluate the extent of a glioma resection is really equivalent among the different imaging modalities remains an area of investigation.

CLINICAL APPLICATIONS

Multimodal definition of tumor margin

Cancer treatment modalities include surgical resection of tumors and their adjacent tissue, which may contain infiltrative neoplastic cells. EoR is increasingly accepted as critical to optimal surgical treatment and patient outcome.[60,61] Incomplete resection of a tumor and its infiltrative elements may increase the risk of recurrence. A surgeon's ability to distinguish between diseased and healthy tissue is limited by visual perception of tumor margins. This definition of the tumor and its borders, whether infiltrative or well circumscribed, is crucial. After an incomplete resection, positive tumor edges are found at the surgical edges of a resection bed. Minimizing the extent of positive tumor edges decreases the recurrence of the primary tumor in breast cancer,[62,63] laryngeal carcinomas,[64,65] rectal cancer,[66] prostate cancer,[67] and glioblastoma multiforme.[68] Surgical approaches that visually distinguish neoplastic tissue from healthy tissue could help surgeons to maximize resection of neoplastic cells from patients and to minimize the amount of residual cells that could form positive edges.[4–6]

For image-guided brain tumor resection, the accurate definition of the tumor margin is essential. Numerous surgical technologies have been developed to facilitate optimal resection, many of which function to guide the surgeon during resection. The primary approach is to define the tumor margin according to the anatomical dataset acquired with CT and MRI. However, since malignant gliomas do not have a distinct margin between the tumor mass and the surrounding brain, achieving gross total tumor resection with anatomical images alone represents a major challenge to the neurosurgeon. To solve this problem, many new imaging techniques were integrated into navigation systems, including metabolic imaging,[21,24,28,29] perfusion imaging,[69] and fluorescence imaging.[37,38]

Metabolic imaging (e.g., PET, MRS, etc.) can be integrated via DICOM format into standard neuronavigation systems to make metabolic imaging-guided brain tumor resection possible. Several studies have reported defining tumor margins using PET datasets for navigated brain tumor resection.[24] However, this technique has many limitations. A more available technique is to define the tumor metabolic margin with MRS. Stadlbauer et al.[20] developed a method to improve the delineation of intrinsic brain tumors based on the changes in metabolism due to tumor infiltration. 1H-MRSI has been used to investigate the spatial distribution of Cho, Cr, and NAA in brain tumors and normal brain. The authors developed a method for coregistration of MRI and MRSI datasets. Integration of the biochemical information into a frameless stereotactic system allowed biopsy sampling from the brain areas that showed normal T2-weighted signal but abnormal 1H-MRSI changes. The histological findings showed tumor infiltration ranging from about 4–17% in areas differentiated from normal tissue by 1H-MRSI only. According to these results, the authors concluded that high spatial resolution 1H-MRSI integrated in navigation could improve delineation of tumor borders compared to routine MRI tumor diagnosis. It remains unknown, however, what the clinical impact of extending resection margins based on such information may be.

In our institution, with standard spectroscopic imaging techniques at 1.5T MRI, we developed a way to import the raw MRS Chemical Shift Images (CSI) into a standard navigation system, followed by full automated segmentation of the tumor including the infiltrative zone in Cho or Cho/NAA ratio map based on the assumption of Gaussian distribution for the Cho/NAA values in normal brain matter.[70] The resulting spectroscopic image of the segmented tumor was used to calculate the area of pathologic brain metabolism due to the lesion. In this way, low-cost, biochemical information guided delineation and resection of selected brain tumors can be achieved (Figures 9.1 and 9.2).

New imaging techniques, like perfusion imaging, can also be used to guide tumor border zone delineation. In a recent report,[69] Roder et al. evaluated the use of intraoperative perfusion-weighted imaging as an aid to detect residual tumor and to guide its resection. Twenty-two patients undergoing intraoperative scanning during the resection of high-grade gliomas were examined with perfusion-weighted ioMRI. In all cases, diagnostic regional cerebral blood volume (rCBV) perfusion maps were generated. Interpretation of perfusion maps demonstrated that gross total resection of

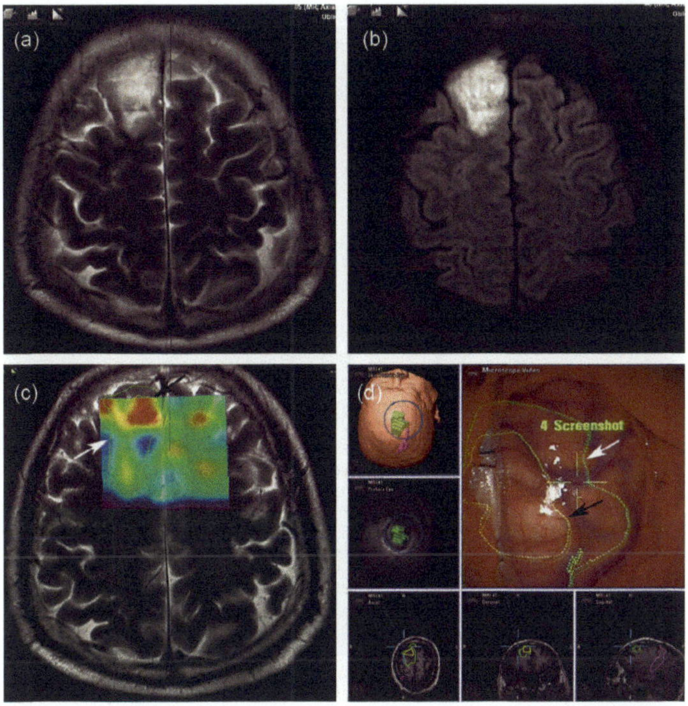

Figure 9.1 Integration of MRS images into neuronavigation. (a, b) T2-weighted and FLAIR MR images reveal a nonenhancing lesion located in right frontal lobe; (c) the MRS images were integrated in multimodal navigation plan. We defined different tumor margins according to conventional MRI images (c, white arrow) and the tumor core with MRS-based metabolic images (c, black arrow), respectively. (d) Intraoperatively, the metabolic margin and T2 image-based margin could be easily overlaid on the surgical field.

gliomas was achieved in 4 of 22 cases (18%), which was macroscopically and histopathologically verified, whereas in 18 of 22 cases (82%), the perfusion-weighted ioMRI revealed hot spots indicating subtotal tumor removal. This study proved the technical feasibility of implementing perfusion-weighted imaging into intraoperative imaging protocols and indicated that this technique may assist the surgeon to detect residual tumor volume and to define the tumor border more precisely, compared with conventional MRI imaging methods.

Another promising approach to such identification is the application of tumor-specific fluorescent probes followed by intraoperative fluorescence imaging to identify residual tumor. Such techniques may allow surgeons to more sensitively identify tumor edges and invasive cells that have infiltrated healthy tissue. Such techniques could lead to more complete tumor resection and thereby improve patient outcomes.[38] Fluorescence imaging has been widely used in the biological sciences to study the structure and function of tissues and individual cells. To date, the fluorescent

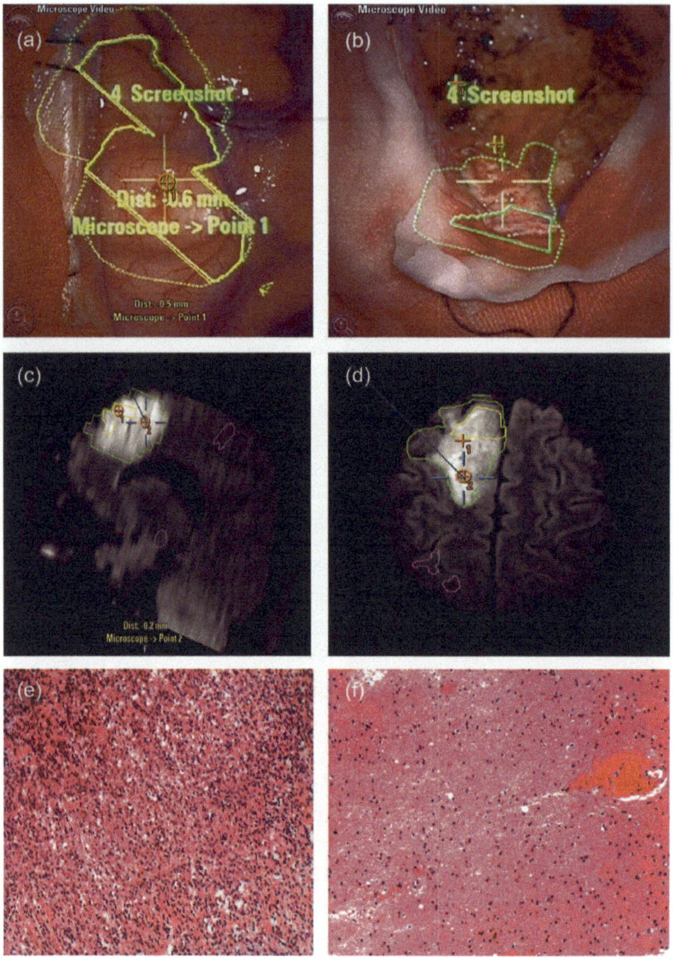

Figure 9.2 (a–d) tissue samples taken from different regions of interest (ROIs) from metabolic imaging based tumor core (ROI 1) and anatomical imaging based tumor margin (ROI 2); while the tissue sample taken from the T2 image-based area (f) had much lower tumor cell concentration than the MRS-depicted tumor core (e).

probes that have been developed are specific to cell types, proteins, genomic information, and ion concentrations.[71,72] Advances in detection hardware now allow nearly real-time in vivo acquisition of images.[73] The speed and sensitivity of fluorescence imaging have fueled interest in identifying intraoperative applications for this technology. Reports from Europe[37,38,46] and China[40] provided high-level evidence that 5-ALA fluorescence-guided surgery is more effective than conventional neuronavigation-guided surgery in increasing diagnostic accuracy and extent of tumor resection, enhancing quality of life, and prolonging survival in patients with high-grade malignant gliomas. However, both the specificity and sensitivity of this approach requires further validation.

In the United States, relatively few fluorescent agents have been approved by the Food and Drug Administration (FDA) for in vivo use in humans. Consequently, these agents are not yet in widespread use. However, as techniques are refined, additional approved fluorescent clinical agents will likely find their way into clinical use.[49,74]

Localization of surrounding eloquent structures

It is well established that EoR is an important prognostic factor in patients with both high-grade[5,6,45] and low-grade [75] gliomas. Another equally critical aspect is the protection of nearby eloquent structures and postoperative health-related quality of life (HRQoL). Multimodal neuronavigation, supported by ioMRI, can not only "visualize" and help to protect eloquent brain structures intraoperatively, but can also influence the extent of tumor resection by helping identify residual tumor and providing imaging data to update neuronavigation, compensating for inaccuracies caused by brain shift.

The usefulness of multimodal imaging guidance for localization of surrounding eloquent structures during brain tumor resection has been demonstrated.[8,10,24] In our series, in all cases, multimodal neuronavigation was successfully integrated into the standard neurosurgical work flow and helped in brain tumor resection, especially for those involving eloquent functional structures. Multimodal neuronavigation can integrate fMRI and DTI images, so that the surrounding functional brain cortex (motor, sensory, visual, and language cortex, etc.) as well as major white matter tracts, can be depicted in the surgical field under the microscope. We have found, in both high- and low-grade glioma cases, that the use of multimodal neuro-navigation with high-field ioMRI seems to be safe and justified for tumor resection control and intraoperative multimodal brain imaging (DTI fiber tracking, MR angiography, etc.).[12,76–78]

For the depiction of functional cortex, task-based BOLD fMRI scans can be used. Well-designed and properly executed motor tasks (e.g., finger tapping) can depict the motor cortex very well. Figure 9.3 shows a 30-year-old right-handed male, who suffered from refractory epilepsy. He was found to have a lesion located in the left frontal lobe (Figure 9.3a and 9.3b). Preoperative fMRI depicted motor cortex activation. The motor cortex for right foot (Figure 9.3d, yellow), right hand (Figure 9.3d, blue, white arrow), and lips (Figure 9.3d, light green) were displayed and marked on the multimodal navigation image. Before the craniotomy, the multimodal functional images, including motor cortex, pyramid and sensory tracts, can be overlaid under the microscope in the surgical field, so that these "invisible" eloquent structures can be "visible" under the microscope. With this technique, the skin incision and bone flap site can be designed according to the contour of tumor and surrounding eloquent structures.

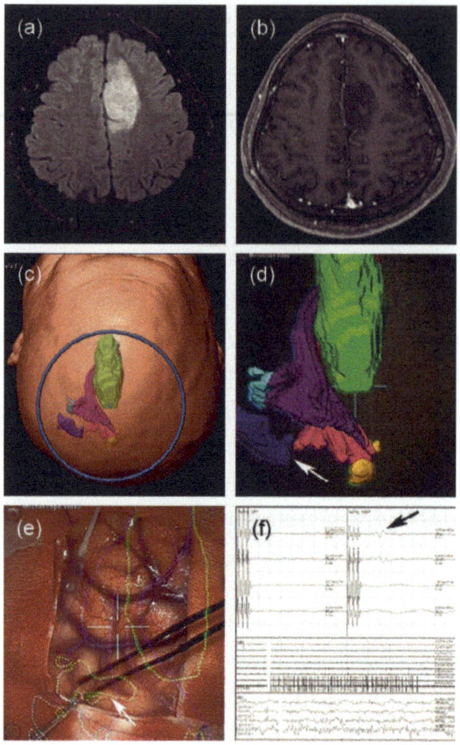

Figure 9.3 (a, b) A 30-year-old right-handed male with refractory epilepsy was found to have a lesion located in the left frontal lobe. (c, d) Preoperative multimodal navigation images. fMRI-depicted motor cortex activation. Motor cortex for right foot (yellow), right hand (blue, white arrow), and lips (aqua) were displayed. Pyramid tract and sensory tract were labeled with purple and pink color, respectively. (e, f) Intraoperative DCS of hand motor cortex.

Intraoperatively, after dural opening, it can also be seen that the tumor itself (Figure 9.3e, green contour) had very similar color and texture with the surrounding normal brain. From this point of view, it was essential to use image-guided navigation to help to localize and resect the tumor, while protecting the surrounding eloquent motor cortex.

The final ioMRI images confirmed the total removal of the tumor, with preservation of the precentral gyrus (Figure 9.4e and 9.4f). The patient has been followed up for 30 months, and showed no tumor progression and no neurologic deficit.

For visual cortex related navigation, standard visual tasks (e.g., black–white alternating checkerboard task) can be used for preoperative fMRI delineation of the visual cortex, followed by importation of these images into the navigation system. For example, Figure 9.5 and 9.6 show a 46-year-old female operated with mainly visual system related navigation. MRI images showed a deep-seated lesion in the right occipital lobe (Figure 9.5a). Because the surgical approach would likely involve the

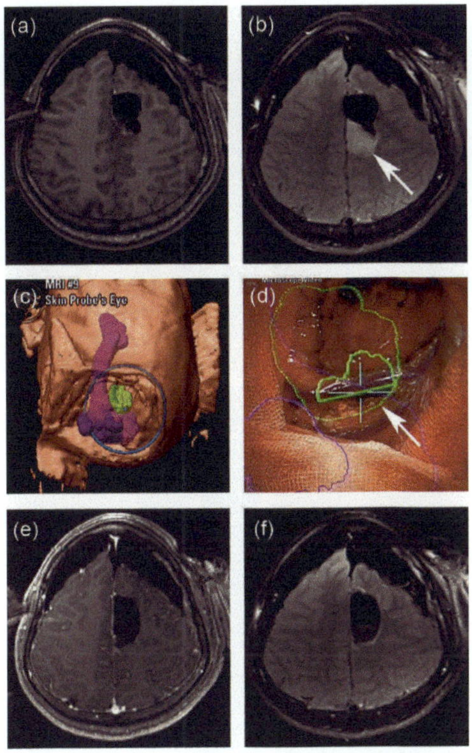

Figure 9.4 Initial ioMRI images revealed tumor remnant in posterior and medial portions of the lesion (a and b, white arrow). The navigation plan was then updated, so that the image of tumor remnant can be overlaid onto the surgical field under the microscope (c and d), guiding resection of the residual of tumor confirmed by the second intraoperative MRI (e and f).

visual cortex and optic radiations, we performed preoperative fMRI with the visual task and depicted the visual cortex, while the right optic radiation was depicted with DTI-based fiber tracking. The multimodal images were then integrated in the navigation system and used for intraoperative navigation (Figure 9.5a). After dural opening, the visual cortex was marked, based on the fMRI findings, and protected, followed by total removal of the lesion. The postoperative course was uneventful. The patient has experienced a very mild and clinically inapparent visual field deficit at 1 month postoperative follow-up (Figure 9.6c and 9.6d). The histological diagnosis was glioblastoma WHO grade IV. The patient then received standard chemotherapy (temozolomide) and radiation therapy. She was followed up for 24 months without tumor progression (Figure 9.6).

For language cortex localization, it is still controversial whether task-based BOLD fMRI has sufficient sensitivity and specificity to depict major language cortex (i.e., Broca and Wernicke cortex).[79–86] However, in recent years, more studies which correlated language fMRI results with DCS results confirmed that a well-designed and

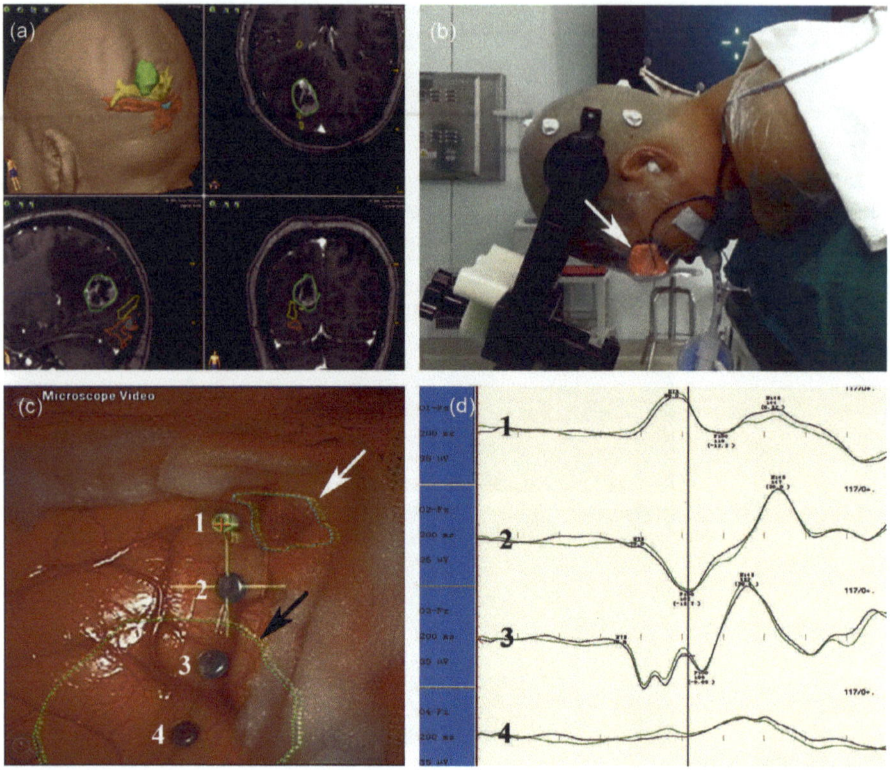

Figure 9.5 Visual cortex-related navigation. (a) Gadolinium-enhanced MR images show an inhomogeneously enhancing lesion in the subcortical right occipital lobe; (b) an LED flash patch (white arrow) was adhered and fixed to the patient's eyes; (c) after dural opening, electrodes were placed around the visual cortex depicted with fMRI based navigation (white arrow); (d) with repeated LED flash stimulation, we successfully recorded visual evoked potential (VEP) from electrode 1.

properly executed fMRI protocol can make language cortex localization more accurate and reliable.[79,81,84] These data suggest that with further technical improvements, imaging of language function, based on BOLD fMRI, may become a good preoperative diagnostic and intraoperative navigation tool for patients with tumors close to language-related brain structures. Figures 9.7 and 9.8 show our procedure for language cortex related navigation. A 48-year-old right-handed male presented with a lesion located in the left frontal operculum lobe (Figure 9.8a and 9.8b). Preoperative fMRI depicted frontal language activations in the putative Broca region just anterior to and in close relationship to the lesion (Figure 9.8c, white arrow). These images were integrated into the navigation system and used for language cortex-related navigation. Immediately after dural opening, the fMRI activations were depicted with navigation under the microscope. In this way, the tumor can be totally removed, while preserving the fMRI-defined functional areas. The intraoperative

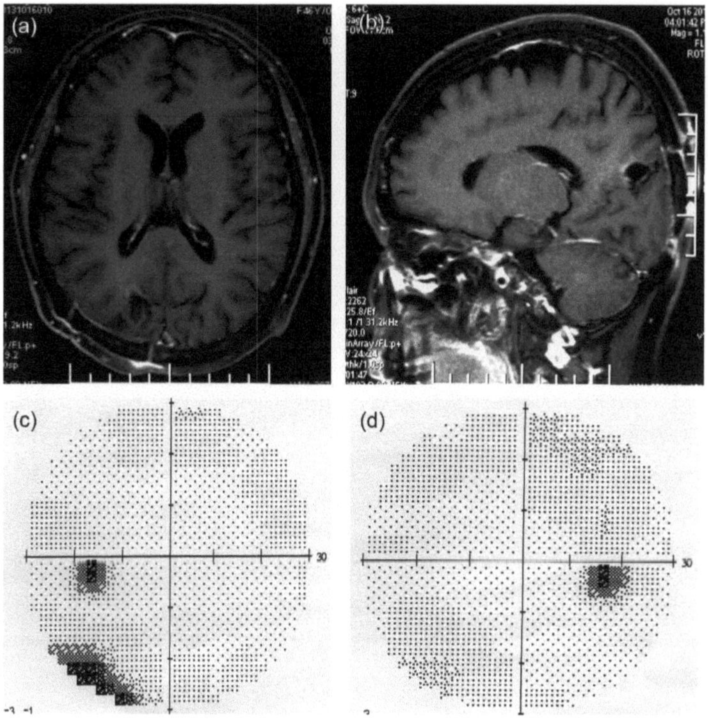

Figure 9.6 (a–d) The patient's 24-month follow-up MRI images and visual field examination results.

and postoperative MRI images confirmed the gross total resection of the tumor (astrocytoma WHO Grade II). The patient had no language deficit and had no seizure. He has been followed for 48 months and showed no tumor progression and no epilepsy.

Protection of the surrounding eloquent major white tracts is also essential. DTI-based fiber navigation can make the usually "invisible" fiber tracts "visible" under the microscope in the surgical field. Presently, the major fiber tracts that are routinely used for intraoperative fiber navigation include pyramid tracts, sensory tracts, optic radiations, and arcuate fasciculus (AF).

Figure 9.9 shows a case with multimodal navigation, including pyramidal tracts, optic radiation, and AF. A 25-year-old right-handed male suffered from refractory epilepsy. Conventional MRI images revealed a nonenhancing tumor located in the left insular lobe (Figure 9.9a–9.9c). Preoperative multimodal imaging revealed that the tumor was surrounded by eloquent structures, including pyramidal tracts, sensory tracts, AF, and fMRI-defined language areas (Figure 9.9d). With microscope-based navigation, the contours of the surrounding eloquent structures can be overlaid in the surgical field (Figure 9.9e and 9.9f), which is useful for the surgeon to plan the cortical incision and the surgical corridor. With ioMRI and multimodal navigation, the

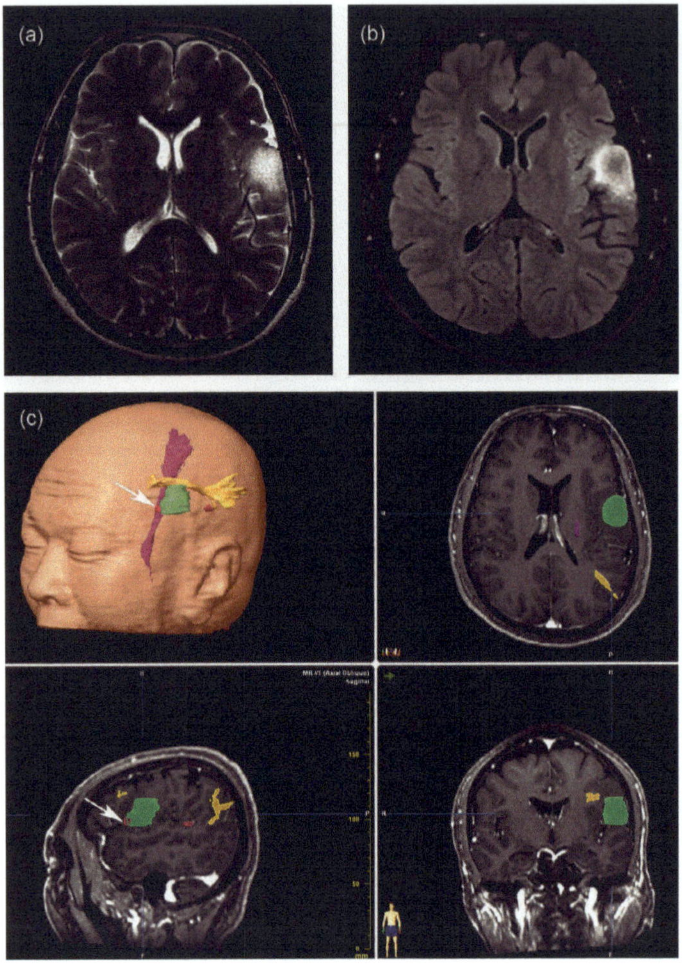

Figure 9.7 Language cortex-related navigation case. (a, b) MRI images revealed a lesion located in the left frontal operculum lobe; (c) preoperative functional brain imaging depicts that frontal language cortex (white arrow) is just anterior to the lesion (green) and has a very close relationship with the lesion. The surrounding arcuate fasciculus (yellow) is also depicted.

insular tumor (astrocytoma WHO II) was totally removed, while the surrounding eloquent fiber tracts were preserved.

The patient has been followed up for 48 months without tumor progression or seizures (Figure 9.9k and 9.9l).

"Brain shift" correction

Tumor removal, brain swelling, the use of brain retractors, and CSF drainage all result in intraoperative brain deformation, which is known as *brain shift*.[87] Thus, in

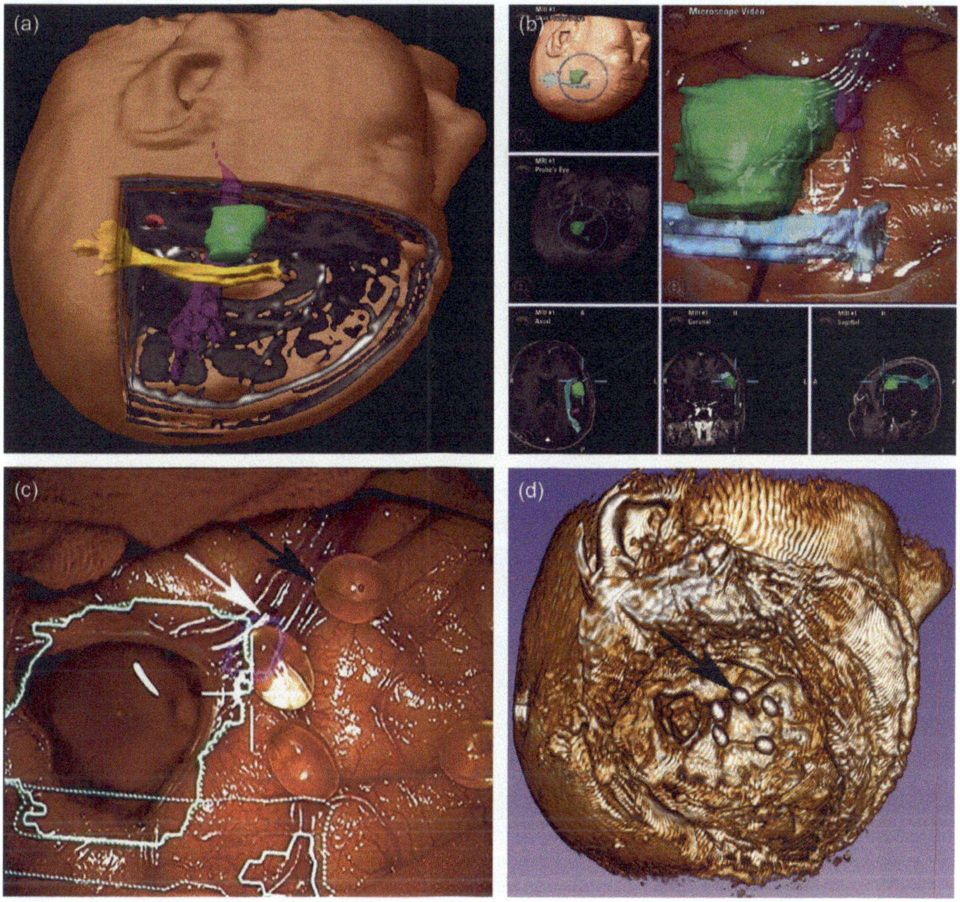

Figure 9.8 Intraoperative pictures of the same patient as Fig 9.7. Immediately after dural opening, when "brain shift" was not very significant, several gas-sterilized vitamin E capsules (c, black arrow) were placed along the margin of Broca cortex (c, white arrow), which was depicted with navigation under the microscope. On the intraoperative MRI images, the markers had very high intensity on both T1- and T2-weighted MRI images. Panel d showed the 3D reconstructed volume rendering Images. The markers can be clearly viewed.

navigation systems relying on preoperative image data only, the accuracy decreases during the course of surgery. Updating the navigation system with high-field ioMRI image data seems to be the most reliable method to compensate for the effects of brain shift. Intraoperative imaging offers the possibility to compensate for the effects of brain shift, because it provides a virtual reproduction of the actual intraoperative physical reality, both in terms of brain deformation and of the actual extent of tumor removal.

Integrating high-field MRI and microscope-based navigation facilitates this intraoperative neuronavigation update procedure.[52] In our institute, navigation updating is

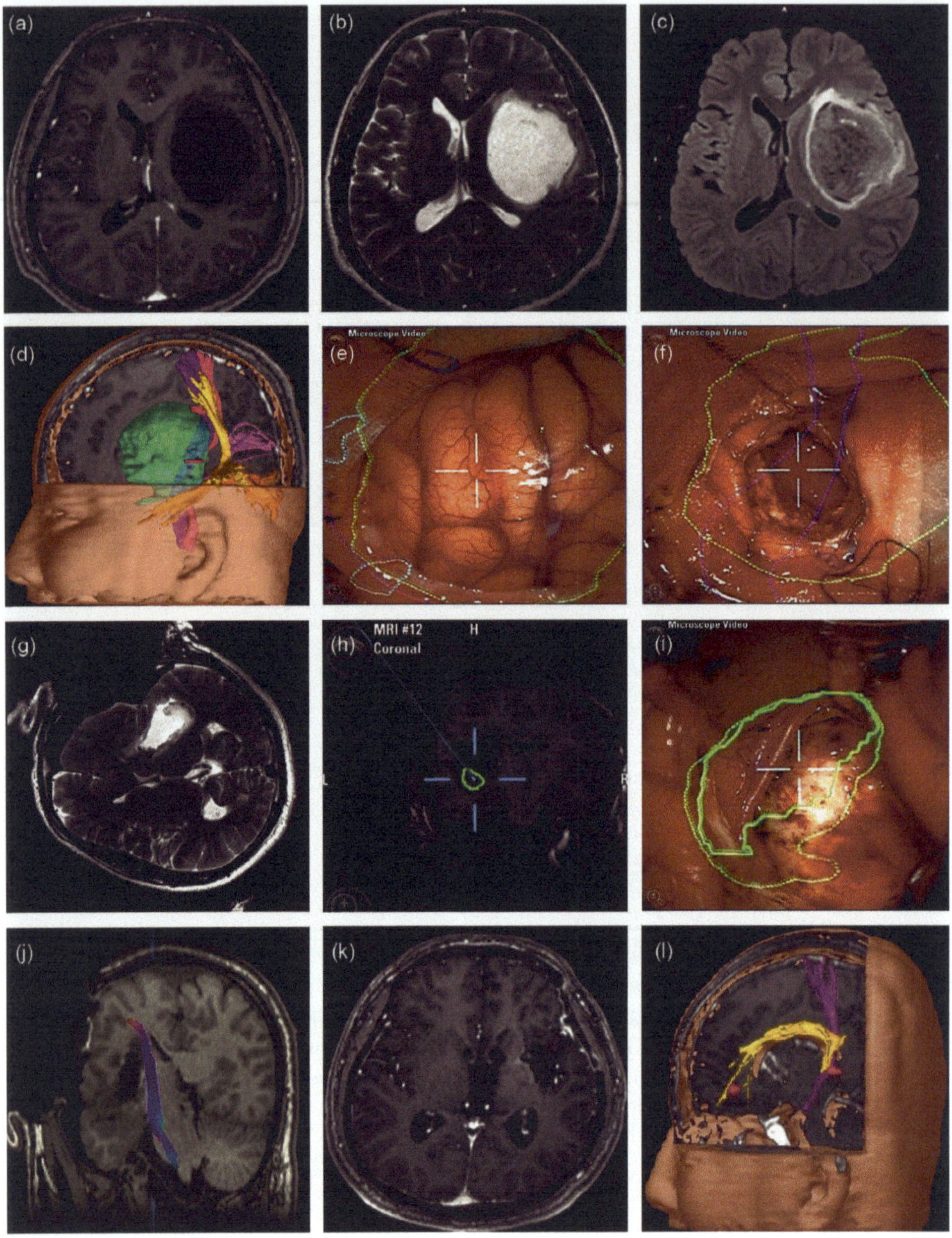

Figure 9.9 Pyramid tract navigation in a left insular glioma case. (a–c) Conventional MRI images; (d) preoperative multimodal imaging revealed that the tumor was surrounded with many eloquent structures, including pyramidal tract, sensory tract, arcuate fasciculus, and frontal and temporal language areas; (e and f) with microscope based navigation, the contours of surrounding eloquent structures can be overlaid in the surgical field; (g–j) the first ioMRI scan depicted tumor remnant in the posterior margin of the resection cavity, while the intraoperative DTI-based fiber tracking confirmed the left pyramid tract and arcuate fasciculus were intact. The contour of tumor remnant was then injected in the surgical field under the microscope; (k and l) 48-month follow-up images revealed no tumor progression.

done with a rigid registration of the intraoperative image data with the preoperative image data, followed by subsequent segmentation of the tumor remnant, and final restoring of the initial patient registration. Alternatively, the intraoperative image registration can be achieved by applying the automatic registration matrix by attaching it to the upper part of the head coil, as is done for the initial patient registration process with preoperative image data.[88] Thus, the registration coordinate system of the preoperative image data is applied to the intraoperative images, serving as an immediate intraoperative image update. Updated image data allow a reliable "brain shift" correction and identification of a tumor remnant.

Microscope-based image injection with the direct visualization of the segmented tumor remnant in the surgical field plays a crucial role in the precise localization and orientation in the resection cavity.

Multimodal navigation patients with preexisting deficits

For patients with preexisting deficits, it may be difficult to perform relevant task-based fMRI for the localization of relevant functional brain cortex. To localize the motor or sensory cortex, phase reversal SSEP can be performed to localize the central sulcus, so that the precentral and postcentral gyri can be distinguished.[89-91] This method is accurate and reliable, even under general anesthesia.

fMRI-based language mapping is difficult to obtain reliably in patients with preexisting language deficits. However, DTI-based fiber tracking can help to localize eloquent language areas (e.g., Broca and Wernicke's areas). Henning et al.[92] reported their method for localization of primary language areas by arcuate fascicle fiber tracing in 13 patients. Their study indicates that tractography of the AF is a useful tool for identification of the parts of the primary language areas. Using tractography as a localization procedure to determine the primary language areas aids in the delineation of patient-specific anatomy and thus may help reduce the risk of postoperative permanent neurological deficit.

Brain mapping: multimodal functional navigation versus awake craniotomy with direct cortical stimulation

In human brain mapping, two basic strategies are commonly used to obtain information about cortical functional representation: (1) active approach: observing the effects of directly eliciting/extinguishing brain activity; and (2) passive approach: recording brain activity during task performance.[93] Techniques using the active approach include direct cortical stimulation (DCS), transcranial electric stimulation (TES), and transcranial magnetic stimulation (TMS).

The first concept of brain mapping with electrical stimulation in humans was generated by Bartholow in 1874.[94] A decade later in 1886, Horsley and his colleague

Jackson applied electrical stimulation to localize the thumb area and decided to excise the area because it was the source of the seizure.[95] In the late 1920s, Penfield developed intraoperative brain mapping techniques while trying to treat patients with intractable epilepsy.[96] His technique was accepted worldwide for epilepsy surgery. He also applied awake craniotomy to some patients with brain tumors.[96] After these seminal achievements, using brain mapping techniques for brain tumor surgery was a logical step from its use in epilepsy surgery and started to be more widely used for this purpose during the last two decades.[97]

Of the various methods used for brain mapping, especially for language areas study, DCS remains the most proven method and is regarded as the "gold standard" for functional brain mapping. This technique allows intraoperative mapping to delineate major eloquent cortex and its relationship with the tumor and facilitates finding the safest transcortical route to the tumor and more aggressive tumor resection while minimizing morbidity by preserving functional tissue.[98] However, as with other tumor resection procedures, it is invasive, and is associated with prolonged operative time, awake craniotomy, and discomfort for the patients. There is even one small randomized study that reported higher blood loss, more neurologic deficits, and less degree of cytoreduction in patients undergoing awake craniotomy for tumors in eloquent cortex.[99] A major limitation of DCS is that it can only be applied intraoperatively once the decision for surgery has already been made. But for the planning of surgical approaches and procedures, preoperative functional maps of eloquent cortex are required as well. Furthermore, in some cases the ability to precisely locate functional areas preoperatively can become a determinant of the feasibility of the neurosurgical intervention. Intraoperatively, the invasive active approaches, like awake craniotomy and brain mapping, are associated with many risks, including seizure, agitation, airway problems, prolonged operative time, discomfort for the patients, etc. Besides the clinical and scientific issues, there may also be cultural differences across the globe in patients' acceptance of the awake craniotomy procedure. Intraoperative brain mapping, moreover, requires a dedicated team with expertise in awake anesthesia, electrophysiology, and neuropsychology; neurosurgeons and anesthetists and all members of the team must be willing to embrace the approach. Noninvasive or less invasive brain mapping methods, which can be also used preoperatively, can be complementary to invasive mapping in centers where it is available and can provide an alternative approach in many centers where awake brain mapping is not performed. This situation may be the reason why, in recent years, brain mapping studies using the passive approach have outnumbered those using the active approach.

Techniques using the passive approach include MEG, electroencephalography (EEG), PET, fMRI, etc. Among them, fMRI has been embraced as an important component of multimodal navigational data used during brain tumor surgery. Multiple case series have proved that fMRI can be used to depict the location of the motor cortex and has very good correlation with "gold standard" DCS results.[100–103] Similarly, several case series have demonstrated that fMRI can also be used to estimate the location of cortical areas involved in speech and language in relationship to brain tumors.[13,104] Though there is less published evidence, fMRI can also be used to map cognitive functions such as calculations in brain tumor patients.[105] All these findings suggest that fMRI is a useful tool in the surgical care of brain tumor patients. Because first, by revealing the relationship of eloquent functional areas to the tumor, fMRI may be used as a strategic tool for preoperative surgical planning.[106] Second, defining the relationship of a lesion to an eloquent region may help determine whether intraoperative mapping via DCS is needed for safe resection. For brain mapping involving language areas, there are several reports suggesting fMRI is less accurate compared to motor area mapping.[82,83,107] For example, Giussani and colleagues report contradictory results when the speech zone was mapped using DCS in comparison with fMRI mapping.[82] This data inconsistency could be because DCS suppresses the projection fibers, whereas fMRI depicts the increased blood flow to areas with an increased neuronal activity.[108] Another possible reason for discrepancies may be related to the difficulty of utilizing an fMRI task protocol that corresponds to those used during intraoperative mapping.[13,83] However, with the advancement of imaging technology and improvement of fMRI algorisms, fMRI for language mapping has been made significant progress in recent years.[109–111]

The active brain mapping approach investigates whether a specific region of the brain is critical for implementing particular cognitive or behavioral functions and therefore is able to answer questions about causal relationships between brain and function. Noninvasive techniques using such an approach include TES and TMS. The noninvasive approaches have the advantage of a greater safety profile and low cost burden for the patients, with TMS being the most important one.

TMS is a noninvasive technique that uses magnetic stimulation to generate electrical current in the cerebral cortex via a device that generates a brief electric current in a coil placed near the patient's head. It has been developed into a technique that allows the closest noninvasive approximation to electrical cortical stimulation. There have been numerous general reviews of the technique and of the potential for TMS in studies and treatment in neurorehabilitation.[112] Recent advances in image processing have allowed the refinement of current TMS-mapping strategies by combining MRI modalities with TMS using a 3D digitizer to measure the position of the stimulating coil and map this position onto an MRI dataset. A frameless stereotactic system that is rigidly fixated to the stimulating coil is used to correlate scalp stimulation sites to the underlying brain anatomy in real time. The accuracy of this new technique has been validated by correlating TMS maps to cortical output maps obtained with DCS output maps.[113,114] The generated multimodal dataset can also be integrated into standard navigation systems and used for multimodal neuronavigation.[113–115] Shamov et al.[116] have recently reported that, using TMS for presurgical planning, the operative time was reduced from an average of 7 hours (using DCS) to an average time of 3 hours and 40 minutes (using TMS) in patients with opercular tumors with a volume more than 50 cm^3.

Neuroimaging techniques, commonly used to acquire functional cortical representation, such as fMRI, PET, EEG, or MEG, have limitations: fMRI and PET provide indirect measures of brain activity with low temporal resolution, and EEG and MEG lack spatial resolution. None of these methods can provide complete insight into causal relations between brain activity and behavior. However, combining such neuroimaging and neurophysiologic methods with TMS offers unique advantages: (1) MRI or PET activation can guide where to stimulate and (2) real-time EEG recording can guide when to stimulate. With the development of newer and more focal TMS systems, the resolution of the stimulation maps can be further enhanced, by moving the coil in small increments and stimulating just over the threshold with a controlled and monitored level of muscle contraction. Using such meticulous mapping protocols, TMS may prove to be an indispensable tool to study cortical connections in humans noninvasively and painlessly. Moreover, integration of TMS for multimodal neuronavigation can be useful not only in the preoperative functional mapping, but also in the preoperative planning of the operative approach and intraoperative planning of the direction of brain retraction and operative corridor for brain tumor resection.

Between noninvasive brain mapping methods (fMRI, MEG, etc.) and invasive "gold standard" DCS, there is a possible middle-ground approach, which is known as intraoperative optical imaging of intrinsic signals (iOIS). This is a brain mapping technique that can visualize brain compartments with micrometer and millisecond resolution. It has been demonstrated to be a potentially useful neurosurgical tool for both functional brain mapping [117–120] and lesion delineation.[121–123] iOIS is usually described as an "invasive" imaging technique because it requires surgery to make the cortex visible. However, as iOIS is performed in the operating room by attaching a charge-coupled device (CCD) camera and optical filter(s) to the operating scope, it is noninvasive compared to the current "gold standard" intraoperative electrophysiological techniques, which require placing electrodes directly onto or into the brain and applying electrical currents. As sensory evoked potentials are robust, even in anesthetized brain, primary sensory cortical functional mapping with iOIS can be performed in fully anesthetized patients. However, language mapping requires a conscious patient, hence for patients with lesions near or in eloquent cortex, intraoperative brain mapping (for any technique, including iOIS) requires the patient to be awake under local anesthesia and cooperative in the operating room in order to perform language mapping. In recent years, with the technology progression, iOIS shows great promise for micrometer intraoperative functional brain mapping and lesion delineation. iOIS can dramatically improve a neurosurgeon's eyesight by creating functional and lesion maps from images from the surgical microscope. More precise individualized intraoperative maps may potentially improve surgical outcomes, especially resections of brain tumors that lay in or near eloquent cortex.

In our institution, we routinely use multimodal neuronavigation, based on images of fMRI, DTI-based fiber tracking, etc., for brain tumor resection. And we are also trying to integrate iOIS images into standard surgical workflow, while experimenting on TMS integration. Table 9.1 summarizes and compares the above mentioned different modalities with traditional "gold standard" DCS under awake craniotomy. We believe that the horizon is not far for this new multimodal approach to work as a promising alternative to DCS.

FUTURE DIRECTIONS

Advances in technology have increasingly improved the precision of intracranial surgery, and image-guided neurosurgery is one of the most promising developments in this field. The potential impact on the management of patients with brain tumors is obvious. The use of multimodal neuronavigation, supported by intraoperative imaging, in brain tumor surgery is safe, reliable and most useful in assessing the EoR of intrinsic, infiltrating brain tumors whose boundaries cannot be clearly visualized during surgery. While several retrospective analyses have documented the benefit of implementing multimodal neuronavigation into the neurosurgical routine in glioma surgery by means of increasing the EoR, prospective randomized studies are still lacking that might prove the benefit of multimodal neuronavigation and ioMRI compared to the standard micro-neurosurgical resection of brain tumors. In recent years, more and more studies, including one prospective randomized clinical trial, has proven that neuronavigation and ioMRI have significant positive impact to prolong survival in patients with glial tumors.[124,125] Although beneficial effects are suggested by previous studies, there is much work to do to promote the use of ioMRI today, and pursuant studies are needed.

One clear advantage of multimodal neuronavigation with ioMRI over the administration of fluorescent porphyrins to visualize tumor tissue intraoperatively is the implementation of functional and advanced structural brain imaging, as well as the discrimination of nonenhancing low-grade gliomas. In the future, the full-spectrum of advanced brain imaging data (e.g., perfusion imaging, metabolic imaging, etc.) is expected to be integrated into neuronavigation systems and be "visualized" intraoperatively. In addition to preoperative functional datasets, in the near future, intraoperative multimodal functional brain imaging data may be able to depict eloquent brain structures, taking into account brain shift or other intraoperative changes.

Table 9.1 Comparison of DCS versus common multimodal navigation methods

	DCS	fMRI navigation	DTI-based fiber navigation	TMS	iOIS
Advantages	1. Intraoperative depiction of the relationship of lesions to eloquent areas 2. Intraoperative localization of functional brain cortex and major subcortical fiber tracts	1. Preoperative depiction of the relationship of lesions to eloquent areas 2. Determining the need for intraoperative electrocortical mapping 3. Preoperative surgical planning 4. Language lateralization defining	1. Preoperative depiction of the relationship of lesions to major fiber tracts 2. Determining the need for intraoperative electrocortical mapping 3. Preoperative surgical planning	1. Preoperative depiction of the relationship of lesions to eloquent areas 2. Preoperative surgical planning 3. Language cortex mapping	1. Intraoperative depiction of the relationship of lesions to eloquent areas 2. Intraoperative localization of functional brain cortex and major subcortical fiber tracts
Limitations	1. Not available for preoperative surgical planning; 2. Risks for intra-operative epilepsy and failure; 3. Intraoperative time consuming; 4. Patients and surgeons' psychological burden	1. Better at demonstrating motor areas than language areas 2. Affected by biological aspects of brain tumors 3. Difficult to acquire in patients with profound neurologic deficits	1. Limited spatial resolution 2. Poor delineation of crossing fibers 3. Image distortion	1. Limited spatial resolution 2. Validation with DCS not fully confirmed	1. Hardware not readily clinically used 2. Complicated data post-processing 3. Validation with DCS not fully confirmed
Invasive	Yes	No	No	No	Partially

Available pre-operatively	No	Yes	Yes	Yes	No
Awake craniotomy needed	Yes	No	No	No	Yes
Overlaid with anatomy	No	Yes	Yes	Yes	Yes
3D visualization	No	Yes	Yes	Yes	No
Patient compliance request	High	High	Low	Low	Moderate
Stress on Surgeons and anesthesiologists	High	Low	Low	Low	Low

Note: DCS: direct cortical stimulation; fMRI: functional magnetic resonance imaging; DTI: diffusion tensor imaging; TMS: transcranial magnetic stimulation; iOIS: intraoperative optical imaging of intrinsic signals; 3D: three dimensional.

CONCLUSIONS

Multimodal image-guided brain tumor resection has been developed during the last three decades. These advances allow neurosurgeons to intraoperatively monitor and alter the surgical procedure assessing residual tumor and nearby eloquent structures, taking into account shift resulting from tumor removal, development of brain edema, and loss of CSF.

Multimodal functional neuronavigation allows more thorough resections of tumors in risk zones with low morbidity. Integration of multimodal functional imaging data with the co-registration of PET data and information from MR spectroscopy (MRS) can add metabolic information leading to true multimodal navigation. Ongoing technology development will continue to add tools to the brain tumor surgeon's armamentarium; ongoing validation will remain a goal.

REFERENCES

1. Kelly PJ, Alker Jr GJ, Goerss S. Computer-assisted stereotactic microsurgery for the treatment of intracranial neoplasms. *Neurosurgery*. 1982;10(3):324–331.
2. Roberts DW, Strohbehn JW, Friets EM, Kettenberger J, Hartov A. The stereotactic operating microscope: accuracy refinement and clinical experience. *Acta Neurochir Suppl*. 1989;46:112–114.
3. Hentschel SJ, Sawaya R. Optimizing outcomes with maximal surgical resection of malignant gliomas. *Cancer Control*. 2003;10(2):109–114.
4. Sanai N, Berger MS. Operative techniques for gliomas and the value of extent of resection. *Neurotherapeutics*. 2009;6(3):478–486.
5. Sanai N, Polley MY, McDermott MW, Parsa AT, Berger MS. An extent of resection threshold for newly diagnosed glioblastomas. *J Neurosurg*. 2011;115(1):3–8.
6. Oppenlander ME, Wolf AB, Snyder LA, et al. An extent of resection threshold for recurrent glioblastoma and its risk for neurological morbidity. *J Neurosurg*. 2014;120(4):846–853.
7. Nimsky C, Ganslandt O, Fahlbusch R. 1.5 T: intraoperative imaging beyond standard anatomic imaging. *Neurosurg Clin N Am*. 2005;16(1):185–200:vii.
8. Nimsky C, Ganslandt O, Fahlbusch R. Functional neuronavigation and intraoperative MRI. *Adv Tech Stand Neurosurg*. 2004;29:229–263.
9. Nimsky C, Ganslandt O, Buchfelder M, Fahlbusch R. Intraoperative visualization for resection of gliomas: the role of functional neuronavigation and intraoperative 1.5 T MRI. *Neurol Res*. 2006;28 (5):482–487.
10. Nimsky C, Ganslandt O, Kober H, et al. Integration of functional magnetic resonance imaging supported by magnetoencephalography in functional neuronavigation. *Neurosurgery*. 1999;44 (6):1249–1255:discussion 1255–1256.
11. Ganslandt O, Buchfelder M, Hastreiter P, Grummich P, Fahlbusch R, Nimsky C. Magnetic source imaging supports clinical decision making in glioma patients. *Clin Neurol Neurosurg*. 2004;107 (1):20–26.
12. Kamada K, Houkin K, Takeuchi F, et al. Visualization of the eloquent motor system by integration of MEG, functional, and anisotropic diffusion-weighted MRI in functional neuronavigation. *Surg Neurol*. 2003;59(5):352–361.
13. Signorelli F, Guyotat J, Schneider F, Isnard J, Bret P. Technical refinements for validating functional MRI-based neuronavigation data by electrical stimulation during cortical language mapping. *Minim Invasive Neurosurg*. 2003;46(5):265–268.
14. Wilkinson ID, Romanowski CA, Jellinek DA, Morris J, Griffiths PD. Motor functional MRI for pre-operative and intraoperative neurosurgical guidance. *Br J Radiol*. 2003;76(902):98–103.

15. Clark CA, Barrick TR, Murphy MM, Bell BA. White matter fiber tracking in patients with space-occupying lesions of the brain: a new technique for neurosurgical planning. *Neuroimage.* 2003;20(3):1601–1608.
16. Hendler T, Pianka P, Sigal M, et al. Delineating gray and white matter involvement in brain lesions: three-dimensional alignment of functional magnetic resonance and diffusion-tensor imaging. *J Neurosurg.* 2003;99(6):1018–1027.
17. Nimsky C, Ganslandt O, Merhof D, Sorensen AG, Fahlbusch R. Intraoperative visualization of the pyramidal tract by diffusion-tensor-imaging-based fiber tracking. *Neuroimage.* 2006;30(4):1219–1229.
18. Coenen VA, Krings T, Mayfrank L, et al. Three-dimensional visualization of the pyramidal tract in a neuronavigation system during brain tumor surgery: first experiences and technical note. *Neurosurgery.* 2001;49(1):86–92:discussion 92–93.
19. Nimsky C, Grummich P, Sorensen AG, Fahlbusch R, Ganslandt O. Visualization of the pyramidal tract in glioma surgery by integrating diffusion tensor imaging in functional neuronavigation. *Zentralbl Neurochir.* 2005;66(3):133–141.
20. Stadlbauer A, Moser E, Gruber S, et al. Improved delineation of brain tumors: an automated method for segmentation based on pathologic changes of 1H-MRSI metabolites in gliomas. *Neuroimage.* 2004;23(2):454–461.
21. Stadlbauer A, Moser E, Gruber S, Nimsky C, Fahlbusch R, Ganslandt O. Integration of biochemical images of a tumor into frameless stereotaxy achieved using a magnetic resonance imaging/magnetic resonance spectroscopy hybrid data set. *J Neurosurg.* 2004;101(2):287–294.
22. Dowling C, Bollen AW, Noworolski SM, et al. Preoperative proton MR spectroscopic imaging of brain tumors: correlation with histopathologic analysis of resection specimens. *Am J Neuroradiol.* 2001;22(4):604–612.
23. Kondziolka D, Lunsford LD, Martinez AJ. Unreliability of contemporary neurodiagnostic imaging in evaluating suspected adult supratentorial (low-grade) astrocytoma. *J Neurosurg.* 1993;79(4):533–536.
24. Braun V, Dempf S, Tomczak R, Wunderlich A, Weller R, Richter HP. Multimodal cranial neuronavigation: direct integration of functional magnetic resonance imaging and positron emission tomography data: technical note. *Neurosurgery.* 2001;48(5):1178–1181.
25. Reithmeier T, Cordeiro J, Mix M, Trippel M, Rottenburger C, Nikkhah G. Impact of automated hotspot detection for (18)FET PET-guided stereotactic biopsy. *Acta Neurochir Suppl.* 2013;117:93–99.
26. Preuss M, Werner P, Barthel H, et al. Integrated PET/MRI for planning navigated biopsies in pediatric brain tumors. *Childs Nerv Syst.* 2014;30(8):1399–1403.
27. Misch M, Guggemos A, Driever PH, et al. F-FET-PET guided surgical biopsy and resection in children and adolescence with brain tumors. *Childs Nerv Syst.* 2015;31(2):261–267.
28. Luyten PR, Marien AJ, Heindel W, et al. Metabolic imaging of patients with intracranial tumors: H-1 MR spectroscopic imaging and PET. *Radiology.* 1990;176(3):791–799.
29. van Der Veen JW, Weinberger DR, Tedeschi G, Frank JA, Duyn JH. Proton MR spectroscopic imaging without water suppression. *Radiology.* 2000;217(1):296–300.
30. Majos C, Alonso J, Aguilera C, et al. Adult primitive neuroectodermal tumor: proton MR spectroscopic findings with possible application for differential diagnosis. *Radiology.* 2002;225(2):556–566.
31. Ott D, Hennig J, Ernst T. Human brain tumors: assessment with in vivo proton MR spectroscopy. *Radiology.* 1993;186(3):745–752.
32. Negendank WG, Sauter R, Brown TR, et al. Proton magnetic resonance spectroscopy in patients with glial tumors: a multicenter study. *J Neurosurg.* 1996;84(3):449–458.
33. Miller BL. A review of chemical issues in 1H NMR spectroscopy: N-acetyl-L-aspartate, creatine and choline. *NMR Biomed.* 1991;4(2):47–52.
34. Michaelis T, Merboldt KD, Bruhn H, Hanicke W, Frahm J. Absolute concentrations of metabolites in the adult human brain in vivo: quantification of localized proton MR spectra. *Radiology.* 1993;187(1):219–227.
35. Urenjak J, Williams SR, Gadian DG, Noble M. Proton nuclear magnetic resonance spectroscopy unambiguously identifies different neural cell types. *J Neurosci.* 1993;13(3):981–989.

36. Croteau D, Scarpace L, Hearshen D, et al. Correlation between magnetic resonance spectroscopy imaging and image-guided biopsies: semiquantitative and qualitative histopathological analyses of patients with untreated glioma. *Neurosurgery*. 2001;49(4):823–829.
37. Stummer W, Pichlmeier U, Meinel T, Wiestler OD, Zanella F, Reulen HJ. Fluorescence-guided surgery with 5-aminolevulinic acid for resection of malignant glioma: a randomised controlled multicentre phase III trial. *Lancet Oncol*. 2006;7(5):392–401.
38. Stummer W, Novotny A, Stepp H, Goetz C, Bise K, Reulen HJ. Fluorescence-guided resection of glioblastoma multiforme by using 5-aminolevulinic acid-induced porphyrins: a prospective study in 52 consecutive patients. *J Neurosurg*. 2000;93(6):1003–1013.
39. Colditz MJ, Jeffree RL. Aminolevulinic acid (ALA)-protoporphyrin IX fluorescence guided tumour resection. Part 1: Clinical, radiological and pathological studies. *J Clin Neurosci*. 2012;19 (11):1471–1474.
40. Zhao S, Wu J, Wang C, et al. Intraoperative fluorescence-guided resection of high-grade malignant gliomas using 5-aminolevulinic acid-induced porphyrins: a systematic review and meta-analysis of prospective studies. *PLoS One*. 2013;8(5):e63682.
41. Behbahaninia M, Martirosyan NL, Georges J, et al. Intraoperative fluorescent imaging of intracranial tumors: a review. *Clin Neurol Neurosurg*. 2013;115(5):517–528.
42. Roberts DW, Valdes PA, Harris BT, et al. Coregistered fluorescence-enhanced tumor resection of malignant glioma: relationships between delta-aminolevulinic acid-induced protoporphyrin IX fluorescence, magnetic resonance imaging enhancement, and neuropathological parameters. Clinical article. *J Neurosurg*. 2011;114(3):595–603.
43. Widhalm G, Wolfsberger S, Minchev G, et al. 5-Aminolevulinic acid is a promising marker for detection of anaplastic foci in diffusely infiltrating gliomas with nonsignificant contrast enhancement. *Cancer*. 2010;116(6):1545–1552.
44. Diez VR, Tejada SS, Idoate GMA, de Eulate RG, Dominguez EP, Aristu MJ. Surgery guided by 5-aminolevulinic fluorescence in glioblastoma: volumetric analysis of extent of resection in single-center experience. *J Neurooncol*. 2011;102(1):105–113.
45. Eyupoglu IY, Buchfelder M, Savaskan NE. Surgical resection of malignant gliomas: role in optimizing patient outcome. *Nat Rev Neurol*. 2013;9(3):141–151.
46. Stummer W, Stocker S, Wagner S, et al. Intraoperative detection of malignant gliomas by 5-aminolevulinic acid-induced porphyrin fluorescence. *Neurosurgery*. 1998;42(3):518–525.
47. Valdes PA, Leblond F, Kim A, et al. Quantitative fluorescence in intracranial tumor: implications for ALA-induced PpIX as an intraoperative biomarker. *J Neurosurg*. 2011;115(1):11–17.
48. Placantonakis DG, Tabaee A, Anand VK, Hiltzik D, Schwartz TH. Safety of low-dose intrathecal fluorescein in endoscopic cranial base surgery. *Neurosurgery*. 2007;61(3 Suppl):161–165.
49. Rey-Dios R, Cohen-Gadol AA. Technical principles and neurosurgical applications of fluorescein fluorescence using a microscope-integrated fluorescence module. *Acta Neurochir (Wien)*. 2013;155 (4):701–706.
50. Schwartz RB, Hsu L, Wong TZ, et al. Intraoperative MR imaging guidance for intracranial neurosurgery: experience with the first 200 cases. *Radiology*. 1999;211(2):477–488.
51. Sun GC, Chen XL, Zhao Y, et al. Intraoperative high-field magnetic resonance imaging combined with fiber tract neuronavigation-guided resection of cerebral lesions involving optic radiation. *Neurosurgery*. 2011;69(5):1070–1084.
52. Nimsky C, Ganslandt O, Von Keller B, Romstock J, Fahlbusch R. Intraoperative high-field-strength MR imaging: implementation and experience in 200 patients. *Radiology*. 2004;233(1): 67–78.
53. Sutherland GR, Kaibara T, Louw D, Hoult DI, Tomanek B, Saunders J. A mobile high-field magnetic resonance system for neurosurgery. *J Neurosurg*. 1999;91(5):804–813.
54. Eyupoglu IY, Hore N, Savaskan NE, et al. Improving the extent of malignant glioma resection by dual intraoperative visualization approach. *PLoS One*. 2012;7(9):e44885.
55. Senft C, Forster MT, Bink A, et al. Optimizing the extent of resection in eloquently located gliomas by combining intraoperative MRI guidance with intraoperative neurophysiological monitoring. *J Neurooncol*. 2012;109(1):81–90.

56. Archip N, Clatz O, Whalen S, et al. Non-rigid alignment of pre-operative MRI, fMRI, and DT-MRI with intra-operative MRI for enhanced visualization and navigation in image-guided neurosurgery. *Neuroimage*. 2007;35(2):609–624.
57. Tirakotai W, Miller D, Heinze S, Benes L, Bertalanffy H, Sure U. A novel platform for image-guided ultrasound. *Neurosurgery*. 2006;58(4):710–718.
58. Comeau RM, Sadikot AF, Fenster A, Peters TM. Intraoperative ultrasound for guidance and tissue shift correction in image-guided neurosurgery. *Med Phys*. 2000;27(4):787–800.
59. Letteboer MM, Willems PW, Viergever MA, Niessen WJ. Brain shift estimation in image-guided neurosurgery using 3-D ultrasound. *IEEE Trans Biomed Eng*. 2005;52(2):268–276.
60. Lacroix M, Abi-Said D, Fourney DR, et al. A multivariate analysis of 416 patients with glioblastoma multiforme: prognosis, extent of resection, and survival. *J Neurosurg*. 2001;95(2):190–198.
61. McGirt MJ, Chaichana KL, Gathinji M, et al. Independent association of extent of resection with survival in patients with malignant brain astrocytoma. *J Neurosurg*. 2009;110(1):156–162.
62. Kreike B, Hart AA, van de Velde T, et al. Continuing risk of ipsilateral breast relapse after breast-conserving therapy at long-term follow-up. *Int J Radiat Oncol Biol Phys*. 2008;71(4):1014–1021.
63. Mannell A. Breast-conserving therapy in breast cancer patients—a 12-year experience. *S Afr J Surg*. 2005;43(2):28–30.
64. Crespo AN, Chone CT, Gripp FM, Spina AL, Altemani A. Role of margin status in recurrence after CO2 laser endoscopic resection of early glottic cancer. *Acta Otolaryngol*. 2006;126(3):306–310.
65. Soudry E, Hadar T, Shvero J, et al. The impact of positive resection margins in partial laryngectomy for advanced laryngeal carcinomas and radiation failures. *Clin Otolaryngol*. 2010;35(5):402–408.
66. Nagtegaal ID, Quirke P. What is the role for the circumferential margin in the modern treatment of rectal cancer. *J Clin Oncol*. 2008;26(2):303–312.
67. Wright JL, Dalkin BL, True LD, et al. Positive surgical margins at radical prostatectomy predict prostate cancer specific mortality. *J Urol*. 2010;183(6):2213–2218.
68. Stummer W, Reulen HJ, Meinel T, et al. Extent of resection and survival in glioblastoma multiforme: identification of and adjustment for bias. *Neurosurgery*. 2008;62(3):564–576.
69. Roder C, Bender B, Ritz R, et al. Intraoperative visualization of residual tumor: the role of perfusion-weighted imaging in a high-field intraoperative magnetic resonance scanner. *Neurosurgery*. 2013;72(2 Suppl Operative):ons151–ons158.
70. Zhu W, Chen X, Zhang J, et al. Automated proton magnetic resonance spectroscopy imaging guided frameless stereotactic biopsy of intracranial lesions. *Zhonghua Wai Ke Za Zhi*. 2014;52(4):280–284.
71. He X, Wang K, Cheng Z. In vivo near-infrared fluorescence imaging of cancer with nanoparticle-based probes. *Wiley Interdiscip Rev Nanomed Nanobiotechnol*. 2010;2(4):349–366.
72. Taraska JW, Zagotta WN. Fluorescence applications in molecular neurobiology. *Neuron*. 2010;66(2):170–189.
73. Gheonea DI, Cartana T, Ciurea T, Popescu C, Badarau A, Saftoiu A. Confocal laser endomicroscopy and immunoendoscopy for real-time assessment of vascularization in gastrointestinal malignancies. *World J Gastroenterol*. 2011;17(1).21–27.
74. Sankar T, Delaney PM, Ryan RW, et al. Miniaturized handheld confocal microscopy for neurosurgery: results in an experimental glioblastoma model. *Neurosurgery*. 2010;66(2):410–417.
75. Sanai N, Chang S, Berger MS. Low-grade gliomas in adults. *J Neurosurg*. 2011;115(5):948–965.
76. Li FY, Chen XL, Sai XY, et al. Application of intraoperative magnetic resonance imaging and multimodal navigation in surgical resection of glioblastoma. *Zhonghua Wai Ke Za Zhi*. 2013;51(6):542–546.
77. Li J, Chen X, Zhang J, et al. Intraoperative diffusion tensor imaging predicts the recovery of motor dysfunction after insular lesions. *Neural Regen Res*. 2013;8(15):1400–1409.
78. Zheng G, Chen X, Xu B, et al. Plasticity of language pathways in patients with low-grade glioma: a diffusion tensor imaging study. *Neural Regen Res*. 2013;8(7):647–654.
79. Kunii N, Kamada K, Ota T, Kawai K, Saito N. A detailed analysis of functional magnetic resonance imaging in the frontal language area: a comparative study with extraoperative electrocortical stimulation. *Neurosurgery*. 2011;69(3):590–596.

80. Hirsch J, Ruge MI, Kim KH, et al. An integrated functional magnetic resonance imaging procedure for preoperative mapping of cortical areas associated with tactile, motor, language, and visual functions. *Neurosurgery.* 2000;47(3):711–721.
81. Kamada K, Sawamura Y, Takeuchi F, et al. Expressive and receptive language areas determined by a non-invasive reliable method using functional magnetic resonance imaging and magnetoencephalography. *Neurosurgery.* 2007;60(2):296–305.
82. Giussani C, Roux FE, Ojemann J, Sganzerla EP, Pirillo D, Papagno C. Is preoperative functional magnetic resonance imaging reliable for language areas mapping in brain tumor surgery? Review of language functional magnetic resonance imaging and direct cortical stimulation correlation studies. *Neurosurgery.* 2010;66(1):113–120.
83. Roux FE, Boulanouar K, Lotterie JA, Mejdoubi M, LeSage JP, Berry I. Language functional magnetic resonance imaging in preoperative assessment of language areas: correlation with direct cortical stimulation. *Neurosurgery.* 2003;52(6):1335–1345.
84. Genetti M, Grouiller F, Vulliemoz S, et al. Noninvasive language mapping in patients with epilepsy or brain tumors. *Neurosurgery.* 2013;72(4):555–565.
85. Herholz K, Reulen HJ, von SHM, et al. Preoperative activation and intraoperative stimulation of language-related areas in patients with glioma. *Neurosurgery.* 1997;41(6):1253–1260.
86. Peck KK, Bradbury M, Petrovich N, et al. Presurgical evaluation of language using functional magnetic resonance imaging in brain tumor patients with previous surgery. *Neurosurgery.* 2009;64(4):644–652.
87. Nabavi A, Black PM, Gering DT, et al. Serial intraoperative magnetic resonance imaging of brain shift. *Neurosurgery.* 2001;48(4):787–797:discussion 797–798.
88. Rachinger J, von KB, Ganslandt O, Fahlbusch R, Nimsky C. Application accuracy of automatic registration in frameless stereotaxy. *Stereotact Funct Neurosurg.* 2006;84(2–3):109–117.
89. Simon MV, Cole AJ, Chang EC, et al. An intraoperative multimodal neurophysiologic approach to successful resection of precentral gyrus epileptogenic lesions. *Epilepsia.* 2012;53(4):e75–e79.
90. Sheth SA, Eckhardt CA, Walcott BP, Eskandar EN, Simon MV. Factors affecting successful localization of the central sulcus using the somatosensory evoked potential phase reversal technique. *Neurosurgery.* 2013;72(5):828–834.
91. Simon MV, Sheth SA, Eckhardt CA, et al. Phase reversal technique decreases cortical stimulation time during motor mapping. *J Clin Neurosci.* 2014;21(6):1011–1017.
92. Henning SL, Seidel K, Wiest R, Beck J, Raabe A. Localization of primary language areas by arcuate fascicle fiber tracking. *Neurosurgery.* 2012;70(1):56–64.
93. Krings T, Chiappa KH, Foltys H, Reinges MH, Cosgrove GR, Thron A. Introducing navigated transcranial magnetic stimulation as a refined brain mapping methodology. *Neurosurg Rev.* 2001;24(4):171–179.
94. Bartholow R. Experimental investigation into the functions of the human brain. *Am J Med Sci.* 1874;67:305–313.
95. Horsley V. Remarks on ten consecutive cases of operations upon the brain and cranial cavity to illustrate the details and safety of the method employed. *Br Med J.* 1887;1(1373):863–865.
96. Penfield W. *No Man Alone. A Neurosurgeon's Life.* Boston, MA: Little Brown; 1977.
97. Keles GE, Lundin DA, Lamborn KR, Chang EF, Ojemann G, Berger MS. Intraoperative subcortical stimulation mapping for hemispherical perirolandic gliomas located within or adjacent to the descending motor pathways: evaluation of morbidity and assessment of functional outcome in 294 patients. *J Neurosurg.* 2004;100(3):369–375.
98. Berger MS. Lesions in functional ("eloquent") cortex and subcortical white matter. *Clin Neurosurg.* 1994;41:444–463.
99. Gupta DK, Chandra PS, Ojha BK, Sharma BS, Mahapatra AK, Mehta VS. Awake craniotomy versus surgery under general anesthesia for resection of intrinsic lesions of eloquent cortex—a prospective randomised study. *Clin Neurol Neurosurg.* 2007;109(4):335–343.
100. Pirotte B, Neugroschl C, Metens T, et al. Comparison of functional MR imaging guidance to electrical cortical mapping for targeting selective motor cortex areas in neuropathic

pain: a study based on intraoperative stereotactic navigation. *Am J Neuroradiol*. 2005; 26(9):2256–2266.
101. Pirotte B, Voordecker P, Neugroschl C, et al. Combination of functional magnetic resonance imaging-guided neuronavigation and intraoperative cortical brain mapping improves targeting of motor cortex stimulation in neuropathic pain. *Neurosurgery*. 2005;56(2 Suppl):344–359.
102. Bartos R, Jech R, Vymazal J, et al. Validity of primary motor area localization with fMRI versus electric cortical stimulation: a comparative study. *Acta Neurochir (Wien)*. 2009;151 (9):1071–1080.
103. Spena G, Nava A, Cassini F, et al. Preoperative and intraoperative brain mapping for the resection of eloquent-area tumors. A prospective analysis of methodology, correlation, and usefulness based on clinical outcomes. *Acta Neurochir (Wien)*. 2010;152(11):1835–1846.
104. Bizzi A, Blasi V, Falini A, et al. Presurgical functional MR imaging of language and motor functions: validation with intraoperative electrocortical mapping. *Radiology*. 2008;248(2):579–589.
105. Ota T, Kamada K, Aoki S, Saito N. Visualization of calculation centres by functional MRI for neurosurgery. *Br J Neurosurg*. 2009;23(4):406–411.
106. Hall WA, Kim P, Truwit CL. Functional magnetic resonance imaging-guided brain tumor resection. *Top Magn Reson Imaging*. 2009;19(4):205–212.
107. Kapsalakis IZ, Kapsalaki EZ, Gotsis ED, et al. Preoperative evaluation with FMRI of patients with intracranial gliomas. *Radiol Res Pract*. 2012;2012:727810.
108. Najib U, Bashir S, Edwards D, Rotenberg A, Pascual-Leone A. Transcranial brain stimulation: clinical applications and future directions. *Neurosurg Clin N Am*. 2011;22(2):233–251:ix.
109. Bauer PR, Reitsma JB, Houweling BM, Ferrier CH, Ramsey NF. Can fMRI safely replace the Wada test for preoperative assessment of language lateralisation? A meta-analysis and systematic review. *J Neurol Neurosurg Psychiatry*. 2014;85(5):581–588.
110. Gutbrod K, Spring D, Degonda N, et al. Determination of language dominance: Wada test and fMRI compared using a novel sentence task. *J Neuroimaging*. 2012;22(3):266–274.
111. Bick AS, Mayer A, Levin N. From research to clinical practice: implementation of functional magnetic imaging and white matter tractography in the clinical environment. *J Neurol Sci*. 2012;312(1–2):158–165.
112. Kobayashi M, Pascual-Leone A. Transcranial magnetic stimulation in neurology. *Lancet Neurol*. 2003;2(3):145–156.
113. Krieg SM, Sollmann N, Hauck T, Ille S, Meyer B, Ringel F. Repeated mapping of cortical language sites by preoperative navigated transcranial magnetic stimulation compared to repeated intraoperative DCS mapping in awake craniotomy. *BMC Neurosci*. 2014;15:20.
114. Picht T, Krieg SM, Sollmann N, et al. A comparison of language mapping by preoperative navigated transcranial magnetic stimulation and direct cortical stimulation during awake surgery. *Neurosurgery*. 2013;72(5):808–819.
115. Saisanen L, Kononen M, Julkunen P, et al. Non-invasive preoperative localization of primary motor cortex in epilepsy surgery by navigated transcranial magnetic stimulation. *Epilepsy Res*. 2010;92 (2–3):134–144.
116. Shamov T, Spiriev T, Tzvetanov P, Petkov A. The combination of neuronavigation with transcranial magnetic stimulation for treatment of opercular gliomas of the dominant brain hemisphere. *Clin Neurol Neurosurg*. 2010;112(8):672–677.
117. Cannestra AF, Bookheimer SY, Pouratian N, et al. Temporal and topographical characterization of language cortices using intraoperative optical intrinsic signals. *Neuroimage*. 2000;12(1):41–54.
118. Cannestra AF, Pouratian N, Bookheimer SY, Martin NA, Beckerand DP, Toga AW. Temporal spatial differences observed by functional MRI and human intraoperative optical imaging. *Cereb Cortex*. 2001;11(8):773–782.
119. Nariai T, Sato K, Hirakawa K, et al. Imaging of somatotopic representation of sensory cortex with intrinsic optical signals as guides for brain tumor surgery. *J Neurosurg*. 2005;103(3):414–423.
120. Schwartz TH. The application of optical recording of intrinsic signals to simultaneously acquire functional, pathological and localizing information and its potential role in neurosurgery. *Stereotact Funct Neurosurg*. 2005;83(1):36–44.

121. Cannestra AF, Pouratian N, Forage J, Bookheimer SY, Martin NA, Toga AW. Functional magnetic resonance imaging and optical imaging for dominant-hemisphere perisylvian arteriovenous malformations. *Neurosurgery*. 2004;55(4):804–812.
122. Popescu MA, Toms SA. In vivo optical imaging using quantum dots for the management of brain tumors. *Expert Rev Mol Diagn*. 2006;6(6):879–890.
123. Toms SA, Lin WC, Weil RJ, Johnson MD, Jansen ED, Mahadevan-Jansen A. Intraoperative optical spectroscopy identifies infiltrating glioma margins with high sensitivity. *Neurosurgery*. 2007;61(1 Suppl):327–335.
124. Senft C, Bink A, Franz K, Vatter H, Gasser T, Seifert V. Intraoperative MRI guidance and extent of resection in glioma surgery: a randomised, controlled trial. *Lancet Oncol*. 2011;12(11):997–1003.
125. Kuhnt D, Becker A, Ganslandt O, Bauer M, Buchfelder M, Nimsky C. Correlation of the extent of tumor volume resection and patient survival in surgery of glioblastoma multiforme with high-field intraoperative MRI guidance. *Neuro Oncol*. 2011;13(12):1339–1348.

CHAPTER 10

Image Guidance and Visualization in Pituitary Surgery

Wenya Linda Bi, Ian F. Dunn, and Edward R. Laws, Jr.
Department of Neurosurgery, Brigham and Women's Hospital, Harvard Medical School, Boston, MA, USA

INTRODUCTION

The evolution of pituitary surgery is intertwined with advances in visualization. Pituitary lesions straddle the boundary between intrinsic and extra-axial brain tumors, depending on their extent and trajectory of growth. Furthermore, the deep seat of the bony sella—cradled by the carotids and four pairs of cranial nerves on either side, the basilar artery posteriorly, and capped by the optic chiasm—provides formidable hurdles to the surgeon and demands a precise understanding of its dimensions. Thus, detailed awareness of anatomic constraints, as provided by static images or maneuverable scopes, is critical to the safety and success of pituitary surgery.

HISTORICAL EVOLUTION

The history of image guidance in pituitary surgery began with the discovery of X-rays (Table 10.1).[14] It is amazing to realize how rapidly Roentgen's discovery in 1895 spread throughout the world (Figure 10.1)—so rapidly that field units with X-ray capability were employed during the Anglo-Boer war in 1899. Dr. Harvey Cushing embraced the utility of radiography immediately upon its discovery, and introduced X-ray imaging to clinical practice by 1896 (Figure 10.2). It was not long before X-rays of the skull in patients with clinical signs of acromegaly, and with visual loss, were used to demonstrate pathological enlargement of the sella.[15] Also noted were abnormal calcifications in the sella and suprasellar region, later associated with craniopharyngiomas. Schloffer used plain film radiography to confirm the presence of sellar pathology prior to the first transsphenoidal surgery in 1907.[1] X-rays in cases of pituitary adenomas were utilized not only to make the diagnosis, but also to guide the placement of radiation sources that were used to treat these tumors as early as 1909 (Figure 10.3).

Walter Dandy of Johns Hopkins introduced ventriculography in 1918,[2] and pneumoencephalography a year later (Figure 10.4).[3] These revolutionary imaging techniques were able to demonstrate suprasellar extension of pituitary and related tumors, and were actually used for targeting of tumors in the early days of proton beam

A. Golby (Ed): Image-Guided Neurosurgery
DOI: http://dx.doi.org/10.1016/B978-0-12-800870-6.00010-8

Table 10.1 The evolution of intraoperative imaging in neurosurgery

Imaging technique	Date of introduction
Roentgenography	1907[1]
Ventriculography, pneumoencephalography	1918[2,3]
Ventricular endoscopy	1922[4]
Angiography	1920s[5]
Fluoroscopy, image intensifier	1960s[6]
Operating microscope	1960s[7]
Computed tomography (CT)	1974
Magnetic resonance imaging (MRI)	1980
Ultrasound imaging	1980s[8]
Metabolic/nuclear imaging	1980s[9]
Fluorescent imaging	1990s[10]
Optical imaging	1990s[11]
iMRI	1994[12]
Pituitary and skull base endoscopy	1996[13]
Nanotechnology	2000s

Figure 10.1 Portrait of Wilhelm Roentgen. The discovery of X-ray electromagnetic radiation by Wilhelm Roentgen in 1895, for which he was awarded the Nobel Prize in physics in 1901, revolutionized medical diagnosis.

radiotherapy and gamma knife radiosurgery. Dandy was also one of the first to use endoscopy for the treatment of hydrocephalus, using the endoscopic image capability for cauterization of the choroid plexus of the ventricles, thus introducing another important tool that would eventually play a critical role in pituitary surgery.

Introduction of cerebral angiography by Egas Moniz (Antonio Caetano de Abreu Freire Egas Moniz) of Portugal in the late 1920s was another imaging advance

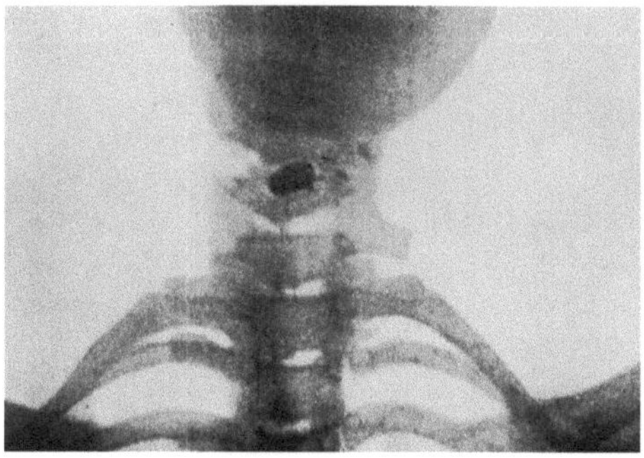

Figure 10.2 Early neurosurgical application of X-ray imaging. X-ray of a bullet in the C6 vertebral body of a patient who presented with a Brown-Séquard syndrome after being shot in the neck, taken by Dr. Harvey Cushing at Johns Hopkins Hospital in November 1896.

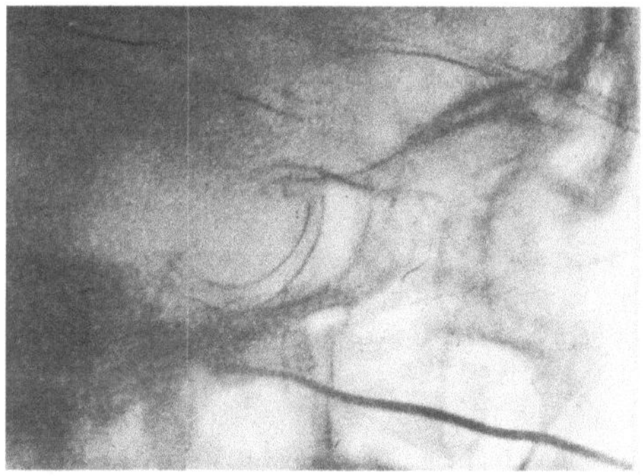

Figure 10.3 Early radiation treatment of pituitary tumor. Skull X-ray taken by Oskar Hirsch in 1910 for transnasal implantation of a radium "bomb" at the end of a probe placed below the sella in the sphenoid sinus. The X-ray shows an expanded sella with a "double floor," diagnostic of a pituitary macroadenoma.

(Figure 10.5).[5] Initially, it could have proved helpful in distinguishing pituitary tumors from intrasellar aneurysms, but did not come into its own as a major adjunct for the diagnosis of pituitary tumors until much later. Indirect angiographic evidence, such as elevation of the anterior cerebral arteries, was helpful in the early diagnosis of pituitary macroadenomas (Figure 10.6). Later, as angiography became much more sophisticated, the actual blood supply of the pituitary and of pituitary adenomas could be delineated, reflecting both the anatomical and physiological correlates of these tumors.[16]

In the 1950s and 60s, polytomography with X-ray showed the details of the sella turcica in much better detail, and was able to demonstrate focal changes in the bony sella produced by intrasellar lesions.[17] However, these images still did not visualize the

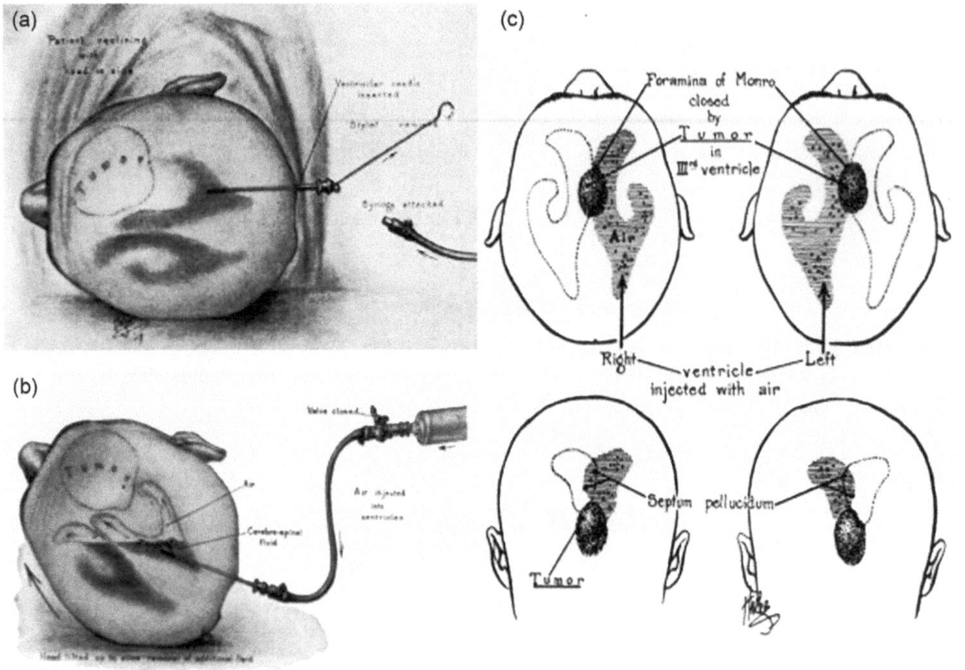

Figure 10.4 Ventriculography and pneumoencephalography. Illustrations of early ventriculography and pneumoencephalography performed by Dr. Walter Dandy. (a) Cerebrospinal fluid egresses from an occipital bur hole, after which a syringe is attached to the end of a blunt needle to introduce air into the ventricles. (b) The head is then tilted so that a new supply of fluid gravitates to the point of the needle and injected air rises to the top. (c) Diagram illustrating failure of air injected into a dilated lateral ventricle to fill the contralateral lateral ventricle and third ventricle, due to obstruction by a third ventricular tumor. *Images from Dandy W,* Surgery of the Brain, *1945.*

pituitary gland or the surrounding neurovascular structures. During the same period, nuclear imaging with Technetium was introduced, and was occasionally useful in diagnosing large pituitary and parasellar tumors. The first attempts at positron emission nuclear scanning occurred at that time as well.

Video fluoroscopic imaging with the X-ray image intensifier was introduced for the real-time monitoring of pituitary surgery by Gerard Guiot of France and Jules Hardy of Canada in the 1960s (Figures 10.7 and 10.8).[6] This advance, along with their introduction of the operating microscope and the concept of the microadenoma, changed the entire course of pituitary surgery.[7,19,20] The transsphenoidal approach, which had been introduced at the turn of the twentieth century, was reintroduced and revitalized. The approach and the technological advances associated with it continued to expand, most recently with the development of the transnasal endoscopic approach to pituitary tumors and a wide variety of other lesions of the anterior skull base.

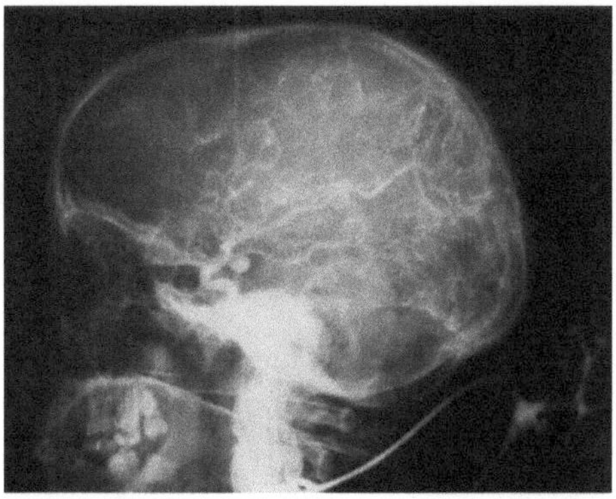

Figure 10.5 Angiography in neurosurgery. An early angiographic demonstration of an intracerebral aneurysm, by Dr. Norman Dott of Edinburgh.

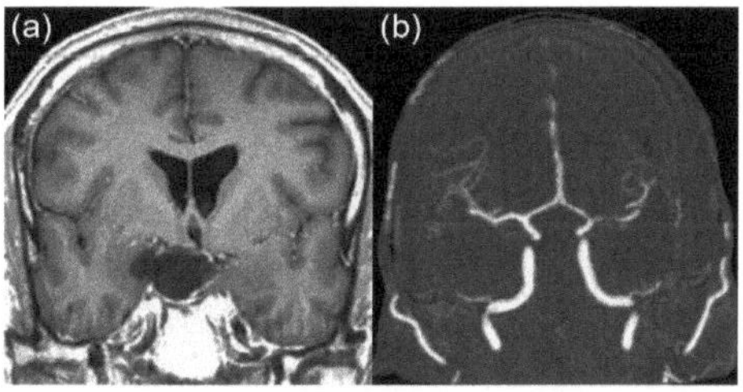

Figure 10.6 Pituitary tumor displacement of anterior cerebral arteries. CT angiography reveals elevation of the anterior cerebral arteries by a pituitary macroadenoma.

The ability to perform intracranial imaging with diagnostic ultrasound had a minor impact in the management of pituitary tumors. Ultrasonic imaging probes with a very low profile were used to image the sellar contents after removal of the bony sellar floor. A truly effective instrument is not yet available, but might be of use in the detection of microadenomas. The innovative use of a diagnostic imaging trans-esophageal probe placed into the sphenoid sinus has proved useful in selected cases, especially as the color Doppler can show blood flow in the carotid arteries and other vascular structures around the sella.[21,22] This has not been developed for general use. The ultrasonic Doppler microvascular probe is useful, however, in identifying vascular structures in the course of pituitary tumor operations.

A pivotal advance in the imaging of all kinds of brain tumors, including pituitary adenomas, was the introduction of CT. The CT scan was revolutionary in detecting the presence of many types of brain tumors, and in delineating the effects of these lesions on surrounding tissues. The CT scan supplanted not only previous diagnostic imaging techniques, but in some cases even the customary details of the physical and

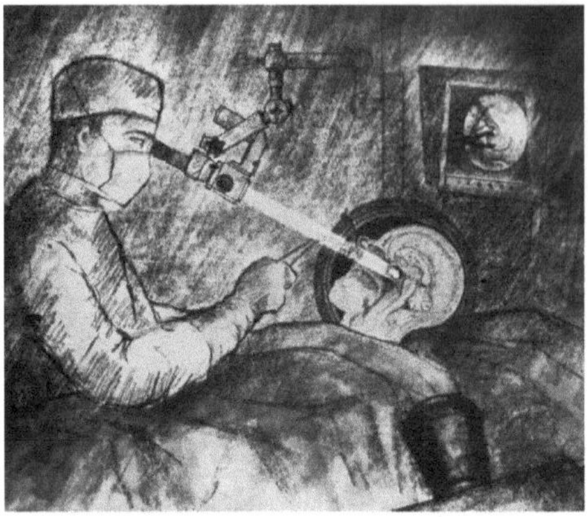

Figure 10.7 Fluoroscopy and the operating microscope for pituitary surgery. Illustration of video fluoroscopic imaging and the operating microscope as used by Dr. Jules Hardy, which revolutionized the course of pituitary surgery. *Image from personal collection of Dr. Edward R. Laws, Jr.*

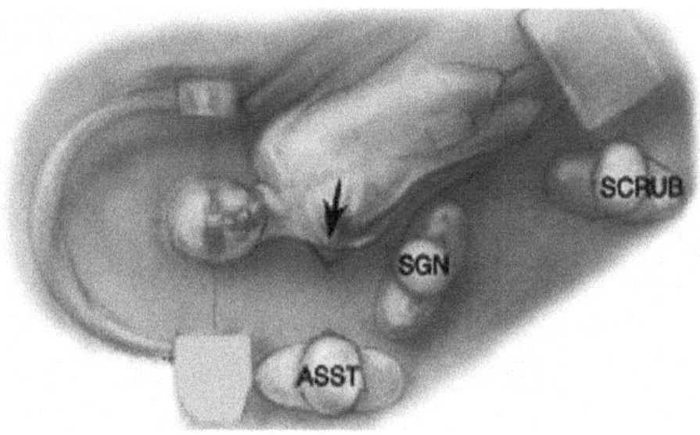

Figure 10.8 Fluoroscopic X-ray in pituitary surgery. Illustration of operating room setup with fluoroscopy positioned at the head of the bed, to assist tumor localization during transnasal anterior skull base surgery. *Image adapted from Jane et al.*[18]

neurological examination! CT-guided surgery, usually for biopsy of intracranial lesions, became a reality, and a harbinger of practical, effective, and accurate image-guided intracranial surgery.

Initial attempts at CT-guided neuronavigation were introduced—first with mechanical articulated arms and pointers, later with infrared detectors, sonic, and electromagnetic methods of registering targets in three-dimensional space. (See Chapter 1 for a detailed discussion of the evolution of these approaches.) These advances were followed by the introduction of MRI, another step forward in improving imaging diagnosis and therapy related to pituitary tumors and other intracranial lesions. The anatomy and physiology of the brain and lesions affecting it could now be imaged in exquisite detail (Figure 10.9). Functional MRI scans shed new light on higher cortical functions as impacted by pituitary disease (Figure 10.10). New

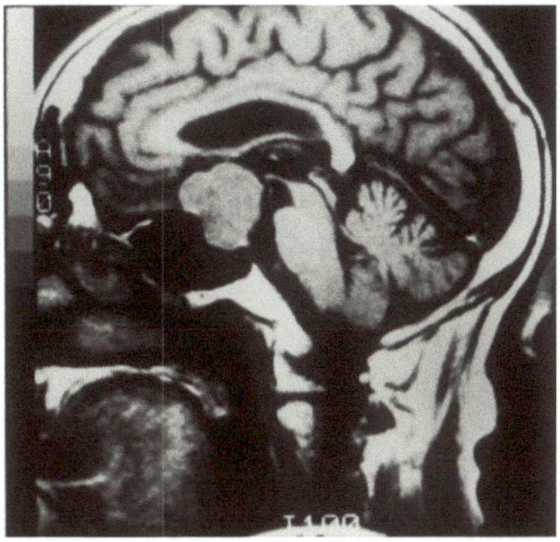

Figure 10.9 Early MRI of pituitary macroadenoma. Incorporation of magnetic resonance imaging into medical practice in the early 1980s revolutionized the diagnosis and planning of pituitary surgery. *MR image from 1984, Mayo Clinic.*

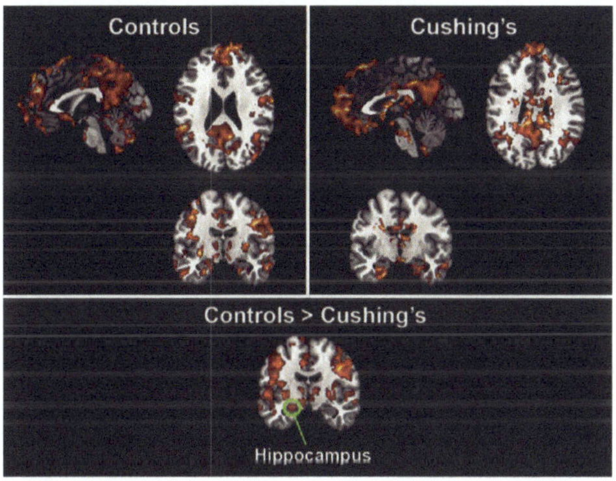

Figure 10.10 Functional imaging as a window into neurocognitive changes in pituitary tumors. Functional MRI of normal controls and Cushing's disease patients demonstrates disorganized localization during a memory recall task, with differential activation of the hippocampus in controls compared to Cushing's disease patients.

technical problems, primarily related to the magnet, made the introduction of MRI-guided surgery difficult and awkward for the surgeon in the early days of development of intraoperative MRI. The expense of intraoperative MRI suites represented another obstacle in their practical development. Currently, these issues have largely been

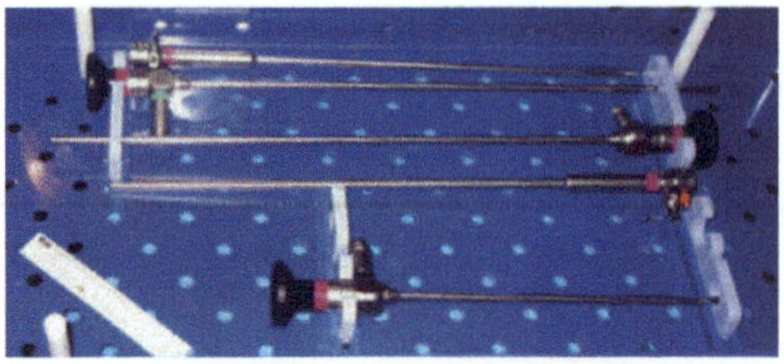

Figure 10.11 Endoscopes. Short, long, straight, and angled endoscopes used for transnasal anterior skull base surgery.

overcome, and highly effective intraoperative MRI (iMRI) units are in regular use, providing great benefit to many patients with a variety of intracranial problems, including pituitary tumors.

CONTEMPORARY IMAGE-GUIDANCE MODALITIES

At present, plain skull X-rays are rarely used in the diagnosis and management of pituitary tumors. Ventriculography, pneumoencephalography, and polytomography are no longer utilized, having been replaced by CT and MRI. Greater resolution of CT and MRI technologies, as well as the ability to overlay these modalities, provide constantly escalating data nuggets to incorporate into perioperative and intraoperative decisions. There may be some evolving uses for nuclear metabolic imaging, particularly in hyperfunctioning pituitary adenomas such as those secreting prolactin, growth hormone, or thyroid secreting hormone.

In addition to preoperative imaging, the operating microscope and endoscopy in and of themselves are advances in image-guided pituitary resection.[13,20] Surgery of pituitary lesions, traditionally done using the operating microscope, is now increasingly done using the operating endoscope, with its unique visualization advantages (Figure 10.11). Endoscopic approaches force closer examination of anatomic relations—the initial aperture, working distances, and safe and optimal corridors between neurovascular structures. The three-dimensional endoscope offers comparable surgical advantages to the 2D high-definition endoscope, but may offer a shorter learning curve.[23] Even more, expanded endonasal approaches have extended the limits of accessible sellar and parasellar lesions to all of the anterior skull base.[24,25] Complementary to the paradigm shift in surgical approach is the integration of skull base surgery principles: to maximize exposure through removal of obstructing bone; preserve normal parenchyma; and create functional and cosmetic reconstructions. Regardless of the surgical approach, image guidance is frequently used, particularly in difficult and reoperative cases, and in academic teaching centers. Image fusion possibilities (CT, MRI, fMRI, PET-CT, digital angiography, ultrasonic Doppler imaging) can further enhance the data set and image matrix for image guidance.

Indications and benefits of image guidance in pituitary surgery

Image guidance in the form of computerized neuronavigation, using either CT or MR images, offers several advantages. In the positioning of the patient's head, it is important to have an optimal and reproducible trajectory through the nostril to the sella. This is helpful from an ergonomic standpoint, and as a way to avoid misdirection in the sagittal plane, either to the anterior fossa or to the clivus and the brainstem. The relationship of the sella to the sphenoid sinus can be accurately determined and followed in real time during the exposure. Enlargement of the sella, thinning of the sellar floor, asymmetries of the sella, and the clinoids, can all be determined and confirmed. This is also true of the optic nerve canals and the parasellar course of the carotid arteries and their respective relationships to the midline and to the cavernous sinuses. The imaging of the surgical midline is critical to avoiding complications of misdirection in the lateral direction. The vomer can be visualized and confirmed with image guidance, as can the position and tortuosity of the carotids.

As surgeons increasingly use the transnasal endoscope for pituitary tumors and other anterior skull base lesions, image guidance also becomes essential to successful surgery and the avoidance of complications. Detailed MRI studies demonstrate the relationship of the surgical target to critical anatomic structures such as the optic nerves, chiasm, and vascular branches of the Circle of Willis. Contact, adherence to, displacement, and distortion of these structures can be appreciated and endoscopic visualization confirmed by image guidance. Navigation within a large tumor can assist and ensure a radical removal of tumor tissue.

Limitations of image guidance

Contemporary methods of image guidance are associated with some difficulties. Because current techniques of computer-based image guidance require registration of the patient's head in relation to the three-dimensional image data set, it may be difficult to obtain the necessary precision, particularly in the anterior—posterior direction. Misregistration can occur, the head may slip or move in relation to the fiducials, and inaccuracies may result that limit the usefulness of the technique.

There may also be problems with the image guidance and visualization afforded by the operating endoscopes. Technical issues with the light path and the camera and its electronic connections can occur. It may be difficult in some cases to keep the lens free of occlusion by blood or mucus. Air bubbles from the endotracheal tube can burst and impair visualization. It is wise to anticipate these issues and to have alternatives such as a backup endoscope or the operating microscope available.

Notably, image guidance does not replace intimate knowledge and familiarity of anatomic relationships. However, increased utilization of image guidance, just like the universal assimilation of the operating microscope, reflects the principles of microneurosurgery—to resect lesions under maximal direct visualization and to optimize safety of the surgical plan of attack.

Preoperative diagnostic imaging

For the pituitary surgeon, the CT scan is often useful for delineating the bony anatomy at the base of the skull, including the approach to the sella where the details of the anatomy of the nasal septum, the turbinates and paranasal sinuses can be critical factors in the transnasal endoscopic approach.[26] Septations in the sphenoid bone frequently lead to the carotid artery and can serve as useful landmarks with preoperative awareness. CT imaging is especially critical in planning repeat surgery or an expanded endonasal ventral skull base approach. CT also is useful in demonstrating calcifications in tumors such as craniopharyngiomas. However, the anatomic diagnoses of the vast majority of pituitary tumors and other parasellar lesions are now being made by MRI, and the precision and utility of this method continue to improve.

A number of MRI sequences have been investigated for optimal delineation of pituitary lesions. MRIs demonstrate distortion of adjacent cavernous sinus, the optic apparatus, and hypothalamic or third ventricular anatomy for macroadenomas. They also allow preoperative determination of the normal pituitary location in a majority of cases, despite compression and distortion by the tumor.[27] Microadenomas remain a diagnostic challenge, even with ever-higher resolution magnets in the current era. One methodology used to detect microadenomas is dynamic contrast MRI, which exploits the preferential vascular supply to the normal hypophysis as opposed to pituitary adenomas. As such, the abnormal lesion will appear initially hypointense after contrast administration compared with the normal gland.[28,29] Some centers have moved away from dynamic MR sequences, given the excellent results obtained from high-resolution static snapshots.[30] Since microadenomas frequently present to medical attention because of the sequelae of hormone excess, another strategy for the diagnosis and follow-up of functional adenomas is metabolic imaging, as discussed below. Inferior petrosal sinus or selective cavernous sinus sampling offers an alternative in cases of difficult microadenoma localization, but carries the attendant risks of a catheter-based procedure.[31]

Metabolic imaging

Hormone-secreting pituitary tumors, especially those producing acromegaly or Cushing's disease, lead to a high burden of medical disease. Patients face a battalion of cardiovascular, respiratory, endocrinologic, arthritic, and cerebrovascular morbidities, often with irreversible physiologic changes, as a consequence of the downstream effects of growth hormone or cortisol. Furthermore, the hormonal effects frequently

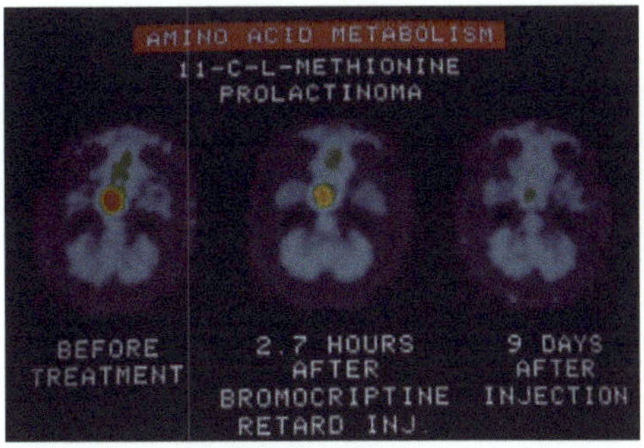

Figure 10.12 Metabolic imaging of prolactinoma. PET imaging demonstrates avid uptake of ^{11}C-L-methionine in a patient with a prolactin-secreting macroadenoma, before bromocriptine injection, with progressive decrease in metabolic uptake 2.7 hrs and 9 days after treatment. *Image courtesy of Dr. Carin Muhr, Akademiska Hospital, Uppsala, Sweden.*

cause these tumors to present in the microadenoma stage. Accurate diagnosis, as well as early detection of recurrence, becomes critical. One adjunct to conventional CT or MR imaging for functional pituitary tumors is metabolic imaging. In particular, positron emission tomography (PET) scans enable more sophisticated measurements of blood flow, glucose metabolism, oxygen utilization, protein synthesis, inflammation, and metabolic changes in response to chemotherapy and radiation therapy (Figure 10.12). Several labeled metabolites, including ^{11}C-L-methionine (MET), ^{18}F-fluorodeoxyglucose (FDG), ^{11}C-L-methylspiperone, ^{11}C-raclopride, and ^{11}C-L-deprenyl, have been studied for the detection of pituitary adenomas using PET.[9,32] The ability to attach radioisotope to receptor-specific ligands allows for tumor-specific targets, such as somatostatin.[33] Metabolic imaging can also enhance the differential diagnosis of the less common metastases to the pituitary gland.

Intraoperative MRI and pituitary surgery

Preoperative imaging, no matter the modality, offers diagnostic and surgical trajectory data, but does not take into account shifts in soft tissue during the course of surgery or provide input on the extent of resection. The development of low-field MR magnets opened the field of intraoperative MRI, with consequent improvements in achieving the goal of surgical resections.[12,34–36] Increased extent of resection, as permitted by safe neurovascular corridors, is critical not only as the standard of treatment for hormone-secreting tumors, aside from medically susceptible prolactinomas, but also for nonfunctional adenomas with mass effect.[37,38] Intraoperative MRI reveals residual tumor in approximately 30–66% of macroadenomas following either microsurgical or endoscopic transsphenoidal resection, usually prompting further resection in the majority of cases.[37,39]

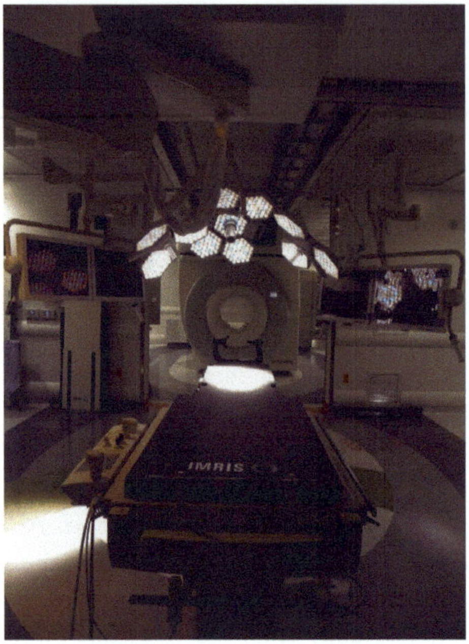

Figure 10.13 Advanced Multimodality Image Guided Operating Suite. The AMIGO suite at Brigham and Women's Hospital offers the ability to utilize image guidance in the form of intraoperative MRI in conjunction with the operating endoscope and computer-assisted neuronavigation, as well as the potential for further incorporation of angiography and PET imaging in one setting.

Our institution has garnered a spectrum of iMRI experience in resecting pituitary lesions, utilizing the initial 0.5 T low-field MR magnet to the current 3 T high-field magnet in the Advanced Multimodality Image Guided Operating Suite (AMIGO). The AMIGO suite offers the ability to utilize image guidance in the form of intraoperative MRI in conjunction with the operating endoscope and computer-assisted neuronavigation (Figure 10.13). It was our feeling that this concept would be best suited in pituitary surgery for the removal of macroadenomas. It was thought that radical resection could more readily be accomplished and confirmed for these lesions, and that necessary decompression of the optic nerves and chiasm could also be documented intraoperatively (Figure 10.14). Another goal was to assure that there were no intraoperative complications such as hemorrhage into the tumor capsule or elsewhere, or excessive intracranial air.

Experience with 22 patients with pituitary lesions has demonstrated the potential of the AMIGO for pituitary surgery (Table 10.2). Our practice is to obtain a baseline MRI study immediately prior to surgery. In most cases, the lesion can be exposed and removed in a standard fashion with the endoscope and image guidance. The tumor cavity is filled with hemostatic gelatin sponge (Gelfoam) material or an abdominal fat graft, and intraoperative MR imaging is carried out and analyzed. T1-weighted

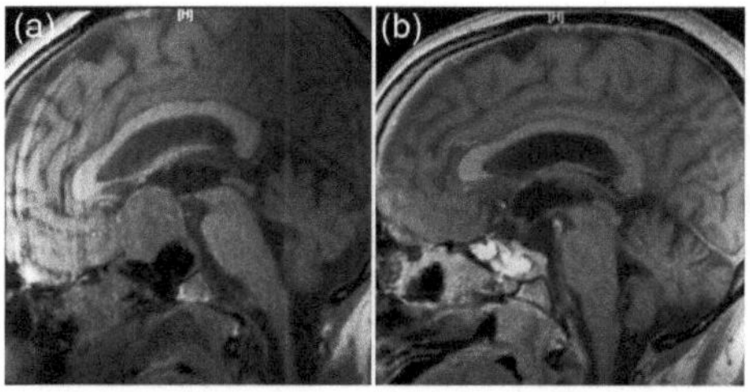

Figure 10.14 Extent of resection of pituitary macroadenoma improves with intraoperative MRI. Sagittal MRI taken intraoperatively during a transnasal endoscopic resection of a pituitary macroadenoma reveals (a) residual tumor above the diaphragm after the initial resection and (b) subsequent total resection.

Table 10.2 Brigham and Women's Hospital AMIGO case series

Adenoma subtype	Number
Null cell (including 1 atypical)	13
GH	3
GH and PRL co-secreting (atypical)	1
Mammosomatotroph	1
ACTH	1
FSH	1
FSH/LH	1
Craniopharyngioma	1

Pituitary lesions ($n = 22$, all > 10 mm) resected with intraoperative MRI surveillance, by histologic subtype.
GH, growth hormone; PRL, prolactin; ACTH, adrenocorticotropic hormone; FSH, follicle-stimulating hormone; LH, luteinizing hormone.

sequences with and without gadolinium and T2-weighted axial images proved to be of highest utility in assessing anatomic residual tumor, as well as differentiating blood products in the surgical field. Dynamic series, which draw upon the preferential vascular uptake of the normal pituitary as compared to adenomas, did not prove to be helpful in the intraoperative setting. If necessary, surgery is resumed, and if additional resection is warranted, another iMRI may be done to confirm the results. In approximately 30% of our patients, the intraoperative MRI demonstrated the need for additional surgery. No complications occurred that were attributable to the AMIGO-based procedure.

Of note, surgical artifacts can impede interpretation of intraoperative MRI at times, leading to both false positive as well as false negative results. Ultimately, the decision to pursue potential residual lesion, as suggested by iMRI, relies upon surgical judgment and experience together with intraoperative neuroradiologic consultation. Experience with intraoperative imaging is critical to interpreting these complex studies.

Although iMRI suites are increasingly available in the contemporary era, routine application to pituitary operative procedures is not always feasible, given institutional case volume, operating costs, and the panoramic views offered by the endoscope during standard transsphenoidal approaches. One comparison of endoscopic evaluation of the extent of resection with iMRI results revealed accurate endoscopic estimates in 85% of cases.[40] Interestingly, residual tumor not detected after endoscopic inspection localized to intrasellar as well as suprasellar compartments. Reasons for misinterpretation of residual tumor despite the angled views provided by the endoscope include a similar appearance of firm tumors to the normal pituitary gland and also progressive discoloration of a hyperemic hypophysis as surgery progresses.[41] Irrefutably, the endoscope offers a superior view of the cavernous sinus and suprasellar compartments in transsphenoidal approaches, as compared to the operating microscope, and serves as a powerful image guidance tool during anterior skull base surgery.

Intraoperative adjunctive imaging modalities

In addition to immediate feedback on the extent of resection provided by iMRI, several molecular adjuncts have been explored to help guide pituitary resection. Disparate capillary densities between the normal pituitary gland and a typically hypovascular adenoma can be leveraged in functional assays. The most frequently studied metabolic agent used in intraoperative image guidance is 5-aminolevulinic acid (5-ALA), especially in intrinsic glial tumors, with some application to pituitary adenomas.[42,43] The photoactive 5-ALA metabolite, protoporphyrin-IX, typically accumulates in tumor cell, thereby allowing optical detection of invasive tumor margins during surgery. In contrast to gliomas, a broad spectrum of fluorescent detection sensitivity has been reported for pituitary adenomas following administration of 5-ALA.[42,43] Practical applications of 5-ALA fluorescence in pituitary tumor resection, including intraoperative identification of microadenomas and detection of invasive margins or residual tumor, remain to be validated.

More recently, injection of indocyanine green (ICG) after sellar exposure has also proved to be a safe and consistent method of differentiating normal pituitary gland from the less fluorescent adenoma under a near-infrared light filter.[44] Furthermore, dural invasion by tumor expressed a higher level of fluorescence compared to normal dura. This technology requires possession of an infrared filter, an amenity that is not ubiquitously available. However, the facile incorporation of this step following sellar exposure is minimally intrusive to the typical workflow, akin to application of the micro-Doppler to verify internal carotid location prior to dural incision.

FUTURE DIRECTIONS

As the limits of optic resolution and noninvasive imaging continue to be explored, the future may portend a greater integration of biologic and functional data with image-guidance systems. In vivo molecular imaging may provide cell level precision in detection of invasive margins or residual nodules of tumor. One prototype in active development is mass spectrometry-based detector probes that can distinguish distinct hormone-secreting cell types in the anterior and posterior pituitary. Following surgery, tractography and optical coherence tomography may provide early prediction of visual recovery after optic nerve and chiasmal decompression. Advances in pituitary surgery have served as a harbinger of technologies applicable to other domains of neurosurgery for the past century. Seamless integration of pre-operative, intraoperative, and postoperative imaging modalities will undoubtedly continue to open new frontiers.

REFERENCES

1. Schloffer H. Erfolgreiche Operationen eines Hypophentamors auf nasalem Wege. *Wien Klin Wochenschr.* 1907;20:621–624.
2. Dandy W. Ventriculography following the injection of air into the cerebral ventricles. *Ann Surg.* 1918;68:5–11.
3. Dandy W. Roentgenography of the brain after the injection of air into the spinal canal. *Ann Surg.* 1919;70:397–403.
4. Dandy W. Cerebral ventriculoscopy. *Johns Hopkins Hosp Bull.* 1922;33:189.
5. Moniz E. Arterial encephalography: importance in the localization of cerebral tumours. *Rev Neurol (Paris).* 1927;34:72–90.
6. Hardy J, Wigser SM. Trans-sphenoidal surgery of pituitary fossa tumors with televised radiofluoroscopic control. *J Neurosurg.* 1965;23(6):612–619.
7. Hardy J. [Surgery of the pituitary gland, using the open trans-sphenoidal approach. Comparative study of 2 technical methods]. *Ann Chir.* 1967;21(15):1011–1022.
8. Chandler WF, Knake JE, McGillicuddy JE, Lillehei KO, Silver TM. Intraoperative use of real-time ultrasonography in neurosurgery. *J Neurosurg.* 1982;57(2):157–163.
9. Bergstrom M, Muhr C, Lundberg PO, Langstrom B. PET as a tool in the clinical evaluation of pituitary adenomas. *J Nucl Med.* 1991;32(4):610–615.
10. Stummer W, Stocker S, Wagner S, et al. Intraoperative detection of malignant gliomas by 5-aminolevulinic acid-induced porphyrin fluorescence. *Neurosurgery.* 1998;42(3):518–525:discussion 525–516.
11. Boppart SA, Brezinski ME, Pitris C, Fujimoto JG. Optical coherence tomography for neurosurgical imaging of human intracortical melanoma. *Neurosurgery.* 1998;43(4):834–841.
12. Black PM, Moriarty T, Alexander 3rd E, et al. Development and implementation of intraoperative magnetic resonance imaging and its neurosurgical applications. *Neurosurgery.* 1997;41(4):831–842.
13. Carrau RL, Jho HD, Ko Y. Transnasal-transsphenoidal endoscopic surgery of the pituitary gland. *Laryngoscope.* 1996;106(7):914–918.
14. Roentgen W. Ueber eine neue Art von Strahlen: vorläufige Mitteilung. *Sitzungsber Physmed Ges Würzburg.* 1895;137:132–141.
15. Medvei VC. The birth of endocrinology: part II. In: Medvei VC, ed. *The History of Clinical Endocrinology.* 2nd ed. New York: Parthenon; 1993:159–194.
16. Powell DF, Baker Jr. HL, Laws Jr. ER. The primary angiographic findings in pituitary adenomas. *Radiology.* 1974;110(3):589–595.
17. Littleton Jr C, Winter F. Polydirectional body section tomography: a new diagnostic method. *Am J Roentgenol.* 1963;89:1179–1193.
18. Jane Jr JA, Thapar K, Kaptain GJ, Maartens N, Laws Jr. ER. Pituitary surgery: transphenoidal approach. *Neurosurgery.* 2002;51(2):435–444.

19. Hardy J, Ciric IS. Selective anterior hypophysectomy in the treatment of diabetic retinopathy. A transsphenoidal microsurgical technique. *JAMA*. 1968;203(2):73–78.
20. Hardy J. Transphenoidal microsurgery of the normal and pathological pituitary. *Clin Neurosurg*. 1969;16:185–217.
21. Yamasaki T, Moritake K, Hatta J, Nagai H. Intraoperative monitoring with pulse Doppler ultrasonography in transsphenoidal surgery: technique application. *Neurosurgery*. 1996;38(1):95–97: discussion 97–98.
22. Dusick JR, Esposito F, Malkasian D, Kelly DF. Avoidance of carotid artery injuries in transsphenoidal surgery with the Doppler probe and micro-hook blades. *Neurosurgery*. 2007;60(4 Suppl 2):322–328: discussion 328–329.
23. Barkhoudarian G, Del Carmen Becerra Romero A, Laws ER. Evaluation of the 3-dimensional endoscope in transsphenoidal surgery. *Neurosurgery*. 2013;73(1 Suppl Operative):ons74–78.
24. Kassam A, Snyderman CH, Mintz A, Gardner P, Carrau RL. Expanded endonasal approach: the rostrocaudal axis. Part I. Crista galli to the sella turcica. *Neurosurg Focus*. 2005;19(1):E3.
25. Kassam A, Snyderman CH, Mintz A, Gardner P, Carrau RL. Expanded endonasal approach: the rostrocaudal axis. Part II. Posterior clinoids to the foramen magnum. *Neurosurg Focus*. 2005;19(1):E4.
26. Miki Y, Kanagaki M, Takahashi JA, et al. Evaluation of pituitary macroadenomas with multidetector-row CT (MDCT): comparison with MR imaging. *Neuroradiology*. 2007;49(4):327–333.
27. Cho CH, Barkhoudarian G, Hsu L, Bi WL, Zamani AA, Laws ER. Magnetic resonance imaging validation of pituitary gland compression and distortion by typical sellar pathology. *J Neurosurg*. 2013;119(6):1461–1466.
28. Bartynski WS, Lin L. Dynamic and conventional spin-echo MR of pituitary microlesions. *Am J Neuroradiol*. 1997;18(5):965–972.
29. Portocarrero-Ortiz L, Bonifacio-Delgadillo D, Sotomayor-Gonzalez A, Garcia-Marquez A, Lopez-Serna R. A modified protocol using half-dose gadolinium in dynamic 3-Tesla magnetic resonance imaging for detection of ACTH-secreting pituitary tumors. *Pituitary*. 2010;13(3): 230–235.
30. Tabarin A, Laurent F, Catargi B, et al. Comparative evaluation of conventional and dynamic magnetic resonance imaging of the pituitary gland for the diagnosis of Cushing's disease. *Clin Endocrinol*. 1998;49(3):293–300.
31. Fujimura M, Ikeda H, Takahashi A, Ezura M, Yoshimoto T, Tominaga T. Diagnostic value of super-selective bilateral cavernous sinus sampling with hypothalamic stimulating hormone loading in patients with ACTH-producing pituitary adenoma. *Neurol Res*. 2005;27(1):11–15.
32. Muhr C. Positron emission tomography in acromegaly and other pituitary adenoma patients. *Neuroendocrinology*. 2006;83(3–4):205–210.
33. Pepe G, Moncayo R, Bombardieri E, Chiti A. Somatostatin receptor SPECT. *Eur J Nucl Med Mol Imaging*. 2012;39(Suppl 1):S41–51.
34. Steinmeier R, Fahlbusch R, Ganslandt O, et al. Intraoperative magnetic resonance imaging with the magnetom open scanner: concepts, neurosurgical indications, and procedures: a preliminary report. *Neurosurgery*. 1998;43(4):739–747:discussion 747–738.
35. Tronnier VM, Wirtz CR, Knauth M, et al. Intraoperative diagnostic and interventional magnetic resonance imaging in neurosurgery. *Neurosurgery*. 1997;40(5):891–900.
36. Fahlbusch R, Ganslandt O, Buchfelder M, Schott W, Nimsky C. Intraoperative magnetic resonance imaging during transsphenoidal surgery. *J Neurosurg*. 2001;95(3):381–390.
37. Nimsky C, von Keller B, Ganslandt O, Fahlbusch R. Intraoperative high-field magnetic resonance imaging in transsphenoidal surgery of hormonally inactive pituitary macroadenomas. *Neurosurgery*. 2006;59(1):105–114:discussion 105–114.
38. Bellut D, Hlavica M, Schmid C, Bernays RL. Intraoperative magnetic resonance imaging-assisted transsphenoidal pituitary surgery in patients with acromegaly. *Neurosurg Focus*. 2010;29(4):E9.
39. Bohinski RJ, Warnick RE, Gaskill-Shipley MF, et al. Intraoperative magnetic resonance imaging to determine the extent of resection of pituitary macroadenomas during transsphenoidal microsurgery. *Neurosurgery*. 2001;49(5):1133–1143.

40. Theodosopoulos PV, Leach J, Kerr RG, et al. Maximizing the extent of tumor resection during transsphenoidal surgery for pituitary macroadenomas: can endoscopy replace intraoperative magnetic resonance imaging? *J Neurosurg*. 2010;112(4):736–743.
41. Jane Jr. JA, Laws Jr. ER. Endoscopy versus MR imaging. *J Neurosurg*. 2010;112(4):734:discussion 735.
42. Eljamel MS, Leese G, Moseley H. Intraoperative optical identification of pituitary adenomas. *J Neurooncol*. 2009;92(3):417–421.
43. Marbacher S, Klinger E, Schwyzer L, et al. Use of fluorescence to guide resection or biopsy of primary brain tumors and brain metastases. *Neurosurg Focus*. 2014;36(2):E10.
44. Litvack ZN, Zada G, Laws Jr. ER. Indocyanine green fluorescence endoscopy for visual differentiation of pituitary tumor from surrounding structures. *J Neurosurg*. 2012;116(5):935–941.

CHAPTER 11

Imaging and Epilepsy: The Key to Surgical Success

Wael Asaad and G. Rees Cosgrove
Department of Neurosurgery, Alpert Medical School of Brown University and Department of Neurosurgery, Brigham and Women's Hospital, Harvard Medical School, Boston, MA, USA

> *"Without doubt, the proper evaluation and selection of patients for surgical management of epilepsy is the most important single factor in determining the success or failure of this form of treatment."*
>
> ***Francis MacNaughton and Theodore Rasmussen, 1975***

INTRODUCTION

Approximately 20–30% of patients with epilepsy have intractable seizures that are poorly controlled with anti-epileptic medications.[1] Focal epilepsy is often the most difficult to control but fortunately these cases are the ones most suitable to a surgical approach. The goal of epilepsy surgery is to accurately localize the seizure focus *and* remove it without causing serious and irreversible neurological deficits.

Before the advent of modern neuroimaging, clinical semiology and electroencephalography (EEG) were the most important methods used to localize the seizure focus. Imaging of the brain was employed but the early radiological techniques of skull X-rays, pneumoencephalography, and angiography provided only indirect evidence of lesions or brain atrophy. Computerized tomography (CT) was able to directly image brain parenchyma and foreign tissue lesions but middle fossa lesions were poorly visualized because of boney artifact.

Magnetic resonance imaging (MRI) is much more sensitive at defining the brain parenchyma itself and is able to distinguish between low grade neoplastic lesions, inflammatory lesions, idiopathic gliosis, post traumatic injury, as well as subtle focal atrophy and dysplasias. This chapter will review the evolving neuroimaging techniques and their application and importance in the modern evaluation and management of patients with intractable focal epilepsy.

BACKGROUND

The major challenge of epilepsy surgery is to accurately localize the zone of seizure onset and its extent. To that end, a presurgical evaluation strategy involves the

A. Golby (Ed): Image-Guided Neurosurgery
DOI: http://dx.doi.org/10.1016/B978-0-12-800870-6.00011-X

accumulation of informative data that includes a history and physical examination, detailed descriptions of the seizure semiology, video/EEG monitoring to capture and characterize typical seizures, advanced neuroimaging and neuropsychological evaluation. Once all of this information is obtained, a hypothesis is derived that implicates a certain region or area of the brain from which the seizures may be arising. Concordance of data heightens the confidence of localization whereas discordance of data suggests the need for additional investigations such as invasive intracranial recordings or inoperability.

The question is always how to value and weigh the acquired information by deciding which information is most important and which might be disregarded. This requires an experienced team with excellent clinical judgment and is the essence of surgical decision-making. The important point to remember is that each modality informs the other. For example, if the clinical semiology of the seizures suggests strongly that they arise in the sensorimotor cortex of the tongue because every seizure begins with a tingling sensation on the right side of the tongue, then one must carefully focus the attention on the left inferior post central gyrus. Similarly, if the patient's seizure always begins with an upraised arm and tonic rotation of the head and upper trunk toward the arm consistent with a supplementary motor onset, then special attention should be turned toward examining the contralateral posterior portion of the superior frontal gyrus.

In the past, clear ictal EEG localization was the most important determinant of seizure onset.[2] Significant limitations of EEG localization are especially evident in foci located deep within the temporal lobe or in parasagittal cortex. Often, demonstration of a clear focal structural lesion on neuroimaging consistent with the seizure semiology is the best predictor of surgical success. For example, complex partial seizures consistent with an EEG abnormality localizing to the ipsilateral temporal lobe where the MRI demonstrates clear volume loss and gliosis in the hippocampus is highly predictive of seizure freedom after surgery. Identification of a clear-cut lesion anywhere in the brain that is at least concordant with other presurgical data is highly suggestive of a favorable outcome after surgery.[3]

Clinical experience from many centers has demonstrated that the best surgical outcomes in intractable epilepsy are those in which a lesion has been identified and removed. The goal therefore of any presurgical evaluation is to "find the lesion." The advent of modern neuroimaging has made this goal ever more possible.

ADVANCED NEUROIMAGING

Advanced neuroimaging can be divided into several categories including structural, metabolic, and functional imaging. The imaging modalities currently available include CT, MRI, PET/SPECT, and magnetoencephalography (MEG). Structural imaging of the brain has largely relied on CT and MRI. While CT can still be useful in some cases, MRI has almost completely supplanted CT as the imaging method of choice, especially for detecting the presence of subtle structural lesions of the brain, i.e., cortical dysplasias. Positron emission tomography (PET) and single photon emission computed tomography (SPECT) have been useful imaging techniques to detect metabolic abnormalities in the brain. Functional imaging can be performed with activation PET (aPET) studies but is mainly performed using functional MRI (fMRI) or MEG.

Each of these different imaging modalities has different spatial resolution and can provide complementary information that may be helpful in interpreting another—i.e., one modality informs the other. MRI clearly provides the best spatial resolution but CT can better visualize boney anatomy and intracranial calcifications while PET/SPECT may demonstrate significant lobar hypometabolism in the face of an apparently structurally normal brain. MRI can also provide both direct and indirect information of a pathological process by either identifying the lesion itself or by demonstrating subtle focal atrophy, gliosis, ventricular asymmetry, or white matter changes.

Structural imaging

MRI is the imaging modality of choice in epilepsy and is used to screen all patients with intractable seizures. CT is no longer used unless a specific question is raised or the patient is unable to undergo MRI because of an incompatible implanted device.

The choice of MRI sequences to screen patients with intractable epilepsy are center specific but typically include 1.5 or 3 Tesla (T) axial whole brain, thin slice, T1-weighted volumetric imaging along with axial T2 FSE and fluid attenuated inversion recovery (FLAIR) sequences.[4] If any cortical abnormality is suspected, additional slices can be obtained in coronal or oblique slices through the areas of interest. Additional susceptibility-weighted sequences, looking for hemosiderin in the case of post traumatic or post-stroke epilepsy, are sometimes warranted as well. If a patient is suspected of having temporal lobe seizures, additional T2 and FLAIR coronal oblique slices are obtained perpendicular to the long axis of the hippocampus. These provide the optimal imaging of the size, signal and internal structure of the hippocampus to diagnose mesial temporal sclerosis (MTS) and atrophy (Figure 11.1).

Before the advent of higher strength magnets, surface coils could be placed over areas of interest to better improve spatial resolution.[5] When a lesion was suspected, a surface coil placed over the area of interest provided much more detailed anatomic information and greatly improved the accuracy of the diagnosis. These coils could only improve visualization in the area directly under the coil; therefore, clinically informed placement of these coils was extremely important (Figure 11.2).

Many academic institutions now have 7 T machines, and 11 T machines are on the horizon. The advantage of MRI at 3 T and beyond is that spatial resolution is doubled over the entire imaging volume so all of the brain parenchyma can be examined (Figure 11.3). These higher-strength MRI scanners provide increased resolution and contrast, potentially enabling the detection of more subtle anatomic variations.[6] Furthermore, images from these scanners can be additionally enhanced

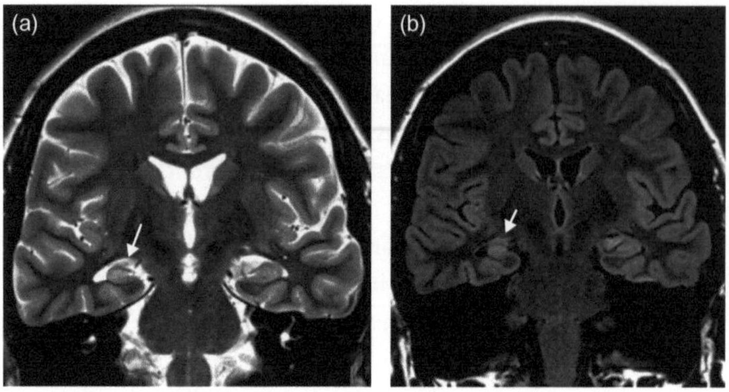

Figure 11.1 T2-weighted (a) and Flair (b) oblique coronal MR images in a patient with intractable complex partial seizures and a history of febrile convulsions as an infant. Note the slightly decreased volume and loss of internal architecture on T2-weighted images as well as the increased signal intensity on Flair images in the right hippocampus.

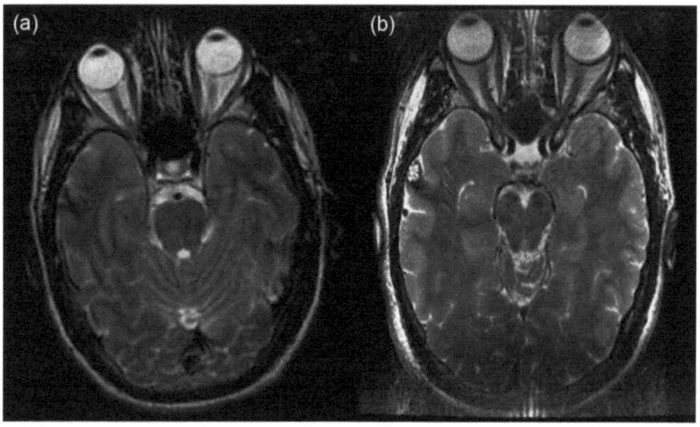

Figure 11.2 Standard T2 fast spin echo axial MR images (a) and images obtained with a surface coil over the temporal regions (b) in a patient with intractable complex partial seizures and a very active interictal spike discharge on scalp EEG in the right temporal leads. The abnormality was clearly identified as a small cavernous angioma and in retrospect there was a suspicious area on standard MRI but this was not clear enough on its own to make a firm diagnosis.

using a new generation of surface coils to produce striking anatomic detail. However, movement artifact is substantially increased at higher field strength and can interfere with image interpretation. The clarity and improved spatial resolution at 3 T and beyond can certainly improve the level of confidence that a lesion has been identified but in most cases, all of these lesions can also be seen at 1.5 T, at least in retrospect. The difference is in the surety of the diagnosis, which is very important in making a surgical decision.

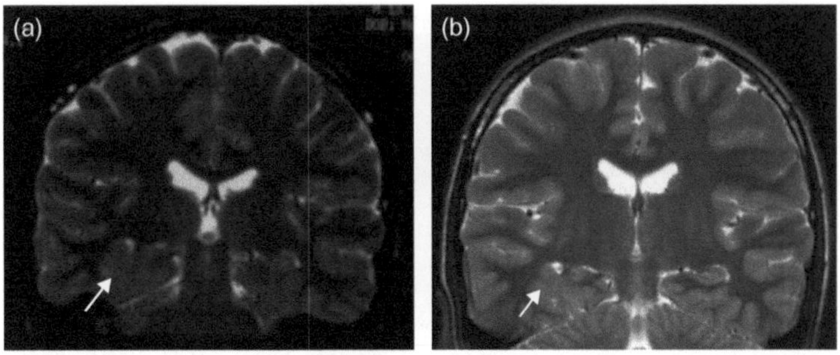

Figure 11.3 T2-weighted coronal MR images at 1.5 T (a) and 3.0 T (b) in a young boy with intractable seizures. The initial MRI at 1.5 T was interpreted as normal but the dysplastic cortex along the collateral sulcus and the periventricular gray matter in the right temporal lobe that is evident on the 3 T images can also be appreciated in retrospect on the 1.5 T images.

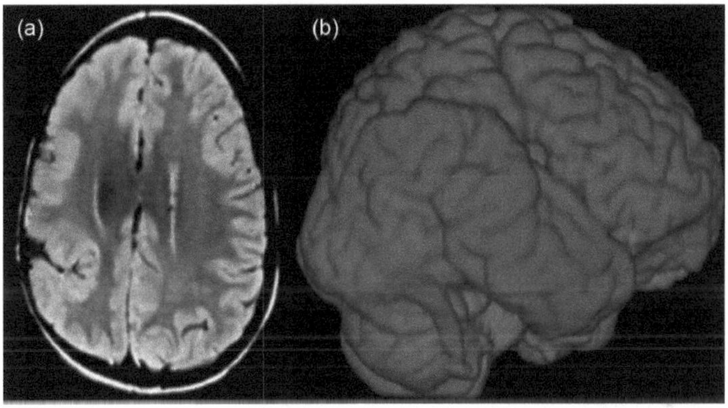

Figure 11.4 Axial T1-weighted MRI (a) of a child with intractable generalized seizures due to right hemispheric schizencephaly and a 3D surface rendering (b) of the same child that demonstrates the extent of the peri-sylvian gyral abnormalities much more clearly. Note the deepened sylvian fissure that extends in to the parietal lobe, the wide flattened gyri and the lack of normal tertiary sulcal patterns.

Volumetric MRI data can also be used to create accurate surface models of the individual patient's brain.[7] Cortical anatomy can be examined to look for subtle abnormalities of gyral and sulcal patternings. These three-dimensional images can also be used to visualize the location and extent of subcortical abnormalities because the underlying gray matter and white matter may be examined using readily available computerized image processing tools (Figure 11.4).

However, even commonly available MRI scanners may have much more information to yield if appropriate signal processing techniques are employed. For example, the application of quantitative measurements can augment visual inspection, such as in cases of decreased hippocampal volume or increased FLAIR signal in patients with

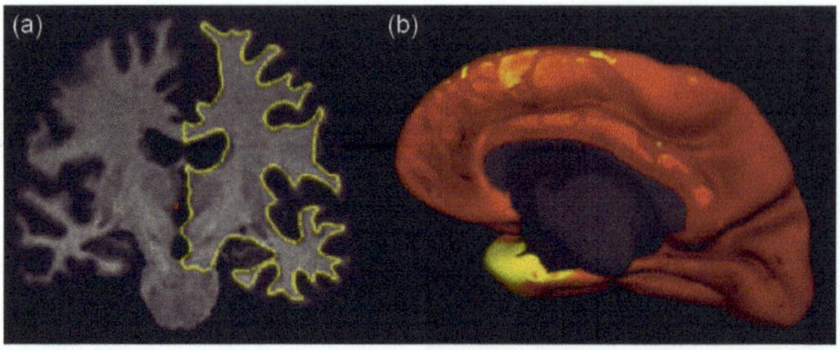

Figure 11.5 Cortical thickness maps (courtesy of Anders Dale, MGH NMR) demonstrating the definition of the gray–white junction (a) and its three-dimensional rendering that displays areas of cortex that are thicker than normal as yellow areas in a patient with right temporal periventricular heterotopia (b). Note the additional areas of cortical thickening suggesting a more diffuse cortical dysplasia that may extend in to the parasagittal frontal cortex.

temporal lobe epilepsy.[8] Semi-automated computer algorithms have been developed to measure regional volumes including the hippocampus, amygdala and parahippocampal cortex in patients with temporal lobe epilepsy. These measurements can often alert the clinician to a subtle loss of volume indicating relative hippocampal atrophy.[9]

Meanwhile, automated morphometry can detect subtle variations in cortical thickness and gyral or sulcal organization that may represent the seizure focus.[10,11] These algorithms can detect the distinct gray–white boundary between cortical gray matter and subcortical white matter with good accuracy and render this information in color-coded cortical thickness maps. The cortical thickness is typically 2–4 mm but varies between Brodmann's areas. Cortical dyplasias are often characterized by thickened and disorganized gray matter and these subtle cortical dyplasias, too difficult to see on standard imaging, can sometimes be detected with these techniques. At the very least, areas that appear thickened on cortical thickness maps can be examined much more carefully and attentively on the original MR image sequences (Figure 11.5). Quantitative imaging can also define the total volumes of gray and white matter in each lobe, which may also inform the trained observer that a subtle lobar abnormality exists that could not be appreciated on direct inspection.[12]

Diffusion tensor imaging (DTI) is a relatively new technique that can better visualize white matter. The directionality of the white matter fiber tracts are color-coded based upon the differential directional mobility of water molecules along versus perpendicular to the fiber bundles, a term known as *anisotropy*. Where axonal bundles are intact and parallel, anisotropy is high; where axonal tracts are spatially disorganized or intermingled, anisotropy is low. These color-coded anisotropy maps can be analyzed to see if the coefficient of anisotropy is maintained or normal in the various lobes of the brain as an indirect measure of dysfunctional connections. The use of this technique in epilepsy surgery remains investigational.[13,14]

An even newer technique for measuring myelin concentration using multi-component driven equilibrium single pulse observation of T_1 and T_2 (McDESPOT)

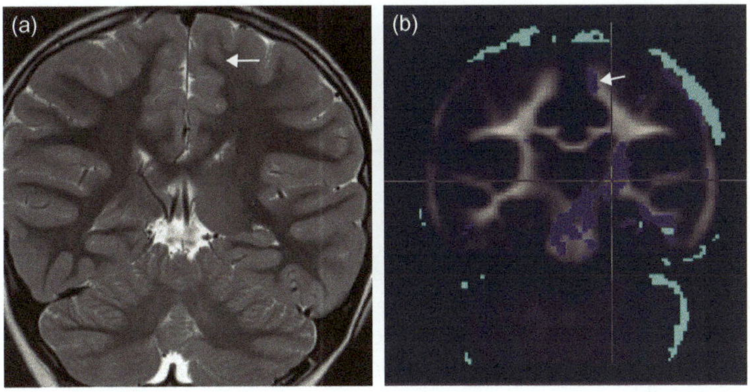

Figure 11.6 T2-weighted coronal images of a 6-year-old girl with severe intractable seizures consisting of right supplementary motor attacks. All MRIs were reportedly normal but a small transmantle dysplasia was identified in the left superior frontal gyrus (a) and McDESPOT imaging demonstrated a focal area of decreased myelin concentration in the same area (b) along with more diffuse myelin loss in the deeper white matter tracts.

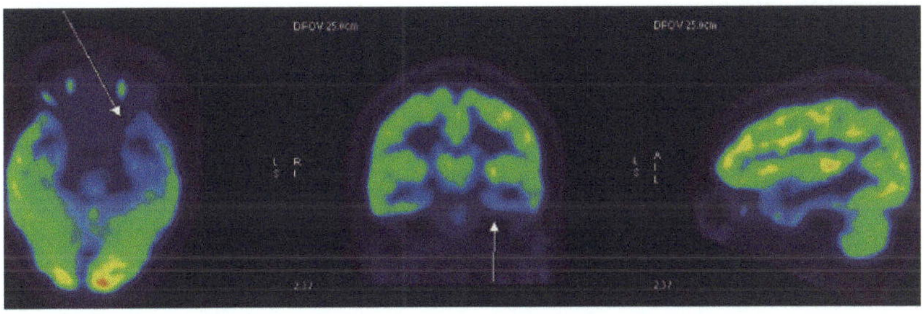

Figure 11.7 Interictal PET images in the axial (a), coronal (b), and sagittal (c) planes in an adult with intractable temporal lobe seizures. Note the decreased metabolism in the left temporal lobe affecting primarily the inferior and mesial temporal cortex.

may also play an important role in evaluating the white matter of patients with epilepsy, especially in pediatric patients.[15] This technique compares the myelin concentration of the subject with an age matched control database and highlights differences. These differences are then presented visually for examination and can be used to focus attention on the cortical area of interest (Figure 11.6).

Metabolic imaging

Metabolic imaging using either PET or SPECT has been used for decades to try to detect abnormalities of metabolism in patients with intractable epilepsy. Both PET and SPECT studies are typically performed in the interictal state and look for areas of cortex that are hypometabolic as compared to the surrounding cortex (Figure 11.7). While heralded as a major method to detect epileptic cortex, they have not been as uniformly helpful as anticipated. PET/SPECT studies can often confirm temporal

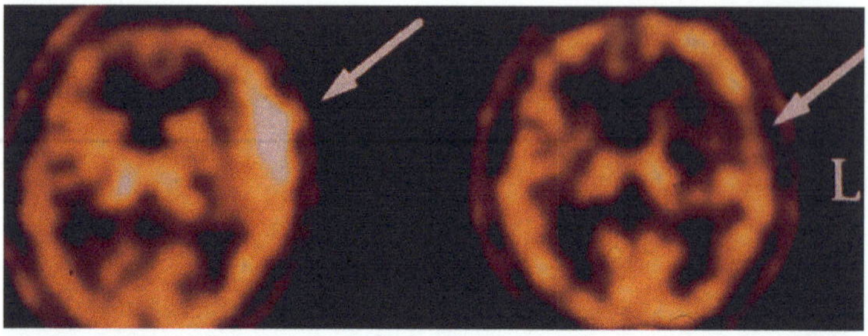

Figure 11.8 Ictal and interictal axial SPECT images in a patient with focal and generalized seizures with a suspected frontal or temporal onset. Peri-ictal injection of the radioisotope reveals increased uptake in the left inferior frontal region (a) confirming a suspicion of focal interictal hypometabolism in the same area (b).

hypometabolism in complex partial seizures of temporal lobe onset in approximately 60–70% of patients.[16,17] They can also be confirmatory in selected extratemporal cases in defining the extent of the abnormality when a structural lesion on MRI has been identified or suspected.

While relatively poor spatial resolution may be responsible for the lack of sensitivity with PET/SPECT, both studies are also more qualitative than quantitative; therefore the observer must be looking for differences in regional cortical metabolism to detect an abnormality. Many centers try to improve sensitivity by preforming ictal SPECT, which entails injecting the radioisotope at the seizure onset and then looking at the difference between this ictal SPECT and an interictal SPECT.[18] This technique is known as subtraction (Figure 11.8). When a radioactive tracer is injected during a seizure, it is concentrated at the site of highest cortical metabolism. At the very beginning of the seizure (first 30–60 seconds), this area is most likely the zone of seizure onset. As the seizure spreads to surrounding cortex, other lobes and contralateral hemisphere, these areas also become hypermetabolic. The problem of ictal SPECT is that it is logistically and practically challenging to administer the tracer in a timely fashion. Therefore the demonstrated area of hypermetabolism may not be in the zone of seizure onset but in those areas to which the seizure has spread, thus providing falsely localizing information. While occasionally very useful, especially in cases of focal status epilepticus, all ictal SPECT data should be interpreted with caution.

Functional imaging

Once the seizure focus has been localized with confidence, the next step is to remove it without causing any irreversible neurological deficit. Functional neuroimaging can be very helpful in this regard and can be performed with aPET, fMRI, or MEG. Activation PET can localize sensorimotor, language and higher cortical function but requires an onsite cyclotron to produce the O^{15} isotope that

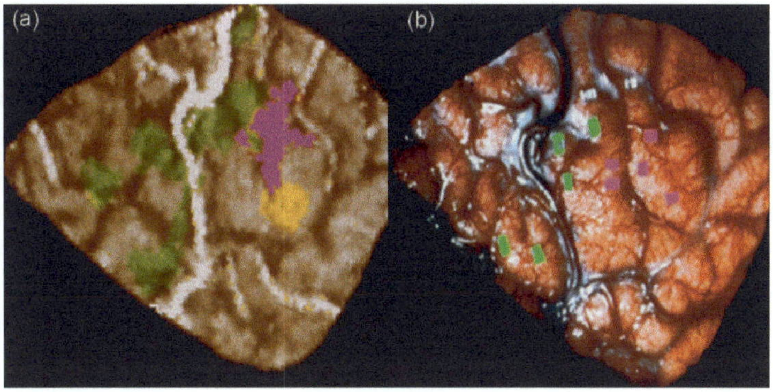

Figure 11.9 Functional MRI of a 26-year-old man with new onset seizures characterized by ictal coughing and speech arrest. MRI demonstrated a small enhancing lesion in the left subcentral gyrus (yellow). The patient underwent fMRI for tongue sensorimotor mapping (mauve) and language mapping (green) and the images were rendered with superimposed cortical venous structures (a). Surgery was performed awake with right tongue contraction and sensory responses marked (mauve) and speech arrest (green) marked with tags on the cortical surface (b). Lesionectomy while testing for speech was performed successfully without neurological deficit. Pathology demonstrated ganglioglioma.

is typically used in such studies. This requirement limits the use of aPET to a handful of centers worldwide and fMRI has almost completely supplanted aPET as the method of choice.

Functional MRI uses the change in T2*-weighted blood oxygen level-dependent (BOLD) signal that occurs in cortical areas during activation tasks that reflect a change in cortical blood flow as a surrogate for neural activity. These BOLD signal changes are quite small and therefore multiple epochs of signal obtained during 30 seconds of an activation task are compared to the same cortical volume during a control or rest state. There are many pragmatic requirements for successful fMRI because these studies have very low signal-to-noise and are movement sensitive, but the design of the task paradigm is crucial to obtain meaningful data. The areas of cortex that have a statistically significant change from baseline are then color coded and registered to a 3D model of the patient's brain for presurgical planning purposes (Figure 11.9). Sensorimotor localization is successful in approximately 90% of subjects and language localization in approximately 60–70% of studies, but all fMRI information must be interpreted with caution. Such caution is particularly advised when expected activations are not seen.[19] The prudent surgeon uses this information to plan out a surgical strategy that incorporates awake cortical mapping and resection techniques when necessary.

MEG can also be used to localize sensorimotor, auditory, and language cortex by calculating a magnetic source dipole based upon repetitive performance of a specific task in much the same way that cortical evoked somatomotor and sensory-evoked potentials are recorded with scalp electrodes. The advantage of MEG is that it has many more detectors than widely spaced scalp electrodes and the magnetic signal recorded is not diminished by the overlying skull and scalp.[20] Commercially available MEG machines are expensive and require dedicated shielded space, however, and thus are not generally available to the practicing neurosurgeon.

MEG can also be used as a tool for seizure onset localization by defining a single source dipole for interictal epileptic activity from the magnetic signals during a scalp-recorded electrical "spike discharge."[21] MEG provides excellent spatial and time resolution and computerized analysis of these small magnetic signals can point to a source for the spike and even trace the propagation of a seizure over time through the adjacent areas of the brain (Figure 11.10).

DISCUSSION

Since the goal of epilepsy surgery is to accurately localize the seizure focus *and* remove it without causing serious and irreversible neurological deficits, it is the responsibility of the epilepsy surgeon to evaluate all presurgical information and formulate a successful surgical plan. It is important to remember that all investigative data informs another and that each modality has a different localizing value. In most cases, finding a lesion on neuroimaging that is concordant with the clinical semiology and EEG data is the most important determinant of a successful surgical outcome. Several clinical studies have confirmed that finding a focal lesion on MRI is the best predictor of outcome.

In one study of 135 patients followed for over 5 years after temporal lobectomy, 69% of patients were seizure free if MRI showed a foreign tissue lesion, 50% were seizure free if MRI showed hippocampal sclerosis and only 21% were seizure free if MRI was normal.[22] The presence of a lesion is even more important in extratemporal epilepsies. In a long-term study of 70 patients with frontal lobe epilepsy, less than 15% were seizure free 5 years after surgery if they had a negative MRI.[23] Similarly, in a review of 51 patients from the Mayo Clinic, less than 50% were seizure free 1–2 years postsurgery without a demonstrable lesion on MRI.[24] The outcome statistics are so discouraging in the extratemporal lobe epilepsies that some centers do not undertake surgery outside the temporal lobe unless an identifiable lesion is seen on MRI.

One important negative predictor of a poor surgical outcome is incomplete resection of the seizure focus or lesion once identified. In this situation, the seizure focus may extend into functionally important cortex and in order to limit the risk of neurological deficit, the resection boundaries are restricted. Functional imaging can often be helpful in this regard by delineating those areas that must be preserved with more confidence, although in most situations, careful resection guided by awake mapping techniques is often the most important determinant.

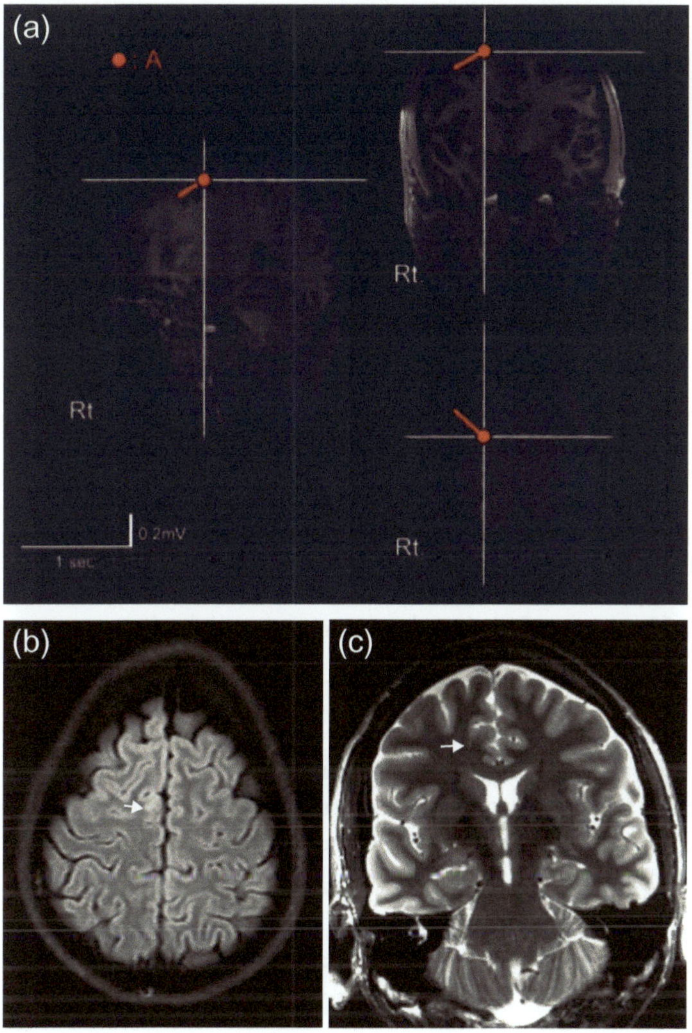

Figure 11.10 This 19-year-old male had mainly nocturnal seizures beginning with tingling in the left thigh followed by tonic extension of the left arm and flexion of the right arm. Diurnal events were associated with frequent falls. Interictal scalp EEG was unhelpful and ictal EEG appeared bilateral from the onset. MEG was performed and revealed a single dipole source in the right frontal lobe (a). Re-interpretation of the MRI demonstrated a small area of high intensity signal in the right parasagittal cortex (b) and coronal MRI revealed a small transmantle dysplasia (c). Focal cortical resection of the dysplasia cured the patient of his seizures. Pathology demonstrated a cortical dysplasia Type IIb.

Although a lesion identified on MRI is often the best indicator for success, there are instances when other presurgical data can be of equal or greater value. For example, a clinical semiology that consists of a very discrete lateralized motor or sensory onset can reliably predict the seizure focus even when no clear MRI lesion is easily

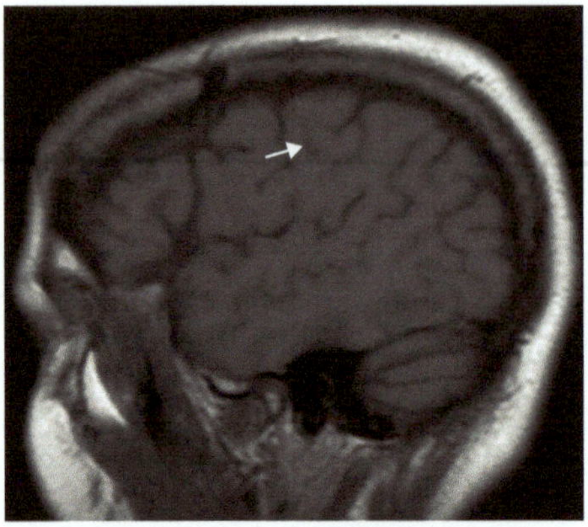

Figure 11.11 This 42-year-old woman had intractable focal seizures beginning with a tingling sensation on the right side of her tongue followed by tonic deviation of the tongue to the left and speech arrest. Seizures occurred up to 10–12 times each day but ictal scalp EEG was nonlocalizing. Both 1.5 T and 3 T MRI obtained on several occasions were completely normal. Careful examination of her MRI in the left inferior Rolandic cortex revealed an unusual sulcus traversing the postcentral gyrus perpendicular to the central sulcus. Craniotomy with subdural grid implantation over the area confirmed frequent seizures originating from this area, which was removed, resulting in complete cessation of her seizures. Pathology demonstrated cortical dysplasia.

identified (Figure 11.11). It is therefore quite clear that all presurgical information is complementary and that all information must be carefully considered before finalizing the surgical plan. Experienced clinicians are able to evaluate which parts of the presurgical evaluation are the most important and which to weight more heavily in the decision-making process.

CONCLUSIONS

Since the most important determinant of a successful surgical outcome in epilepsy surgery is identification of a lesion that explains the patient's seizures, it is imperative that the surgical epilepsy team makes every effort to "find the lesion." All presurgical evaluation data is considered in order to accurately define the focal zone of seizure onset, the extent of the epileptic lesion and the boundaries of important cortical function to avoid neurological deficits because incomplete resection bodes failure. Ultimately, it is the surgeon's responsibility within this multidisciplinary team to make the final decision regarding all aspects of the planned surgery.

Improvements in imaging devices and novel applications of existing devices hold the promise of revealing structural abnormalities in cases once thought to be "nonlesional," thereby greatly increasing the prospects for a surgical cure. While we have many new and powerful neuroimaging tools available to us that clearly improve our chances for a successful surgical outcome, it is important to remember the old adage from Lars Leksell that *A fool with a tool is still a fool.* Guided by this principle, it is important to make the best possible decisions to maximize the patient's chances for surgical success.

REFERENCES

1. Kwan P, Brodie MJ. Early identification of refractory epilepsy. *N Engl J Med*. 2000;342:314–319.
2. Cascino GD. Video-EEG monitoring in adults. *Epilepsia*. 2002;43(Suppl 3):80–93.
3. Tellez-Zenteno JF, Dhar R, Wiebe S. Long-term seizure outcomes following epilepsy surgery: a systematic review and meta-analysis. *Brain*. 2005;128:1188–1198.
4. Lessor RP, Modic MT, Weinstein MA, et al. MRI in patients with intractable epilepsy. *Arch Neurol*. 1986;43:367–371.
5. Grant PE, Barkovich AJ, Wald LL, Dillon WP, Laxer KD, Vigneron DB. High-resolution surface-coil MR of cortical lesions in medically refractory epilepsy: a prospective study. *Am J Neuroradiol*. 1997;18(2):291–301.
6. Knake S, Triantafyllou C, Wald LL, et al. 3 T phased array MRI improves the presurgical evaluation in focal epilepsies: a prospective study. *Neurology*. 2005;65:1026–1031.
7. Xu C, Pham DL, Rettmann ME, Yu DN, Prince JL. Reconstruction of the human cerebral cortex from magnetic resonance images. *IEEE Trans Med Imaging*. 1999;18(6):467–480.
8. Keller SS, Roberts N. Voxel-based morphometry of temporal lobe epilepsy: an introduction and review of the literature. *Epilepsia*. 2008;49(5):741–757.
9. Bernasconi N, Bernasconi A, Caramanos Z, Antel SB, Andermann F, Arnold DL. Mesial temporal damage in temporal lobe epilepsy: a volumetric MRI study of the hippocampus, amygdala and para-hippocampal region. *Brain*. 2003;126:462–469.
10. Bernasconi A, Bernasconi N, Bernhardt BC, Schrader D. Advances in MRI for 'cryptogenic' epilepsies. *Nat Rev Neurol*. 2011;7(2):99–108.
11. Hong SJ, Kim H, Schrader D, Bernasconi N, Bernhardt BC, Bernasconi A. Automated detection of cortical dysplasia type II in MRI-negative epilepsy. *Neurology*. 2014;83(1):48–55.
12. Peng SJ, Harnod T, Tsai JZ, et al. Evaluation of subcortical grey matter abnormalities in patients with MRI-negative cortical epilepsy determined through structural and tensor magnetic resonance imaging. *BMC Neurol*. 2014;14:104:May 14.
13. Ahmadi ME, Hagler Jr DJ, McDonald CR, et al. Side matters: diffusion tensor imaging tractography in left and right temporal lobe epilepsy. *Am J Neuroradiol*. 2009;30(9):1740–1747.
14. Chen Q, Lui S, Li CX, et al. MRI-negative refractory partial epilepsy: Role for diffusion tensor imaging in high field MRI. *Epilepsy Res*. 2008;80(1):83–89.
15. Spader HS, Ellermeier A, O'Muircheartaigh J, et al. Advances in myelin imaging with potential clinical application to pediatric imaging. *Neurosurg Focus*. 2013;34(4):E9.
16. Engel Jr J, Kuhl DE, Phelps ME, Crandall PH. Comparative localization of epileptic foci in partial epilepsy by PCT and EEG. *Ann Neurol*. 1982;12:529–537.
17. Lee BI, Marklan ON, Siddiqui AR, et al. Single photon emission computed tomography (SPECT) brain imaging, intractable complex partial seizures. *Neurology*. 1986;36:1471–1477.
18. Marks DA, Katz A, Hoffer P, Spencer SS. Localization of extratemporal epileptic foci during ictal single photon emission computed tomography. *Ann Neurol*. 1992;31:250–255.
19. Cosgrove GR, Buchbinder BR, Jiang H. Functional magnetic resonance imaging for intracranial navigation. *Neurosurg Clin N Am*. 1996;7(2):313–322.
20. Lau M, Yam D, Burneo JG. A systematic review on MEG and its use in the presurgical evaluation of localization-related epilepsy. *Epilepsy Res*. 2008;79:97–104.

21. Knake S, Halgren E, Shiraishi H, et al. The value of multichannel MEG and EEG in the presurgical evaluation of 70 epilepsy patients. *Epilepsy Res*. 2006;69(1):80–86.
22. Berkovic SF, McIntosh AM, Kalnins RM, et al. Preoperative MRI predicts outcome of temporal lobectomy: an actuarial analysis. *Neurology*. 1995;45(7):1358–1363.
23. McIntosh AM, Averill CA, Kalnins RM, et al. Long-term seizure outcome and risk factors for recurrence after extratemporal epilepsy surgery. *Epilepsia*. 2012;53(6):970–978.
24. Wetjen NM, Marsh WR, Meyer FB, et al. Intracranial electroencephalography seizure onset patterns and surgical outcomes in nonlesional extratemporal epilepsy. *J Neurosurg*. 2009;110(6):1147–1152.

CHAPTER 12

Image-Guided Open Cerebrovascular Surgery

Rajiv Khajuria, Bradley A. Gross, and Rose Du
Department of Neurosurgery, Brigham and Women's Hospital, Harvard Medical School, Boston, MA, USA

INTRODUCTION

Image guidance is an important tool in the armamentarium of any cerebrovascular neurosurgeon. From the time-honored mainstay of intraoperative digital subtraction angiography (DSA) to modern advances with indocyanine green videoangiography (ICG-VA), effective obliteration of an aneurysm, resection of a vascular malformation, and patency of direct bypasses can all be verified with intraoperative image guidance. In this chapter, we review image guidance adjuncts used in cerebrovascular neurosurgery—from treatment of aneurysms, to arteriovenous shunts, to cavernous malformations (CMs), to moyamoya.

ANEURYSMS

Intracranial aneurysms are abnormal focal outpouchings of cerebral arteries that most commonly develop at branching points of the major arteries of the circle of Willis, where vessel walls are especially structurally vulnerable.[1,2] Over 80% of intracranial aneurysms are located in the anterior circulation, most frequently at the junction of the internal carotid artery and the posterior communicating artery, the anterior communicating artery complex, or the bifurcation of the middle cerebral artery.[1–4] Postmortem examinations suggest the prevalence of intracranial aneurysms in adults is 1–5%, the majority being small and incidental.[1,2,5] A subarachnoid hemorrhage (SAH) resulting from rupture of an aneurysm is associated with high rates of morbidity and mortality.[1,2,6–10] Patients typically present with severe headache of acute onset, often accompanied by nausea or vomiting and possible loss of consciousness.[1,2] The outcome is often devastating; approximately 12% of patients die before receiving medical care, 40% of hospitalized patients do not survive longer than one month after the event, and more than 30% of the survivors have significant neurological deficits.[1,2,6–10] Even patients considered to have an otherwise good outcome often suffer from persistent cognitive deficits.[9]

Diagnostic imaging modalities that enable detection of intracranial aneurysms include conventional angiography, magnetic resonance angiography (MRA), and computed tomography angiography (CTA).[1,11–15] Conventional catheter-based angiography is the

A. Golby (Ed): Image-Guided Neurosurgery
DOI: http://dx.doi.org/10.1016/B978-0-12-800870-6.00012-1

gold standard methodology for detecting intracranial aneurysms and determining important anatomical properties, particularly neck anatomy, relation to adjacent branch vessels, presence of daughter domes, and the location of the rupture site.[16–20]

Treatment options include observation, endovascular coiling, and surgical clipping.[1,2] Microsurgical clipping is used for aneurysms unamenable to endovascular coiling due to complex neck/dome anatomy, symptomatic aneurysms and typically aneurysms of the middle cerebral artery bifurcation. Its ultimate goal is the complete occlusion of the aneurysm from the intracranial circulation without any residual neck while preserving the blood flow in parent, branching, and perforating vessels.[1,2] However, this goal is not always achieved. In postoperative angiographic studies, the incidence of residual filling of aneurysms has ranged from 2 to 8% and the incidence of parent or branching artery occlusion from 4 to 12%.[21–25] The consequences of suboptimal surgical results are significant, as aneurysm remnants are associated with a significant risk of regrowth and rupture.[7,26–28] In addition, aneurysmal rebleeding and unintended vessel occlusion is associated with a marked risk of disabling postoperative stroke.[29] If an unintended neck remnant or vessel occlusion is detected, reexploration may be required. Thus, intraoperative verification of aneurysm occlusion and parent and branch vessel patency is crucial.

Digital subtraction angiography—the gold standard

Historically, digital subtraction angiography (DSA) was the gold standard for monitoring the treatment of cerebrovascular neurosurgical disorders intraoperatively and postoperatively. Diagnostic angiography entails catheterization of the common femoral artery, typically with a 4 French or 5 French diagnostic catheter. Using fluoroscopic guidance, the catheter is advanced over a guidewire into the parent artery of interest (often an internal carotid artery or vertebral artery), and an angiographic run of the head is carried out. Though occurring rarely, this approach does have associated risks, including potential access complications (groin hematoma, access vessel dissection, and retroperitoneal hematoma) as well as complications associated with catheterization (vessel dissection and embolus dislodgment, each potentially resulting in stroke).[21,30–39]

DSA electronically subtracts background structures such as bone in real time, leaving only the opacified blood vessels in the rendered images.[40] In DSA, the first image acquired prior to contrast administration is used to generate a digital mask that is subtracted from images acquired during both the contrast angiographic run and also fluoroscopy.[40,41] The subtracted images are generated by a hardware component called the digital image processing system.[40] Multiple studies have shown that intraoperative angiography is a highly accurate method for detection of inadequately placed clips and hence enables correction of the clip position during surgery.[30–32,34] In the majority of cases, findings of intraoperative DSA concords with results of postoperative DSA.[30–32,34]

Aside from the aforementioned general, though rare, complications of diagnostic angiography, the addition of an average of 30 minutes to the procedure may exceed the ischemic tolerance of brain tissue in critical cases. In fact, clip readjustment following DSA quality control is associated with a stroke rate of up to 33%.[29] Moreover, DSA does not allow observation of small perforating vessels and the spatial resolution is limited, though it does have high contrast resolution.[42–44] An additional issue to consider is the radiation exposure.[40] It therefore stands to reason that an intraoperative assessment tool that is easier, more expedient and safer to use than DSA that provides useful information with respect to vascular flow and aneurysm patency would be a highly desirable alternative to the routine use of DSA during aneurysm surgery.

Indocyanine green videoangiography visualizes angioarchitecture in high resolution

Indocyanine green videoangiography (ICG-VA) (Figure 12.1) is a simple, reliable, and comparatively safe and cost-effective method that allows intraoperative observation and documentation of blood flow in vessels of any size with high resolution.[45,46] ICG is a near-infrared (NIR) fluorescent nontoxic tricarbocyanine dye that was approved by the Food and Drug Administration in 1956 for cardiocirculatory and liver function diagnostic uses and in 1975 for ophthalmic angiography.[45–47] The absorption and emission peaks of ICG (805 and 835 nm, respectively) lie within the tissue optical window where absorption attributable to endogenous chromophores is low.[45–47] NIR light can therefore penetrate tissue to depths of several millimeters to a few centimeters. After intravenous injection, ICG quickly binds to lipoproteins (within 1 to 2 seconds) and remains in the circulation.[45–47] ICG is not metabolized in the body and is excreted rapidly and exclusively by the liver into bile.[45–47] Its short plasma half-life of 3 to 4 minutes allows repeated intraoperative injections. It is not reabsorbed from the intestine nor does it undergo enterohepatic recirculation.[45–47] The recommended dose for ICG video angiography is 0.2 to 0.5 mg/kg; the maximal daily dose should not exceed 5 mg/kg.[46]

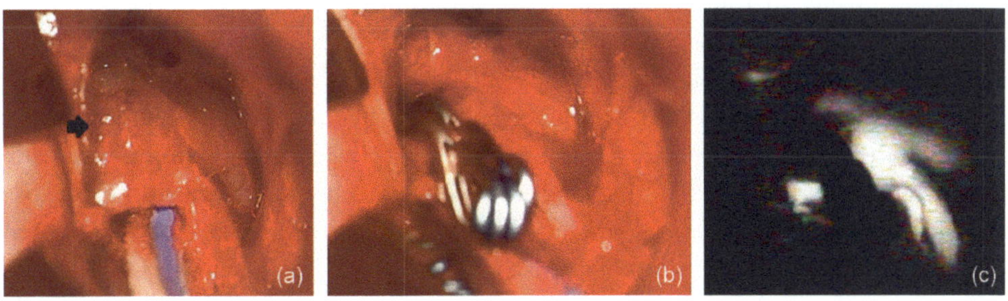

Figure 12.1 ICG-VA in aneurysm clipping. This anterior communicating artery aneurysm (a, arrow) was clipped with two fenestrated clips (b). ICG-VA confirmed patency of the parent vessels (including the anterior cerebral artery within the fenestrations) and occlusion of the aneurysm (c).

The technique uses an NIR camera integrated into the operating microscope for detection and acquisition of real-time high spatial resolution and high-contrast images.[45,46] Following intravenous injection of the ICG dye, the surgical field of interest is illuminated with NIR light at the excitation wavelength (about 750 to 800 nm) while fluorescence emission is observed at longer wavelengths (over 800 nm).[45,46] An optical filter blocks both the ambient and the excitation light so that only ICG-induced fluorescence is collected. The result is real-time angiographic images providing arterial, capillary, and venous phases, which can be seen on a video screen and recorded by a nonintensified video camera for further analysis.[45,46] This also allows the repetition of playback loops as needed during surgery. ICG-VA additionally has the important potential advantage over DSA of enabling visualization of the blood flow in small perforating vessels after microneurosurgical clipping.[48,49] Although there are other options to assess the blood flow including microvascular Doppler and ultrasonic perivascular flow probe, these methods, like DSA, lack the ability to detect patency of perforating arteries.[50–54]

Following the introduction of ICG-VA into cerebrovascular surgery, multiple studies have reported the utility of this technique in the intraoperative detection of aneurysmal remnants and parent or branch vessel compromise following aneurysm surgery.[44–46,55–57] No side effects are associated with the dye.[44–47,49,55–58] Using ICG-VA during aneurysm surgery, parent vessel stenosis or occlusions of small perforating arteries can be detected and resolved in a period of about 3–5 minutes by replacing the clips intraoperatively.[44–46,48,55–57] Therefore, postoperative ischemic deficits may be reduced. Multiple studies reported concordance rates of 90%–100% between ICG-VA and post- and intraoperative DSA.[44,45,55–57] Thus, clip adjustment rates are similar to those reported in multiple case series in which DSA was the sole intraoperative technique for assessing adequacy of aneurysm clipping.[30,32,33,37] Factors that may contribute to ICG-VA-DSA discordance are deep aneurysm location (anterior communicating, basilar) and those with complex flow patterns.[44,56,58]

Although ICG-VA has become the method of choice for intraoperative assessment of clipping quality in routine use, there are some limitations to the technique which may require the use of intraoperative DSA in select cases. Such cases include those where neck residuals are located behind the aneurysm or are very small, as these cannot always be detected by ICG-VA.[44,56,58] The restricted field of view and limited area of observation offered in accordance with the chosen surgical approach are the major limitations of ICG-VA.[44–46,55,56,58] As a result, ICG-VA can only visualize the proximal aspect of distal branching vessels, whereas during DSA, the distal flow can be compared between different branches remote from the clipping site.[44] Furthermore, vessels obscured by blood clots, aneurysms, or brain tissue cannot be observed using this technique.[44,45,55,56] Consequently, ICG-VA is less effective for deep-seated aneurysms, and these have been reported to be associated with neck remnants more often than surface aneurysms.[44,56,58] Thus, for these as well as for giant, thick-walled and complex aneurysms, there may be a need for verification of the findings by intraoperative DSA. Particularly in cases of complex aneurysmal reconstructions with bypass, though ICG-VA is used routinely intraoperatively, we regularly perform formal DSA to confirm bypass patency and aneurysm occlusion (Figure 12.2). Another limitation of ICG-VA is that calcifications, atherosclerotic plaque and partially or completely thrombosed

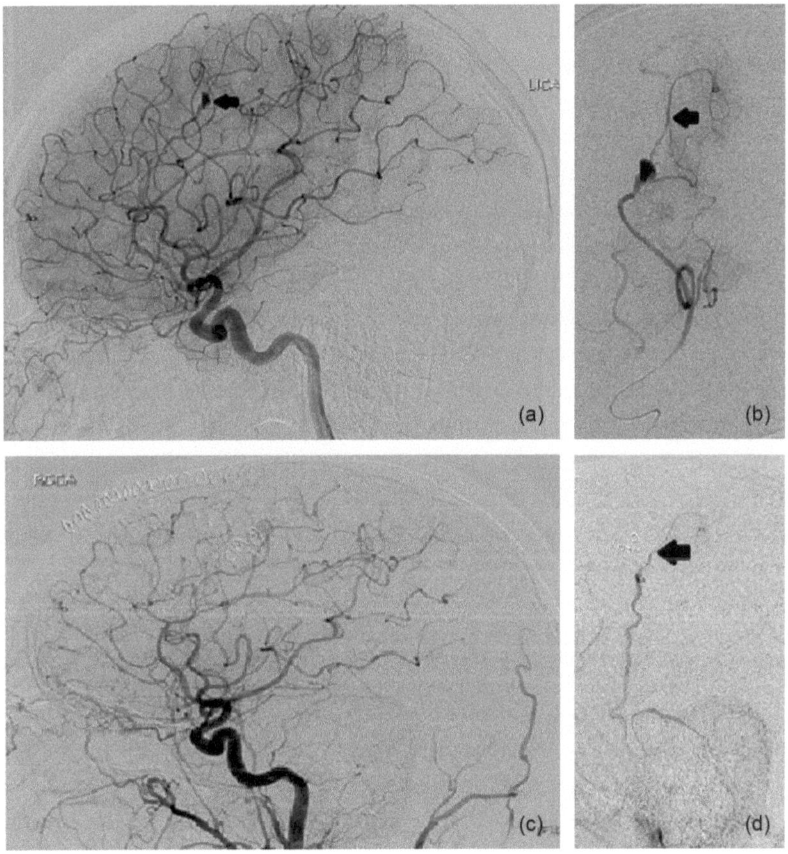

Figure 12.2 DSA to evaluate bypass patency after trapping a complex mycotic aneurysm. This distal MCA mycotic aneurysm (a, lateral internal carotid injection, arrow) was located on a precentral branch supplying the motor cortex, confirmed by superselective injection of sodium amytal (b, superselective distal MCA injection, arrow denotes distal precentral branch). The aneurysm was trapped and EC–IC bypass was performed, anastomosing the end of the parietal branch of the superficial temporal artery to the precentral branch distal to the mycotic aneurysm. As confirmed by the lateral internal carotid artery injection in (c), the aneurysm was obliterated. A selective external carotid artery injection demonstrated filling of the precentral branch via the EC–IC bypass (arrow, d). The patient remained neurologically intact postoperatively.

aneurysms may attenuate the fluorescent signal and affect the ICG angiographic results.[45,55,56] In such cases, it may be difficult for the surgeon to identify any residual filling. Moreover, repeated application of ICG over short intervals can cause false positive findings.[42,44] In addition, filling of the arteries from proximal to distal should be undertaken carefully to avoid misdiagnosing the retrograde filling of the branches distal to aneurysm.[42–44] As ICG-VA is not a quantitative method, any suspicious or delayed detection of fluorescence should be verified by other techniques such as microvascular Doppler or intraoperative DSA.[42–44]

Despite these limitations, the advantages of ICG-VA over DSA make it more suitable for use on a routine basis. Results of ICG-VA are available within 3 minutes, which allows immediate removal or correction of the position of an aneurysm clip before critical cerebral ischemia can occur.[42–46,49,55,56,58] Furthermore, this imaging modality can easily be repeated as needed. Consequently, ICG-VA is a simple tool for intraoperative quality assessment and documentation of surgical outcomes. The images can be stored on the microscope video recording for further analysis after surgery.[43,45] Moreover, ICG-VA has much higher spatial resolution than DSA, allowing the surgeon to observe the patency of all vessels within the surgical field of view, including small perforating or cortical arteries of submillimeter diameter.[44,45,48,49] Hence, even in complex cases requiring intraoperative DSA, the ICG-VA technique may be used as a supplement to assess the patency of perforating vessels that cannot be assessed with intraoperative DSA. In addition to benefits to the patient, ICG-VA is safer for the surgeon and surgical staff as no potentially harmful radiation is involved.[45,46] Finally, ICG-VA is also simpler, faster, less invasive and much more cost-effective than DSA, and thus is a crucial tool for the routine use in monitoring and improving outcome in cerebrovascular surgery.[45,46]

Frameless stereotaxy as an adjunct to localize mycotic aneurysms

Although saccular aneurysms of the circle of Willis are approached without the need for intraoperative stereotaxy, complex distal mycotic aneurysms are often approached through craniotomies guided by frameless stereotactic navigation. Mycotic aneurysms most often develop as a result of septic emboli from infective endocarditis lodging into these distal vessels, though contiguous spread as a result of cavernous sinus thrombophlebitis or meningitis may result in infectious aneurysms of the proximal intracranial circulations.[59–61] Most commonly they are treated with endovascular takedown of the vessel if noneloquent or proximal clip occlusion with bypass if the vessel supplies eloquent territory (Figure 12.2).[59,62]

Even though larger study series are missing from the literature, several reports suggest the usefulness of frameless stereotactic navigation for localization of distal mycotic aneurysms where open surgery is recommended.[62–65] Both CT angiographic and angiographic MR imaging guided frameless stereotaxy have been shown to be useful adjuncts in the management of mycotic aneurysms, improving patient outcome by decreasing morbidity and mortality.[63,66,67] This imaging technique enables a smaller, precisely placed craniotomy. 3D representation of the aneurysm and the adjacent arteries in correct orientation facilitate identification and dissection of the aneurysm.[63,64,67]

ARTERIOVENOUS MALFORMATIONS

Arteriovenous malformations (AVMs) are arteriovenous shunts supplied by pial cerebral arteries with direct drainage into the venous system without intermediate capillaries. Studies suggest the prevalence of AVMs in the general population is approximately 0.01%.[68–70] In about 50% of cases, AVMs present clinically with intracerebral hemorrhage, most commonly occurring for the first time in individuals in the third decade of life.[68,71–74] Seizures, mass effect, and ischemic steal caused by AVM are other possible clinical signs.[72] The risk of rupture of an AVM is estimated at 2–4% per year; deep AVMs, those with exclusively deep venous drainage, those associated with aneurysms, and ruptured lesions possess a greater risk.[75,76] Hemorrhage due to a ruptured AVM is a devastating event; 5–10% of affected patients die and approximately 40% suffer from persistent neurologic deficits.[68,72] Angiography prior to treatment remains the mandatory gold standard for analyzing the vascular anatomy of the AVM, including the evaluation of the presence or absence of associated aneurysms, presence or absence of obstruction of venous outflow, and pattern of venous drainage.[72,77] The ultimate goal of treatment is the complete obliteration of the AVM to prevent hemorrhage while preserving functional status.[72,77] Subtotal treatment does not confer protection against future hemorrhage and can worsen disease history. Surgical resection results in the immediate cure of the AVM and is associated with the highest overall rates of obliteration as compared to other treatment modalities.[77–81] In the course of surgical treatment, a circumferential dissection of arterial feeders is performed, and each is disconnected prior to disconnecting the AVM's venous outflow.[77,81–83] After lesion removal, formal angiography must be performed to evaluate for residual shunting.[77,81–83]

Frameless stereotaxy in arteriovenous malformation surgery

We typically perform preoperative MRI and fuse with a preoperative CTA to be used for neuronavigation. Contrast-enhanced CT scans and CT angiograms not only enable the presentation of three-dimensional anatomical AVM characteristics but also indicate the relation of the lesion to osseous structures in proximity to the cranial base.[84] In addition, CTA can simultaneously visualize feeding and draining vessels of an AVM.[84] Moreover, CTA gives good vessel contrast even in presence of hemorrhage of an AVM, and is therefore often superior in localizing AVMs in such cases compared to

MR-based imaging.[84] Contrast-enhanced MRI and MRA also enable visualization of anatomical AVM characteristics, but additionally display parenchymal anatomy in superior resolution compared with CT scans.[66,85]

Neuronavigation is a helpful adjunct to identify the optimal site for skin incision and tailoring the craniotomy.[66,84,85] In addition to ICG-VA, mentioned below, frameless stereotaxy can be used after the craniotomy as well to help distinguish feeding arteries from draining veins.[66,84,85] This is of particular importance during the final stages of resection along the deep periventricular margins, where observation is limited by the AVM nidus and hemorrhage may occur. In cases in which preoperative embolization is performed, directly visualized, embolized vessels can serve as intraprocedural fiducials in correlation with neuronavigation to further facilitate intraprocedural localization and account for brain shift. Overall, frameless stereotaxy has been shown to reduce operative time and blood loss during AVM resection.[66,84,85]

Although frameless stereotaxy is used to facilitate localization and early portions of the dissection, it is not used as a means to confirm AVM resection. High quality DSA remains the "gold standard" for evaluating vascular flow in AVM surgery and is mandatory at the end of surgery to ensure complete resection; many centers perform DSA while the patient remains under the same anesthesia from surgery but has been moved to a dedicated angiography suite.[30,32,34,77,86]

Indocyanine green videoangiography to guide AVM resection

ICG-VA provides a rapid means of intraoperative real-time analysis of arterial, early venous, capillary, and venous phases, making it very helpful for AVM vessel identification.[87–90] Identification of en-passage vessels is also facilitated. It can be used in the early phases of the dissection to help distinguish arterial feeders from veins and in late phases to qualitatively detect early filling of identified draining veins. However, ICG-VA has several limitations in AVM surgery. Visualization may be limited by the surgical approach. Consequently, ICG-VA is frequently of limited use for deep-seated AVMs that must be approached through a long, narrow corridor, where deeply located parts of the AVM are only accessible when partial removal is achieved.[87–90] In addition, ICG fluorescence is obscured by tissue, blood clots, or calcifications within the vessels. The precondition of evacuating as much hematoma as possible before dye injection in order to attain visualization could be problematic particularly in ruptured lesions.[87–90] Furthermore, ICG-VA is not reliable for assessing absence of flow in the main draining vein postresection because a remnant nidus may be draining into an unexposed deep draining vein. Also detection of residual AVM nidus is not reliable using this technique, especially if the nidus is fragmented by clots, diffuse, or covered by overlying brain parenchyma.[87–90] Identification of remnant nidus is only possible

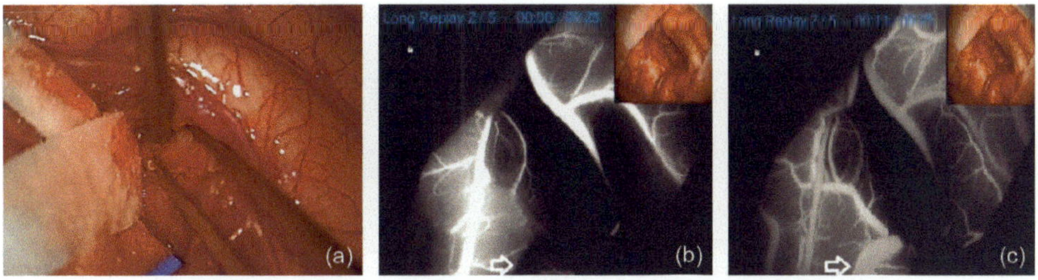

Figure 12.3 ICG-VA in AVM surgery. This superficial AVM (a) was resected with intraprocedural ICG-VA, which demonstrated no further early filling of the draining vein (b, arrow); pane c illustrates the vein filling in the subsequent normal venous phase suggestive of AVM obliteration but requiring formal digital subtraction angiography confirmation.

in the case of superficial venous drainage, and provided that the residual nidus is connected to this superficial vein (Figure 12.3). It is generally useful if it demonstrates residual AVM nidus or early filling of draining vein; however, the absence of either is not definitive in confirming successful AVM obliteration. Finally, ICG-VA alone does not improve clinical outcomes.[87–90]

Despite these limitations, ICG-VA might still be beneficial for a selected subset of superficial AVMs, which are amenable to this technique. Advantages of using ICG-VA include the rapidly acquired information that is immediately integrated into the surgical view and also the possibility of detecting incompletely resected AVM nidus before DSA is performed.[87–90] This may decrease the number of intraoperative and postoperative angiograms needed. Hence, while ICG-VA is not suitable as the sole imaging modality to confirm residual disease, it may present a helpful adjunct to intraoperative DSA and shorten procedure duration.[87–90] Formal angiography remains the gold standard for the evaluation of complete AVM resection.[77,87–90]

Role of indocyanine green videoangiography in arteriovenous fistulas

ICG-VA can also be used in the treatment of cerebral and spinal arteriovenous fistulas. In a fashion analogous to cerebral AVMs, ICG-VA can be used prior to and after obliterating a fistula, providing preliminary confirmatory evidence of successful obliteration prior to formal cerebral angiography (Figure 12.4). DSA is the gold standard for the diagnosis of AVFs.[77,91] The technique provides clearly distinguishable visualization of the fistulous point(s), early venous, capillary, feeding arteries, and draining vein phases and also allows identification of abnormal retrograde drainage. Intraoperative findings on ICG-VA correlate with findings on postoperative angiography.[92–97] Particularly in the case of spinal AVFs, intraoperative DSA is challenging with the patient in the prone position with the requisite need to catheterize small segmental vessels.

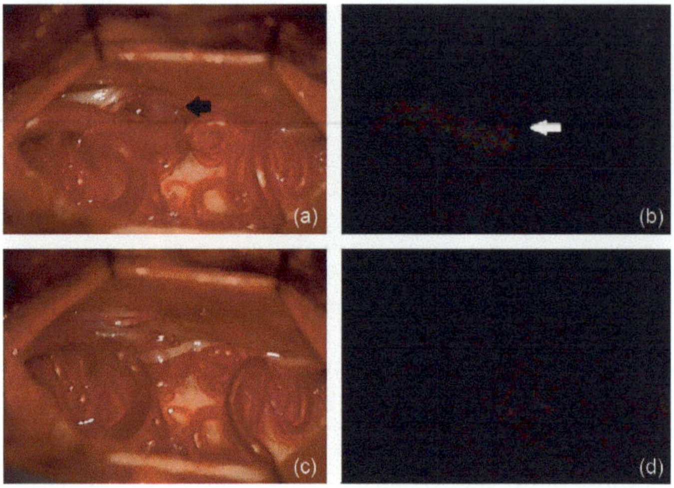

Figure 12.4 ICG-VA in spinal dural AVF surgery. This spinal dural AVF (type I) is seen with a typical fistula point at the interface of the thecal sac dura and nerve root sleeve (arrow). ICG-VA illustrates an early filling vein (b, arrow). After disconnection (c), no shunting is seen (d, venous phase).

ICG-VA can be used to precisely detect the AVFs in real time during surgery. Additionally, it is able to rapidly detect incomplete fistula obliteration if early venous shunting is demonstrated. Disconnection of AVFs is determined by the absence of an immediate arteriovenous shunt and by delayed filling of the venous plexus.[92–97] However, limitations mentioned previously also apply to the use of ICG-VA for surgery of AVFs. Only structures visualized under the operating microscope are evaluable. The vessels need to be fully exposed for observation and the penetration of the fluorescence is obscured by blood clots, calcifications or atherosclerosis.[92–97] As with AVMs, cerebral angiography is necessary to determine complete obliteration.

CAVERNOUS MALFORMATIONS

Cavernous malformations (CMs), also known as cavernous hemangiomas, cavernous angiomas or cavernomas, are vascular lesions found throughout the body including the central nervous system. CMs are composed of a collection of endothelium-lined sinusoids without intervening brain parenchyma, resembling the appearance of a purple lobulated mulberry.[98,99] These lesions are prone to small hemorrhages that are rarely symptomatic.[98,99] Autopsy and MRI studies suggest the prevalence of CMs in the general population is approximately 0.3–0.6%.[98,99] While symptomatic lesions manifest in all age groups, the peak incidence of detected CMs is usually between the third and fifth decade of life with females and males being equally affected.[98,100,101]

The majority of cases are sporadic, but CMs can also be inherited.[98–100] Familial forms are often associated with multiple cavernomas, with a high incidence of epilepsy and neurologic deficits.[98,102] Common symptoms such as seizure or focal neurologic deficit result from low-pressure hemorrhages that exert a mass effect on the surrounding brain lesions. MRI is the radiographic tool of choice for detection of CMs, which present on T2-weighted sequences with a classic surrounding hypointense hemosiderin rim.[77,98,99] The lesions are angiographically occult.[98,103]

Patients with symptomatic CMs should be considered for surgical resection.[77,98,99] Outcomes of surgical therapy have been remarkably good.[104,105] The goal of CM surgery is gross total resection of the entire lesion to eliminate the risk of bleeding while preserving normal surrounding vasculature and associated developmental venous anomalies.[77,98,99,105]

Frameless stereotaxy and intraoperative MRI to guide CM resection

Preoperative planning and mapping of eloquent areas adjacent to the CM are the most critical part of the procedure, as any inaccuracy in direction of the approach can lead to significant difficulties in finding small lesions within the parenchyma. The highest accuracy is provided by combining knowledge of anatomical landmarks in the affected region and using stereotactic navigation.[66,106–110] Frameless stereotaxy based on MR-imaging is routinely applied to aid in the localization of CM as it is a useful adjunct for planning the craniotomy, choosing the best approach, and designing the skin incision. This technique provides excellent anatomic orientation during dissection and is particularly useful for deep lesions to facilitate lowest possible risk of permanent neurologic deficits caused by damage of surrounding eloquent brain structures.[66,106–110]

Intraoperative MRI allows the surgical outcome to be evaluated immediately and reliably.[106] In case further resection is necessary, navigation can be updated and guided until the greatest possible extent of resection is achieved.

Surgical results using intraoperative MRI and neuronavigation are excellent and can reduce surgery-related morbidity in comparison with conventional surgery.[66,106–110] Hence, CMs are among the most common indications for cranial frameless stereotaxy in open cerebrovascular surgery. The limitation is the availability of intraoperative MR-imaging. If not available, the technique lacks the ability to provide real-time information.[66,107–110] However, intraoperative MRI is used with increasing frequency and even functional MRI or DTI are increasingly available, allowing the surgeon to identify eloquent cortex accurately and reassess the surgical strategy based on high precision real-time data.[106] Especially in cases where multiple lesions are present and being addressed in a single operative sitting, intraoperative MR-imaging is indispensable as resection of each lesion can result in relevant brain shift and anatomical changes, making preoperative imaging data potentially unreliable.[106]

MOYAMOYA

Moyamoya, referring to progressive intracranial stenosis and formation of abnormal collateral "moyamoya" vessels, is a considerable source of morbidity as a result of both ischemic and hemorrhagic sequelae. In the majority of cases, moyamoya inevitably progresses.[111–114] The outcome without treatment is poor.[115–117] Diagnosis of the condition is based on characteristic radiographic findings, detectable on CTA, MRA or DSA.[111,118–122] Early recognition with prompt therapeutic intervention is crucial in order to achieve the best outcome in patients.[111]

The goal of treatment is to prevent strokes by improving blood flow to the affected cerebral hemisphere and by reduction of moyamoya-associated collaterals.[111] Surgical treatment uses the external carotid artery as a source of new blood flow to the ischemic hemisphere through anastomosis to an intracranial artery, usually the middle cerebral artery (extracranial to intracranial arterial bypass, EC–IC bypass).[111,123–126] Direct bypass surgery is an effective method to prevent stroke in affected patients; indirect bypass via onlay of the superficial temporal artery and/or the temporalis muscle is a potentially efficacious approach in children.[127] In cases where direct bypass is employed (adults), intraoperative assessment of bypass patency is critical.

ICG-VA to assess bypass patency

Direct intraoperative inspection lacks sufficient reliability in EC–IC bypass patency assessment.[128–130] Historically, DSA was the gold-standard technique for intraoperative assessment of graft patency.[131] However, as mentioned before, DSA is a technique with several limitations, including its invasiveness, spatial resolution, high costs, need of advanced expertise, time consumption and exposure to radiation.[131,132] A technique overcoming all of these limitations is ICG-VA. ICG-VA is a simple, cost-effective and well-tolerated technique for the rapid intraoperative assessment of bypass patency with high image quality and spatial resolution, which facilitates the identification of stenoses at the anastomotic site or vessel obstructions.[132,133] Bypass patency can be evaluated from the filling of the graft with ICG following injection of the fluorescent dye (Figure 12.5).[132,133] Findings of intraoperative ICG-VA concur with results on postoperative DSA or CTA.[132,133] If needed, the imaging study can be repeated within 15 minutes following clearance of the dye.[132,133] These repeated injections do not affect patient safety and image quality.[132,133] In addition to its use for assessing bypass patency, ICG-VA is also a useful and reliable adjunct to identify suitable recipient vessels for the anastomosis for EC–IC bypass surgery.[134] While ICG-VA is a valuable intraoperative tool, CTA and MRA remain ideal noninvasive techniques to assess bypass patency during the early postoperative phase. In addition, conventional angiography remains a valuable postoperative imaging study in patients

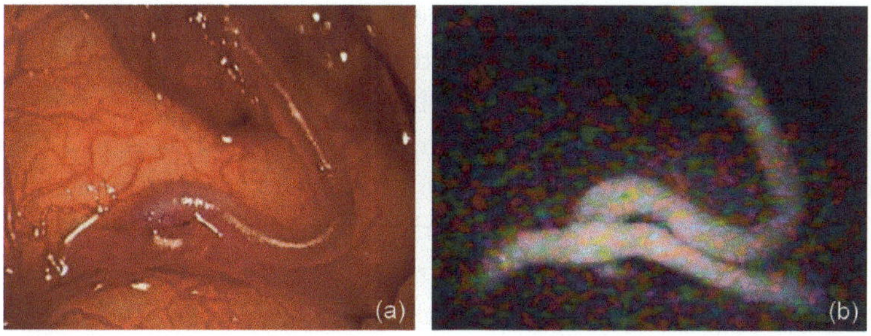

Figure 12.5 ICG-VA in bypass surgery. Following EC–IC bypass (a), ICG-VA confirms patency (b).

undergoing EC–IC bypass surgery, because it not only provides morphological information on bypass patency but also dynamic information on the extent of intracranial filling through the bypass.[131]

DISCUSSION

ICG-VA, although relatively new, is already an indispensable adjunct to cerebrovascular neurosurgery, providing real-time angiographic images of arterial, capillary and venous phases. These are recorded and can be repeatedly displayed as needed. In aneurysm surgery, the technique is of use in the intraoperative detection of aneurysmal remnants and parent or branch vessel compromise following aneurysm surgery, enabling intraoperative clip adjustment and eliminating the need of a potential subsequent operation.[44–48,55–57] It is also advantageous over the historically used DSA by enabling visualization of the blood flow in small perforating vessels after microneurosurgical clipping, which is crucial as perforating arteries may be distorted or occluded during stages of approach, dissection, or clipping of the aneurysm.[48,49] During surgery of superficial AVMs, ICG-VA can help distinguish feeding arteries from draining veins; en passage vessels are also readily visualized. In selected cases, evaluation of a potential residual nidus might be possible by assessment of early filling of a draining vein. Detection of incompletely resected AVM nidus may decrease the number of intraoperative and postoperative angiograms needed and shorten surgery time.[87–90] In the course of treatment of AVFs, ICG-VA can be used to precisely detect the fistula in real time and visualize the draining vein during surgery. Additionally, it is a useful aid for detecting incomplete fistula obliteration.[92–97] After EC–IC bypass, ICG-VA provides high quality imaging of adjacent cerebral arteries and the graft, allowing the identification of nonfunctioning bypasses, which can be revised intraoperatively to avoid subsequent additional surgery and bypass surgery related morbidity.[132,133]

However, ICG-VA also has limitations including the restricted field of view and obscuration of the fluorescence by calcifications, atherosclerotic plaque, tissue, and blood clots. These restrictions make the technique less useful in aneurysms where neck residuals are obscured by the clip and surrounding tissue. ICG-VA is also less helpful for deep-seated aneurysms and AVMs.[44–49,55–57,87,92–97] Despite these limitations, ICG-VA is widely applied in cerebrovascular surgery as it is a relatively safe, time and cost-effective method, making it advantageous over angiography. In cases where ICG-VA lacks sufficient reliability, the use of formal cerebral angiography is still mandatory and the gold standard. This includes complex aneurysms and arteriovenous shunts where obliteration should be confirmed by formal DSA.

Frameless stereotaxy is routinely used to localize AVMs, CMs, and also mycotic aneurysms. It can guide the location and size of the craniotomy while also providing intraprocedural localization during vascular malformation resection. Intraoperative MRI, though time consuming and not always available, is a useful adjunct during resection of CMs and can account for brain shift, a limitation of frameless stereotaxy based on preoperative imaging.

CONCLUSION

DSA, ICG-VA, and frameless stereotaxy are the primary image-guidance tools used by the cerebrovascular neurosurgeon. DSA remains the gold standard in assessing arteriovenous shunt obliteration, complex aneurysm obliteration and dynamics after EC–IC bypass. ICG-VA can be used as a single modality to verify simple aneurysm obliteration, early EC–IC bypass patency and to facilitate AVM and AVF resection. Frameless stereotaxy is used in vascular malformation neurosurgery ubiquitously and can be augmented with intraoperative MRI during resection of CMs.

REFERENCES

1. Schievink WI. Intracranial aneurysms. *N Engl J Med.* 1997;336(1):28–40.
2. Brisman JL, Song JK, Newell DW. Cerebral aneurysms. *N Engl J Med.* 2006;355(9):928–939.
3. Kassell NF, Torner JC, Haley Jr. EC, Jane JA, Adams HP, Kongable GL. The international cooperative study on the timing of aneurysm surgery. Part 1: overall management results. *J Neurosurg.* 1990;73(1): 18–36.
4. Fox JL. *Intracranial Aneurysms.* New York: Springer-Verlag; 1983.
5. Wiebers DO, Whisnant JP, Huston III J, et al. Unruptured intracranial aneurysms: natural history, clinical outcome, and risks of surgical and endovascular treatment. *Lancet.* 2003;362(9378):103–110.
6. Schievink WI, Wijdicks EF, Parisi JE, Piepgras DG, Whisnant JP. Sudden death from aneurysmal subarachnoid hemorrhage. *Neurology.* 1995;45(5):871–874.
7. Fogelholm R, Hernesniemi J, Vapalahti M. Impact of early surgery on outcome after aneurysmal subarachnoid hemorrhage. A population-based study. *Stroke.* 1993;24(11):1649–1654.
8. Inagawa T, Tokuda Y, Ohbayashi N, Takaya M, Moritake K. Study of aneurysmal subarachnoid hemorrhage in Izumo City, Japan. *Stroke.* 1995;26(5):761–766.
9. Tidswell P, Dias PS, Sagar HJ, Mayes AR, Battersby RD. Cognitive outcome after aneurysm rupture: relationship to aneurysm site and perioperative complications. *Neurology.* 1995;45(5):875–882.

10. Johnston SC, Selvin S, Gress DR. The burden, trends, and demographics of mortality from subarachnoid hemorrhage. *Neurology.* 1998;50(5):1413–1418.
11. Huston III J, Nichols DA, Luetmer PH, et al. Blinded prospective evaluation of sensitivity of MR angiography to known intracranial aneurysms: importance of aneurysm size. *Am J Neuroradiol.* 1994;15(9):1607–1614.
12. Pertuiset B, Haisa T, Bordi L, Abou Ouf S, Eissa M. Detection of a ruptured aneurysmal sac by MRI in a case of negative angiogram. Successful clipping of an anterior communicating artery aneurysm. Case report. *Acta Neurochir (Wien).* 1989;100(1–2):84–86.
13. Curnes JT, Shogry ME, Clark DC, Elsner HJ. MR angiographic demonstration of an intracranial aneurysm not seen on conventional angiography. *Am J Neuroradiol.* 1993;14(4):971–973.
14. Schwartz RB, Tice HM, Hooten SM, Hsu L, Stieg PE. Evaluation of cerebral aneurysms with helical CT: correlation with conventional angiography and MR angiography. *Radiology.* 1994;192(3):717–722.
15. Hope JK, Wilson JL, Thomson FJ. Three-dimensional CT angiography in the detection and characterization of intracranial berry aneurysms. *Am J Neuroradiol.* 1996;17(3):439–445.
16. Dammert S, Krings T, Moller-Hartmann W, et al. Detection of intracranial aneurysms with multislice CT: comparison with conventional angiography. *Neuroradiology.* 2004;46(6):427–434.
17. Chappell ET, Moure FC, Good MC. Comparison of computed tomographic angiography with digital subtraction angiography in the diagnosis of cerebral aneurysms: a meta-analysis. *Neurosurgery.* 2003;52(3):624–631.
18. White PM, Wardlaw JM, Easton V. Can noninvasive imaging accurately depict intracranial aneurysms? a systematic review. *Radiology.* 2000;217(2):361–370.
19. Kouskouras C, Charitanti A, Giavroglou C, et al. Intracranial aneurysms: evaluation using CTA and MRA. Correlation with DSA and intraoperative findings. *Neuroradiology.* 2004;46(10):842–850.
20. Okahara M, Kiyosue H, Yamashita M, et al. Diagnostic accuracy of magnetic resonance angiography for cerebral aneurysms in correlation with 3D-digital subtraction angiographic images: a study of 133 aneurysms. *Stroke.* 2002;33(7):1803–1808.
21. Alexander TD, Macdonald RL, Weir B, Kowalczuk A. Intraoperative angiography in cerebral aneurysm surgery: a prospective study of 100 craniotomies. *Neurosurgery.* 1996;39(1):10–17.
22. Drake CG, Allcock JM. Postoperative angiography and the "slipped" clip. *J Neurosurg.* 1973;39(6): 683–689.
23. Macdonald RL, Wallace MC, Kestle JR. Role of angiography following aneurysm surgery. *J Neurosurg.* 1993;79(6):826–832.
24. Feuerberg I, Lindquist C, Lindqvist M, Steiner L. Natural history of postoperative aneurysm rests. *J Neurosurg.* 1987;66(1):30–34.
25. Rauzzino MJ, Quinn CM, Fisher III WS. Angiography after aneurysm surgery: indications for "selective" angiography. *Surg Neurol.* 1998;49(1):32–40.
26. David CA, Vishteh AG, Spetzler RF, Lemole M, Lawton MT, Partovi S. Late angiographic follow-up review of surgically treated aneurysms. *J Neurosurg.* 1999;91(3):396–401.
27. Hernesniemi J, Vapalahti M, Niskanen M, et al. One-year outcome in early aneurysm surgery: a 14 years experience. *Acta Neurochir (Wien).* 1993;122(1–2):1–10.
28. Ohman J, Heiskanen O. Timing of operation for ruptured supratentorial aneurysms: a prospective randomized study. *J Neurosurg.* 1989;70(1):55–60.
29. Batjer HH, Frankfurt AI, Purdy PD, Smith SS, Samson DS. Use of etomidate, temporary arterial occlusion, and intraoperative angiography in surgical treatment of large and giant cerebral aneurysms. *J Neurosurg.* 1988;68(2):234–240.
30. Barrow DL, Boyer KL, Joseph GJ. Intraoperative angiography in the management of neurovascular disorders. *Neurosurgery.* 1992;30(2):153–159.
31. Chiang VL, Gailloud P, Murphy KJ, Rigamonti D, Tamargo RJ. Routine intraoperative angiography during aneurysm surgery. *J Neurosurg.* 2002;96(6):988–992.
32. Derdeyn CP, Moran CJ, Cross DT, Grubb Jr. RL, Dacey Jr. RG. Intraoperative digital subtraction angiography: a review of 112 consecutive examinations. *Am J Neuroradiol.* 1995;16(2):307–318.
33. Klopfenstein JD, Spetzler RF, Kim LJ, et al. Comparison of routine and selective use of intraoperative angiography during aneurysm surgery: a prospective assessment. *J Neurosurg.* 2004;100(2):230–235.

34. Martin NA, Bentson J, Vinuela F, et al. Intraoperative digital subtraction angiography and the surgical treatment of intracranial aneurysms and vascular malformations. *J Neurosurg*. 1990;73(4): 526–533.
35. Origitano TC, Schwartz K, Anderson D, Azar-Kia B, Reichman OH. Optimal clip application and intraoperative angiography for intracranial aneurysms. *Surg Neurol*. 1999;51(2):117–124.
36. Payner TD, Horner TG, Leipzig TJ, Scott JA, Gilmor RL, DeNardo AJ. Role of intraoperative angiography in the surgical treatment of cerebral aneurysms. *J Neurosurg*. 1998;88(3):441–448.
37. Tang G, Cawley CM, Dion JE, Barrow DL. Intraoperative angiography during aneurysm surgery: a prospective evaluation of efficacy. *J Neurosurg*. 2002;96(6):993–999.
38. Yanaka K, Asakawa H, Noguchi S, et al. Intraoperative angiography evaluation of the microsurgical clipping of unruptured cerebral aneurysms. *Neurol Med Chir (Tokyo)*. 2002;42(5):193–200.
39. Kivisaari RP, Porras M, Ohman J, Siironen J, Ishii K, Hernesniemi J. Routine cerebral angiography after surgery for saccular aneurysms: is it worth it? *Neurosurgery*. 2004;55(5):1015–1024.
40. Castillo M, Ramalho J. *Vascular Imaging of the Central Nervous System: Physical Principles, Clinical Applications, and Emerging Techniques*. Hoboken, NJ: John Wiley & Sons Inc; 2013.
41. Menken M, United States Congress Office of Technology Assessment. For sale by the Supt of Docs., U.S. G.P.O. *The Cost Effectiveness of Digital Subtraction Angiography in the Diagnosis of Cerebrovascular Disease*. Washington, DC: Congress of the U.S. 1985.
42. Dashti R, Laakso A, Niemela M, Porras M, Hernesniemi J. Microscope integrated indocyanine green video-angiography in cerebrovascular surgery. *Acta Neurochir Suppl*. 2011;109:247–250.
43. Dashti R, Laakso A, Niemela M, et al. Application of microscope integrated indocyanine green video-angiography during microneurosurgical treatment of intracranial aneurysms: a review. *Acta Neurochir Suppl*. 2010;107:107–109.
44. Dashti R, Laakso A, Niemela M, Porras M, Hernesniemi J. Microscope-integrated near-infrared indocyanine green videoangiography during surgery of intracranial aneurysms: the Helsinki experience. *Surg Neurol*. 2009;71(5):543–550.
45. Raabe A, Nakaji P, Beck J, et al. Prospective evaluation of surgical microscope-integrated intraoperative near-infrared indocyanine green videoangiography during aneurysm surgery. *J Neurosurg*. 2005;103(6):982–989.
46. Raabe A, Beck J, Gerlach R, Zimmermann M, Seifert V. Near-infrared indocyanine green video angiography: a new method for intraoperative assessment of vascular flow. *Neurosurgery*. 2003;52(1): 132–139.
47. Alander JT, Kaartinen I, Laakso A, et al. A review of indocyanine green fluorescent imaging in surgery. *Int J Biomed Imaging*. 2012;2012:940585.
48. de Oliveira JG, Beck J, Seifert V, Teixeira MJ, Raabe A. Assessment of flow in perforating arteries during intracranial aneurysm surgery using intraoperative near-infrared indocyanine green videoangiography. *Neurosurgery*. 2007;61(3 Suppl):63–72.
49. de Oliveira JG, Beck J, Seifert V, Teixeira MJ, Raabe A. Assessment of flow in perforating arteries during intracranial aneurysm surgery using intraoperative near-infrared indocyanine green videoangiography. *Neurosurgery*. 2008;62(6 Suppl 3):1300–1310.
50. Bailes JE, Tantuwaya LS, Fukushima T, Schurman GW, Davis D. Intraoperative microvascular Doppler sonography in aneurysm surgery. *Neurosurgery*. 1997;40(5):965–970:discussion 970–962.
51. Charbel FT, Hoffman WE, Misra M, Hannigan K, Ausman JI. Role of a perivascular ultrasonic micro-flow probe in aneurysm surgery. *Neurol Med Chir*. 1998;38(Suppl):35–38.
52. Firsching R, Synowitz HJ, Hanebeck J. Practicability of intraoperative microvascular Doppler sonography in aneurysm surgery. *Minim Invasive Neurosurg*. 2000;43(3):144–148.
53. Marchese E, Albanese A, Denaro L, Vignati A, Fernandez E, Maira G. Intraoperative microvascular Doppler in intracranial aneurysm surgery. *Surg Neurol*. 2005;63(4):336–342.
54. Stendel R, Pietila T, Al Hassan AA, Schilling A, Brock M. Intraoperative microvascular Doppler ultrasonography in cerebral aneurysm surgery. *J Neurol Neurosurg Psychiatr*. 2000;68(1):29–35.
55. Roessler K, Krawagna M, Dorfler A, Buchfelder M, Ganslandt O. Essentials in intraoperative indocyanine green videoangiography assessment for intracranial aneurysm surgery: conclusions from 295 consecutively clipped aneurysms and review of the literature. *Neurosurg Focus*. 2014;36(2):E7.

56. Gruber A, Dorfer C, Standhardt H, Bavinzski G, Knosp E. Prospective comparison of intraoperative vascular monitoring technologies during cerebral aneurysm surgery. *Neurosurgery.* 2011;68(3):657–673.
57. Li J, Lan Z, He M, You C. Assessment of microscope-integrated indocyanine green angiography during intracranial aneurysm surgery: a retrospective study of 120 patients. *Neurol India.* 2009;57(4):453–459.
58. Washington CW, Zipfel GJ, Chicoine MR, et al. Comparing indocyanine green videoangiography to the gold standard of intraoperative digital subtraction angiography used in aneurysm surgery. *J Neurosurg.* 2013;118(2):420–427.
59. Ducruet AF, Hickman ZL, Zacharia BE, et al. Intracranial infectious aneurysms: a comprehensive review. *Neurosurg Rev.* 2010;33(1):37–46.
60. Nakahara I, Taha MM, Higashi T, et al. Different modalities of treatment of intracranial mycotic aneurysms: report of 4 cases. *Surg Neurol.* 2006;66(4):405–409.
61. Dhomne S, Rao C, Shrivastava M, Sidhartha W, Limaye U. Endovascular management of ruptured cerebral mycotic aneurysms. *Br J Neurosurg.* 2008;22(1):46–52.
62. Peters PJ, Harrison T, Lennox JL. A dangerous dilemma: management of infectious intracranial aneurysms complicating endocarditis. *Lancet Infect Dis.* 2006;6(11):742–748.
63. Carvalho FG, Godoy BL, Reis M, Gasparetto EL, Wajnberg E, de Souza JM. Frameless stereotactic navigation for intraoperative localization of infectious intracranial aneurysm. *Arq Neuropsiquiatr.* 2009;67(3B):911–913.
64. Dashti R, Hernesniemi J, Niemela M, et al. Microneurosurgical management of distal middle cerebral artery aneurysms. *Surg Neurol.* 2007;67(6):553–563.
65. Carlson JD, Liu JK, Dogan A, Sincoff E, Anderson GJ, Delashaw Jr. JB. Use of frameless stereotactic computed tomography venography for intraoperative localization of dural arterial venous fistulas: case report. *Surg Neurol.* 2008;70(5):521–525.
66. Golfinos JG, Fitzpatrick BC, Smith LR, Spetzler RF. Clinical use of a frameless stereotactic arm: results of 325 cases. *J Neurosurg.* 1995;83(2):197–205.
67. Origitano TC, Anderson DE. CT angiographic-guided frameless stereotactic-assisted clipping of a distal posterior inferior cerebellar artery aneurysm: technical case report. *Surg Neurol.* 1996;46(5):450–453:discussion 453–454.
68. ApSimon HT, Reef H, Phadke RV, Popovic EA. A population-based study of brain arteriovenous malformation: long-term treatment outcomes. *Stroke.* 2002;33(12):2794–2800.
69. Brown Jr. RD, Wiebers DO, Torner JC, O'Fallon WM. Incidence and prevalence of intracranial vascular malformations in Olmsted County, Minnesota, 1965 to 1992. *Neurology.* 1996;46(4):949–952.
70. Al-Shahi R, Fang JS, Lewis SC, Warlow CP. Prevalence of adults with brain arteriovenous malformations: a community based study in Scotland using capture-recapture analysis. *J Neurol Neurosurg Psychiatr.* 2002;73(5):547–551.
71. Brown Jr. RD, Wiebers DO, Torner JC, O'Fallon WM. Frequency of intracranial hemorrhage as a presenting symptom and subtype analysis: a population-based study of intracranial vascular malformations in Olmsted Country, Minnesota. *J Neurosurg.* 1996;85(1):29–32.
72. Friedlander RM. Clinical practice. Arteriovenous malformations of the brain. *N Engl J Med.* 2007;356(26):2704–2712.
73. Ondra SL, Troupp H, George ED, Schwab K. The natural history of symptomatic arteriovenous malformations of the brain: a 24-year follow-up assessment. *J Neurosurg.* 1990;73(3):387–391.
74. Stapf C, Mast H, Sciacca RR, et al. Predictors of hemorrhage in patients with untreated brain arteriovenous malformation. *Neurology.* 2006;66(9):1350–1355.
75. Gross BA, Du R. Natural history of cerebral arteriovenous malformations: a meta-analysis. *J Neurosurg.* 2013;118(2):437–443.
76. Gross BA, Du R. Rate of re-bleeding of arteriovenous malformations in the first year after rupture. *J Clin Neurosci.* 2012;19(8):1087–1088.
77. Gross BA, Du R. Diagnosis and treatment of vascular malformations of the brain. *Curr Treat Options Neurol.* 2014;16(1):279.

78. Heros RC, Korosue K, Diebold PM. Surgical excision of cerebral arteriovenous malformations: late results. *Neurosurgery.* 1990;26(4):570–577:discussion 577–578.
79. Spetzler RF, Martin NA. A proposed grading system for arteriovenous malformations. *J Neurosurg.* 1986;65(4):476–483.
80. Pikus HJ, Beach ML, Harbaugh RE. Microsurgical treatment of arteriovenous malformations: analysis and comparison with stereotactic radiosurgery. *J Neurosurg.* 1998;88(4):641–646.
81. Gross BA, Du R. Surgical and radiosurgical results of the treatment of cerebral arteriovenous malformations. *J Clin Neurosci.* 2012;19(7):1001–1004.
82. Clatterbuck RE, Hsu FP, Spetzler RF. Supratentorial arteriovenous malformations. *Neurosurgery.* 2005;57(1 Suppl):164–167.
83. Hashimoto N. Microsurgery for cerebral arteriovenous malformations: a dissection technique and its theoretical implications. *Neurosurgery.* 2001;48(6):1278–1281.
84. Coenen VA, Dammert S, Reinges MH, Mull M, Gilsbach JM, Rohde V. Image-guided microneurosurgical management of small cerebral arteriovenous malformations: the value of navigated computed tomographic angiography. *Neuroradiology.* 2005;47(1):66–72.
85. Russell SM, Woo HH, Joseffer SS, Jafar JJ. Role of frameless stereotaxy in the surgical treatment of cerebral arteriovenous malformations: technique and outcomes in a controlled study of 44 consecutive patients. *Neurosurgery.* 2002;51(5):1108–1116:discussion 1116–1108.
86. Munshi I, Macdonald RL, Weir BK. Intraoperative angiography of brain arteriovenous malformations. *Neurosurgery.* 1999;45(3):491–497.
87. Hanggi D, Etminan N, Steiger HJ. The impact of microscope-integrated intraoperative near-infrared indocyanine green videoangiography on surgery of arteriovenous malformations and dural arteriovenous fistulae. *Neurosurgery.* 2010;67(4):1094–1103:discussion 1103–1094.
88. Killory BD, Nakaji P, Gonzales LF, Ponce FA, Wait SD, Spetzler RF. Prospective evaluation of surgical microscope-integrated intraoperative near-infrared indocyanine green angiography during cerebral arteriovenous malformation surgery. *Neurosurgery.* 2009;65(3):456–462.
89. Ng YP, King NK, Wan KR, Wang E, Ng I. Uses and limitations of indocyanine green videoangiography for flow analysis in arteriovenous malformation surgery. *J Clin Neurosci.* 2013;20(2):224–232.
90. Zaidi HA, Abla AA, Nakaji P, Chowdhry SA, Albuquerque FC, Spetzler RF. Indocyanine green angiography in the surgical management of cerebral arteriovenous malformations: lessons learned in 130 consecutive cases. *Neurosurgery.* 2014;10(Suppl 2):246–251.
91. Nagata S, Morioka T, Natori Y, Matsukado K, Sasaki T, Yamada T. Factors that affect the surgical outcomes of spinal dural arteriovenous fistulas. *Surg Neurol.* 2006;65(6):563–568.
92. Colby GP, Coon AL, Sciubba DM, Bydon A, Gailloud P, Tamargo RJ. Intraoperative indocyanine green angiography for obliteration of a spinal dural arteriovenous fistula. *J Neurosurg Spine.* 2009;11(6):705–709.
93. Hanel RA, Nakaji P, Spetzler RF. Use of microscope-integrated near-infrared indocyanine green videoangiography in the surgical treatment of spinal dural arteriovenous fistulae. *Neurosurgery.* 2010;66(5):978–984.
94. Kato N, Tanaka T, Suzuki Y, et al. Multistage indocyanine green videoangiography for the convexity dural arteriovenous fistula with angiographically occult pial fistula. *J Stroke Cerebrovasc Dis.* 2012;21(8):918:e911–e915.
95. Oh JK, Shin HC, Kim TY, et al. Intraoperative indocyanine green video-angiography: spinal dural arteriovenous fistula. *Spine.* 2011;36(24):E1578–E1580.
96. Schuette AJ, Cawley CM, Barrow DL. Indocyanine green videoangiography in the management of dural arteriovenous fistulae. *Neurosurgery.* 2010;67(3):658–662.
97. Wang G, Ma G, Ma J, et al. Surgical treatment of spinal vascular malformations performed using intraoperative indocyanine green videoangiography. *J Clin Neurosci.* 2013;20(6):831–836.
98. Smith ER, Scott RM. Cavernous malformations. *Neurosurg Clin N Am.* 2010;21(3):483–490.
99. Zabramski JM, Wascher TM, Spetzler RF, et al. The natural history of familial cavernous malformations: results of an ongoing study. *J Neurosurg.* 1994;80(3):422–432.

100. Gault J, Sarin H, Awadallah NA, Shenkar R, Awad IA. Pathobiology of human cerebrovascular malformations: basic mechanisms and clinical relevance. *Neurosurgery.* 2004;55(1):1–16.
101. Baumann SB, Noll DC, Kondziolka DS, et al. Comparison of functional magnetic resonance imaging with positron emission tomography and magnetoencephalography to identify the motor cortex in a patient with an arteriovenous malformation. *J Image Guid Surg.* 1995;1(4):191–197.
102. Labauge P, Laberge S, Brunereau L, Levy C, Tournier-Lasserve E. Hereditary cerebral cavernous angiomas: clinical and genetic features in 57 French families. Societe Francaise de Neurochirurgie. *Lancet.* 1998;352(9144):1892–1897.
103. Kesava PP, Turski PA. MR angiography of vascular malformations. *Neuroimaging Clin N Am.* 1998; 8(2):349–370.
104. Scott RM, Barnes P, Kupsky W, Adelman LS. Cavernous angiomas of the central nervous system in children. *J Neurosurg.* 1992;76(1):38–46.
105. Amin-Hanjani S, Ogilvy CS, Ojemann RG, Crowell RM. Risks of surgical management for cavernous malformations of the nervous system. *Neurosurgery.* 1998;42(6):1220–1227.
106. Sun GC, Chen XL, Zhao Y, et al. Intraoperative MRI with integrated functional neuronavigation-guided resection of supratentorial cavernous malformations in eloquent brain areas. *J Clin Neurosci.* 2011;18(10):1350–1354.
107. Zhao J, Wang Y, Kang S, et al. The benefit of neuronavigation for the treatment of patients with intracerebral cavernous malformations. *Neurosurg Rev.* 2007;30(4):313–318.
108. Leal PR, Houtteville JP, Etard O, Emery E. Surgical strategy for insular cavernomas. *Acta Neurochir.* 2010;152(10):1653–1659.
109. Grunert P, Charalampaki K, Kassem M, Boecher-Schwarz H, Filippi R, Grunert Jr. P. Frame-based and frameless stereotaxy in the localization of cavernous angiomas. *Neurosurg Rev.* 2003;26(1):53–61.
110. Tirakotai W, Sure U, Benes L, Krischek B, Bien S, Bertalanffy H. Image-guided transsylvian, transinsular approach for insular cavernous angiomas. *Neurosurgery.* 2003;53(6):1299–1304.
111. Scott RM, Smith ER. Moyamoya disease and moyamoya syndrome. *N Engl J Med.* 2009;360(12): 1226–1237.
112. Suzuki J, Takaku A. Cerebrovascular "moyamoya" disease. Disease showing abnormal net-like vessels in base of brain. *Arch Neurol.* 1969;20(3):288–299.
113. Imaizumi T, Hayashi K, Saito K, Osawa M, Fukuyama Y. Long-term outcomes of pediatric moyamoya disease monitored to adulthood. *Pediatr Neurol.* 1998;18(4):321–325.
114. Kuroda S, Ishikawa T, Houkin K, Nanba R, Hokari M, Iwasaki Y. Incidence and clinical features of disease progression in adult moyamoya disease. *Stroke.* 2005;36(10):2148–2153.
115. Choi JU, Kim DS, Kim EY, Lee KC. Natural history of moyamoya disease: comparison of activity of daily living in surgery and non surgery groups. *Clin Neurol Neurosurg.* 1997;99(Suppl 2):S11–S18.
116. Kurokawa T, Chen YJ, Tomita S, Kishikawa T, Kitamura K. Cerebrovascular occlusive disease with and without the moyamoya vascular network in children. *Neuropediatrics.* 1985;16(1):29–32.
117. Ezura M, Takahashi A, Yoshimoto T. Successful treatment of an arteriovenous malformation by chemical embolization with estrogen followed by conventional radiotherapy. *Neurosurgery.* 1992;31 (6):1105–1107.
118. Yamada I, Suzuki S, Matsushima Y. Moyamoya disease: comparison of assessment with MR angiography and MR imaging versus conventional angiography. *Radiology.* 1995;196(1):211–218.
119. Katz DA, Marks MP, Napel SA, Bracci PM, Roberts SL. Circle of Willis: evaluation with spiral CT angiography, MR angiography, and conventional angiography. *Radiology.* 1995;195(2):445–449.
120. Takanashi JI, Sugita K, Niimi H. Evaluation of magnetic resonance angiography with selective maximum intensity projection in patients with childhood moyamoya disease. *Eur J Paediatr Neurol.* 1998;2(2):83–89.
121. Fujiwara H, Momoshima S, Kuribayashi S. Leptomeningeal high signal intensity (ivy sign) on fluid-attenuated inversion-recovery (FLAIR) MR images in moyamoya disease. *Eur J Radiol.* 2005; 55(2):224–230.
122. Yamada I, Matsushima Y, Suzuki S. Moyamoya disease: diagnosis with three-dimensional time-of-flight MR angiography. *Radiology.* 1992;184(3):773–778.

123. Smith ER, Scott RM. Surgical management of moyamoya syndrome. *Skull Base*. 2005;15(1):15–26.
124. Isono M, Ishii K, Kobayashi H, Kaga A, Kamida T, Fujiki M. Effects of indirect bypass surgery for occlusive cerebrovascular diseases in adults. *J Clin Neurosci*. 2002;9(6):644–647.
125. Veeravagu A, Guzman R, Patil CG, Hou LC, Lee M, Steinberg GK. Moyamoya disease in pediatric patients: outcomes of neurosurgical interventions. *Neurosurg Focus*. 2008;24(2):E16.
126. Czabanka M, Pena-Tapia P, Scharf J, et al. Characterization of direct and indirect cerebral revascularization for the treatment of European patients with moyamoya disease. *Cerebrovasc Dis*. 2011; 32(4):361–369.
127. Scott RM, Smith JL, Robertson RL, Madsen JR, Soriano SG, Rockoff MA. Long-term outcome in children with moyamoya syndrome after cranial revascularization by pial synangiosis. *J Neurosurg*. 2004;100(2 Suppl Pediatrics):142–149.
128. Mendelowitsch A, Taussky P, Rem JA, Gratzl O. Clinical outcome of standard extracranial-intracranial bypass surgery in patients with symptomatic atherosclerotic occlusion of the internal carotid artery. *Acta Neurochir*. 2004;146(2):95–101.
129. Schmiedek P, Piepgras A, Leinsinger G, Kirsch CM, Einhupl K. Improvement of cerebrovascular reserve capacity by EC-IC arterial bypass surgery in patients with ICA occlusion and hemodynamic cerebral ischemia. *J Neurosurg*. 1994;81(2):236–244.
130. Sundt Jr. TM, Siekert RG, Piepgras DG, Sharbrough FW, Houser OW. Bypass surgery for vascular disease of the carotid system. *Mayo Clin Proc*. 1976;51(11):677–692.
131. Yanaka K, Fujita K, Noguchi S, et al. Intraoperative angiographic assessment of graft patency during extracranial-intracranial bypass procedures. *Neurol Med Chir*. 2003;43(10):509–512.
132. Woitzik J, Horn P, Vajkoczy P, Schmiedek P. Intraoperative control of extracranial-intracranial bypass patency by near-infrared indocyanine green videoangiography. *J Neurosurg*. 2005;102(4):692–698.
133. Awano T, Sakatani K, Yokose N, et al. Intraoperative EC-IC bypass blood flow assessment with indocyanine green angiography in moyamoya and non-moyamoya ischemic stroke. *World Neurosurg*. 2010; 73(6):668–674.
134. Pena-Tapia PG, Kemmling A, Czabanka M, Vajkoczy P, Schmiedek P. Identification of the optimal cortical target point for extracranial-intracranial bypass surgery in patients with hemodynamic cerebrovascular insufficiency. *J Neurosurg*. 2008;108(4):655–661.

CHAPTER 13

Functional Neurosurgery: Deep Brain Stimulation and Gene Therapy

R. Mark Richardson
Epilepsy and Movement Disorders Surgery, University of Pittsburgh, Pittsburgh, PA, USA

INTRODUCTION

Frame-based stereotaxis has been a well-established technique for a variety of neurosurgical procedures since the development of stereotactic frames in the first half of the last century.[1,2] Although initially brain atlas-based, deep brain stimulation (DBS) using a stereotactic frame has been image guided since its inception and translation from ablative procedures, beginning first with pneumoencephalograms during DBS procedures for pain in the 1960s and continuing with the advent of magnetic resonance imaging (MRI) and modern DBS in the late 1980s. Image-guided surgical navigation systems now allow DBS procedures to be performed using frameless stereotaxis with skin or bone fiducials and optical or electromagnetic instrument tracking technology.[3–5] Regardless of the platform, targeting for DBS electrode implantation is based on the interpretation of preoperative MRI, with most centers employing intraoperative electrophysiological confirmation via microelectrode recordings (MERs) to verify actual lead location. Recent advances in real-time MRI guidance, however, are changing this paradigm. This chapter covers advances in image guidance for functional neurosurgery in three important areas: (1) optimization of preoperative techniques for visualizing the target structure and related tractography to optimize target selection in DBS, (2) development of interventional MRI (iMRI) for real-time visualization of DBS electrode placement in MRI-visible anatomic targets, and (3) development of iMRI gene therapy delivery strategies for the treatment of Parkinson's disease (PD).

DEEP BRAIN STIMULATION

In the United States, DBS received full U.S. Food and Drug Administration (FDA) approval for the treatment of essential tremor (ET) in 1997 and for PD in 2002. Subsequently, under humanitarian device exemptions (HDEs), DBS was approved for dystonia in 2003 and for obsessive–compulsive disorder (OCD) in 2009. The European Union additionally approved DBS for the treatment of epilepsy in 2010. Although targets for movement disorders and epilepsy have thus far been gray matter structures, emerging targets for psychiatric indications such as OCD and depression are white matter fiber

A. Golby (Ed): Image-Guided Neurosurgery
DOI: http://dx.doi.org/10.1016/B978-0-12-800870-6.00013-3

bundles. Methods for localizing the FDA-approved targets on MRI will be described with further discussion on the use of tractography both to refine traditional gray matter targeting as well as to define the optimal target site for neuropsychiatric indications.

GENERAL CONSIDERATIONS IN FRAME-BASED DBS

In stereotactic surgery, the surgeon must perform a registration process to make MRI space match the real-time physical space as closely as possible. This is accomplished most commonly using rigid frame fixation and any of several neuro-navigation software packages now available. The preoperative MRI may be obtained on the day of surgery with an MRI localizer attached to the stereotactic frame. Alternately, a volumetric CT scan may be obtained in the frame using a CT localizer and merged with a recent MRI. The author prefers the latter method, because it reduces the scan time for the patient on the day of surgery, as well as the time in the stereotactic frame. Because many patients with movement disorders require conscious sedation, or even general anesthesia, in order to obtain the highest quality MRI, this scan may be obtained on an outpatient basis prior to surgery, avoiding the potential prolonged effects of anesthetics on motor symptoms or cognitive function that can affect participation in awake surgery. Although merging the CT to an MRI introduces a small possible source of error in stereotactic targeting, in practice this is typically not consequential as long as the automatic software merge is carefully double-checked by the surgeon. If MRI is used alone, distortion of the localizer fiducials is a potential source of error, because magnetic fields are heterogeneous, with the greatest potential distortion at the periphery of the scanning field.[6]

Targeting is accomplished through a combination of indirect and direct methods. Indirect targeting relies on the visualization of the anterior commissure (AC) and posterior commissure (PC) and the known relationship of these structures to the surgical target. Direct targeting relies on the visualization of the actual surgical target for stereotactic coordinate generation. In order to best visualize the commissures, an axial T1 volumetric scan is obtained. We use a magnetization-prepared rapid gradient echo (MPRAGE) sequence, a scan type designed to minimize distortion errors,[7] and align the slice plane parallel to the intercommissural line. This sequence is a 3D volumetric gradient-echo image set covering the entire brain in 1.5-mm-thick slices with a zero interslice distance. A contrast enhanced scan is also obtained for use in avoiding vasculature during trajectory planning.

It is efficient to set the target based on standard AC—PC coordinates in the neuronavigation software using the T1 image, in order to plan the trajectories prior to fine-tuning the target location in a subsequent step on the T2 targeting scan. In this way, the trajectory through the target nucleus in oblique, in-plane views can be taken into consideration during the final target selection. The entry point is set as close to the coronal suture as possible given the patient's sulcal anatomy, which provides a favorable sagittal trajectory angle for electrode placement in the subthalamic nucleus (STN), globus pallidus internus (GPi), and ventral intermediate nucleus of the thalamus (Vim). This location also has the benefit of providing an opportunity to double-check the frame setup during the procedure by visually confirming that the entry point on the skull is at the expected location relative to the suture. The coronal suture is not too frontal to produce cosmetic issues, but is well rostral of the primary motor cortex (typically two gyri in front), which helps to minimize morbidity in the rare instance of a subcortical hemorrhage at the entry point. If the frame scan is a CT scan, identifying the coronal suture is simple. In planning the trajectory, the author leaves a minimum distance of the width of the standard guide tube (2.1 mm) between the trajectory and any vessel clearly identified on imaging, in addition to avoiding passing through the ventricle. It is best to avoid transgressing the ventricle during electrode placement, due to increased risk of postoperative confusion,[8] loss of targeting accuracy,[9,10] and hemorrhage.[11]

For visualization of basal ganglia targets, a volumetric axial T2 scan is obtained, which can be acquired as a slab through the target region, in a plane parallel to the AC–PC line. T2-fast spin echo (T2-FSE) and inversion recovery (FSE/IR) techniques have been used most commonly for over a decade.[12] The increased accuracy of targeting resulting from direct visualization on MRI, as opposed to solely indirect or atlas-based targeting, has been reported in multiple studies (see Brunenberg et al.[13] for review). Every MRI scanner is different, however, and optimization of targeting sequences by an experienced neuroradiologist or MR physicist is critical. Choosing the appropriate sequences on a given scanner and regular shimming of the MRI scanner can minimize the degree of distortion, and the use of frequent phantom testing can determine if any errors are present. Images obtained on a 1.5 T scanner typically are adequate for stereotactic planning, but the author prefers 3 T scans for target visualization due to the better resolution that results from the increased signal-to-noise ratio. Although increasing magnetic field strength also increases the potential for image distortion, error due to MR distortion at 3 T has been estimated to be less than 0.3 mm within a 10-mm cubic region surrounding the deep brain targets.[14] Alterman and Tagliati have advocated using only axial FSE/IR due to its resistance to magnetic susceptibility artifact and ease in identifying the AC and PC on inverted images.[15] These authors solely employ the indirect method to target for DBS implantation, finding that even in targeting the STN, which is not well visualized on FSE/IR, the target could be localized with one or two MER trajectories in 80% of cases.[16]

Even greater structural detail of brain targets can be achieved with ultra-high field (7 T) MRI (Figure 13.1). Anatomical images of the human brain acquired at 7 T demonstrate superior image resolution and contrast that significantly improves anatomic delineation of DBS targets, allowing the identification of internal architecture within these targets.[17] For example, 7 T images demonstrate the clear boundary dividing the STN from the substantia nigra, the medial medullary lamina between GPe and GPi, and direct visualization of internal thalamic nuclei, including Vim. Duchin

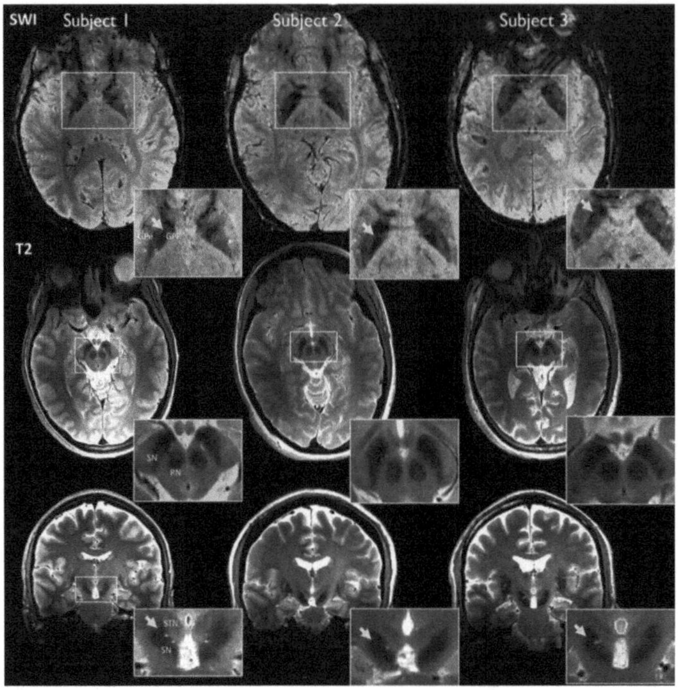

Figure 13.1 Structural imaging at 7 T. (Top row) Axial high-resolution susceptibility-weighted 7 T images (SWIs) in three subjects at the level of the globus pallidus (GP), putamen, and thalamus. The white arrows indicate the border between GPe and GPi, known as the lamina pallidi medialis (uppermost insets). (Middle row) Axial high-resolution T2-weighted images in three subjects at the level of the substantia nigra (SN) and red nucleus (RN). (Bottom row) Coronal T2-weighted images in three subjects at the level of the STN and SN. Coronal images provide good contrast, enabling differentiation between SN and STN along the lateral–medial axis, as indicated by the white arrow in each inset. *Reprinted with permission.*[19]

et al. characterized the amount of geometrical distortion present at 7 T relative to standard clinical imaging obtained on a 1.5 T scanner, in subjects undergoing preoperative evaluations for DBS surgery.[18] 7 T images could be coregistered to CT images as typically occuring with 1.5 T scans and exhibited minimal additional distortions in the midbrain region, compared to 1.5 T images. These distortions could be corrected by regional-based linear registrations, indicating the feasibility of using 7 T MRI for DBS targeting. In another report from this group, diffusion imaging at 7 T was used to produce subject-specific parcellations of the basal ganglia and thalamus into subterritories based on their distinct connectivity patterns, which demonstrated overlap with functional connectivity maps generated by resting-state functional MRI.[19] Although 7 T MRI machines are not widely available, and currently no 7 T MRI scanner is approved for clinical use by the FDA, the advantages of ultra-high field imaging may prove invaluable for refining DBS targets in the future.

PREOPERATIVE IMAGING AND TARGET SELECTION

Subthalamic nucleus

The STN is the most common target used in the treatment of PD. The typical imaging sequence used to target the STN is a T2-weighted FSE. To reduce the time required for imaging, only a slab through the target area is obtained.[20] Multiple groups have described the use of susceptibility-weighted imaging (SWI) and other T2 sequences to delineate the boundaries of the STN, taking advantage of the increased sensitivity of these methods to iron deposition.[13,21,22] At 3 T, T2*-FLASH2D images have been shown to provide superior reliability for visualizing the STN, particularly with respect to the coronal view, although SWI also provides significantly higher contrast-to-noise ratio values for the STN than standard T2-weighted imaging.[23] (Note that FLASH is an acronym for fast low angle shot. Different manufacturers of MRI equipment use different names for this type of sequence: Siemens uses the name FLASH, General Electric uses the name SPGR (spoiled gradient echo), and Philips uses the name CE-FFE-T1 for contrast-enhanced fast field echo.) This improved visualization is particularly obvious for the most posterior portions of the STN. High-resolution three-dimensional T2*-weighted angiography (HR 3-D SWAN) is another sequence that has been shown to have superior resolution to standard T2-FSE.[24] Lefranc et al. used the position of microelectrodes determined by intraoperative CT to show an excellent correspondence of the electrophysiologic intraoperative findings to the anatomic MRI data. To date, whether the use of higher resolution images for target planning results in better clinical outcomes has not been reported. In addition, the use of diffusion-weighted imaging to visualize the relationship of white matter pathways in the basal ganglia to planned electrode locations[25] may well play an increasing role in STN targeting.

Precise targeting of the dorsolateral (sensorimotor) part of the STN is critical for obtaining the best possible effect on motor symptoms, as well as for minimizing undesirable motor and nonmotor side effects. Targeting is initiated on the T2 axial slice 4 mm below the AC–PC line, although it is possible to vary this up to 1 mm either superiorly or inferiorly if better STN visualization is obtained due to differences in an individual patient's anatomy (indirect coordinates are $x = \pm 12$, $y = -3$, and $z = -4$). In this plane, targeting begins with identification of the red nucleus and the medial border of the STN. The target is the center of the STN in the medial–lateral direction, aligned in the anterior–posterior plane with the anterior border of the red nucleus (Figure 13.2), a location that is the presumed center of the sensorimotor territory of the STN.[26] The medial border is typically better visualized than the lateral border, and if the latter is not well visualized, the target can be placed 2 mm from the medial border of the STN.[27] The trajectory can then be adjusted, as permitted by the individual brain anatomy, in the sagittal and coronal planes to optimize the distance

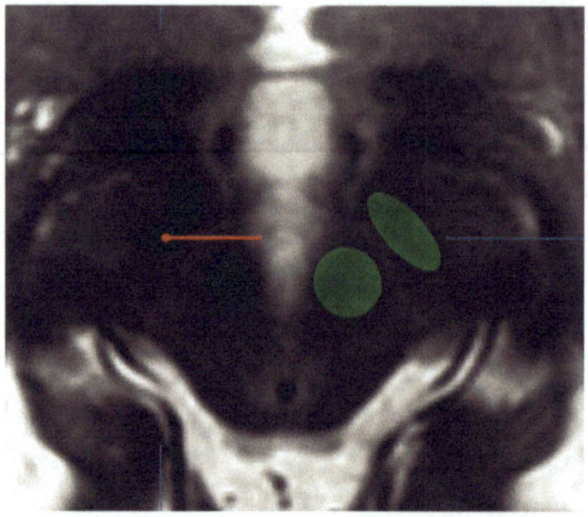

Figure 13.2 STN targeting. The region of the STN is seen on a T2-weighted sequence 4 mm below and parallel to the AC–PC plane. A line is drawn from the midline to the posterior-lateral STN that is aligned with the anterior border of the red nucleus, and the target is placed on this line in the center of the STN.

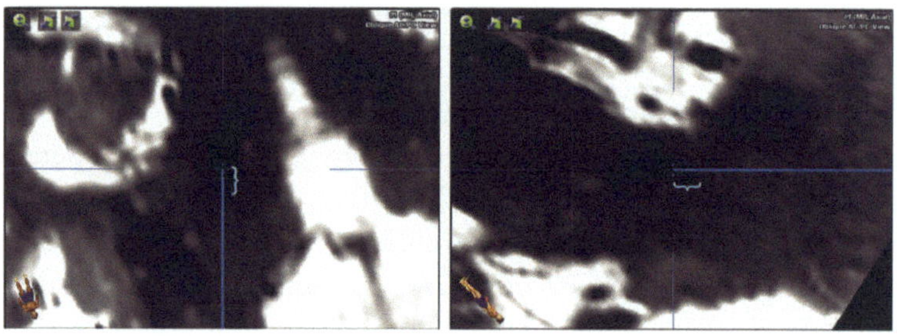

Figure 13.3 In-plane trajectory views. The planning software can be used to predict the depth at which entry into the target nucleus will occur during MER. The brace indicates the distance from STN entry to target in coronal-oblique (left) and sagittal-oblique (right) in-plane views.

traveled through the dorsolateral STN. In-plane trajectory views in the neuronavigation software then can also be used to predict the depth at which the STN will be entered during microelectrode recording (Figure 13.3).

An alternate target in the subthalamic region for improving motor symptoms in PD[28] and ET[29,30] is the caudal zona incerta nucleus (cZI), a horizontally elongated area of gray matter cells separating the lenticular fasciculus from the thalamic fasciculus. This structure, however, is not visualized on standard T2 imaging, and electrode placements therefore rely on placing the electrode posteromedial to the postero-dorsal STN[28], based on the expected location of cZI from postmortem studies. At 3 T, T2*-FLASH2D images however do visualize the zona incerta, although only the rostral portion[31]. Targeting of cZI, therefore, currently requires the knowledge of this region garnered from anatomic atlases.

Globus pallidus

The internal segment of the globus pallidus (GPi) is the most widely used stimulation target for the treatment of dystonia. The GPi is also frequently targeted in the treatment of PD and has been shown to have similar efficacy to STN stimulation.[32] Like the STN, the GPi has a functional topography that dictates the placement of the electrode in the sensorimotor territory or posterior lateral portion of GPi.[33–35] Typical scans used to target GPi include T2-FSE and FSE/IR, and it can be advantageous to obtain both scans and use the one in which the GPi is better visualized. Similar to results reported in imaging the STN, it has been suggested that at 3 T the GPi is best visualized with T2*-FLASH2D images and that SWI offers a significantly higher contrast-to-noise ratio for the GPi compared to standard T2-weighted imaging[36] (Figure 13.4). We find, however, that inverting the IR image often gives quite adequate visualization. In addition, probabilistic tractography has been applied to explore connectivity patterns of the dorsal and ventral GPi, with ventral GPi showing stronger measures of connectivity to primary sensory cortex and posterior motor cortical regions and dorsal GPi showing stronger measures of connectivity to motor and premotor regions.[37] This result suggests that diffusion-weighted imaging increasingly may be used to visualize different functional and anatomic subsystems of basal ganglia nuclei for prospective targeting.

The target is selected in the plane of the AC–PC line (indirect coordinates are $x = 21$, $y = 2$, and $z = 0$). A line is measured along the border of the GPi and the internal capsule and divided into thirds. The target is placed 3–4 mm lateral to the junction of the posterior 1/3 and anterior 2/3 of the pallidocapsular border, which is 1 mm medial to the internal medullary lamina (the white matter boundary between the GPi and the GPe) (Figure 13.5).[27] The probe's eye view in the neuronavigation

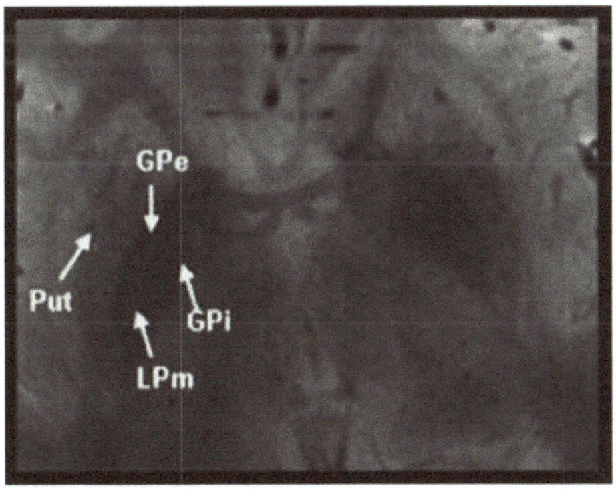

Figure 13.4 Axial views of the pallidal region in T2*-FLASH2D-sequences. The GPi and the surrounding structures are indicated (GPi: internal globus pallidus; GPe: external globus pallidus; Put: putamen; LPm: lamina palladi medialis). *Reprinted with permission.*[36]

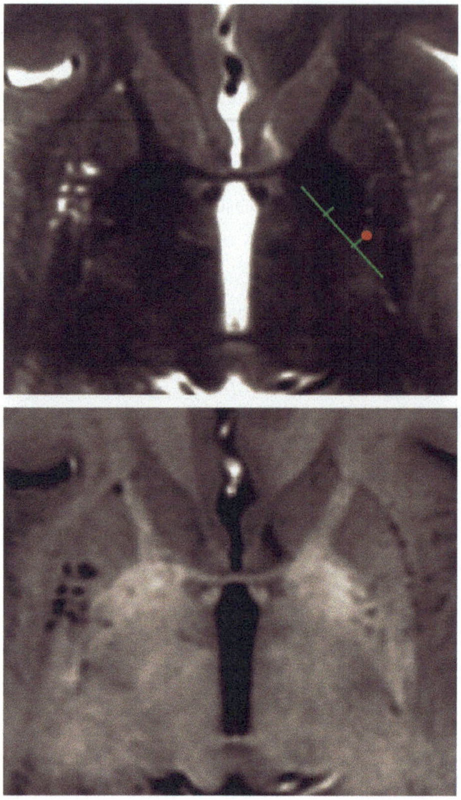

Figure 13.5 GPi targeting. The region of the GPi seen on an inversion recovery sequence at the AC–PC plane. The pallidocapsular border is divided into thirds. The target is one-third the distance forward from the posterior end of the pallidocapsular border and 3–4 mm lateral along a line perpendicular to the pallidocapsular border.

software can then be used to scroll in the planned trajectory beyond the target to the base of the pallidum, to verify that the trajectory is terminating over the lateral aspect of, or just lateral to, the optic tract.

Thalamus

Individual nuclei of the thalamus cannot be delineated visually on standard T1- and T2-weighted MR images obtained at either 1.5 or 3 T. Other groups have reported the use of specialized sequences to visualize the Vim for the treatment of ET. The use of an optimized 3D MPRAGE protocol at 3 T for maximizing the contrast between the gray matter and the white matter in the thalamus has been described.[38] This optimization resulted in reliable identification of the four large thalamic nuclei groups: anterior, lateral, medial, and posterior. Multiple groups have applied probabilistic diffusion tractography imaging (DTI) to the retrospective analysis of effective and ineffective lead locations in the Vim. Tractography initiated from the effective stimulation site has suggested that the most efficacious DBS contact for controlling tremor may colocalize with the thalamic region having the highest probability of connection with the primary motor cortex,[39] or with the premotor cortex.[40] The use of DTI prospectively in DBS targeting of the Vim has not been reported. The use of resting BOLD MRI to discriminate the Vim based on functional connectivity measurements between the motor cortex, the superior cerebellum, and the thalamus has also been described.[41]

Routine targeting of the Vim nucleus in the treatment of ET, therefore, is accomplished indirectly using stereotactic coordinates and a standard volumetric T1 gradient echo scan, followed by direct targeting that takes into account the width of the third ventricle. The author begins with coordinates of $x = \pm 12$ mm, $y = +6$ mm from PC, and $z = 0$, and modifies the target to lie 10 mm from the lateral edge of the third ventricle on the MR slice in the AC–PC plane.[42]

Ventral capsule/ventral striatum

In contrast to the brain targets in movement disorders being gray matter nuclei, there is increasing recognition of the importance of direct stimulation of white matter tracts in the treatment of psychiatric disorders. The approved target for OCD is now termed the ventral capsule/ventral striatum (VC/VS) to reflect the fact that the target includes the anterior limb of the internal capsule in addition to the adjacent ventral striatum.[43] Specific scan sequences vary between centers but, in general, T2-weighted sequences are used to optimize gray-white differentiation. In the worldwide experience with DBS in OCD, the target systematically has moved more posterior to that from the original target for anterior capsulotomy, primarily in response to clinical results. The current target is at the junction of the anterior capsule and the ventral striatum, within 1–2 mm of the posterior border of the AC, with only the most distal contact in the caudal nucleus accumbens[43] (Figure 13.6). Nuttin et al. have further described that targeting occurs on the axial slice in the AC–PC plane, selecting the posterior edge of the anterior commissure at about 6 mm from midline[44] (Figure 13.7).

McIntyre and colleagues have led the field in analyzing white matter tractography related to the efficacy of DBS for OCD and treatment-resistant depression (TRD).[45] By combining clinical data, diffusion tensor tractography, and computer models of patient-specific neurostimulation to identify particular axonal pathways activated by VC/VS DBS, they found that clinical responders exhibited pathways coursing lateral and medial to the VS or dorsal and lateral to the nucleus accumbens that were activated in the model.[46] Similarly, one common pathway was identified across non-responders. More sophisticated, patient-specific tractography activation models were subsequently applied to the study of subcallosal cingulate white matter DBS in

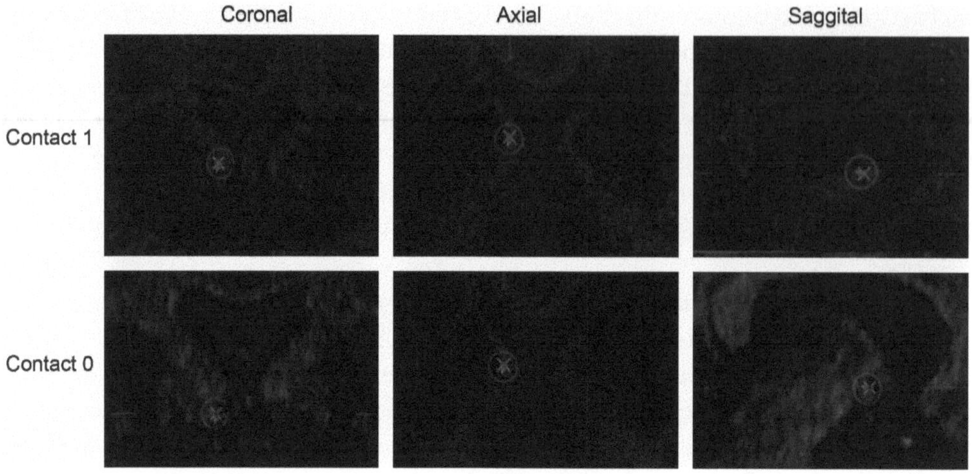

Figure 13.6 VC/VS targeting. Post implant locations (X within circles) of the two lower electrode contacts in an OCD patient. The centers of contacts 0 (bottom) and 1 (top) in the coronal, axial, and sagittal planes are shown on the right side. *Reprinted with permission.*[43]

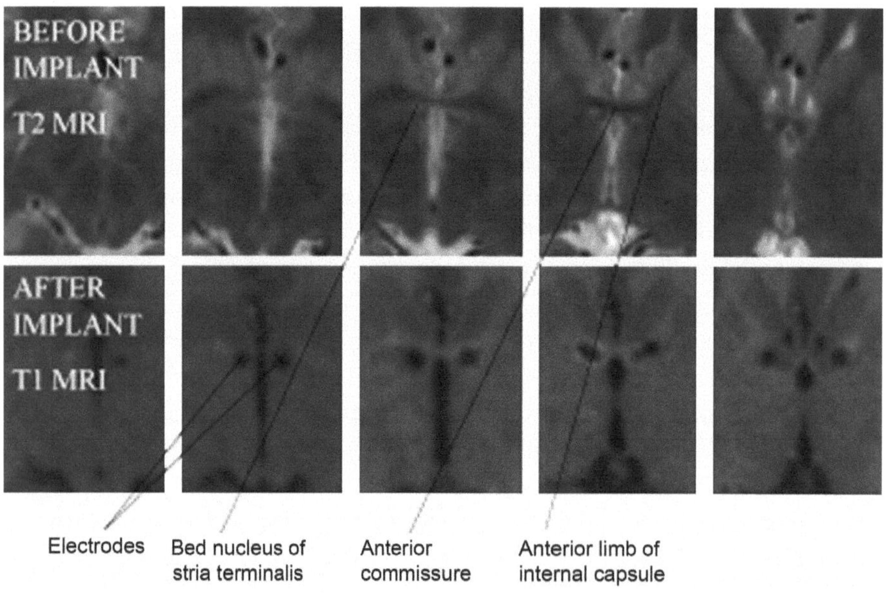

Figure 13.7 VC/VS targeting. Preoperative (top row) T2-weighted MRI, axial adjacent brain slices, parallel to AC–PC, and perpendicular to the midsagittal plane. Postoperative (bottom row) T1-weighted MRI, which demonstrates the position of the lead at the stria terminalis. *Reprinted with permission.*[44]

TRD.[47] Whole-brain tractography combined with models of the volume of tissue activated by clinical stimulation parameters suggested that all DBS responders at 2 years shared bilateral pathways from their activation volumes to (1) the medial frontal cortex via forceps minor and uncinate fasciculus; (2) rostral and dorsal cingulate cortex via the cingulum bundle; and (3) subcortical nuclei, whereas nonresponders did not consistently show these connections.[48] This ongoing work suggests that the activation of a critical mass of a unique combination of cortical, subcortical, and cingulate pathways might be necessary for therapeutic benefit, and that small differences in electrode location could generate substantial differences in the directly activated pathways.

Postoperative imaging

There are three common methods for verifying the location of the implanted electrode: intraoperative CT, postoperative CT, and postoperative MRI. We prefer to obtain postoperative MRI in all patients, as these scans have been shown to be quite safe, despite overrestrictive manufacturer's guidelines related to MRI.[49] Although the artifact is larger than that obtained with CT, and slightly eccentric, detected lead location varies less than a millimeter between MRI and CT imaging.[50] The value of postoperative MRI is that there is no required fusion to the preoperative MRI, so any potential residual brain shift that occurred intraoperatively is taken into account. If desired however, the pre- and postoperative images can be fused to estimate the lead placement error. Thousands of implanted leads, at multiple centers, have been scanned in 1.5 T magnets with no significant complications reported.[51]

Several groups have published their approach to the use of intraoperative CT scanning in DBS implantation.[52–55] This remains a method for checking lead location but does not provide real-time visualization of lead placement in the anatomic target, which cannot be visualized on CT and must be inferred following fusion to preoperative MRI. Nonetheless, this method has high fidelity for placing electrodes in the anatomically selected target, with reported errors on the order of 1.5 mm.[56]

INTERVENTIONAL MRI-GUIDED DBS

Prospective stereotaxy

As discussed in the previous sections, frame-based and frameless stereotaxis both rely on imaging obtained before surgery. Image registration to operative space requires image fusion and setting of frame coordinates (frame-based stereotaxy) or fiducial registration using a neuro-navigation system (frameless stereotaxy). Opening the dura, however, allows a combination of cerebrospinal fluid egress and air entry that may combine to create unpredictable, varying degrees of brain shift, even in deep brain structures.[57,58] As these stereotactic procedures do not allow true intraoperative visualization of the brain and target area during device implantation, they are inherently susceptible to inaccuracies resulting from potential errors in registration, brain shift, and technique that are not detectable intraoperatively.

An interventional MRI (iMRI) approach using a skull-mounted aiming device was developed over the past decade at the University of California, San Francisco.[59] The goal of this approach is to improve the safety and accuracy of DBS placement to allow intraoperative visualization of both the brain target and the DBS electrode. This strategy is based on the concept of prospective stereotaxy, the alignment of a skull-mounted trajectory guide within an MRI system.[60] Initially applied to perform brain biopsies[61,62] and subsequently to place DBS leads,[61,63,64] this approach provides an immediate detection of complications, eliminates the need for microelectrode mapping, and reduces brain penetrations. The key features of this strategy are (1) integration of planning, insertion, and real-time MRI confirmation of DBS lead placement during a single procedure with patients positioned on the MRI gantry; (2) trajectory alignment and DBS lead insertion via a burr hole-mounted trajectory guide rather than traditional stereotactic frame and arc system; (3) definition of target coordinates with respect to the MRI isocenter rather than to a separate stereotactic space using fiducial markers; (4) patient positioning supine under general anesthesia without the performance of MER or stimulation testing; and (5) acquisition of target images after burr hole creation and intracranial air entry.[64]

In a first-generation platform, a commercially available skull-mounted aiming device (Nexframe MR, Medtronic, Minneapolis, Minnesota) was used to implant DBS electrodes into the STN of patients with PD using 1.5 T iMRI.[64] The accuracy was found to be superior to that of frame-based and frameless stereotaxis for DBS with comparable clinical outcomes and a low complication rate. Subsequently, this strategy was expanded with the development of a second-generation platform, the ClearPoint system (MRI Interventions, Inc., Irvine, California), which is based on a specially designed trajectory guide (SMARTFrame) with control software specifically dedicated to this application.[59] This platform is approved for use in 3 T scanners, though at this field strength the potential for image distortion should be carefully evaluated, and the contribution of an MR physicist to the initial establishment of scan parameters on a site's specific 3 T scanner is currently warranted. The 3 T environment, however, may become the optimal imaging environment for iMRI-guided procedures as this technique matures.

ClearPoint system

DBS placement using the ClearPoint system can occur in a truly intraoperative MRI suite, or in a diagnostic MRI scanner. The platform is compatible with all currently available MRI scanners. Most of the following description pertains to performing simultaneous, bilateral lead placement in a diagnostic MRI scanner. The procedure begins with shaving of the head and infiltration of epinephrine-containing local anesthetic into the scalp at the level of the coronal suture, prior to transfer of the patient into the MRI scanner room. Because Bovie electrocautery cannot be used in the MRI environment, advanced infiltration of the scalp facilitates hemostasis on opening. The head is secured in a fixation device, either the one manufactured by MRI Interventions with an integrated imaging coil, or a device that allows positioning of two loop coils on either side of the head. The patient is prepped and draped at the head of the MRI bore, such that if the procedure is performed in a diagnostic scanner, this area is transformed into a miniature sterile operating theater. A stick-on, MRI visible grid is placed on each side of the head, centered near the coronal suture. The patient is then moved to the magnetic isocenter, and a volumetric T1 scan with contrast is obtained. The superior aspect of the forehead is optimally placed at the center of the magnet to position the isocenter at a point that will be approximately equidistant from both the external trajectory guide and the deep brain targets.

The control software (ClearPoint) exists on a standalone workstation in the MRI control room and communicates with the host computer of the MR system via a network link over which digital imaging and communications in medicine (DICOM) data are sent. The general workflow is divided into three different stages in the software: burr-hole planning (entry), target selection and trajectory visualization (target), and alignment of trajectory guide and insertion monitoring (navigate). The software is designed for either unilateral or bilateral procedures, and a second monitor within the magnet room provides the surgeon with visualization of the sequential steps in the workflow. Using ClearPoint, the target is selected based on standard AC–PC coordinates, without taking into consideration direct visualization of target anatomy at this stage. Appropriate trajectories are chosen, and the entry points for these trajectories are then projected onto reconstructions of the marking grids, which are automatically recognized by the control software. The software provides the grid coordinates at which the burr holes should be created for the established trajectories (Figure 13.8, upper panel). In comparison to frame-based surgery, lateral entry points are limited if using an integrated coil/head fixation system, because this current hardware limits the placement of the aiming device, in addition to increasing the risk for bore collision in a smaller bore magnet. Therefore, patients with large ventricles that force the trajectory more laterally may be better candidates for frame-based surgery, if the surgeon does not intend to transgress the ventricle. A customized sharp screwdriver is then passed through the skin at the indicated grid coordinate and used to mark the skull (Figure 13.8, lower panel). Care must be taken that the marking tool does not deviate as a result of inadvertent scalp movement. A coronal incision is then made directly through these marks. Although simultaneous GPi targeting is possible if the chosen trajectories are planned lateral to medial, traditional parasagittal trajectories require unilateral, sequential implantation of GPi due to the required proximity of the burr holes that precludes mounting two trajectory guide bases at the same time.

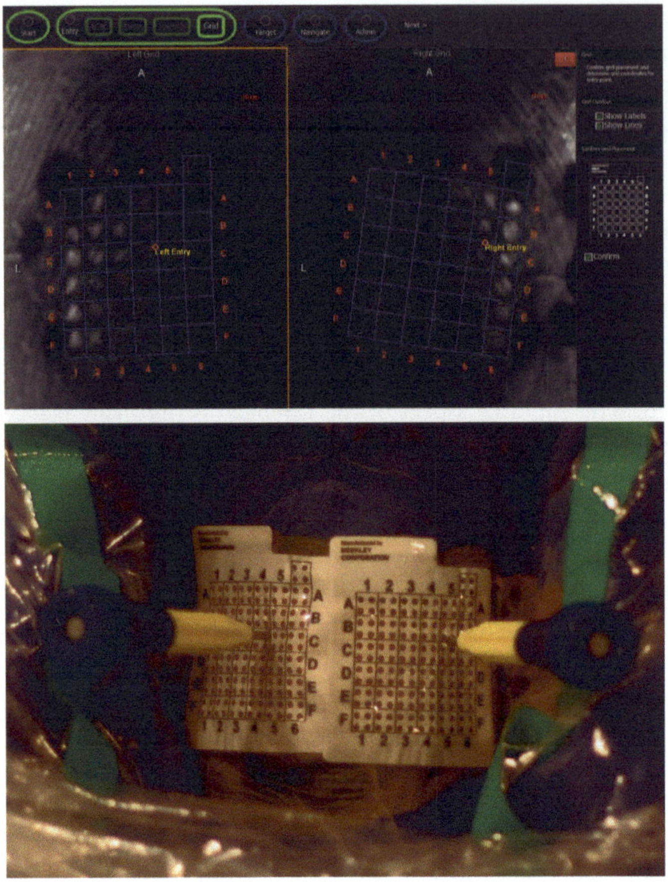

Figure 13.8 Marking the skull entry site using the ClearPoint interventional-MRI platform. (Top panel) The software segments the contrast-filled grid placed on the scalp and projects the entry site location for the chosen trajectory onto the appropriate grid location. (Bottom panel) Marking tools are used to pierce the grids and to penetrate the skull. After opening, the site where the burr hole should be centered after opening is indicated by the skull marking.

The central component of the ClearPoint system is the skull-mounted trajectory guide, or SMARTFrame, which contains an MR-visible targeting cannula. Two degrees of motion (pitch and roll) pivot the frame around the spherical tip, which is a fixed point in space. Thus, once the SMARTFrame is mounted, the entry point is fixed. The pitch and roll adjustments allow alignment of the targeting cannula along a trajectory parallel to the planned trajectory. Two other sets of controls (X−Y stage) permit the user to align the parallel trajectory to the planned trajectory. The X−Y stage, however, offers limited range, with a maximum offset in either orthogonal direction of 2.5 mm. Imaging of three MR-visible fiducial markers built into the base of the frame in a plane that is perpendicular to the default null pitch and roll angulations allows the software to

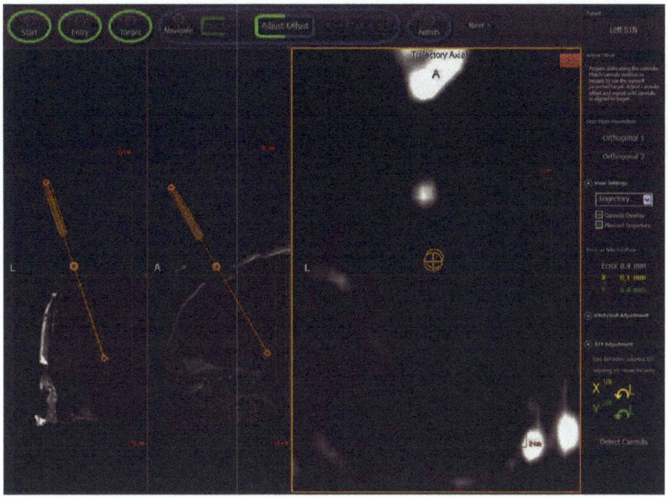

Figure 13.9 Aligning the trajectory and predicting the error in ClearPoint. Screen shot following the final X–Y translational maneuver, showing a predicted error of 0.4 mm. The open circle depicting the trajectory is very slightly anterior to the target, which is a preferred error in this case, given the slight posterior bias in this particular scanner.

detect the angulation of the frame with respect to the patient and target. Subsequently, instructions are provided to the surgeon for adjusting the entry point. Adjustments of the SMARTFrame are made by turning knobs on the trajectory guide via a controller that attaches to the frame and is long enough to permit adjustments from the head of the magnet while the patient remains at isocenter.

Once the burr holes have been made, it is important to decide whether the dura will be opened widely, prior to mounting the SMARTFrames, to allow any brain shift to occur before the anatomic targeting sequences are repeated. The author prefers to use a sharp stylet to puncture the dura in a subsequent step, once trajectories are aligned. However, one may prefer to see the cortical surface during stylet penetration, especially if there is any question about the proximity of cortical vascula ture to the cortical entry point. If the dura is to be opened on both sides, this should be done at this stage, to allow any brain shift to occur, prior to repeating the anatomic targeting scans.

Once the dura has been opened widely or a decision has been made to use the sharp stylet at a later step, and the SMARTFrames have been mounted over the burr holes, the patient is moved into the magnet. A new volumetric T1 image is then obtained, as well as a T2-FSE and/or IR slab, roughly centered on the AC–PC plane. These sequences are the same as would be obtained with routine preoperative MRI for MER-based surgery. The target is then selected based on direct visualization of the anatomy (Figure 13.9). The final trajectory, now dependent on the

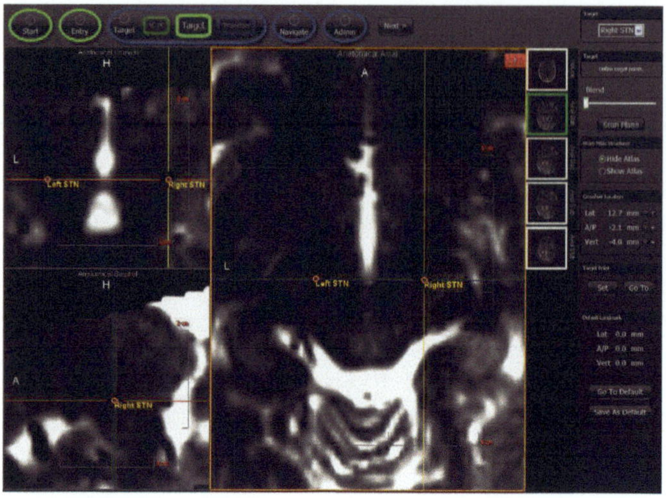

Figure 13.10 STN targeting using ClearPoint software. Target points are selected based on real-time imaging of brain structures.

mounted position of the trajectory guide, is then confirmed to be similar to that originally intended. If the SMARTFrame was not mounted ideally with respect to the intended cortical entry point, the software allows the surgeon to offset the tip of the target cannula using the X–Y stage up to 2 mm. If this maneuver is undertaken, one must anticipate having to rely more on the pitch and roll adjustments in subsequent steps, because little room may remain on the X–Y stage. At this point, the software will give a warning if the target is outside of the angular range of the device or if a bore collision during probe insertion is anticipated.

In the navigation step, the software generates the scan planes required for acquiring imaging data through the base of the trajectory guide, and these are entered manually in the scanner software. This image is sent back to the workstation, which detects the position of the targeting cannula and provides instructions for adjusting the pitch and roll controls to bring the targeting cannula collinear with the desired trajectory. After an adjustment is made, the same scan can be acquired, and this process is repeated until the software predicts that the trajectory guide is aimed within 2 mm of the target. Next, the software provides new parameters for acquiring an oblique coronal and oblique sagittal plane along the specified trajectory. When these scans are sent to the workstation, the software autodetects the actual trajectory of the targeting cannula, predicts a linear fit at target depth, and determines the distances from target in the X–Y planes (Figure 13.10). The surgeon then adjusts the X–Y translation stage to correct for this predicted error, based on instructions from the software, and the process is repeated iteratively until an acceptable predicted error is achieved. With the ClearPoint platform, one should achieve a predicted 2-D vector error of less than 0.5 mm. However, different MRI scanners may have a reproducible bias in terms of the location of the implanted ceramic stylet in relation to the chosen target. For instance, the author typically leaves the trajectory such that there is always a predicted error in the anterior direction, given a propensity for the ceramic stylet to fall posterior to the target when the stylet is implanted with a predicted error close to zero on the Siemens Magnetom scanner.

At this stage, the software provides the length on the electrode at which to set the depth stop so that it terminates at the selected target when inserted. The SMARTFrame comes with a rigid, nonconductive, blunt ceramic stylet that is laminated in plastic (diameter, 1.4 mm) and produces minimal MR artifact. Before inserting the DBS electrode, a depth stop is set on the ceramic stylet, which is secured in a peel-away plastic sheath. The stylet is inserted, and scans are obtained to verify correct targeting and absence of complications, prior to the electrode implantation. If the dura has not been widely opened prior to mounting the SMARTFrames, a sharp ceramic stylet can be used to sharply pierce the dura and pia in one maneuver. When the blunt stylet is subsequently inserted in the peel-away sheath, an MR image can be acquired after partial insertion to rule out subcortical hemorrhage from brain deflection, or the blunt stylet can be inserted to target in one step. Note that if the dura was opened in a previous stage, it is critical that the surgeon opens the pia in a traditional fashion with scalpel and cautery, or with the sharp stylet, and that the stylet is visualized to enter the pia with minimal cortical deflection in order to minimize the risk of subcortical hemorrhage. Once the rigid stylet has been placed to the target depth, a final evaluation step is used to determine the targeting error by comparing the distance from the stylet artifact to the intended target. If targeting accuracy is unacceptable, the software can calculate the adjustment necessary on the X–Y stage, and the surgeon can remove the blunt stylet and sheath, make the adjustment, and reinsert.

When the surgeon is satisfied that the ceramic stylet has been placed within an acceptable distance from the intended target, the stylet is removed from the plastic peel-away sheath and the DBS electrode is inserted (Figure 13.11). The sheath is then peeled away, and a final scan is obtained to verify electrode position. When measuring the length of insertion on the DBS electrode, the author typically adds 2 mm when targeting the STN (target plane typically 4 mm below the AC–PC line) or 4 mm when targeting the GPi (target plane typically at the level of the AC–PC line). This takes into account the intention to leave the bottom of contact 0 at the ventral border of the STN or just superiolateral to the optic tract when targeting GPi.

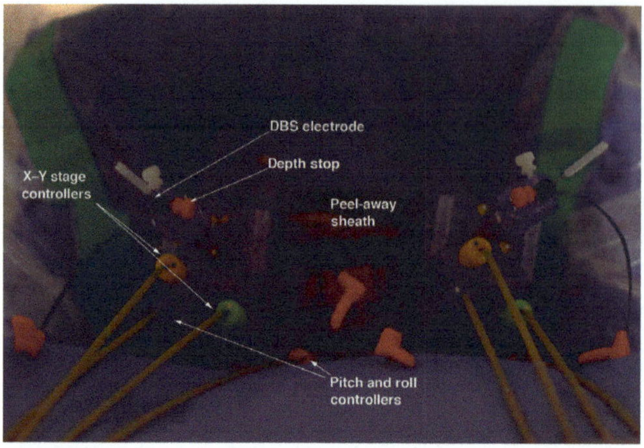

Figure 13.11 DBS electrodes inserted through bilateral SmartFrames. The electrodes have been placed to target depth, and the peel-away sheaths will be removed once electrode locations have been verified by MRI.

Microelectrode vs. iMRI-guided DBS

Although some authors have suggested that DBS electrodes may be implanted successfully without the aid of intraoperative microelectrode recording, based solely on preoperative MRI[56] or combined with impendance monitoring,[65–67] the use of near real-time MRI to guide placement ensures that the electrode is placed in the intended target. The reported accuracy for frame-based and frameless stereotactic placement of DBS leads is typically between 2 and 3 mm.[4] Accuracy with the ClearPoint system results in probe placement less than 1 mm from the intended target in cadaver and nonhuman primate studies[59,68]; at the time of publication, similar errors in clinical use have been reported in abstract form. Submillimetric accuracy is greater than that which can be obtained using standard frame-based stereotaxy in conjunction with preoperative MRI, although errors on the order of 1.5 mm have been reported using 3 T MRI fused to intraoperative CT[56] or MRI alone,[65] without microelectrode recording.

Whether the intended MRI-visible anatomic target is equivalent to that which would be identified based on intraoperative physiology, with regard to therapeutic outcome in patients, will eventually be shown by long-term studies, the data for which are currently unavailable. The rationale for expecting equivalent outcomes, however, is based on the prior experience of many groups with frame-based, microelectrode-guided DBS. The STN and GPi are "mature" targets for iMRI-based targeting due to their easy visibility on MRI, well-characterized internal and surrounding anatomy, and the extensive literature correlating lead locations with clinical outcomes.[69] In this collective experience, the intraoperative criteria for successful lead placement has been physiological, either by detection of STN or GPi cells with movement-related responses, or by reduction of Parkinsonian symptoms with macro-stimulation. Many groups have performed posthoc correlation of lead location on postoperative MRI with single-unit physiology, thresholds for stimulation-induced adverse events, and clinical success, demonstrating that the dorsolateral region of the MR imaging-defined STN, and the posteriolateral region of GPi, reliably correspond to sensorimotor territory. As tractography methods improve and it becomes possible to include high-resolution patient-specific imaging of white matter pathways in target planning, other targets may emerge as appropriate for iMRI-guided localization.

INTERVENTIONAL MRI-GUIDED DELIVERY OF GENE THERAPY VECTORS

Expectations for the eventual success of gene therapy and direct drug delivery approaches to treat PD remain high despite recent failed trials in the field. Currently, there are two clinical trials under way in the United States to study the delivery of AAV2 vector-based therapeutics: AADC, the final rate-limiting enzyme in dopamine synthesis (NCT01973543), and GDNF, a potent neurotropic factor (NCT01621581). These trials are employing convection-enhanced delivery via iMRI for real-time imaging of vector distribution by convection-enhanced delivery (CED), in an attempt to overcome obstacles that may have prevented previous PD drug delivery trials from reaching efficacy goals.[70,71]

CED is a term that denotes the use of a pressure gradient to generate bulk flow within the brain parenchyma, i.e., convection of macromolecules within the interstitial fluid driven by infusing a solution through a cannula placed directly in the targeted structure.[72] This method allows therapeutic agents to be homogenously distributed through large volumes of brain tissue by bypassing the blood–brain barrier and surpassing simple diffusion. Following the successful use of CED in preclinical studies of AAV2–AADC in Parkinsonian nonhuman primates (NHPs),[73–76] CED has been used to deliver this gene therapy to PD disease patients with encouraging results in an initial phase I trial.[77,78] The use of CED alone, however, is not adequate for ensuring optimal translation of preclinical to clinical results. An extensive analysis of the different infusion techniques used in the chronic GDNF protein infusion trials underscored the importance of variable cannula design and placement in producing discrepant outcomes.[79] In fact, postoperative imaging analysis of infusions from the AAV2–AADC trial delivered in patients by CED, without real-time visualization, demonstrated poor coverage of the putamen in some cases.[80]

To increase the translation of positive results from preclinical intracerebral drug delivery studies into successful clinical trials, a method for real-time image-guided CED has evolved over several years with extensive modeling in NHP.[81] Visualizing infusions in real time provides the neurosurgeon with rapid feedback on the physical and anatomic diffusion parameters important for optimizing gene transfer and reduces the potential for adverse effects.[82] Initially described by Oldfield and colleagues using albumin-linked surrogate tracers,[83] the current technique of real-time CED employs iMRI to monitor the distribution of therapeutic agents that are coinfused with gadolinium-related tracers.[84] Coinfusion of therapeutic agents and gadolinium-diethylenetriamene pentaacetic acid has been used by Lonser et al. to treat patients with intrinsic brainstem lesions,[85] whereas gadoteridol has been shown to be an accurate tracer for visualizing the distribution of AAV2 vectors and subsequent transgene expression in nonhuman primates.[70,86]

Infusion studies in nonhuman primates demonstrate that differences in cytoarchitecture between target structures, particularly between gray matter and white matter, are an anatomic determinant of infusate distribution. For instance, corona radiata infusions clearly show distribution following white matter fibers, whereas the limited size of the putamen restricts the allowable volume of infusion in that nucleus. Important potential pitfalls for putamenal infusion have been identified, including anterior lateral leakage that can be visualized by signal enhancement that follows the perivascular space of lateral striate arteries connecting to perivascular space of the medial cerebral artery, terminating in the Sylvian fissure and insular cortex.[87] Despite this potential for leakage, the real-time monitoring technique allows the surgeon to stop the infusion at any point, thereby permitting the filling of the putamen or similar structures with some precision. Additionally, real-time MRI is required to verify that cannula step and tip are both located within a safe zone for gray matter structures, no closer than 3 mm from any white matter border, for minimizing leakage.[88]

The safety and success of iMRI CED in initial clinical use in neuro-oncology, combined with preclinical data demonstrating improved control of infusate delivery and prediction of therapeutic distribution, have allowed current clinical trials of intracerebral direct drug delivery to employ this treatment strategy. The ClearPoint system, described above, was designed with this intention in mind, and a custom-designed CED catheter is available for this purpose. Initially validated in nonhuman primates,[68] this platform was used specifically to model the parameters for delivery of a gene therapy vector to the putamen.[70] Within this paradigm, a previous finding that real-time imaging of coinfused Gd/AAV2–GDNF predicts the anatomic distribution of subsequent GDNF expression within the target structure was confirmed in an NHP model.[86] The predicted locations in the putamen for cannula placement and infusate volume most likely to achieve GDNF transgene expression have now been applied to the clinical delivery of AAV2–GDNF. Precise delivery to the human putamen has been observed in initial patients (Figure 13.12). The ability to incorporate tractography studies into iMRI drug delivery procedures[89] in the future should further improve and expand this clinical application, by improving the predication of drug distribution

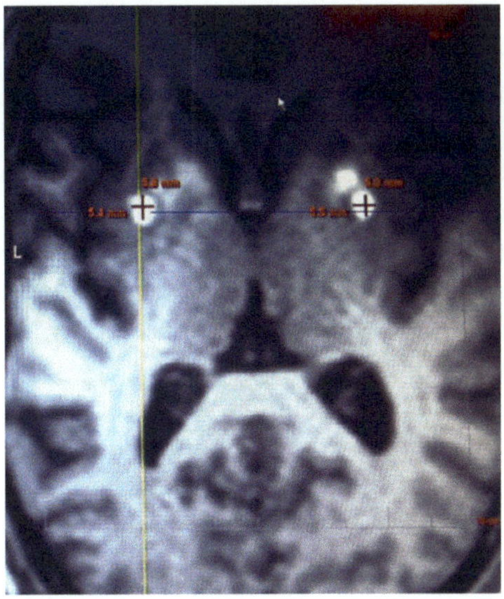

Figure 13.12 iMRI-guided AAV2–GDNF infusions. The co infusion of contrast agent with viral vector allows the real-time visualization of the therapeutic agent in the bilateral putamen of a PD patient. *Image courtesy of Dr. John Heiss, Surgical Neurology Branch, National Institute of Neurological Disorders and Stroke.*

along white matter pathways. For instance, an automated algorithm for predicting drug distribution within gray matter structures has been developed that autosegments the target and leakage-risk structures and predicts the optimal cannula location to maximize coverage and containment.[90] Evidence for leakage of the infusate can be monitored with serial MRI scans during the infusion, and the infusion can be terminated if leakage into cerebrospinal fluid (CSF) spaces or along perivascular routes is noted. The main concern with leakage is that, once it occurs, one loses the ability to effectively deliver the therapeutic to the target in the predicted anatomic distribution and with the predicted concentration, in addition to any concerns regarding widespread distribution of the viral vector, or other agent, through the CSF.

DISCUSSION

Image guidance in functional neurosurgery has evolved from atlas-based approaches with minimal patient-specific imaging information available to single-modality preoperative MRI using standard T1- and T2-weighted sequences to real-time intraoperative imaging of the brain target and electrode trajectory. As diffusion-weighted tractography imaging continues to improve, these sequences will play an increasing role in the targeting of white matter pathways involved in movement and neuropsychiatric disorders. Although current tractography methods are not validated at the fine resolution required for reliably visualizing fiber tracts through the basal ganglia (e.g., pallidofugal fibers), efforts in this regard are under way.[19] The incorporation of functional data, such as microelectrode recording, into the surgical imaging plan in real time is also becoming possible with the advent of software tools that enable interactive 3D visualization of coregistered MRI scans, 3D brain atlases, MER data, and postimplantation CT scans with the predicted volume of tissue activated as a function of stimulation parameters[91] or with a central repository housing physiological data from hundreds of patients.[92] The limited ability of currently available clinical navigation systems for automating the fusion of multiple data types, however, has thus far restricted these applications to the research domain.

Robotic systems are a logical evolution for incorporating automated fusion of imaging and functional modalities with the automation of actual steps in the stereotactic procedure itself. The NeuroMate robotic arm, whose use has been pioneered by Dr. Benabid and his group in Grenoble, is the foremost example of this technology. This group has implanted over 1500 electrodes using evolving prototypes of this system, demonstrating its effectiveness in integrating data, performing complex calculations, and performing repetitive tasks.[93] Stereotactic approaches are well suited for the application of robotic assistance, due to the relatively stable position of the brain within the rigid cranial vault, and the ability to target brain regions based on imaging data that can be localized in stereotactic space. Image-linked robotic systems do not replace any of the roles of the neurosurgeon, but provide assistance by increasing precision and speed. The cost of robotic systems, however, thus far has been cost-prohibitive for many centers. Although the technology currently exists for the creation of truly automated, image-guided, supervisory-controlled robotic systems, their safe and effective use always will require an experienced neurosurgeon to evaluate the dynamic work environment in real time. In order to continue improving the contribution of image guidance to neuromodulation, better automation platforms for the real-time combination of multimodal data, on commercially available workstations that are widely accessible, are essential.

In summary, the advent of real-time MRI guidance in functional neurosurgery has opened a new era in this field. The ability to visualize brain targets in real time during DBS surgery has increased access to this procedure for patients, and the incorporation of tractography data in the future may further improve this technique. Long-term follow-up data are likely to support the notion that direct real-time MRI targeting of the STN and GPi leads to clinical outcomes for patients that are at least equal to that obtained with traditional MER-guided approaches. The caveat is that iMRI–DBS should continue to be utilized by functional neurosurgeons trained in MER-guided techniques, who have the experience in both patient and target selection to optimize chances for clinical success. With regard to intracerebral drug delivery, it is now clear that iMRI guidance is an absolute must for any new clinical trial. As multimodality imaging is further incorporated into iMRI platforms, our ability to effectively deliver therapeutics via CED will continue to improve. Finally, the advantages of multimodality real-time MRI guidance will need to be weighed against the need for MER in the evaluation of new targets and indications for DBS, such as neuropsychiatric disorders.

REFERENCES

1. Spiegel EA, Wycis HT, Marks M, Lee AJ. Stereotaxic apparatus for operations on the human brain. *Science*. 1947;106(2754):349–350.
2. Rahman M, Murad GJA, Mocco J. Early history of the stereotactic apparatus in neurosurgery. *Neurosurg Focus*. 2009;27(3):E12.
3. Henderson JM, Holloway KL, Gaede SE, Rosenow JM. The application accuracy of a skull-mounted trajectory guide system for image-guided functional neurosurgery. *Comput Aided Surg*. 2004;9(4):155–160.
4. Holloway KL, Gaede SE, Starr PA, Rosenow JM, Ramakrishnan V, Henderson JM. Frameless stereotaxy using bone fiducial markers for deep brain stimulation. *J Neurosurg*. 2005;103(3):404–413.
5. Tai CH, Wu RM, Lin C-H, et al. Deep brain stimulation therapy for Parkinson's disease using frameless stereotaxy: comparison with frame-based surgery. *Eur J Neurol*. 2010;17(11):1377–1385.
6. Sumanaweera TS, Adler JR, Napel S, Glover GH. Characterization of spatial distortion in magnetic resonance imaging and its implications for stereotactic surgery. *Neurosurgery*. 1994;35(4):696–703.
7. diPierro CG, Francel PC, Jackson TR, Kamiryo T, Laws ER. Optimizing accuracy in magnetic resonance imaging-guided stereotaxis: a technique with validation based on the anterior commissure-posterior commissure line. *J Neurosurg*. 1999;90(1):94–100.
8. Gologorsky Y, Ben-Haim S, Moshier EL, et al. Transgressing the ventricular wall during subthalamic deep brain stimulation surgery for Parkinson disease increases the risk of adverse neurological sequelae. *Neurosurgery*. 2011;69(2):294–300.
9. Zrinzo L, van Hulzen ALJ, Gorgulho AA, et al. Avoiding the ventricle: a simple step to improve accuracy of anatomical targeting during deep brain stimulation. *J Neurosurg*. 2009;110(6):1283–1290.
10. Khan MF, Mewes K, Gross RE, Skrinjar O. Assessment of brain shift related to deep brain stimulation surgery. *Stereotact Funct Neurosurg*. 2008;86(1):44–53.
11. Ben-Haim S, Asaad WF, Gale JT, Eskandar EN. Risk factors for hemorrhage during microelectrode-guided deep brain stimulation and the introduction of an improved microelectrode design. *Neurosurgery*. 2009;64(4):754–762.
12. Starr PA. Placement of deep brain stimulators into the subthalamic nucleus or globus pallidus internus: technical approach. *Stereotact Funct Neurosurg*. 2002;79(3–4):118–145.
13. Brunenberg EJL, Platel B, Hofman PAM, Haar Romeny Ter BM, Visser-Vandewalle V. Magnetic resonance imaging techniques for visualization of the subthalamic nucleus. *J Neurosurg*. 2011;115(5):971–984.
14. Balachandran R, Welch EB, Dawant BM, Fitzpatrick JM. Effect of MR distortion on targeting for deep-brain stimulation. *IEEE Trans Biomed Eng*. 2010;57(7):1729–1735.
15. Alterman RL, Tagliati M. Preparation for movement disorder surgery. In: Bakay RAE, ed. *Movement Disorder Surgery: The Essentials*. New York: Thieme; 2008:58–69.
16. Ben-Haim S, Gologorsky Y, Monahan A, Weisz D, Alterman RL. Fiducial registration with spoiled gradient-echo magnetic resonance imaging enhances the accuracy of subthalamic nucleus targeting. *Neurosurgery*. 2011;69(4):870–875.
17. Abosch A, Yacoub E, Ugurbil K, Harel N. An assessment of current brain targets for deep brain stimulation surgery with susceptibility-weighted imaging at 7 Tesla. *Neurosurgery*. 2010;67(6):1745–1756.
18. Duchin Y, Abosch A, Yacoub E, Sapiro G, Harel N. Feasibility of using ultra-high field (7 T) MRI for clinical surgical targeting. Bankiewicz K, ed. *PLoS One*. 2012;7(5):e37328.

19. Lenglet C, Abosch A, Yacoub E, De Martino F, Sapiro G, Harel N. Comprehensive in vivo mapping of the human basal ganglia and thalamic connectome in individuals using 7 T MRI. *PLoS One.* 2012;7(1):e29153.
20. Starr PA, Christine CW, Theodosopoulos PV, et al. Implantation of deep brain stimulators into the subthalamic nucleus: technical approach and magnetic resonance imaging-verified lead locations. *J Neurosurg.* 2002;97(2):370–387.
21. Slavin KV, Thulborn KR, Wess C, Nersesyan H. Direct visualization of the human subthalamic nucleus with 3 T MR imaging. *Am J Neuroradiol.* 2006;27(1):80–84.
22. Dormont D, Ricciardi KG, Tandé D, et al. Is the subthalamic nucleus hypointense on T2-weighted images? A correlation study using MR imaging and stereotactic atlas data. *Am J Neuroradiol.* 2004;25(9):1516–1523.
23. Kerl HU, Gerigk L, Pechlivanis I, Al-Zghloul M, Groden C, Nölte I. The subthalamic nucleus at 3.0 Tesla: choice of optimal sequence and orientation for deep brain stimulation using a standard installation protocol—clinical article. *J Neurosurg.* 2012;117(6):1155–1165.
24. Lefranc M, Derrey S, Merle P, et al. High-resolution three-dimensional T2 Star Weighted Angiography (HR 3-D SWAN). *Neurosurgery.* 2014;74(6):615–627.
25. Sweet JA, Walter BL, Gunalan K, Chaturvedi A, McIntyre CC, Miller JP. Fiber tractography of the axonal pathways linking the basal ganglia and cerebellum in Parkinson disease: implications for targeting in deep brain stimulation. *J Neurosurg.* 2014;120(4):988–996.
26. Starr PA, Theodosopoulos PV, Turner R. Surgery of the subthalamic nucleus: use of movement-related neuronal activity for surgical navigation. *Neurosurgery.* 2003;53(5):1146–1149.
27. Larson PS, Starr PA, Martin AJ. *Interventional MRI-Guided DBS: a Practical Atlas.* 2013. Paul Larson Publishers.
28. Plaha P, Ben-Shlomo Y, Patel NK, Gill SS. Stimulation of the caudal zona incerta is superior to stimulation of the subthalamic nucleus in improving contralateral parkinsonism. *Brain.* 2006;129 (Pt 7):1732–1747.
29. Plaha P, Javed S, Agombar D, et al. Bilateral caudal zona incerta nucleus stimulation for essential tremor: outcome and quality of life. *J Neurol Neurosur Ps.* 2011;82(8):899–904.
30. Plaha P, Khan S, Gill SS. Bilateral stimulation of the caudal zona incerta nucleus for tremor control. *J Neurol Neurosur Ps.* 2008;79(5):504–513.
31. Kerl HU, Gerigk L, Huck S, Al-Zghloul M, Groden C, Nölte IS. Visualisation of the zona incerta for deep brain stimulation at 3.0 Tesla. *Clin Neuroradiol.* 2012;22(1):55–68.
32. Follett KA, Weaver FM, Stern M, et al. Pallidal versus subthalamic deep-brain stimulation for Parkinson's disease. *N Engl J Med.* 2010;362(22):2077–2091.
33. Chang EF, Turner RS, Ostrem JL, Davis VR, Starr PA. Neuronal responses to passive movement in the globus pallidus internus in primary dystonia. *J Neurophysiol.* 2007;98(6):3696–3707.
34. Guridi J, Gorospe A, Ramos E, Linazasoro G, Rodriguez MC, Obeso JA. Stereotactic targeting of the globus pallidus internus in Parkinson's disease: imaging versus electrophysiological mapping. *Neurosurgery.* 1999;45(2):278–287.
35. Vitek JL, Chockkan V, Zhang JY, et al. Neuronal activity in the basal ganglia in patients with generalized dystonia and hemiballismus. *Ann Neurol.* 1999;46(1):22–35.
36. Nölte IS, Gerigk L, Al-Zghloul M, Groden C, Kerl HU. Visualization of the internal globus pallidus: sequence and orientation for deep brain stimulation using a standard installation protocol at 3.0 Tesla. *Acta Neurochir.* 2011;154(3):481–494.
37. Rozanski VE, Vollmar C, Cunha JP, et al. Connectivity patterns of pallidal DBS electrodes in focal dystonia: a diffusion tensor tractography study. *Neuroimage.* 2014;84(C):435–442.
38. Bender B, Manz C, Korn A, Nagele T, Klose U. Optimized 3D magnetization-prepared rapid acquisition of gradient echo: identification of thalamus substructures at 3 T. *Am J Neuroradiol.* 2011; 32(11):2110–2115.
39. Klein JC, Barbe MT, Seifried C, Baudrexel S, Runge M. The tremor network targeted by successful VIM deep brain stimulation in humans. *Neurology.* 2012;78(11):787–795.
40. Pouratian N, Zheng Z, Bari AA, Behnke E, Elias WJ, DeSalles AAF. Multi-institutional evaluation of deep brain stimulation targeting using probabilistic connectivity-based thalamic segmentation. *J Neurosurg.* 2011;115(5):995–1004.

41. Anderson JS, Dhatt HS, Ferguson MA, et al. Functional connectivity targeting for deep brain stimulation in essential tremor. *Am J Neuroradiol.* 2011;32(10):1963–1968.
42. Papavassiliou E, Rau G, Heath S, et al. Thalamic deep brain stimulation for essential tremor: relation of lead location to outcome. *Neurosurgery.* 2004;54(5):1120–1129.
43. Greenberg BD, Gabriels LA, Malone DA, et al. Deep brain stimulation of the ventral internal capsule/ventral striatum for obsessive-compulsive disorder: worldwide experience. *Mol Psychiatry.* 2010;15(1):64–79.
44. Nuttin B, Gielen F, van Kuyck K, et al. Targeting bed nucleus of the stria terminalis for severe obsessive-compulsive disorder: more unexpected lead placement in obsessive-compulsive disorder than in surgery for movement disorders. *WNEU.* 2013;80(3–4):S30:e11–e16.
45. Lujan JL, Chaturvedi A, McIntyre CC. Tracking the mechanisms of deep brain stimulation for neuropsychiatric disorders. *Front Biosci.* 2008;13:5892–5904.
46. Lujan JL, Chaturvedi A, Malone DA, Rezai AR, Machado AG, McIntyre CC. Axonal pathways linked to therapeutic and nontherapeutic outcomes during psychiatric deep brain stimulation. *Hum Brain Mapp.* 2012;33(4):958–968.
47. Lujan JL, Chaturvedi A, Choi KS, et al. Tractography-activation models applied to subcallosal cingulate deep brain stimulation. *Brain Stimul.* 2013;6(5):737–739.
48. Riva-Posse P, Choi KS, Holtzheimer PE, et al. Defining critical white matter pathways mediating successful subcallosal cingulate deep brain stimulation for treatment-resistant depression. *Biol Psychiatry.* 2014;76(12):963–969.
49. Larson PS, Richardson RM, Starr PA, Martin AJ. Magnetic resonance imaging of implanted deep brain stimulators: experience in a large series. *Stereotact Funct Neurosurg.* 2008;86(2):92–100.
50. Pinsker MO, Herzog J, Falk D, Volkmann J, Deuschl G, Mehdorn M. Accuracy and distortion of deep brain stimulation electrodes on postoperative MRI and CT. *Zentralbl Neurochir.* 2008;69(3):144–147.
51. Chhabra V, Sung E, Mewes K, Bakay RA, Abosch A, Gross RE. Safety of magnetic resonance imaging of deep brain stimulator systems: a serial imaging and clinical retrospective study. *J Neurosurg.* 2010;112(3):497–502.
52. Smith AP, Bakay RAE. Frameless deep brain stimulation using intraoperative O-arm technology. *J Neurosurg.* 2011;115(2):301–309.
53. Burchiel KJ, McCartney S, Lee A, Raslan AM. Accuracy of deep brain stimulation electrode placement using intraoperative computed tomography without microelectrode recording. *J Neurosurg.* 2013; 119(2):301–306.
54. Shahlaie K, Larson PS, Starr PA. Intraoperative computed tomography for deep brain stimulation surgery: technique and accuracy assessment. *Neurosurgery.* 2011;68:ONS114–ONS124.
55. Holloway K, Docef A. A quantitative assessment of the accuracy and reliability of O-arm images for deep brain stimulation surgery. *Neurosurgery.* 2013;72(1 Suppl Operative):47–57.
56. Burchiel KJ, McCartney S, Lee A, Raslan AM. Accuracy of deep brain stimulation electrode placement using intraoperative computed tomography without microelectrode recording. *J Neurosurg.* 2013;119(2):301–306.
57. Ivan ME, Yarlagadda J, Saxena AP, et al. Brain shift during bur hole-based procedures using interventional MRI. *J Neurosurg.* 2014;121(1):149–160.
58. Hunsche S, Sauner D, Maarouf M, et al. Intraoperative X-ray detection and MRI-based quantification of brain shift effects subsequent to implantation of the first electrode in bilateral implantation of deep brain stimulation electrodes. *Stereotact Funct Neurosurg.* 2009;87(5):322–329.
59. Larson PS, Starr PA, Bates G, Tansey L, Richardson RM, Martin AJ. An optimized system for interventional mri guided stereotactic surgery: preliminary evaluation of targeting accuracy. *Neurosurgery.* 2012;70(1 Suppl Operative):95–103.
60. Prospective stereotaxy—a novel method of trajectory alignment using real-time image guidance. 2001;13(3):452–457.
61. Martin AJ, Hall WA, Roark C, Starr PA, Larson PS, Truwit CL. Minimally invasive precision brain access using prospective stereotaxy and a trajectory guide. *J Magn Reson Imaging.* 2008;27(4): 737–743.
62. Hall WA, Truwit CL. Intraoperative MR-guided neurosurgery. *J Magn Reson Imaging.* 2008;27(2): 368–375.

63. Martin AJ, Larson PS, Ostrem JL, et al. Placement of deep brain stimulator electrodes using real-time high-field interventional magnetic resonance imaging. *Magn Reson Med.* 2005;54(5):1107–1114.
64. Starr PA, Martin AJ, Ostrem JL, Talke P, Levesque N, Larson PS. Subthalamic nucleus deep brain stimulator placement using high-field interventional magnetic resonance imaging and a skull-mounted aiming device: technique and application accuracy. *J Neurosurg.* 2010;112(3):479–490.
65. Foltynie T, Zrinzo L, Martinez-Torres I, et al. MRI-guided STN DBS in Parkinson's disease without microelectrode recording: efficacy and safety. *J Neurol Neurosur Ps.* 2011;82(4):358–363.
66. Nakajima T, Zrinzo L, Foltynie T, et al. MRI-guided subthalamic nucleus deep brain stimulation without microelectrode recording: can we dispense with surgery under local anaesthesia? *Stereotact Funct Neurosurg.* 2011;89(5):318–325.
67. Aviles-Olmos I, Aviles-Olmos I, Kefalopoulou Z, et al. Long-term outcome of subthalamic nucleus deep brain stimulation for Parkinson's disease using an MRI-guided and MRI-verified approach. *J Neurol Neurosur Ps.* 2014;85(12):1419–1425.
68. Richardson RM, Kells AP, Martin AJ, et al. Novel platform for MRI-guided convection-enhanced delivery of therapeutics: preclinical validation in nonhuman primate brain. *Stereotact Funct Neurosurg.* 2011;89(3):141–151.
69. Vega RA, Holloway KL, Larson PS. Image-guided deep brain stimulation. *Neurosurgery Clinics of NA.* 2014;25(1):159–172.
70. Richardson RM, Kells AP, Rosenbluth KH, et al. Interventional MRI-guided putaminal delivery of AAV2-GDNF for a planned clinical trial in Parkinson's disease. *Mol Ther.* 2011;19(6):1048–1057.
71. San Sebastian W, Richardson RM, Kells AP, et al. Safety and tolerability of magnetic resonance imaging-guided convection-enhanced delivery of AAV2-hAADC with a novel delivery platform in nonhuman primate striatum. *Hum Gene Ther.* 2012;23(2):210–217.
72. Bobo RH, Laske DW, Akbasak A, Morrison PF, Dedrick RL, Oldfield EH. Convection-enhanced delivery of macromolecules in the brain. *Proc Natl Acad Sci USA.* 1994;91(6):2076–2080.
73. Bankiewicz KS, Eberling JL, Kohutnicka M, et al. Convection-enhanced delivery of AAV vector in parkinsonian monkeys: in vivo detection of gene expression and restoration of dopaminergic function using pro-drug approach. *Exp Neurol.* 2000;164(1):2–14.
74. Bankiewicz KS, Forsayeth J, Eberling JL, et al. Long-term clinical improvement in MPTP-lesioned primates after gene therapy with AAV-hAADC. *Nature.* 2006;14(4):564–570.
75. Forsayeth JR, Eberling JL, Sanftner LM, et al. A dose-ranging study of AAV-hAADC therapy in Parkinsonian monkeys. *Nature.* 2006;14(4):571–577.
76. Daadi MM, Pivirotto P, Bringas J, et al. Distribution of AAV2-hAADC-transduced cells after 3 years in Parkinsonian monkeys. *Neuroreport.* 2006;17(2):201–204.
77. Christine CW, Starr PA, Larson PS, et al. Safety and tolerability of putaminal AADC gene therapy for Parkinson disease. *Neurology.* 2009;73(20):1662–1669.
78. Eberling JL, Jagust WJ, Christine CW, et al. Results from a phase I safety trial of hAADC gene therapy for Parkinson disease. *Neurology.* 2008;70(21):1980–1983.
79. Morrison PF, Lonser RR, Oldfield EH. Convective delivery of glial cell line-derived neurotrophic factor in the human putamen. *J Neurosurg.* 2007;107(1):74–83.
80. Valles F, Fiandaca MS, Eberling JL, et al. Qualitative imaging of adeno-associated virus serotype 2-human aromatic L-amino acid decarboxylase gene therapy in a phase I study for the treatment of Parkinson disease. *Neurosurgery.* 2010;67(5):1377–1385.
81. Richardson RM, Varenika V, Forsayeth JR, Bankiewicz KS. Future applications: gene therapy. *Neurosurg Clin N Am.* 2009;20(2):205–210.
82. Varenika V, Dickinson P, Bringas J, et al. Detection of infusate leakage in the brain using real-time imaging of convection-enhanced delivery. *J Neurosurg.* 2008;109(5):874–880.
83. Nguyen TT, Pannu YS, Sung C, et al. Convective distribution of macromolecules in the primate brain demonstrated using computerized tomography and magnetic resonance imaging. *J Neurosurg.* 2003;98(3):584–590.
84. Fiandaca MS, Varenika V, Eberling J, et al. Real-time MR imaging of adeno-associated viral vector delivery to the primate brain. *Neuroimage.* 2009;47(Suppl 2):T27–T35.

85. Lonser RR, Warren KE, Butman JA, et al. Real-time image-guided direct convective perfusion of intrinsic brainstem lesions. Technical note. *J Neurosurg.* 2007;107(1):190–197.
86. Su X, Kells AP, Aguilar Salegio EA, et al. Real-time MR imaging with gadoteridol predicts distribution of transgenes after convection-enhanced delivery of AAV2 vectors. *Mol Ther.* 2010;18(8):1490–1495.
87. Krauze MT, Saito R, Noble C, et al. Effects of the perivascular space on convection-enhanced delivery of liposomes in primate putamen. *Exp Neurol.* 2005;196(1):104–111.
88. Yin D, Valles FE, Fiandaca MS, et al. Optimal region of the putamen for image-guided convection-enhanced delivery of therapeutics in human and non-human primates. *Neuroimage.* 2011;54(Suppl 1): S196–S203.
89. Rosenbluth KH, Eschermann JF, Mittermeyer G, Thomson R, Mittermeyer S, Bankiewicz KS. Analysis of a simulation algorithm for direct brain drug delivery. *Neuroimage.* 2011;1–7.
90. Rosenbluth KH, Martin AJ, Mittermeyer S, Eschermann J, Dickinson PJ, Bankiewicz KS. Rapid inverse planning for pressure-driven drug infusions in the brain. Chen M, ed. *PLoS One.* 2013;8(2): e56397.
91. Miocinovic S, Noecker AM, Maks CB, Butson CR, McIntyre CC. Cicerone: stereotactic neurophysiological recording and deep brain stimulation electrode placement software system. *Acta Neurochir Suppl.* 2007;97(Pt 2):561–567.
92. D'Haese P-F, Pallavaram S, Li R, et al. CranialVault and its CRAVE tools: a clinical computer assistance system for deep brain stimulation (DBS) therapy. *Med Image Anal.* 2012;16(3):744–753.
93. Benabid A-L, Wallace B, Hoffman DS, et al. Implantation of multiple electrodes and robotic techniques. In: Bakay RAE, ed. *Movement Disorder Surgery: The Essentials.* New York: Thieme Medical Pub. Inc; 2009.

CHAPTER 14

Image Guidance for Spine Surgery

Viren Vasudeva[1], Ziev Moses[1], Tyler Cole[2], Yakov Gologorsky[3], and Yi Lu[1]
[1]Department of Neurosurgery, Brigham and Women's Hospital, Boston, MA, USA
[2]Stanford University School of Medicine, Stanford, CA, USA
[3]Department of Neurosurgery, Mount Sinai School of Medicine, New York, NY, USA

INTRODUCTION

Image-guided spine surgery was first introduced in the mid-1990s.[1] This advancement drew on technology that had initially been developed for intracranial neurosurgical procedures.[2] However, the anatomy of the spine makes the application inherently more complex. External fiducials placed on the skin can provide accurate registration for the brain; however, this is not feasible for registration in the spine as the skin and underlying soft tissue can move in relation to spinal bony structures.[3–6] Furthermore, the dynamic intersegmental motion of the spine makes registration even more challenging as the overall alignment does not remain constant with changes in positioning. Despite these difficulties, the use of image guidance in spine surgery has clear theoretical benefits for the improvement of accuracy of hardware placement in complex cases or patients with abnormal anatomy.

Without navigation, the trajectory of pedicle screw placement is determined based on the surgeon's understanding of local anatomic landmarks and intraoperative X-rays.[7] Because this skill varies with the experience and expertise of each surgeon, the rates of pedicle screw misplacement have been reported as high as 21% to 40% in the lumbar spine[8,9] and 43–55% in the thoracic spine.[10,11] This naturally inspired surgeons to explore intraoperative navigation as a method to enhance outcomes. Concurrent advancement in computer and imaging technology over the past 20 years has allowed the development of several progressively advanced systems for interactive surgical navigation. These systems in turn have been shown to benefit both spine surgeons and patients alike.

TYPES OF IMAGE GUIDANCE IN SPINE SURGERY

In general, the process of computer-navigated spine surgery entails image acquisition, surgical planning, registration to patient anatomy in the operating room, use of navigation to guide surgery or the placement of implants, and finally acquisition of additional imaging for confirmation as necessary.[12,13] The method of image acquisition and registration are the cornerstones of any image-guidance system [14] and, as technology has improved, this has evolved over time.

A. Golby (Ed): Image-Guided Neurosurgery
DOI: http://dx.doi.org/10.1016/B978-0-12-800870-6.00014-5

The earliest form of image-guided navigation in the spine was based on thin-cut preoperative computed tomography (CT) and registration with point matching and surface matching techniques. This then gave way to intraoperative image acquisition with 2D fluoroscopy or 3D C-arm cone beam CT. These systems allowed for automatic registration and, because imaging was obtained with the patient already in the surgical position, minimized the inaccuracies associated with shifting anatomy. More recently, some centers have started to use intraoperative CT (iCT) and magnetic resonance imaging (MRI) (ioMRI) as advanced methods of image guidance, but their widespread use is still limited by cost and the large size.

Conventional image guidance for spine surgery: plain radiography and C-Arm fluoroscopy

Before the advent of intraoperative navigation, surgeons relied on the knowledge of spinal anatomy as well as serial radiography or C-arm fluoroscopy to guide the placement of spinal instrumentation.[5] Intraoperative neurophysiologic monitoring could also be used as a supportive method.[14]

Plain radiography, initially developed by Wilhelm Roentgen in 1895,[12] proved useful for localizing the skin incision, determining the correct level of surgery, and confirming acceptable position of implants prior to leaving the operating room.[15] However, a considerable period of time could elapse during the development of X-rays and only static images could be obtained.

C-arm fluoroscopy improved upon plain radiography by allowing surgeons to view images immediately and to obtain serial images in rapid succession. The unique advantage of C-arm fluoroscopy is the ability to view images in real time. This means that with any surgical manipulation the surgeon can immediately obtain an image that will accurately reflect any changes in the patient's anatomy or in the position of surgical instruments. By comparison, other forms of image-guided spinal navigation currently in use depend on an imaging study that is acquired prior to surgery. After image registration, the surgery is essentially performed based on the "stored" image in the computer. Therefore, any errors in image registration or shift of the patient's anatomy after the image acquisition will result in misleading information, potentially leading to inaccurate surgical manipulations. The real-time fluoroscopic C-arm image essentially eliminates this problem. For this reason, many surgeons continue to use fluoroscopy as the primary source of imaging in the operating room.

C-arm fluoroscopy has been crucial for the development of minimally invasive spine surgery. One example is the use of fluoroscopy for accurate percutaneous pedicle screw placement, using either a one or two (biplanar) C-arm technique to view anteroposterior (AP) and lateral images (Figure 14.1).

The use of C-arm fluoroscopy for spine surgery requires an experienced X-ray technician to accurately position the C-arm into true AP and lateral projections. The spine surgeon must then correctly interpret the two-dimensional fluoroscopic images

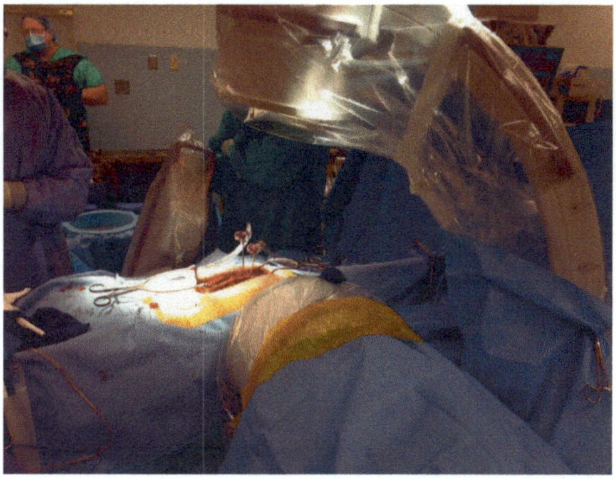

Figure 14.1 The setup of biplanar C-arm fluoroscopy for pedicle screw placement.

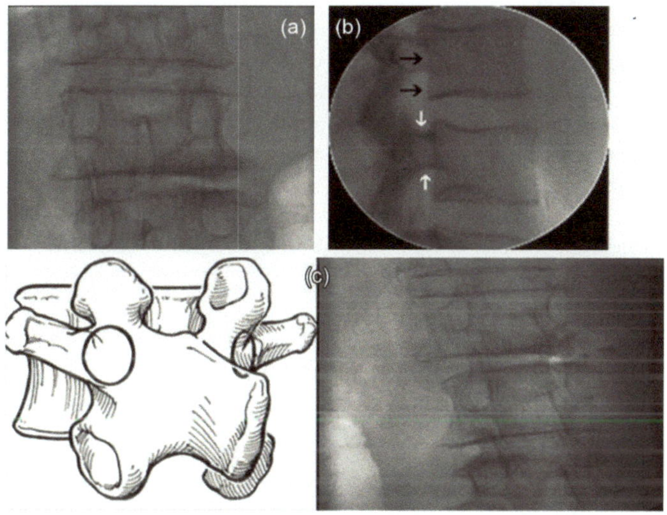

Figure 14.2 The properly aligned AP and lateral fluoroscopic spinal images.

and mentally reconstruct these images into a three-dimensional structure. In a properly aligned AP view, the superior endplate appears as a sharp, single line and the pedicle shadows are located just caudal to the superior endplate. The spinous process shadow is centered between the two pedicle shadows. In a true lateral view, the superior endplate must also be a sharp, single line. The shadow from both the left and right pedicles should be superimposed and the posterior cortex of the vertebral body should appear as a single line, indicating no rotation of the vertebrae. Given the changing lordotic and kyphotic angles at each vertebral level and possible scoliosis in some patients, to obtain true AP and lateral fluoroscopic images may require adjustment of the C-arm at each level (Figure 14.2).

Table 14.1 Comparison of use of fluoroscopy and image navigation in spine surgery

C-arm fluoroscopy	Computer-assisted image navigation
Real-time image	Image obtained prior to surgical manipulation
No concern for anatomy shift during the surgery	Anatomy change after the image acquisition may lead to inaccuracy
No need for registration	Potential for registration error
No need for reference base	Reference base necessary which sometimes requires a separate incision
Inexpensive	Expensive initial capital investment
No need for special surgical instruments	Need to use special instrumentations that can be "tracked"
No ability to visualize axial images	Modern systems allow 3D and multiplanar reconstructions (including axial plane)
Increased radiation exposure to surgeon and OR staff	Minimal occupational radiation exposure
Surgeon and OR staff must wear lead aprons, thyroid shields, etc.	No need to wear protective gear
Surgeon must interpret the 2D image data to understand 3D anatomy	More straightforward and easy to understand virtual 3D computer images for navigation
Working around the C-arm can be ergonomically challenging	The images are acquired at the start of surgery, and then the imaging device can be removed
May be difficult to use in complex instrumentation scenarios	More versatile instrumentation possibilities
Fluoroscopy is often inadequate to check the accuracy of instrumentation in OR	Ability to check the accuracy of instrumentation immediately after placement before closure

One of the major disadvantages of fluoroscopy is increased radiation exposure to the surgeon and operating room staff, especially when continuous fluoroscopy is used.[16] The surgeon is at particularly high risk of radiation exposure due to close proximity to the field. Also, using one C-arm, it is only possible to obtain images in a single plane. If biplanar imaging is required, then two independent C-arms must be used and this can be ergonomically challenging.[5,17] Another important limitation of using fluoroscopy in spine surgery is the difficulty obtaining spine images in the axial plane. Yet, axial images are very important in confirming the accuracy of instrumentations (Table 14.1).

Preoperative CT-based navigation

The earliest form of image-guided navigation in spine surgery was described by Nolte et al. in 1995 and relied on a thin-cut preoperative CT with 1–2-mm slices for navigation.[18,2] This landmark achievement opened up a new dimension for spine surgery, allowing greater preoperative planning and simulation of surgery and enabling surgeons for the first time to visualize anatomy in the spine that was not under direct sight.[19]

Like the system developed by Nolte et al., all preoperative CT-based spinal navigation systems require a thin-cut preoperative CT scan of the relevant portion of the spine.[20] This data is then transferred to a navigation workstation that creates 3D and multiplanar reconstructions that allow the surgeon to plan the trajectory of screws and simulate the positioning of other implants before surgery, thereby optimizing accuracy. As patients are often in the prone or lateral position for surgery, there is likely to be some shift in the relative position of each vertebrae compared to the preoperative CT scan, which is usually obtained supine. It is therefore necessary to select anatomical landmarks on each level of interest and to register each level individually prior to using navigation in the operating room.[21,1,22,4] In order to accomplish this, the surgeon needs to select four to five anatomical landmarks on the surface of each level of interest within the preoperative data set for registration once the exposure has been completed. Once the correct level to be navigated is identified and exposed, a dynamic reference base (DRB) is firmly attached to that bony segment in order to begin registration.[23] The points that were identified preoperatively are located on the surface of the bone to which the DRB is attached and matched to the data on the workstation computer. This process is referred to as point matching. To enhance the registration, the surgeon may also touch random additional points on the exposed bone, which were not selected preoperatively in a process called surface matching. The accuracy of the registration is then tested using a tracked instrument to identify anatomical landmarks and ensuring that they correlate with the projected images on the navigation workstation. It is essential that the registration is accurate. If there is a mismatch, the process must be repeated. Once the accuracy of the registration has been verified, navigation can be used for placement of hardware such as pedicle screws. To navigate other vertebral segments, the registration process must be repeated at each additional level.[20]

Although preoperative CT-based navigation systems are able to provide surgeons with greater ability for preoperative planning and better intraoperative support compared to conventional fluoroscopy, there are several limitations. First, this navigation system requires an image-guided protocol thin-cut preoperative CT, which adds time and cost to the preparation of the procedure and increases radiation exposure to the patient.[5] As discussed above, another limitation is that registration is performed using a preoperative CT scan in the supine position, whereas the surgical procedure is usually performed on a patient in the prone position. As the alignment of vertebral segments shift with different body position, each segment needs to be registered individually prior to navigation and this process is time consuming and inefficient, especially in large multilevel surgeries.[21,1] Another disadvantage is that it can be difficult to obtain adequate reproducible anatomical landmarks for point matching registration, particularly in patients with abnormal anatomy or who have undergone previous laminectomy. In some of these situations, it may be possible to place percutaneous spinal fiducial markers prior to the preoperative CT scan, which could then serve as more

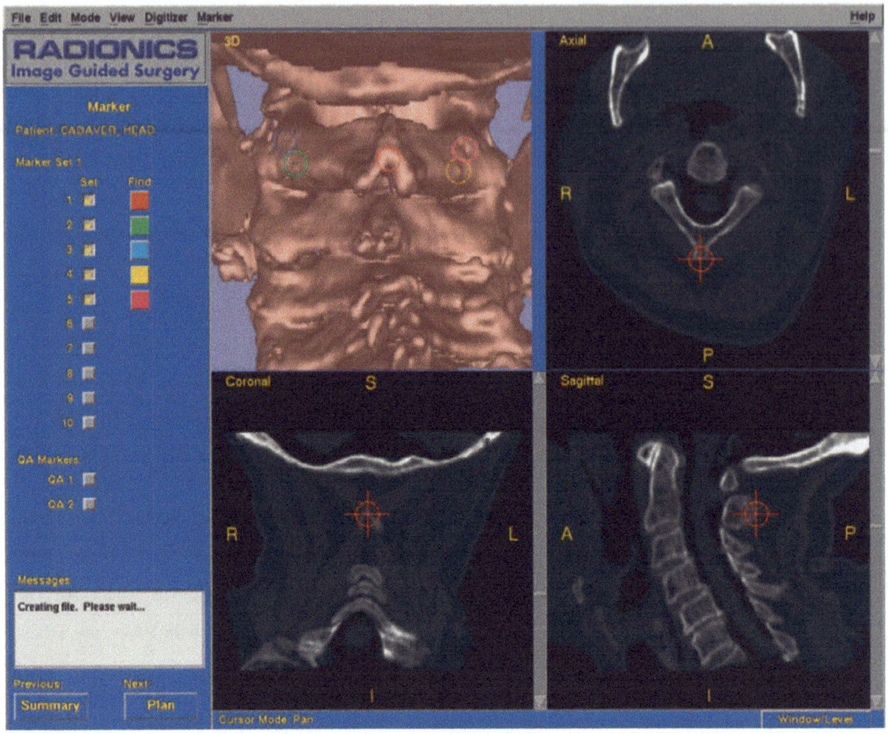

Figure 14.3 Three-dimensional reconstruction and axial, coronal, and sagittal views are seen on the computer workstation using preoperative CT for image guidance. The image on the upper left illustrates five selected points for point matching registration (*used with permission from Holly*[5]).

reliable points for point matching registration; however, this is not widely used due to its time-consuming process.[19] Even in patients with normal anatomy who have not undergone previous surgery, an extensive and meticulous bony exposure is essential to avoid registration inaccuracy and this can be a tedious process.[1] Also, because exposure of the bony spinal anatomy is key to the registration process, this system is of little use in minimally invasive procedures. A semi automatic method of registration that does not require point matching or surface matching was developed by merging intraoperative AP and lateral fluoroscopic images to the preoperative CT for registration.[13] Even though this could be used in minimally invasive procedures, it was inferior to the more sophisticated emerging technologies (Figure 14.3).

Two-dimensional fluoroscopy-based navigation

Due to the extensive preoperative preparation required and the difficulty with accurate registration, preoperative CT-based, computer-assisted navigation has not been widely adopted.[24] In large part, registration inaccuracy was due to shift of anatomy

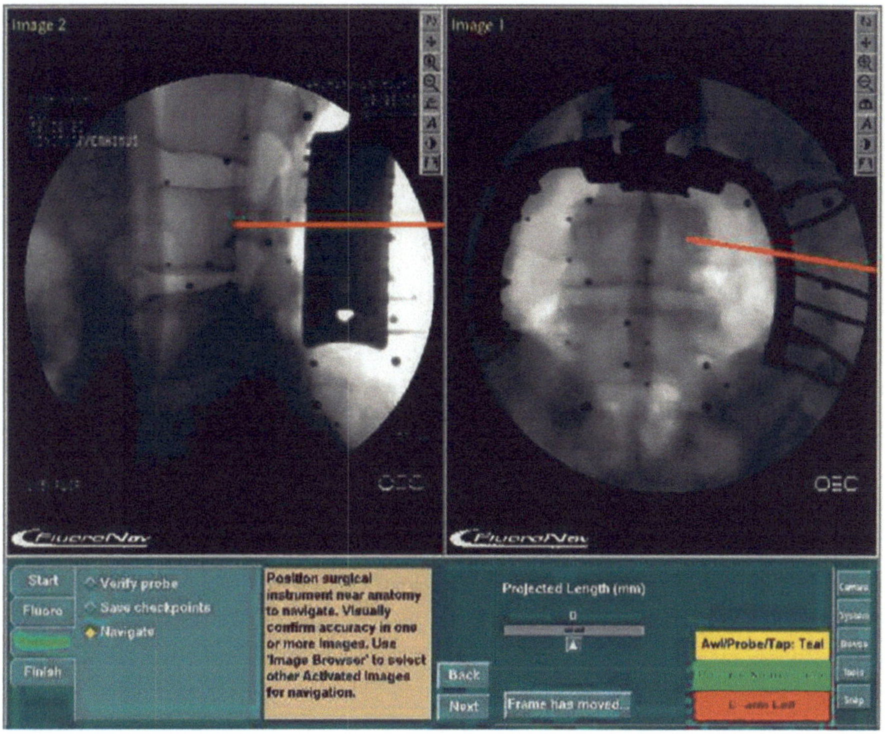

Figure 14.4 AP and lateral images obtained using virtual fluoroscopy demonstrating the trajectory for lumbar pedicle screw placement (*used with permission from Holly*[5]).

from surgical positioning. This led to the development of new navigation technologies relying instead on intraoperative image acquisition, which minimized the problem of anatomic shift as the patient was in the surgical position while the imaging was obtained on which registration would be based. The first such system was created using 2D fluoroscopy.

In this system, the C arm is equipped with a calibration target that is recognized by the camera of the navigation system. Because the system is oriented to the geometric configuration of the C-arm, it is able to ascertain the space that an object occupies within the scanned area.[20] To use this system, a DRB is first attached to the patient, and both AP and lateral intraoperative fluoroscopic images are obtained. These images are transferred to the image guidance system for processing and automatic registration. The navigation system is then able to track the location of a surgical instrument and construct an overlay of the instrument onto the previously obtained fluoroscopic images.[5] (See Figures 14.4 and 14.5.)

Two-dimensional fluoroscopy-based navigation has many advantages over preoperative CT-based computer-assisted navigation. One major benefit is the use of

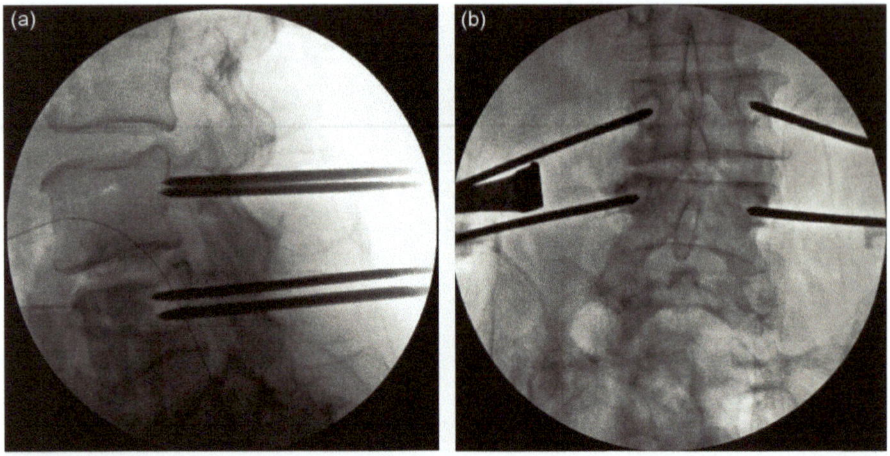

Figure 14.5 Similar AP and lateral images obtained using real-time fluoroscopy with Jamshidi needle for lumbar pedicle screw placement.

intraoperative image acquisition, which reduces inaccuracy due to anatomic shift from positioning. Another major advantage is that registration of the spine occurs automatically after image acquisition. This eliminates the process of point matching and surface matching and further increases the accuracy of registration.[24] Also, there is no need for a preoperative CT with special image guidance protocol, which is more convenient and helps to minimize radiation exposure for the patient. Two-dimensional fluoroscopy allows for improved ergonomics compared to routine fluoroscopy, as the C-arm can be removed once registration is complete. This is particularly helpful in cases where multiplanar imaging is required.[1] Although this was a remarkable improvement, there were still significant limitations. For example, accurate registration and navigation using 2D fluoroscopy-based systems rely on the ability to obtain a high-quality fluoroscopic projection in the operating room. Unfortunately, this could be difficult in certain areas of the spine such as the low cervical or high thoracic regions or in patients with obesity or low bone density.[5,20,14] If intraoperative fluoroscopy is of poor quality, this type of navigation cannot be used. Perhaps the most significant disadvantage of 2D fluoroscopy is that this system does not allow surgeons to view 3D or multiplanar reconstructions of the imaging.[1] Without axial images, the theoretical benefit of instrumentation accuracy with image navigation compared to real-time fluoroscopy is likely minimal.

Computer-assisted navigation: cone beam CT

The development of cone beam CT (cbCT) imaging technology was a major breakthrough for the field of image-guided spine surgery. This technology utilizes a cone-shaped X-ray beam to obtain multiple intraoperative fluoroscopic images as the device rotates around the patient. These rotational images are then used to generate a CT-like 3D data set, which can be reconstructed to provide axial, coronal, and sagittal views of the anatomy.[5] These can then be used for navigation. As in 2D fluoroscopy-based navigation systems, the position of the X-ray source is tracked during the "spin," and the system is able to determine the location of the patient's anatomy and DRB within the scanned area. This allows the navigation system to perform registration automatically following intraoperative image acquisition.[1,20] If a preoperative CT or MRI is available, it can be coregistered to the intraoperative imaging for improved image quality and enhanced visual information.

Cone beam CT-based navigation systems maintained many of the features that had been introduced with 2D fluoroscopy-based navigation, but eliminated many of the disadvantages. Similar to 2D fluoroscopy-based systems, cbCT-based navigation relies on intraoperative image acquisition, eliminating registration error due to anatomic shift in position. Registration is still automatic, time efficient, and user-friendly.[14] Additionally, because exposure of the posterior elements is not necessary for registration, this method of navigation supports minimally invasive surgeries in which the DRB could be attached to a spinous process or iliac crest through a very small incision. For operations in the high cervical spine, the DRB may also be attached to a Mayfield head holder. Like 2D fluoroscopy systems, the cbCT device can be removed from the operating room after image acquisition. This improves ergonomics in the operating room and allows other surgeons to share the same device.

Cone beam CT has several advantages over 2D fluoroscopy primarily because image quality is much greater with cbCT and the surgeons are able to view 3D and multiplanar reconstructions in the operating room. This is especially important in the lower cervical, upper thoracic spine as well as in obese and osteopenic patients in whom it may be difficult to obtain high-quality images using 2D fluoroscopy. Although cbCT devices have lower image quality and a smaller field of view compared to standard CT scanners, they are adequate in most situations to confirm acceptable placement of spinal instrumentation before leaving the operating room. The major disadvantage to this system, however, is that each spin on a conventional cbCT device has a limited scan volume that can only fully image three to four segments at a time. Therefore, for large, multiple level surgeries, additional "spins" may be required.

Several cbCT devices that are commonly used today are the ARCADIS® Orbic 3D (Siemens), the Ziehm Vision Vario 3D (Ziehm), the Ziehm Vision FD Vario 3D (Ziehm), and the O-arm® (Medtronic) (Table 14.2). The O-arm and Ziehm Vision FD Vario 3D use flat panel detectors rather than a conventional C-arm image intensifier to acquire images, resulting in higher image quality and larger scan volume compared to standard cbCT devices. The O-arm, for example, is able to create a 3D cylindrical image, which is 15 cm high and 20 cm in diameter. Because more vertebral

Table 14.2 Comparison of common cbCT devices

	Size (footprint/weight)	Gantry opening (cm)	Flat panel detector	Degrees of rotation	3D image volume
Arcadis Orbic 3D	215 × 80 cm/745lbs	78	No	190°	Cube volume 12 × 12 × 12 cm
Ziehm Vision Vario 3D	184 × 80 cm/595lbs	76	No	135°	Cube volume 12.8 × 12.8 × 12.8 cm
Ziehm Vision FD Vario 3D	184 × 80 cm/595lbs	89.5	Yes	135°	Cube volume 12.8 × 12.8 × 12.8 cm
Medtronic O-Arm	249 cm × 81.3 cm/ 1950lbs	69.9	Yes	360°	Cylindrical volume 20 cm diameter × 15 cm length

levels can be visualized in a single scan, there is a reduced need for recentering the device and re-registration. The O-arm also acquires images over a full 360°, resulting in further improvement in image quality. (See Figures 14.6 and 14.7.)

Computer-assisted navigation: intraoperative CT and MRI

Use of intraoperative CT (iCT) in neurosurgery began in the late 1970s.[25–28] This method of imaging, however, did not gain popularity due to cost, limited image quality, inefficiency, time-consuming modifications to work flow, and difficulty with patient positioning in the operating room. However, with recent improvements in the quality of iCT devices as well as the ability to couple iCT with modern neuronavigation systems, there has been a renewed interest in exploring its applications within spine surgery.[25,1,29,30] Barsa et al. found that using a portable 32-slice intraoperative CT scanner was safe and effective for spinal instrumentation navigation. Using iCT-based navigation, they placed 571 spinal implants at all spinal levels, including cervical pedicle screws with 99.13% accuracy. (See Figure 14.8 for plan of placement.) There were no neurovascular complications, and there were no situations in which the patient's body habitus or positioning caused degradation of the intraoperative imaging. They did, however, find that using iCT created several extra steps adding 18–40 minutes to the operation time.[31]

Due to larger field of view and the ability of the device to translate longitudinally over the patient, the entire spine can be imaged with a single scan. This is a dramatic improvement over cbCT devices, which can only visualize a portion of the spine and may require recentering and re-registration for large multilevel surgeries. The improved image quality of iCT devices allows the evaluation of soft tissue better than

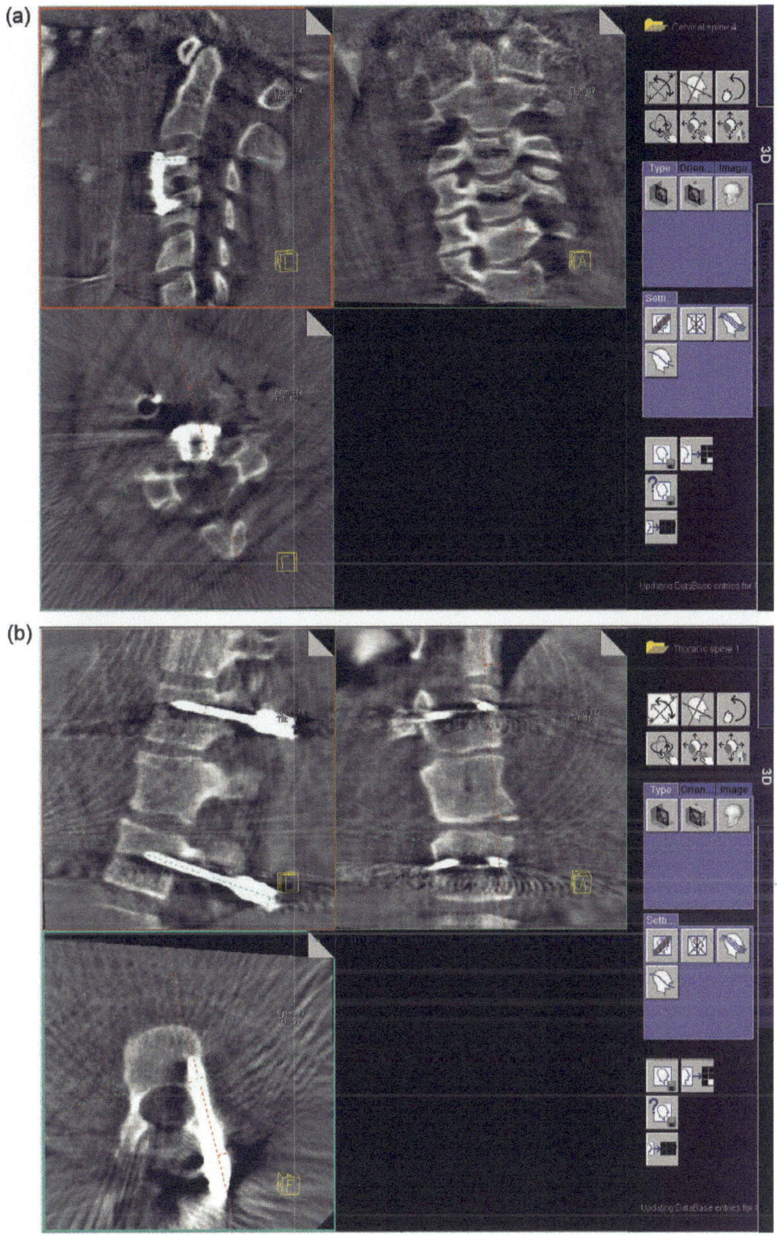

Figure 14.6 Images obtained with Arcadis Orbic 3D system. (*Images provided by Siemens Healthcare USA and used with permission.*)

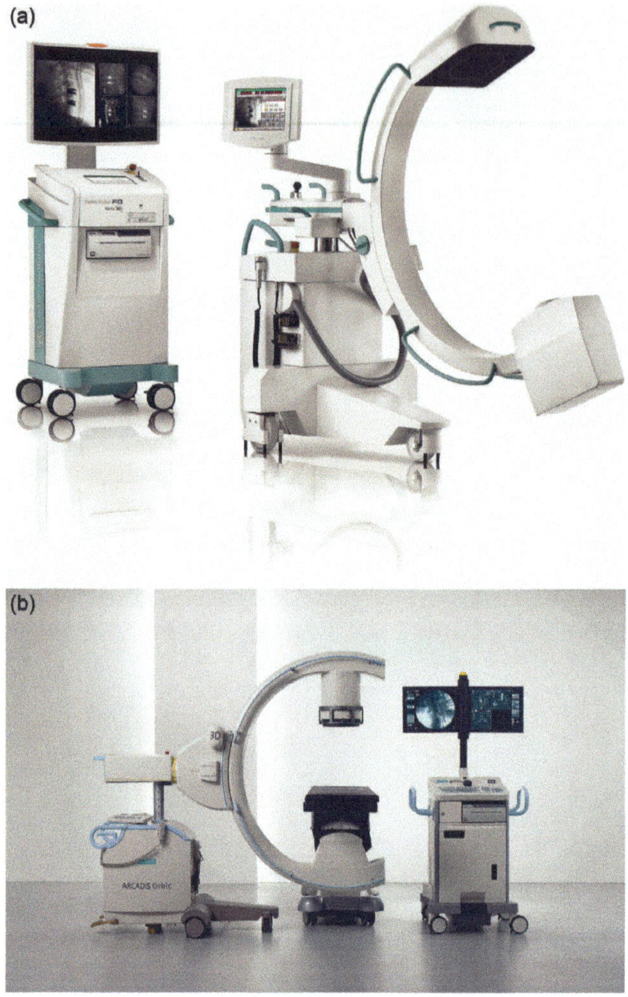

Figure 14.7 Two common cbCT devices in use today: (a) Ziehm Vision FD Vario 3D (*image provided by Ziehm Imaging and used with permission*) and (b) Arcadis Orbic 3D (*image provided by Siemens Healthcare USA and used with permission*). The O-arm, which is another common cbCT in use today, is not pictured.

cbCT, making it a more versatile imaging modality with applications in cranial neurosurgery as well as other surgical subspecialties. (See Figure 14.9.)

There are, however, several disadvantages of the use of iCT. This technology is significantly more expensive than cbCT, which typically provides adequate image quality and scanning volume for most spinal operations. In addition, iCT devices are larger and more difficult to maneuver. Being a true CT scanner, there is no break in the gantry, making it more difficult and time consuming to position the device around

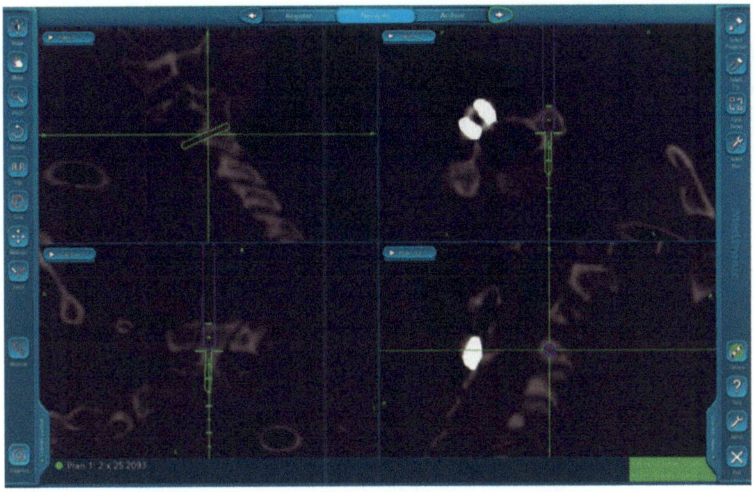

Figure 14.8 Plan for placement of cervical pedicle screw using a portable intraoperative CT scanner.[31] (*Figure 3 from Ref. 31, used with kind permission from Springer Science and Business Media.*)

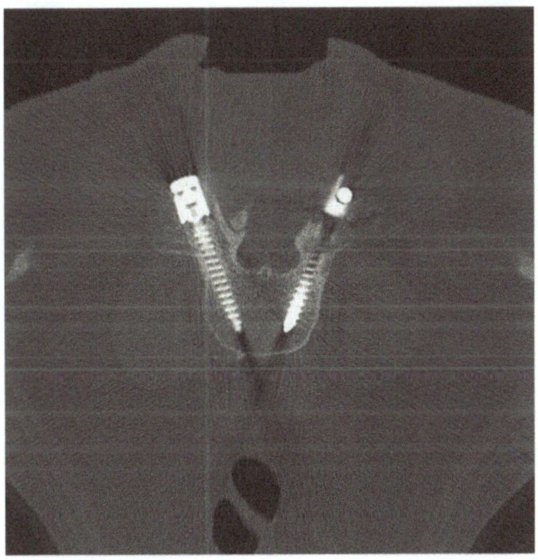

Figure 14.9 Axial image showing placement of thoracic pedicle screws obtained on a portable iCT device showing improved image quality compared to cbCT (as seen in Figure 14.7). (*Image provided by NeuroLogica Corporation and used with permission.*)

the patient. Additionally, a special radiolucent operating table must be used with iCT devices as the traditional Jackson table will not allow the closed bore of the iCT devices to pass through. As a result, in order for the device to be used by different surgeons at the same time, multiple radiolucent operating tables must be purchased and moving the iCT device between rooms could be a challenge.

There are currently two commercially available portable iCT scanners: BrainLab Airo and Neurologica BodyTom (Table 14.3 and Figure 14.10). Both of these devices are portable 32-slice CT scanners. Although these two devices are fundamentally very similar, one key difference is that the BrainLab Airo has an integrated table system,

Table 14.3 Comparison of the available portable iCT devices

	Size (footprint/ weight)	Bore size (cm)	Detector type	Scan sequences	3D Image volume
Neurologica BodyTom	256.5 × 104 cm/ 3510lbs	85	32-slice CT	Helical, axial, dynamic	Cylindrical volume 60 cm diameter × 200 cm length
BrainLab Airo	195 × 38 cm/ 2150lbs	107	32-slice CT	Helical	Cylindrical volume 51.5 cm × 100 cm

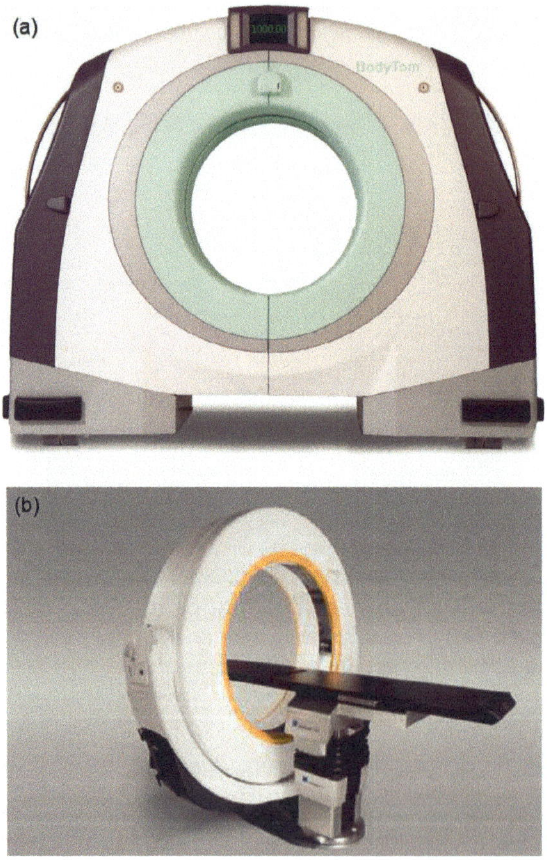

Figure 14.10 Two portable iCT devices in use today: (a) Neurologica BodyTom (*image provided by NeuroLogica Corporation and used with permission*) and (b) BrainLab Airo (*image provided by Brainlab, Inc. and used with permission*).

which is mechanically separate but attached to the base of the device. During the scan, the ring translates along rails on the mobile base so the device is not dependent on floor flatness to obtain images. This may provide a better image quality, yet it is harder for the scanner to be detached from the operating table and be used in another operating room. The BodyTom, however, functions by translating automatically on the floor, independently from the operating table.

The use of ioMRI has become increasingly popular in cranial neurosurgery, especially in the resection of brain tumors; however, the use of this technology in spine surgery is very limited. This is likely due to the high cost and the poor resolution of bony structures on MRI compared to cbCT or iCT. In some situations such as resection of tumors in the spine where high-quality soft tissue imaging is necessary, it may be more advantageous to fuse a preoperative MRI with the iCT or cbCT used for registration.

Table 14.4 compares the different imaging navigation modalities used in image-guided spine surgery.

APPLICATIONS OF IMAGE-GUIDED NAVIGATION IN SPINE SURGERY

Accuracy of pedicle screw placement

Several studies have compared the accuracy of screw placement with freehand techniques to those using more modern image-guided techniques. One of the initial randomized controlled trials in this field, published in 2000, demonstrated the accuracy of pedicle screws placed using the freehand technique compared to those using 3D point matching image guidance and reported a breach rate of 13.4% and 4.6%, respectively.[32] In a study investigating the use of 3D fluoroscopy-based navigation, it was found that of 188 screws placed on 25 patients, 2.6% breached medially and 2.6% were misaligned with the pilot trajectory.[33] This was a significant improvement compared with other studies demonstrating 86.6% accuracy in adult nonnavigated screw placement and 93.7% accuracy in adult navigated screw placement. In a meta-analysis of literature prior to 2007 on the accuracy of pedicle screw placement, there was a median accuracy of 90.3% in 12,299 pedicle screws placed without navigation versus a median accuracy of 95.2% in 3059 pedicle screws placed with navigation.[1] Another meta-analysis of 23 studies investigating pedicle screw placement with computer-assisted navigation demonstrated an accuracy of 84.7% in 2437 compared to 93.3% accuracy with navigation.[34] However, these studies did not differentiate navigation technique. In a meta-analysis of 20 studies from 2000 to 2011 investigating primarily CT-based computer-navigated versus freehand pedicle screw placement, the overall relative risk for pedicle perforation was determined to be 0.39 ($p < 0.001$) in favor of navigated compared to nonnavigated screw placements.[35] The overall risk for perforation with navigation was 6%, in comparison with 15% without imaging assistance. The anatomic area found to have the most decreased risk with navigated screw

Table 14.4 Comparison of different imaging navigation modalities in image-guided spine surgery

	Fluoroscopy	Preoperative CT	2D Fluoroscopy	cbCT	iCT/ioMRI
Ability to use with navigation system	No	Yes	Yes	Yes	Yes
Registration method	N/A	Point matching/surface matching	Automatic	Automatic	Automatic
Compatibility with MIS	Yes	No	Yes	Yes	Yes
Need for special preoperative imaging	No	Yes	No	No	No
Intraoperative image acquisition in surgical position	Yes	No	Yes	Yes	Yes
Real-time imaging	Yes	No	No, but can update navigation intraoperative	No, but can update navigation intraoperative	No, but can update navigation intraoperative
Image quality	Only 2D imaging is possible. Low image quality in low cervical/high thoracic spine and in obese and osteopenic patients	3D dimensional reconstructions as well as axial/coronal/sagittal views available	Only 2D imaging is possible. Low image quality in low cervical/high thoracic spine and in obese and osteopenic patients	Image quality less than routine CT. 3D dimensional reconstructions as well as axial/coronal/sagittal views available	Highest quality intraoperative imaging. 3D dimensional reconstructions as well as axial/coronal/sagittal views available.
Ability to confirm adequate placement of instrumentation in OR	No	No	No	Yes	Yes

placement was in the thoracic spine (RR 0.32). In lumbar and cervical spine, the relative risks for perforation using navigation were 0.38 and 0.49, respectively. There were no neurological complications in the navigated group (4814 screws total), but there were three in the nonnavigated group (3725 screws total). They concluded that the number of pedicle screws that need to be inserted with neuronavigation to prevent one perforation was 11.1. However, it is still unclear whether there is a clinically apparent benefit in cases in which pedicle malposition would most likely be asymptomatic. A meta-analysis comparing CT-based navigation to 2D fluoroscopy-based navigation showed higher accuracy with CT-based techniques (90.76% vs. 85.48%) in all procedural subgroups tested.[36] Another meta-analysis of screw placement accuracy that differentiated between preoperative CT-based, 2D fluoroscopy-based, and 3D fluoroscopy-based screw placement accuracy showed a decreased in vivo risk of pedicle violation in all groups, with relative risks of 0.32–0.60, 0.27–0.48, and 0.09–0.36, respectively.[24] Among navigated techniques, however, there was no statistically significant difference noted. A more recent meta-analysis, however, showed 3D fluoroscopic guidance to be superior to both conventional and 2D fluoroscopic guidance in all spinal levels, but did not compare these modalities to CT-based guidance.[37] Accurate screw placement for each group was 68.1%, 84.3%, and 95.5% for conventional, 2D, and 3D fluoroscopic image guidance, respectively.[37]

Correction of spinal deformity

The placement of instrumentation in patients with spinal deformity, such as congenital pediatric scoliosis or adult *de novo* degenerative scoliosis, is a particularly challenging situation in which the use of image navigation is of great benefit. In these patients, the normal spinal anatomy is obscured by deformity in the coronal and sagittal planes and significant rotatory listhesis.[38,30] The morphometry of vertebrae in patients with scoliosis is also substantially different compared to that in patients with normal spinal anatomy, which further complicates the placement of spinal instrumentation.[39] Most significantly, the pedicle width at the concave side is usually smaller. In addition, the spinal cord is usually closer to the concave side and the aorta is often displaced posterolaterally. Therefore, the risk of significant neurological or vascular damage from misplacement of pedicle screws is much higher. Using the traditional freehand technique, it may be difficult for the surgeon to recognize the location and trajectory of the pedicles, and therefore, higher rates of hardware malposition have been observed. Without image guidance, Liljenqvist et al. demonstrated a 25% perforation rate in thoracic pedicle screws for the treatment of idiopathic scoliosis in a series of 122 pedicle screws in 32 patients. Medial perforation was observed in 8.3% of screws and lateral perforation in 14.2% of screws.[40] Similarly, Belmont et al. retrospectively examined the placement of 399 freehand thoracic pedicle screws in 47 patients and found that patients with coronal plane deformity had a significantly higher rate of pedicle breach (41% vs 27%; $p = 0.04$).[41]

There have been several studies demonstrating the benefit of using image guidance for navigation in deformity surgery.[42–44] Kotani et al. demonstrated that using a preoperative CT-based navigation in adolescent scoliosis surgery, the pedicle screw perforation rate decreased from 11% in the freehand control group to 1.8% in the navigation group. The perforations in the control group were usually medial (89%), which implies that use of navigation may be more useful in that situation.[42] Similarly, Sakai et al. retrospectively examined 40 patients with idiopathic scoliosis and found a pedicle perforation rate of 11.4% when preoperative CT-based navigation was used, compared to 28% in the control group.[43]

Three-dimensional fluoroscopy-based navigation has also been used in surgery for spinal deformity. This navigation system is at least as accurate as preoperative CT-based systems in deformity surgery but is more time efficient. Kotani et al. found that the time to position one screw using CT-based navigation was 10.9 minutes; however, with 3D fluoroscopy-based navigation, this decreased to 5.4 minutes.[45] In pediatric patients with congenital spinal deformity such as scoliosis, kyphosis, and spinal dysgenesis, 3D fluoroscopy-based image guidance has also shown promising results. Larson et al. reported a 99.3% accuracy rate of pedicle screw placement in such cases.[46] Ughwanogho et al. noted more optimal placement of pedicle screws in patients with adolescent idiopathic scoliosis compared to conventional freehand screw insertion with fluoroscopy confirmation.[47]

Due to the large field of view, iCT-based navigation systems are particularly helpful in deformity surgery. Although cbCT systems have a limited field of view that only allows the imaging of up to 3–5 vertebral bodies, iCT systems allow long segment (>10 segments) image acquisition and navigation. The correction of long segmental deformity can be evaluated at the end of the case via another iCT scan before closure. Tormenti et al. placed 164 thoracolumbar pedicle screws in 12 patients undergoing surgery for correction of scoliosis using iCT-based navigation. With iCT they observed a 1.2% rate of pedicle breach compared to 5.2% in the control group using standard fluoroscopy-based techniques.[48] Scheufler et al. operated on 30 individuals with adult scoliosis and demonstrated that use of iCT for image acquisition and navigation was superior to routine multiplanar fluoroscopy and led to a major change in intraoperative workflow. They were able to successfully place up to 8 levels of thoracolumbar instrumentation with a single registration scan and obtained excellent position in >99% of screws.[30] In addition, because patients with adult scoliosis have major rotational and coronal deformity, use of multiplanar fluoroscopy required frequent C-arm adjustments at each level resulting in increased radiation exposure for the surgeon and the patient. This problem was effectively avoided with the use of iCT-based navigation.

These studies together demonstrated that the use of image guidance technology is particularly advantageous in spinal deformity surgery. The benefits include optimization of hardware placement, reduced occupational radiation exposure, reduced time in the operating room, and the ability to check hardware placement and overall deformity correction with a final iCT scan at the end of the case.

Minimally invasive spine surgery

Minimally invasive spine surgery, especially when using techniques focused on the preservation of normal muscular structure, leads to less surgical pain, risk of infection, blood loss, and faster recovery. However, muscle preservation unavoidably leads to less direct visualization of the bony reference points used for instrumentation. The use of image-guided navigation effectively overcomes the lack of direct visualization in minimally invasive spine surgery and has facilitated the improvement and development of minimally invasive techniques.

A good example is minimally invasive pedicle screw placement. The traditional method of lumbar pedicle screw placement requires wide dissection of paraspinal muscles and detachment of the muscle from the spinous processes, laminae, facets, and the transverse processes. The screw entry point is then determined by the identification of the junction of the transverse process and the superior articular process. With image guidance, the entry point and angle of the screw insertion can be determined without the need for the direct visualization of anatomic reference points, thereby eliminating the need for muscle dissection and allowing the preservation of the normal neurovascular structures that support the paraspinal muscles. Minimally invasive pedicle screw placement can be achieved with the use of biplanar fluoroscopy. Generally, two C-arms are required to provide simultaneous AP and lateral fluoroscopic views for the accurate placement of the pedicle screws. This requires frequent beam angle adjustments at each level for optimal screw insertion, which increases time and radiation exposure for the case. It also requires the use of Jamshidi needles and Kirschner wires. Although infrequent, there is a risk of nerve root damage, visceral, or vascular injury associated with the use of Jamshidi needle or Kirschner wire. It is also possible that the K-wire could break during the procedure.[49]

The ability to use image-guided navigation for minimally invasive pedicle screw placement has changed screw insertion technique. With accurate image registration, the insertion of pedicle screws does not require either the anatomic identification of screw entry point or the identification of accurate pedicle shadows on fluoroscopy. The screw placement is simplified to direct drilling, tapping, and insertion of screws using the virtual 3D navigation images on the computer. This eliminates a lot of guesswork involved in finding the proper pedicle trajectory. The pedicle screw diameter and length can be optimized with fewer concerns about pedicle wall violation or

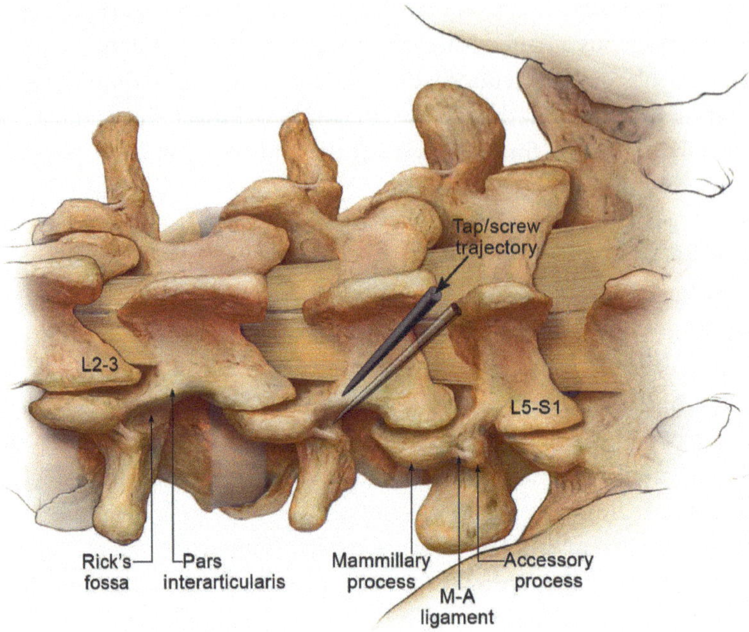

Figure 14.11 Medial to lateral pedicle screw insertion entry point and trajectory. By entering more medially on the surface of the pars compared to a conventional pedicle screw, the screw trajectory traverses more cortical bone and the construct, which sits adjacent to the spinous process, better underlies the paraspinal muscles. (*Used with kind permission from Springer Science and Business Media. © 2014 Springer: "Minimally Invasive Wiltse Approaches for Posterolateral Fusion" by S Ritland in* Minimally Invasive Spinal Deformity Surgery, *Figure 34.9.*)

anterior cortex violation, thereby increasing the screw pullout strength and biomechanical stability. In many cases, there is also no need for use of Jamshidi needles or Kirschner wires, eliminating the risk of complications associated with these tools.

Use of image-guided navigation also creates new possibilities for hardware insertion, which may provide further advantage in preserving normal physiologic structures. Several screw trajectories that are not able to be performed safely with anatomical guidance or fluoroscopic guidance alone could be achieved with the help of stereotactic navigation. In fact, as long as the screws are placed in bony elements that provide strong purchase and pullout strength, they do not necessarily have to be placed along the longitudinal axis of the pedicles. Several screw trajectories have been suggested to further reduce the need for soft tissue dissection. For example, a medial to lateral lumbar pedicle screw trajectory (Figures 14.11 and 14.12) allows less muscle dissection for the placement of pedicle screws due to the more medial screw entry point on the pars than traditional pedicle screws. Using this technique, the screws aim laterally and superiorly toward the lateral superior corner of the vertebral body and are generally

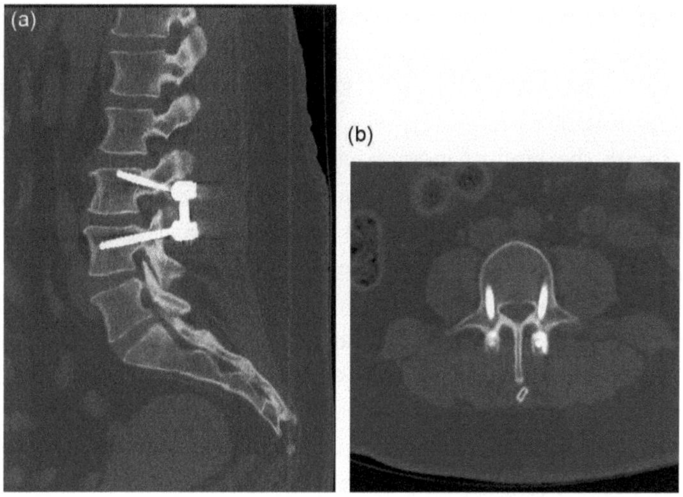

Figure 14.12 Example of the medial to lateral pedicle screws insertion at the L3 level.

smaller in both diameter and length compared to traditional pedicle screws. The margin of error is smaller than with traditional pedicle screws, and the screws traverse mostly cortical bone so it is not as easy to probe through the planned trajectory based on tactile feedback. The screw trajectory is also not along the longitudinal axis of the pedicle so fluoroscopy is less helpful in guiding the screw placement. Medial breach of the screws would place the traversing nerve roots that are just medial to the lateral recess in danger. Lateral breach during tapping or screw insertion would ruin the possibility of using the trajectory for screw insertion. Use of intraoperative navigation effectively solves these problems and makes the medial to lateral pedicle screw insertion a viable alternative in minimally invasive lumbar spine surgery. Similarly, other posterior instrumentation possibilities can be achieved with image guidance with better accuracy. For example, transfacet pedicle screws have been shown to provide similar stability to traditional pedicle screw-rod construct.[50,51] Translaminar facet screws have also been used as an alternative method for posterior stabilization.[52] Neither of these screw placement trajectories is intuitive, even with fluoroscopy. Therefore, use of navigation might prove to be helpful in placing those screws in the intended locations, thereby maximizing the potential benefit.

Minimally invasive spine: cervical spine

Holly and Foley were among the first to describe the placement of minimally invasive percutaneous screws throughout the cervical region using image guidance.[53] In a 2006 study, using 3D fluoroscopy-based navigation, they demonstrated accurate placement of 41 percutaneous pedicle screws (97.6%) in three cadaver specimens. Trajectories included C1–C2 transarticular placement, C3–6 lateral masses, and C7 pedicle placement. There was only one minor cortical breech at C7. In a large group of acute spinal trauma, investigators introduced an intraoperative O-arm for assistance.[54] In this setting, investigators successfully placed C1–C2 percutaneous transarticular screws in one patient. This was of particular benefit given a traumatic skin ulceration that may have interfered with the kind of incision used for an open approach. In addition, the authors successfully treated a Hangman's fracture using C2 pars osteosynthesis screws.

Minimally invasive anterior approaches to the cervical spine have also benefited from image guidance. In addition to treatment of basilar invagination via transcervical odontoidectomy as described below, the group has treated a recurrent cervical chordoma with endoscope-assisted transcervical resection. Accurate identification of the tumor, vertebral arteries, and delineation of the dural sac all benefited from image-based visualization.[55]

There have also been advancements in the minimally invasive treatment of degenerative cervical disease through the use of intraoperative navigation. In one such technical report, a modified transcorporeal anterior cervical microforaminotomy (MTACM) is described. Although already a MIS procedure aimed to directly decompress the pathology and avoid a larger procedure such as traditional ACDF, the modified approach uses a 3D fluoroscopy-based navigation platform.[56] Eight patients with foraminal disc herniation underwent MTACM with good results. The authors discuss the benefits of intraoperative navigation in selecting a skin entry point and minimizing its size as well as the drill hole size. Although times were longer compared with their nonnavigated experience, they felt that the improved safety and efficacy outweighed the increased operative time.

Anterior screw placement for odontoid fracture screw placement is another area in which image-guided MIS has developed. Traditional screw placement using biplanar fluoroscopy was compared with an image-guided approach using a 3D fluoroscopy-based navigation system.[57] Not only did fusion across the fracture increase from 88% to 95% in the image-guided group, but also the total length of surgery was decreased despite the longer time needed to prepare the operating room.

Another interesting application of navigated MIS includes the use of real-time MR combined with percutaneous technique. In one case report, the authors describe a cervical intradural fenestration of a cyst anterior to the spinal cord.[58] Avoiding the extensive dissection otherwise required by an open approach, the authors used an MR-compatible spinal needle to puncture the cyst under MR guidance. The needle coursed through the C1–2 interlaminar space into the dural sac and the cyst contents were aspirated. The patient was free of postoperative complications, and at 9 month follow-up, complete disappearance of the cyst was confirmed.

Minimally invasive spine: thoracic spine

Pedicle screw placement in the thoracic spine can be particularly challenging given the small pedicle size compared with other regions of the spine and the complex

3D anatomy. In addition, a freehand technique relies heavily on tactile feedback and anatomic landmarks, skills that tend to improve with experience. Misplacement rates have been quoted as high as 55% in previous cadaveric studies.[59] Consequences can include nerve root damage, hemorrhage, and spinal cord injury. Human cadaveric studies of thoracic pedicle screw placement have led to reductions in structurally significant violations.[60] Minimal access techniques including percutaneous screw placement coupled with navigation in the thoracic spine are being explored. Holly et al. placed 94 percutaneous screws into cadavers using a 3D fluoroscopy device with an accuracy of 94.7%.[61] The average pedicle diameter was 6 mm, and of the five pedicles with a cortical breech, the diameter was 4.6 mm.

In another study of 50 patients undergoing 3D fluoroscopy-assisted pedicle screw placement, 20% of patients underwent percutaneous placement[62] and 5 of 40 pedicle screws (12.5%) breached the pedicle medially. There were no pedicle wall violations in any other direction. All breeches were less than 2 mm of the pedicle and no revisions were required.

In a study of six patients with unstable thoracic spine fractures, Kakarla et al. placed 37 pedicle screws with the assistance of a 3D fluoroscopy-based navigation system[63] and a total of seven screws were misplaced, six of which had a cortical violation less than 2 mm and the other was greater than 2 mm. All misplacements were clinically asymptomatic and none of the screws required revision. In the large trauma series above by Schouten et al., 2 of 27 patients underwent attempted placement of percutaneous thoracic pedicle screws.[54] One patient with diffuse idiopathic skeletal hyperostosis and a T10–11 hyperextension injury underwent successful placement of pedicle screws from T7–L2. However, a preexisting kyphosis prevented the placement of the upper thoracic screws using navigation, and instead they were placed fluoroscopically. The other patient underwent attempted fixation of T2–10, but this was aborted as the intraoperative imaging suggested suboptimal placement and was converted to an open approach using anatomic landmarks.

Image guidance has also been combined with minimally invasive percutaneous kyphoplasty to treat patients with vertebral compression fractures. In a series of patients undergoing lumbar and thoracic kyphoplasties, the authors report on 11 thoracic (T4–12) procedures using 3D fluoroscopy-based navigation.[64] No complications occurred and the authors report a reduction in radiation time from an average of 293.2 seconds in nine historical controls compared with 41.3 seconds in the study cohort.

Minimally invasive spine: lumbosacral spine

The accuracy of freehand pedicle screw placement in the lumbar spine has been reported around 79.0% in at least one meta-analysis and even lower during deformity surgery, with misplacement levels as high as 30% in pedicle screws at the thoracolumbar region.[65,30] Baaj et al. reported on 14 patients undergoing 3D fluoroscopy-assisted placement of minimally invasive percutaneous screws in the thoracolumbar region.[66] Eleven patients underwent procedures for degenerative diagnoses, whereas three patients were operated on for traumatic instability. Of 110 placed screws, 68 were available for retrospective review. Four lateral breaches occurred (<2 mm) and two medial breaches occurred (one >2 mm, one >4 mm), with only one medial breech undergoing immediate revision. No patients suffered from postoperative neurological deficits attributable to misplaced screws.

As described above, image guidance in correction of spinal deformity has many benefits, including improved accuracy of pedicle screw placement. The study by Scheuffler and colleagues on a series of 30 patients with adult onset scoliosis describes an approach utilizing minimal access techniques combined with image guidance.[30] Pedicle screw navigation was provided by biplanar fluoroscopy in 16 patients, whereas iCT-based navigation was used in the other 14 most recent patients. Unilateral mini-open transforaminal lumbar interbody fusions (TLIFs) were undertaken on the side of the concavity. A total of 415 screws were placed using the two imaging techniques. Intraoperative CT navigation required on average 17.5 minutes for automatic registration across all segments. No revision implants were needed and no grade II or higher misplacements were made. Another variant of the lumbar interbody fusion, the AxiaLIF procedure, which is a minimally invasive presacral approach for L4–S1 fusion, has benefited from intraoperative navigation. Luther et al. operated on six patients with refractory low back pain, performing L5–S1 procedures on five patients and an L4–S1 procedure on one patient.[67] A 3D fluoroscopy-based system was used for intraoperative navigation. No patients required reoperation and all implanted screws were in the correct position.

Navigated percutaneous iliosacral screws have also been studied in the setting of pelvic injury resulting in sacroiliac instability.[68] A wide variation in sacroiliac joint morphology in addition to the close association of the L5 nerve root has been associated with the relatively higher rates of screw misplacement and neurological injury following nonnavigated percutaneous approaches. Smith and colleagues studied three various navigation systems compared with fluoroscopy in a cadaver model. All navigation systems achieved higher accuracy scores compared with standard nonnavigated fluoroscopy. Another group studied percutaneous iliosacral screw fixation using assistance with an O-arm.[69] Forty pelvic ring fractures were treated by surgery using both open and percutaneous techniques as well as fluoroscopic and CT-guided navigation. There were no neurological complications or malpositioning of screws in the navigated group.

Image guidance in difficult surgical scenarios

In addition to the instrumentation in deformity cases and/or to facilitate minimally invasive spine surgery, image guidance has been found to be a useful adjunct for surgeons in multiple other difficult instrumentation scenarios. This is especially true in situations where there are poor anatomical landmarks or where hardware must be placed in close relation to neurovascular structures that are not under direct vision. It is also useful in circumstances where hardware is placed through a difficult or unusual approach in which the operating surgeon has limited experience.

Posterior upper cervical spine surgery

Given the unique anatomic constraints of the cervical region, including the close relationship between the vertebral arteries and the bony elements, image guidance is of particular interest in this area. Indeed, vertebral artery injury has been reported in as many as 4.1% of procedures using a conventional transarticular approach for C1–C2 screw placement.[70] Other potential complications include spinal cord and nerve root injuries. In patients with rheumatoid arthritis, normal anatomical landmarks may be destroyed at the C1–C2 junction, making instrumentation even more hazardous.[71] Because of this low margin for error, use of image navigation systems has been found to be a helpful adjunct for surgeons operating in this region.

The placement of C1–C2 transarticular screws using image guidance has been well described. Early studies demonstrated that the use of preoperative CT-based navigation systems decreased but did not eliminate the risk of screw misplacement.[72–74,71] Acosta et al. found that in some cases in which transarticular screw placement was thought to be anatomically difficult or impossible based on routine preoperative imaging, they were able to successfully place the screw using surgical navigation.[74] Uehera et al. performed a retrospective study of 20 patients who underwent C1–C2 fusion with transarticular screws using preoperative CT-based navigation (Figure 14.13). They observed a perforation rate of 2.6% and had no major complications, including vertebral artery injury or pseudoarthrosis.[75]

Another option for C1–C2 stabilization is placement of lateral mass screws at C1 and either laminar or pedicle screw fixation at C2. Rajasekaran et al. describe a case

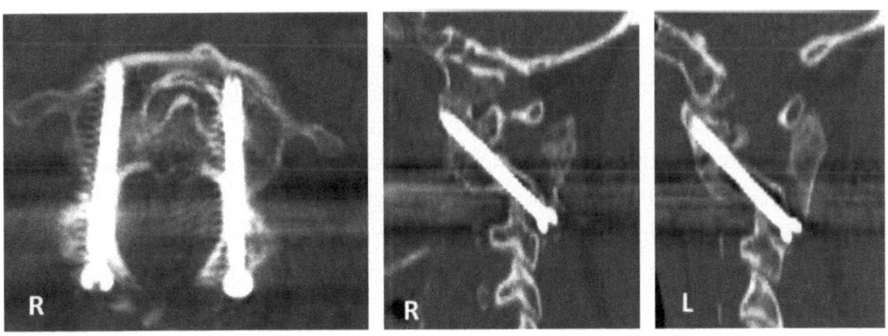

Figure 14.13 Postoperative CT scan after transarticular screw fixation performed using CT-based navigation demonstrates accurate screw placement.[75] (*Used with permission from M. Uehara.*[75])

of direct pedicle screw fixation of C2 using 3D fluoroscopy-based navigation for successful treatment of a displaced Hangman's fracture.[76] Nottmeier et al. described a series of 18 patients who underwent placement of instrumentation into the occipital and C1–C2 levels using 3D fluoroscopy-based navigation and found that this technique was safe and accurate, with only 1 out of 78 screws (a C2 laminar screw) demonstrating a minor lateral breach on postoperative CT.[77] Yang et al. found similarly that placement of C1 lateral mass and C2 pedicle screws using 3D fluoroscopy-based navigation resulted in fewer perforations compared to conventional fluoroscopy. They also found that the use of navigation resulted in shorter operative time, less blood loss, and decreased radiation exposure.[78] More recently, the use of iCT-based navigation has shown promising results for posterior fixation of C1–C2.

Additionally, use of navigation allows surgeons to consider the possibility of placing pedicle screws in the subaxial cervical spine. Traditionally, many surgeons have avoided using cervical pedicle screws due to the narrow trajectory and close proximity to critical neurovascular structures despite their biomechanical advantage over the more commonly placed lateral mass screws.[79,80] However, advanced navigation techniques may prove to be useful in the safe placement of pedicle screws in the subaxial cervical spine.

Transoral and transcervical odontoidectomy

Image navigation may also be useful in approaching the ventral surface of the craniocervical junction, which is significant for its complicated anatomy and proximity to multiple critical structures. Using a cranial neuronavigation system and fiducial-based registration, Veres et al. performed a transoral odontoidectomy in three patients with either osteomyelitis or atlantoaxial subluxation with basilar impression. Because the fiducial points were located on the skull, HALO fixation was required in these patients to prevent intersegmental motion at occiput–C1 and C1–C2 and to ensure the accuracy of registration.[81] The authors found that this navigation system provided an excellent 3D guidance of the C1–C2 complex and allowed for safe and efficient resection of the odontoid process. In order to avoid the morbidity of a transoral approach, Dasenbrock et al. described an endoscopic, image-guided approach for odontoidectomy in 15 patients with basilar invagination using 3D fluoroscopy-based navigation.[82,83] In this approach, a tubular retractor is placed on the ventral aspect of the atlantoaxial complex through a midcervical incision exposing an area from the top of the anterior arch of C1 to the caudal end of the C2–C3 disc space. The odontoid is then resected using endoscopy for visual control and 3D fluoroscopy for navigation and to obtain intraoperative imaging to assess completeness of resection. Although they encountered incidental durotomy resulting in cerebrospinal fluid leak in 20% of patients, the surgeons found that image guidance was useful in guiding the surgical trajectory and evaluating whether decompression was adequate. In another case study,

Leng et al. removed an os odontoideum that was compressing the cervicomedullary junction through an image-guided endoscopic endonasal approach following posterior fixation. The authors reported a short recovery and also avoided the morbidity associated with a traditional transoral approach.[84]

Oblique corpectomy for cervical spondylitic myelopathy

Cervical spondylitic myelopathy (CSM) is a common condition treated by spine surgeons. Because the compressive pathology in CSM is often located ventral to the spinal cord, anterior approaches to the cervical spine are commonly used for decompression. These anterior surgeries, however, cause instability requiring instrumentation and fusion, which exposes the patient to graft and hardware-related complications. Within the past 20 years, surgeons have explored an oblique corpectomy via an anterolateral approach for CSM. In selected patients, this operation may be used to decompress the cervical spinal cord while preserving more than half of the vertebral bodies so as not to cause instability requiring hardware placement or fusion.[85] Despite several publications demonstrating good clinical results, this surgery remains uncommon and is performed only by a handful of spine surgeons[86] (Figure 14.14). One reason this surgery has not gained popularity may be that the use of an oblique drilling plane is not familiar to many surgeons and the lack of bony anatomical landmarks may make it easy to lose orientation during surgery. Because of this, it conceptually makes sense that image guidance during surgery would be a useful tool. Lee et al. evaluated the use of a C-arm-based navigation system in 11 patients who underwent multilevel oblique corpectomy. When compared to the control group, they found that the use of image navigation resulted in more complete bony resection, faster operative times, and improved Japanese Orthopaedic Association Scores.[87]

Transvertebral pedicle screws

Another uncommon operation that may be more practical with the use of image navigation is the placement of transvertebral pedicle screws. This technique, which was first described in 1994,[88] has traditionally been used in cases of high-grade spondylolisthesis at L5–S1 and is useful for patients with osteoporosis as each screw crosses more cortical boundaries. Transvertebral pedicle screws have also been found to have a biomechanical advantage over traditional pedicle screws at that level.[89] Nottmeier et al. reported the use of 3D fluoroscopy-based image guidance for the placement of transvertebral pedicle screws in the thoracic spine in 12 patients. They found that they were able to achieve high rates of fusion and only experienced one minor pedicle breach (out of 35 screws), which was without clinical significance.[77] The use of 3D fluoroscopy-based navigation in this surgery proved useful because of the long and narrow trajectory that each screw must follow for adequate placement. Thoracic

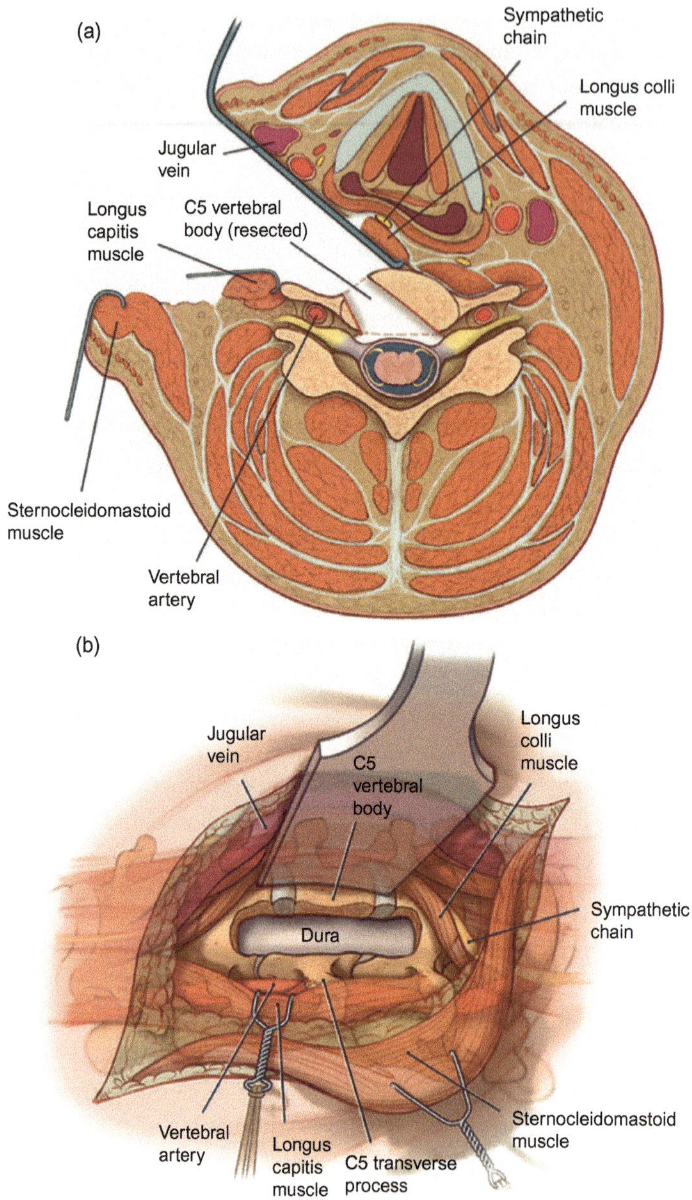

Figure 14.14 (a) The approach for oblique cervical corpectomy. (b) The surgeon's view at completion of surgery.[86] (*Used with permission from T. Kiris,[86] p. 676.*)

pedicles tend to be narrow in the axial plane; however, the screws must be medialized enough that the tip remains within the vertebral body above. In the sagittal plane, the screw must pass from the inferodorsal aspect of the pedicle to the superoventral portion in order to cross both endplates of the disc space above. This requires a precise trajectory that would be difficult to obtain without navigation.

Artificial disc placement

Intraoperative navigation may also prove beneficial in artificial disc replacement (ADR) surgery. Recently, lumbar ADR has been shown to be a viable alternative to fusion for patients with discogenic pain, otherwise unresponsive to nonoperative management. However, outcomes after ADR have been directly influenced by position of the implant in the intervertebral disc space.[90] Correct placement is considered critical to ensuring long-term function and optimal clinical outcome with total disc arthroplasties. Off-center misplacement potentially predisposes implants to asymmetric loading with subsequent shifts in axes of rotation leading to premature implant wear, implant loosening, and nonphysiologic stresses on adjacent segments and facet joints leading to accelerated adjacent-level degeneration.

Although most disc arthroplasties are inserted and analyzed using fluoroscopy, this method carries a risk of misplacement because of errors relating to parallax and skewed spinous processes. To that end, Rauschmann et al. compared the insertion of 15 lumbar intervertebral disc prostheses using preoperative CT-based navigation by an inexperienced ADR surgeon, with 15 prostheses placed with exclusive use of fluoroscopy by an experienced ADR surgeon in 10 human cadaver spine specimens. After insertion, CT scans were analyzed using computer software to assess placement accuracy of each disc prosthesis. Only three navigated disc prostheses were suboptimal and none was poorly placed, compared with two discs suboptimally placed and one poorly placed in the manual group. They concluded that navigated placement of the implants was more accurate, especially in the coronal plane.[91]

Application of image guidance in spinal oncology

The use of intraoperative navigation systems has been shown to be particularly helpful in oncological spine surgery, especially in localizing and determining the margins of resection of complex tumors.[92] Many of these cases can be difficult due to abnormal regional anatomy and biomechanics.[93] Using navigation technology, surgeons can better understand the boundaries of the tumor and its relationship to neurovascular structures—enabling the surgeon to safely maximize the extent of resection. In addition, navigation can be very useful in reconstruction and stabilization of the spine following tumor resection.

There are very few publications reporting on the use of image guidance in surgery for spinal tumors. Bandeira et al. used preoperative CT-based navigation for the resection of various tumors, including osteoid osteoma, osteoblastoma, angioma, and metastatic disease, and felt that the use of navigation improved tumor localization and excision.[92] Campos et al. used a preoperative CT-based navigation system to localize an osteoid osteoma within the T9 vertebral body, which was curettaged using video-assisted thoracoscopic surgery[94] with good results as well. In this case, the use of navigation allowed the surgeons to avoid sacrificing additional bone, which may have caused instability requiring fusion.[95] Similarly, Nagashima et al. report the use of preoperative CT-based navigation for resection of an osteoid osteoma of the C2 pedicle without creating instability.[96] Using 3D fluoroscopy, Rajasekaran et al. described the resection of three spinal osteoid osteomas. In addition to localizing and excising the lesion without the removal of excess bone, the use of 3D fluoroscopy allowed for easier registration and confirmation of adequacy of resection prior to leaving the operating room.[97]

The use of image guidance may be particularly helpful in the resection of tumors such as chordomas or sarcomas where *en bloc* surgical resection with negative margins is correlated with improved outcomes.[98–100] Dasenbrock et al. described the usefulness of preoperative CT-based navigation for *en bloc* resection of sacral chordomas in three patients. Because of complex anatomy in this region, tumor resection can be difficult and it is not always possible to achieve negative margins. However, they found that the use of navigation helped to visualize the 3D anatomy of the tumor, minimize the risk of injury to adjacent structures, decrease the risk of violating the tumor capsule, and increase the likelihood of achieving negative margins.[100] Dasenbrock et al. and Fubiyashi et al. also felt that the use of navigation in complex spinal tumors could in some situations allow surgeons to perform resection in a single-stage operation; otherwise, a combined anterior and posterior approach would be necessary.[99,100]

Guppy et al. provide an excellent example of the application of image guidance in oncological spine surgery. They performed *en bloc* resection of a C2–3 chordoma with instrumented fusion from the occiput to C6 using an O-arm-based navigation system.[101] In this case, the patient, a 49-year-old woman who presented with difficulty breathing, neck pain, and radicular pain, underwent MRI demonstrating an enhancing mass centered at C2 with epidural extension and severe narrowing of the spinal canal. The tumor also encased the right vertebral artery (which was embolized prior to surgery) and displaced the right internal carotid artery. Using 3D fluoroscopy-based navigation, the surgeons were able to successfully perform a two-stage posterior–anterior surgery. In the first stage, the surgeons performed a C2–3 laminectomy and facetectomy and placed lateral mass screws and pedicle screws in the subaxial cervical spine. The navigation system helped to identify the lateral masses and location of the vertebral artery allowing for safe bone removal. In the second stage of the surgery, *en bloc* resection of the tumor was performed from a transcervical, transmandibular, transpalatal approach. In this stage, image guidance helped to define the margins of the tumor for resection, protect the left vertebral artery, and place anterior C1 lateral mass screws.[101] This case illustrates many of the benefits of navigation in tumor surgery. Using navigation as an adjunct, the authors were able to (1) gain better understanding of the complex, abnormal, 3D anatomy of the tumor; (2) protect vital neurovascular structures, including the spinal cord, nerve roots, carotid arteries, and left vertebral artery, which alone was supplying the posterior circulation; (3) accurately place instrumentation in a difficult scenario; and (4) achieve *en bloc* resection of the tumor.

Another strategy that may be utilized in tumor surgery is fusion of preoperative MRI or CT data to the intraoperative imaging modality that is being used for navigation. This technique creates an overlay that allows the surgeon to integrate and analyze multiple imaging modalities simultaneously.[102] This is particularly useful in determining the soft tissue extent of tumors or location of vascular structures, which are not well seen on routine CT or fluoroscopy.

SPECIFICS OF IMAGE GUIDANCE

Time efficiency

With the advancement of navigation systems in spine surgery, many surgeons have been concerned that the use of this technology might increase the length of each operation. In fact, the early image-guided navigation systems utilizing preoperative CT scan and point matching techniques were considerably time consuming. Standard techniques of pedicle screw insertion employing fluoroscopy have reported fluoroscopy time in pedicle screw placement varying from 3.4 to 66 s per screw with significant associated radiation; yet overall screw placement time can be largely practitioner dependent.[103]

Early studies on the time required for a registration procedure on one vertebral level using preoperative CT-based image guidance was 6–8 minutes, whereas the time required for application of a pedicle screw using these computer-assisted techniques was an additional 6–10 minutes.[4] More recent studies demonstrate the average time required for registration of a vertebral segment using preoperative imaging as less than 2 minutes, even with insufficient initial registrations 13% of the time, with an accuracy obtained of 0.9 mm.[22] Girardi et al., in a study of 62 patients who underwent lumbar decompression and pedicle screw fixation using a preoperative CT-based navigation system, reported the time required for screw placement using a single, multilevel registration. The mean time required to insert each screw was 6.6 (3.3–12.5) minutes. When studying L5–S1 instrumented fusions with the use of fluoroscopy-based image guidance, a significant decrease in OR time has been reported compared to standard fluoroscopy.[103] For 3D fluoroscopy-guided lumbar procedures, significantly reduced mean operative times have been demonstrated (231 minutes) compared to procedures with conventional fluoroscopy being used (321 minutes).[104] For one-level procedures, the difference remained significant; however, there was no statistically significant difference in multilevel procedures. For odontoid fractures, a study comparing 2D to 3D fluoroscopy guidance demonstrated significantly shortened length of surgery in the 3D group (83 vs. 115 minutes); however, the total operating room time (149 vs. 162 minutes) was the same given the preincision preparation time needed (66 vs. 47 minutes).[57] During operations in which intraoperative CT is used for navigation, interruption for image acquisition has been shown to be in the range of 9 minutes for spinal surgery.[25] In another study, intraoperative CT scanning for navigation was again shown to be 9 minutes, with most scan durations falling within 7.5 and 10.5 minutes.[29] Using a portable intraoperative CT scanner, it was estimated that 18–40 minutes was added to operative time and varied based on the extent of instrumented area and anatomical complexity.[31,105]

Overall, these estimates are specific to the experience of the surgeons and OR staff of each study, and the time to accomplish the necessary tasks can depend largely on prior surgeon experience and staff efficiency, which are not well studied in the literature. For each individual practitioner, important considerations regarding the effect on the operative time should include preincision preparation time, intraoperative scanning and registration time, need for rescanning prior to closure, OR staff training, and surgeon experience with a particular image guidance system.

Intraoperative radiation exposure

As additional imaging techniques enter into practice for spinal surgeons, secondary exposure to ionizing radiation is a common concern. Known risks of radiation exposure include increased risk of cataracts, lymphoma, and thyroid cancer.[106] A 2005 study reported a five fold higher risk of lifetime cancer rates among orthopedic surgeons compared to nonorthopedic surgeons[106]; however, there is significant debate within the literature regarding overall and long-term effects of low-dose radiation, with no firm causal relationship established.[107] With less invasive techniques that tend to employ greater amounts of radiation, it is important for surgeons to have a general understanding of the degree of radiation involved. Barrier techniques, positioning, radiation–distance relationships, and understanding modality-specific radiation levels are important in minimizing exposure to ionizing radiation. Traditionally, minimally invasive spine surgery relies heavily on the fluoroscopy, thereby leading to significantly greater radiation exposure for both patients and surgeons.[106] The use of fluoroscopic time for minimally invasive spine procedures on average is about 3 times that of the equivalent open procedures. In a study of 31 minimally invasive transforaminal lumbar interbody fusions, radiation to the surgeon was found to be highest at the middle finger and did not differ significantly from the surgical assistant or radiographic technologist; overall, radiation levels fell within the International Commission of Radiological Protection's guidelines.[108]

The advancement of neuronavigation has significantly decreased the radiation exposure for both the operating room staff and the patients. With the modern navigation systems, the intraoperative imaging is acquired with the patient in surgical position prior to the start of the surgery. Most of the time, this imaging is completed with the operating room staff and the surgeons either out of the operating room or behind a lead shield, thereby minimizing the radiation exposure. After the initial imaging study, there is little need for fluoroscopy, so neither the staff nor the surgeons need to wear a lead apron or thyroid shield during the case. Several studies have confirmed decreased occupational radiation exposure with the use of navigation. Use of navigation with 3D fluoroscopy, for example, has been shown to reduce radiation exposure to the surgeon during pedicle screw placement by a factor of 10 compared to freehand fluoroscopy, with the only radiation exposure being from preoperative-level determination and positioning of the fluoroscopy device before image acquisition.[109] When using 3D fluoroscopy-based navigation for spinal fusion procedures, one study found that the duration of radiation was reduced to 72 seconds compared to 105 seconds in a nonnavigated procedure.[110] Smith et al. demonstrated the mean radiation exposure to the torso of procedures performed with standard fluoroscopy as 13 times higher than procedures performed with computer-assisted image guidance.[111] In a clinical series of TLIF procedures, radiation exposure was undetectable when navigation-assisted fluoroscopy was used compared to 12.4 mREM of radiation delivered to the surgeon during unilateral MIS TLIF without navigation; the total fluoroscopy time for the navigation-assisted group was 57.1 seconds compared with 147.2 seconds using fluoroscopy.[112]

Radiation exposure also varies by physical relation to instrument, with 3D fluoroscopy and CT typically demonstrating the highest radiation exposure to the patient, and C-arm providing the most radiation scatter to the surgeon and surgical assistant.[113] Using anthropomorphic models, Zhang et al. calculated that the radiation exposure to patients using 3D fluoroscopy is about 50% of that of a standard 64-slice body CT.[114]

Cost

Considering the evolving health care environment in the United States and increasing focus on cost-effectiveness in surgical care, there has been little research published regarding the comparative financial impact of using image navigation in spine surgery. Although clinically significant complications and revisions secondary to misplaced screws are fairly uncommon regardless of imaging modality, the costs associated with subsequent procedures are significant. Overall, imaging techniques have substantially improved misplacement rates of hardware and decreased revision rates.

In a small institutional study comparing standard fluoroscopy to computer-assisted image guidance for pedicle screw placement, it was found that the revision surgery rate decreased from 3% to 0%, with an overall cost saving of $71,286 per 100 cases.[115] The difference in revision rate was, however, not statistically significant [therefore, the accuracy of these cost savings on a larger scale is unclear.] In a separate Italian study comparing intraoperative O-arm imaging to preoperative CT-based image guidance, there was significant shortening of the procedure time; however, there was no significant cost difference from the hospital perspective.[116] They found that a financial "break-even" point to be 154 procedures using image guidance per year; however, in practice, this number would likely be dependent on the specific case mix and costs specific to individual hospitals, as well as interdepartmental utilization of imaging devices to offset purchasing costs. With upfront costs of image navigation systems in the range of $225,000 to $1 million, it may be financially beneficial for surgical centers that perform high-volume spine procedures or centers performing complex spinal deformities. Another consideration is that the use of intraoperative image systems such as O-arm will decrease the need for postoperative CT imaging to confirm hardware placement.

Larger, multi-institutional studies are needed to assess cost-effectiveness of image guidance systems for spine surgery, and consideration of local practice patterns and costs is essential in the decision of any individual institution to adopt imaging navigation into the practice.

CONCLUSION

The rapid advancement of imaging and navigation technology has made spine surgery safer and more efficient than ever, especially for cases requiring difficult surgical approaches or instrumentation. The ability to create three-dimensional reconstructions of patient anatomy in the operating room and to use this data for interactive navigation has allowed spine surgeons to more confidently approach complex surgeries. This technology has also been crucial to the continued advancement of minimally invasive spine surgery. In turn, patients benefit from faster recovery time, less postsurgical pain, and lower infection rates. Image-guided spine surgery also benefits the surgeons and operating room staff with decreased occupational radiation exposure. Currently, image-guided navigation technology is limited by the lack of real-time image feedback, unlike the traditional fluoroscopy. Therefore, the accuracy and safety of image guidance is largely dependent on the registration process. In addition, the size, cost, and maneuverability of these units still limit the widespread usage of larger navigation systems such as iCT, whereas smaller systems are limited by the reduced image quality and low scan volume, which only allow the surgeon to see a few segments of the spine at a time.

REFERENCES

1. Nottmeier EW. A review of image-guided spinal surgery. *J Neurosurg Sci*. 2012;56(1):35–47.
2. Nolte LP, Zamorano L, Visarius H, et al. Clinical evaluation of a system for precision enhancement in spine surgery. *Clin Biomech*. 1995;10(6):293–303.
3. Brodwater BK, Roberts DW, Nakajima T, Friets EM, Strohbehn JW. Extracranial application of the frameless stereotactic operating microscope: experience with lumbar spine. *Neurosurgery*. 1993;32 (2):209–213:discussion 213.
4. Lee TC, Yang LC, Liliang PC, Su TM, Rau CS, Chen HJ. Single versus separate registration for computer-assisted lumbar pedicle screw placement. *Spine*. 2004;29(14):1585–1589.
5. Holly LT. Image-guided spinal surgery. *Int J Med Robot*. 2006;2(1):7–15.
6. Roessler K, Ungersboeck K, Dietrich W, et al. Frameless stereotactic guided neurosurgery: clinical experience with an infrared based pointer device navigation system. *Acta Neurochir*. 1997;139 (6):551–559.
7. Steinmann JC, Herkowitz HN, el-Kommos H, Wesolowski DP. Spinal pedicle fixation. Confirmation of an image-based technique for screw placement. *Spine*. 1993;18(13):1856–1861.
8. Laine T, Makitalo K, Schlenzka D, Tallroth K, Poussa M, Alho A. Accuracy of pedicle screw insertion: a prospective CT study in 30 low back patients. *Eur Spine J*. 1997;6(6):402–405.
9. Castro WH, Halm H, Jerosch J, Malms J, Steinbeck J, Blasius S. Accuracy of pedicle screw placement in lumbar vertebrae. *Spine*. 1996;21(11):1320–1324.
10. Belmont Jr. PJ, Klemme WR, Dhawan A, Polly Jr. DW. In vivo accuracy of thoracic pedicle screws. *Spine*. 2001;26(21):2340–2346.
11. Xu R, Ebraheim NA, Ou Y, Yeasting RA. Anatomic considerations of pedicle screw placement in the thoracic spine. Roy-Camille technique versus open-lamina technique. *Spine*. 1998;23(9):1065–1068.
12. Mezger U, Jendrewski C, Bartels M. Navigation in surgery. *Langenbecks Arch Surg*. 2013;398 (4):501–514.
13. Wong KC, Kumta SM. Use of computer navigation in orthopedic oncology. *Curr Surg Rep*. 2014;2:47.
14. Tjardes T, Shafizadeh S, Rixen D, et al. Image-guided spine surgery: state of the art and future directions. *Eur Spine J*. 2010;19(1):25–45.
15. Hsiang J. Wrong-level surgery: a unique problem in spine surgery. *Surg Neurol Int*. 2011;2:47.
16. Sanders R, Koval KJ, DiPasquale T, Schmelling G, Stenzler S, Ross E. Exposure of the orthopaedic surgeon to radiation. *J Bone Joint Surg Am*. 1993;75(3):326–330.
17. Thambiraj S, Quraishi NA. Intra-operative localisation of thoracic spine level: a simple "'K'-wire in pedicle" technique. *Eur Spine J*. 2012;21(Suppl 2):S221–S224.
18. Nolte LP, Visarius H, Arm E, Langlotz F, Schwarzenbach O, Zamorano L. Computer-aided fixation of spinal implants. *J Image Guid Surg*. 1995;1(2):88–93.
19. Winkler D, Vitzthum HE, Seifert V. Spinal markers: a new method for increasing accuracy in spinal navigation. *Comput Aided Surg*. 1999;4(2):101–104.
20. Gebhard F, Weidner A, Liener UC, Stockle U, Arand M. Navigation at the spine. *Injury*. 2004;35 (Suppl 1):S-A35–S-45.
21. Moses ZB, Mayer RR, Strickland BA, et al. Neuronavigation in minimally invasive spine surgery. *Neurosurg Focus*. 2013;35(2):E12.
22. Nottmeier EW, Crosby TL. Timing of paired points and surface matching registration in three-dimensional (3D) image-guided spinal surgery. *J Spinal Disord Tech*. 2007;20(4):268–270.
23. Foley KT, Smith MM. Image-guided spine surgery. *Neurosurg Clin N Am*. 1996;7(2):171–186.
24. Tian NF, Huang QS, Zhou P, et al. Pedicle screw insertion accuracy with different assisted methods: a systematic review and meta-analysis of comparative studies. *Eur Spine J*. 2011;20(6):846–859.
25. Uhl E, Zausinger S, Morhard D, et al. Intraoperative computed tomography with integrated navigation system in a multidisciplinary operating suite. *Neurosurgery*. 2009;64(5 Suppl 2):231–239: discussion 239–240.
26. Engle DJ, Lunsford LD. Brain tumor resection guided by intraoperative computed tomography. *J Neurooncol*. 1987;4(4):361–370.

27. Lunsford LD, Parrish R, Albright L. Intraoperative imaging with a therapeutic computed tomographic scanner. *Neurosurgery.* 1984;15(4):559–561.
28. Shalit MN, Israeli Y, Matz S, Cohen ML. Intra-operative computerized axial tomography. *Surg Neurol.* 1979;11(5):382–384.
29. Zausinger S, Scheder B, Uhl E, Heigl T, Morhard D, Tonn JC. Intraoperative computed tomography with integrated navigation system in spinal stabilizations. *Spine.* 2009;34(26):2919–2926.
30. Scheufler KM, Cyron D, Dohmen H, Eckardt A. Less invasive surgical correction of adult degenerative scoliosis, part I: technique and radiographic results. *Neurosurgery.* 2010;67(3):696–710.
31. Barsa P, Fröhlich R, Benes 3rd V, Suchomel P. Intraoperative portable CT-scanner based spinal navigation—a feasibility and safety study. *Acta Neurochir.* 2014;156(9):1807–1812.
32. Laine T, Lund T, Ylikoski M, Lohikoski J, Schlenzka D. Accuracy of pedicle screw insertion with and without computer assistance: a randomised controlled clinical study in 100 consecutive patients. *Eur Spine J.* 2000;9(3):235–240.
33. Patil S, Lindley EM, Burger EL, Yoshihara H, Patel VV. Pedicle screw placement with O-arm and stealth navigation. *Orthopedics.* 2012;35(1):e61–e65.
34. Verma R, Krishan S, Haendlmayer K, Mohsen A. Functional outcome of computer-assisted spinal pedicle screw placement: a systematic review and meta-analysis of 23 studies including 5,992 pedicle screws. *Eur Spine J.* 2010;19(3):370–375.
35. Shin BJ, James AR, Njoku IU, Hartl R. Pedicle screw navigation: a systematic review and meta-analysis of perforation risk for computer-navigated versus freehand insertion. *J Neurosurg Spine.* 2012;17(2):113–122.
36. Tian NF, Xu HZ. Image-guided pedicle screw insertion accuracy: a meta-analysis. *Int Orthop.* 2009;33(4):895–903.
37. Mason A, Paulsen R, Babuska JM, et al. The accuracy of pedicle screw placement using intraoperative image guidance systems. *J Neurosurg Spine.* 2014;20(2):196–203.
38. Perennou D, Marcelli C, Herisson C, Simon L. Adult lumbar scoliosis. Epidemiologic aspects in a low-back pain population. *Spine.* 1994;19(2):123–128.
39. Liljenqvist UR, Link TM, Halm HF. Morphometric analysis of thoracic and lumbar vertebrae in idiopathic scoliosis. *Spine.* 2000;25(10):1247–1253.
40. Liljenqvist UR, Halm HF, Link TM. Pedicle screw instrumentation of the thoracic spine in idiopathic scoliosis. *Spine.* 1997;22(19):2239–2245.
41. Belmont Jr. PJ, Klemme WR, Robinson M, Polly Jr. DW. Accuracy of thoracic pedicle screws in patients with and without coronal plane spinal deformities. *Spine.* 2002;27(14):1558–1566.
42. Kotani Y, Abumi K, Ito M, et al. Accuracy analysis of pedicle screw placement in posterior scoliosis surgery: comparison between conventional fluoroscopic and computer-assisted technique. *Spine.* 2007;32(14):1543–1550.
43. Sakai Y, Matsuyama Y, Nakamura H, et al. Segmental pedicle screwing for idiopathic scoliosis using computer-assisted surgery. *J Spinal Disord Tech.* 2008;21(3):181–186.
44. Nakanishi K, Tanaka M, Misawa H, Sugimoto Y, Takigawa T, Ozaki T. Usefulness of a navigation system in surgery for scoliosis: segmental pedicle screw fixation in the treatment. *Arch Orthop Trauma Surg.* 2009;129(9):1211–1218.
45. Kotani T, Akazawa T, Sakuma T, et al. Accuracy of pedicle screw placement in scoliosis surgery: a comparison between conventional computed tomography-based and O-arm-based navigation techniques. *Asian Spine J.* 2014;8(3):331–338.
46. Larson AN, Polly Jr. DW, Guidera KJ, et al. The accuracy of navigation and 3D image-guided placement for the placement of pedicle screws in congenital spine deformity. *J Pediatr Orthop.* 2012;32(6):e23–e29.
47. Ughwanogho E, Patel NM, Baldwin KD, Sampson NR, Flynn JM. Computed tomography-guided navigation of thoracic pedicle screws for adolescent idiopathic scoliosis results in more accurate placement and less screw removal. *Spine.* 2012;37(8):E473–E478.
48. Tormenti MJ, Kostov DB, Gardner PA, Kanter AS, Spiro RM, Okonkwo DO. Intraoperative computed tomography image-guided navigation for posterior thoracolumbar spinal instrumentation in spinal deformity surgery. *Neurosurg Focus.* 2010;28(3):E11.

49. Nottmeier EW, Fenton D. Three-dimensional image-guided placement of percutaneous pedicle screws without the use of biplanar fluoroscopy or Kirschner wires: technical note. *Int J Med Robot.* 2010;6(4):483–488.
50. Chen C-J, Saulle D, Fu K-M, Smith JS, Shaffrey CI. Dysphagia following combined anterior-posterior cervical spine surgeries. *J Neurosurg Spine.* 2013;19(3):279–287.
51. Agarwala A, Bucklen B, Muzumdar A, Moldavsky M, Khalil S. Do facet screws provide the required stability in lumbar fixation? A biomechanical comparison of the Boucher technique and pedicular fixation in primary and circumferential fusions. *Clin Biomech.* 2012;27(1):64–70.
52. Amoretti N, Amoretti ME, Hovorka I, Hauger O, Boileau P, Huwart L. Percutaneous facet screw fixation of lumbar spine with CT and fluoroscopic guidance: a feasibility study. *Radiology.* 2013;268 (2):548–555.
53. Holly LT, Foley KT. Percutaneous placement of posterior cervical screws using three-dimensional fluoroscopy. *Spine.* 2006;31(5):536–540.
54. Schouten R, Lee R, Boyd M, et al. Intra-operative cone-beam CT (O-arm) and stereotactic navigation in acute spinal trauma surgery. *J Clin Neurosci.* 2012;19(8):1137–1143.
55. Hsu W, Kosztowski TA, Zaidi HA, Gokaslan ZL, Wolinsky JP. Image-guided, endoscopic, transcervical resection of cervical chordoma. *J Neurosurg Spine.* 2010;12(4):431–435.
56. Kim JS, Eun SS, Prada N, Choi G, Lee SH. Modified transcorporeal anterior cervical microforaminotomy assisted by O-arm-based navigation: a technical case report. *Eur Spine J.* 2011;20(Suppl 2): S147–S152.
57. Martirosyan NL, Kalb S, Cavalcanti DD, et al. Comparative analysis of isocentric 3-dimensional C-arm fluoroscopy and biplanar fluoroscopy for anterior screw fixation in odontoid fractures. *J Spinal Disord Tech.* 2013;26(4):189–193.
58. Takahashi S, Morikawa S, Egawa M, Saruhashi Y, Matsusue Y. Magnetic resonance imaging-guided percutaneous fenestration of a cervical intradural cyst. Case report. *J Neurosurg.* 2003;99 (3 Suppl):313–315.
59. Gelalis ID, Paschos NK, Pakos EE, et al. Accuracy of pedicle screw placement: a systematic review of prospective in vivo studies comparing free hand, fluoroscopy guidance and navigation techniques. *Eur Spine J.* 2012;21(2):247–255.
60. Youkilis AS, Quint DJ, McGillicuddy JE, Papadopoulos SM. Stereotactic navigation for placement of pedicle screws in the thoracic spine. *Neurosurgery.* 2001;48(4):771–778.
61. Holly LT, Foley KT. Three-dimensional fluoroscopy-guided percutaneous thoracolumbar pedicle screw placement. Technical note. *J Neurosurg.* 2003;99(3 Suppl):324–329.
62. Oertel MF, Hobart J, Stein M, Schreiber V, Scharbrodt W. Clinical and methodological precision of spinal navigation assisted by 3D intraoperative O-arm radiographic imaging. *J Neurosurg Spine.* 2011;14(4):532–536.
63. Kakarla UK, Little AS, Chang SW, Sonntag VK, Theodore N. Placement of percutaneous thoracic pedicle screws using neuronavigation. *World Neurosurg.* 2010;74(6):606–610.
64. Villavicencio AT, Burneikiene S, Bulsara KR, Thramann JJ. Utility of computerized isocentric fluoroscopy for minimally invasive spinal surgical techniques. *J Spinal Disord Tech.* 2005;18(4):369–375.
65. Kosmopoulos V, Schizas C. Pedicle screw placement accuracy: a meta-analysis. *Spine.* 2007;32(3): E111–E120.
66. Baaj AA, Beckman J, Smith DA. O-Arm-based image guidance in minimally invasive spine surgery: technical note. *Clin Neurol Neurosurg.* 2013;115(3):342–345.
67. Luther N, Tomasino A, Parikh K, Hartl R. Neuronavigation in the minimally invasive presacral approach for lumbosacral fusion. *Minim Invasive Neurosurg.* 2009;52(4):196–200.
68. Smith HE, Yuan PS, Sasso R, Papadopolous S, Vaccaro AR. An evaluation of image-guided technologies in the placement of percutaneous iliosacral screws. *Spine.* 2006;31(2):234–238.
69. Coste C, Asloum Y, Marcheix PS, Dijoux P, Charissoux JL, Mabit C. Percutaneous iliosacral screw fixation in unstable pelvic ring lesions: the interest of O-arm CT-guided navigation. *Orthop Traumatol Surg Res.* 2013;99(4 Suppl):S273–S278.
70. Madawi AA, Casey AT, Solanki GA, Tuite G, Veres R, Crockard HA. Radiological and anatomical evaluation of the atlantoaxial transarticular screw fixation technique. *J Neurosurg.* 1997;86 (6):961–968.

71. Kelleher MO, Tan G, Sarjeant R, Fehlings MG. Predictive value of intraoperative neurophysiological monitoring during cervical spine surgery: a prospective analysis of 1055 consecutive patients. *J Neurosurg Spine*. 2008;8(3):215–221.
72. Weidner A, Wahler M, Chiu ST, Ullrich CG. Modification of C1–C2 transarticular screw fixation by image-guided surgery. *Spine*. 2000;25(20):2668–2673.
73. Borm W, Konig RW, Albrecht A, Richter HP, Kast E. Percutaneous transarticular atlantoaxial screw fixation using a cannulated screw system and image guidance. *Minim Invasive Neurosurg*. 2004;47 (2):111–114.
74. Acosta Jr. FL, Quinones-Hinojosa A, Gadkary CA, et al. Frameless stereotactic image-guided C1-C2 transarticular screw fixation for atlantoaxial instability: review of 20 patients. *J Spinal Disord Tech*. 2005;18(5):385–391.
75. Uehara M, Takahashi J, Hirabayashi H, et al. Computer-assisted C1-C2 transarticular screw fixation "Magerl technique" for atlantoaxial instability. *Asian Spine J*. 2012;6(3):168–177.
76. Rajasekaran S, Vidyadhara S, Ramesh P, Shetty AP. Randomized clinical study to compare the accuracy of navigated and non-navigated thoracic pedicle screws in deformity correction surgeries. *Spine*. 2007;32(2):E56–E64.
77. Nottmeier EW, Pirris SM. Placement of thoracic transvertebral pedicle screws using 3D image guidance. *J Neurosurg Spine*. 2013;18(5):479–483.
78. Yang YL, Zhou DS, He JL. Comparison of isocentric C-arm 3-dimensional navigation and conventional fluoroscopy for C1 lateral mass and C2 pedicle screw placement for atlantoaxial instability. *J Spinal Disord Tech*. 2013;26(3):127–134.
79. Ling JM, Dinesh SK, Pang BC, et al. Routine spinal navigation for thoraco-lumbar pedicle screw insertion using the O-arm three-dimensional imaging system improves placement accuracy. *J Clin Neurosci*. 2014;21(3):493–498.
80. Yu FB, Chen DY, Wang XW, Liu XW. [Radiographic comparison of anterior cervical fusion after two-level discectomy or single-level corpectomy for two-level cervical spondylotic myelopathy]. *Zhonghua yi xue za zhi*. 2012;92(37):2636–2640.
81. Veres R, Bago A, Fedorcsak I. Early experiences with image-guided transoral surgery for the pathologies of the upper cervical spine. *Spine*. 2001;26(12):1385–1388.
82. Wolinsky JP, Sciubba DM, Suk I, Gokaslan ZL. Endoscopic image-guided odontoidectomy for decompression of basilar invagination via a standard anterior cervical approach. Technical note. *J Neurosurg Spine*. 2007;6(2):184–191.
83. Dasenbrock HH, Clarke MJ, Bydon A, et al. Endoscopic image-guided transcervical odontoidectomy: outcomes of 15 patients with basilar invagination. *Neurosurgery*. 2012;70(2):351–359:discussion 359–360.
84. Leng LZ, Anand VK, Hartl R, Schwartz TH. Endonasal endoscopic resection of an os odontoideum to decompress the cervicomedullary junction: a minimal access surgical technique. *Spine*. 2009;34 (4):E139–E143.
85. George B, Gauthier N, Lot G. Multisegmental cervical spondylotic myelopathy and radiculopathy treated by multilevel oblique corpectomies without fusion. *Neurosurgery*. 1999;44(1):81–90.
86. Kiris T, Kilincer C. Cervical spondylotic myelopathy treated by oblique corpectomy: a prospective study. *Neurosurgery*. 2008;62(3):674–682.
87. Lee HY, Lee SH, Son HK, et al. Comparison of multilevel oblique corpectomy with and without image guided navigation for multi-segmental cervical spondylotic myelopathy. *Comput Aided Surg*. 2011;16(1):32–37.
88. Abdu WA, Wilber RG, Emery SE. Pedicular transvertebral screw fixation of the lumbosacral spine in spondylolisthesis. A new technique for stabilization. *Spine*. 1994;19(6):710–715.
89. Minamide A, Akamaru T, Yoon ST, Tamaki T, Rhee JM, Hutton WC. Transdiscal L5–S1 screws for the fixation of isthmic spondylolisthesis: a biomechanical evaluation. *J Spinal Disord Tech*. 2003;16(2):144–149.
90. Goel VK, Faizan A, Palepu V, Bhattacharya S. Parameters that effect spine biomechanics following cervical disc replacement. *Eur Spine J*. 2012;21(Suppl 5):S688–S699.

91. Rauschmann MA, Thalgott J, Fogarty M, et al. Insertion of the artificial disc replacement: a cadaver study comparing the conventional surgical technique and the use of a navigation system. *Spine*. 2009;34(10):1110–1115.
92. Bandiera S, Ghermandi R, Gasbarrini A, Barbanti Brodano G, Colangeli S, Boriani S. Navigation-assisted surgery for tumors of the spine. *Eur Spine J*. 2013;22(Suppl 6):S919–S924.
93. Errani C, Kreshak J, Ruggieri P, Alberghini M, Picci P, Vanel D. Imaging of bone tumors for the musculoskeletal oncologic surgeon. *Eur J Radiol*. 2013;82(12):2083–2091.
94. Lee JY, Hilibrand AS, Lim MR, et al. Characterization of neurophysiologic alerts during anterior cervical spine surgery. *Spine*. 2006;31(17):1916–1922.
95. Campos WK, Gasbarrini A, Boriani S. Case report: curetting osteoid osteoma of the spine using combined video-assisted thoracoscopic surgery and navigation. *Clin Orthop Relat Res*. 2013;471 (2):680–685.
96. Nagashima H, Nishi T, Yamane K, Tanida A. Case report: osteoid osteoma of the C2 pedicle—surgical technique using a navigation system. *Clin Orthop Relat Res*. 2010;468(1):283–288.
97. Rajasekaran S, Kamath V, Shetty AP. Intraoperative Iso-C three-dimensional navigation in excision of spinal osteoid osteomas. *Spine*. 2008;33(1):E25–E29.
98. Smitherman SM, Tatsui CE, Rao G, Walsh G, Rhines LD. Image-guided multilevel vertebral osteotomies for en bloc resection of giant cell tumor of the thoracic spine: case report and description of operative technique. *Eur Spine J*. 2010;19(6):1021–1028.
99. Fujibayashi S, Neo M, Takemoto M, et al. Computer-assisted spinal osteotomy: a technical note and report of four cases. *Spine*. 2010;35(18):E895–E903.
100. Dasenbrock HH, Clarke MJ, Bydon A, et al. En bloc resection of sacral chordomas aided by frameless stereotactic image guidance: a technical note. *Neurosurgery*. 2012;70(1 Suppl Operative):82–87: discussion 87–88.
101. Guppy KH, Chakrabarti I, Banerjee A. The use of intraoperative navigation for complex upper cervical spine surgery. *Neurosurg Focus*. 2014;36(3):E5.
102. Wong KC, Kumta SM, Antonio GE, Tse LF. Image fusion for computer-assisted bone tumor surgery. *Clin Orthop Relat Res*. 2008;466(10):2533–2541.
103. Sasso RC, Garrido BJ. Computer-assisted spinal navigation versus serial radiography and operative time for posterior spinal fusion at L5-S1. *J Spinal Disord Tech*. 2007;20(2):118–122.
104. Fraser J, Gebhard H, Irie D, Parikh K, Hartl R. Iso-C/3-dimensional neuronavigation versus conventional fluoroscopy for minimally invasive pedicle screw placement in lumbar fusion. *Minim Invasive Neurosurg*. 2010;53(4):184–190.
105. Barsa P, Suchomel P. Portable CT scanner-based navigation in lumbar pedicle screw insertion. *Eur Spine J*. 2013;22(6):1446–1450.
106. Yu E, Khan SN. Does less invasive spine surgery result in increased radiation exposure? A systematic review. *Clin Orthop Relat Res*. 2014;472(6):1738–1748.
107. Nelson EM, Monazzam SM, Kim KD, Seibert JA, Klineberg EO. Intraoperative fluoroscopy, portable X-ray, and CT: patient and operating room personnel radiation exposure in spinal surgery. *Spine J*. 2014;14(12):2992–2994.
108. Funao H, Ishii K, Momoshima S, et al. Surgeons' exposure to radiation in single- and multi-level minimally invasive transforaminal lumbar interbody fusion: a prospective study. *PLoS One*. 2014;9 (4):e95233.
109. Villard J, Ryang YM, Demetriades AK, et al. Radiation exposure to the surgeon and the patient during posterior lumbar spinal instrumentation: a prospective randomized comparison of navigated versus non-navigated freehand techniques. *Spine*. 2014;39(13):1004–1009.
110. Gebhard FT, Kraus MD, Schneider E, Liener UC, Kinzl L, Arand M. Does computer-assisted spine surgery reduce intraoperative radiation doses? *Spine*. 2006;31(17):2024–2027.
111. Smith HE, Welsch MD, Sasso RC, Vaccaro AR. Comparison of radiation exposure in lumbar pedicle screw placement with fluoroscopy vs computer-assisted image guidance with intraoperative three-dimensional imaging. *J Spinal Cord Med*. 2008;31(5):532–537.

112. Kim CW, Lee YP, Taylor W, Oygar A, Kim WK. Use of navigation-assisted fluoroscopy to decrease radiation exposure during minimally invasive spine surgery. *Spine J.* 2008;8(4):584–590.
113. Abdullah KG, Bishop FS, Lubelski D, Steinmetz MP, Benzel EC, Mroz TE. Radiation exposure to the spine surgeon in lumbar and thoracolumbar fusions with the use of an intraoperative computed tomographic 3-dimensional imaging system. *Spine.* 2012;37(17):E1074–E1078.
114. Zhang D, Savandi AS, Demarco JJ, et al. Variability of surface and center position radiation dose in MDCT: Monte Carlo simulations using CTDI and anthropomorphic phantoms. *Med Phys.* 2009;36(3):1025–1038.
115. Watkins RG, Gupta A, Watkins RG. Cost-effectiveness of image-guided spine surgery. *Open Orthop J.* 2010;4:228–233.
116. Costa F, Porazzi E, Restelli U, et al. Economic study: a cost-effectiveness analysis of an intraoperative compared with a preoperative image-guided system in lumbar pedicle screw fixation in patients with degenerative spondylolisthesis. *Spine J.* 2013;14(8):1790–1796.

CHAPTER 15

Image-Guided Radiosurgery and Stereotactic Radiotherapy

Agam Sharda[1] **and John R. Adler, Jr.**[1,2]
[1]Varian Medical Systems, Inc., Palo Alto, CA, USA
[2]Department of Neurosurgery, Stanford University, Stanford, CA, USA

IMAGE-GUIDED RADIOSURGERY: DEVELOPMENT AND APPLICATIONS OF LINEAR ACCELERATOR-BASED SRS

For over four decades, Lars Leksell's concept of a multitude of radiation beams converging on and ablating a target has been the guiding principle for the field of radiosurgery.[1] Frame-based targeting and immobilization of the skull, not surprisingly using the type of stereotactic frames that Leksell himself invented, was critical to this initial concept of radiosurgery. In 1985, the first glimmerings of image-guided targeting appeared to a young neurosurgical fellow at the Karolinska Institute (see Adler et al.[2]), who was observing the early days of radiosurgery under the guidance of Lars Leksell. The conceptual breakthrough Adler made, and which subsequently has been at the heart of all image-guided navigation, was that the patient's skeleton can provide a unique reference frame[3] for the soft tissue contained within. To implement this idea within the domain of radiosurgery, a method of targeting which utilized X-ray image-to-image correlation was developed; of note, this technique mimicked in many ways the use of orthogonal projection of X-rays that was commonly used in 1985 during conventional stereotactic procedures. Eventually, these image-guided radiosurgical techniques were brought into the commercial realm in 1994, with the creation of the CyberKnife® Robotic Radiosurgery System (Accuray Incorporated, Sunnyvale, CA, USA).

Over the past two decades, further advances in machine control, intratreatment imaging and image processing have made it possible to administer all radiosurgery without rigid skeletal fixation, and thereby ushered in the era of image-guided therapeutic radiation that has profoundly transformed the entire field of radiation oncology. Of note, image-guided targeting has been absolutely critical to expanding the principles of stereotactic radiosurgery (SRS) beyond the brain to previously unimaginable extracranial targets, e.g., spine, lung, liver, prostate, and kidney. This progression in the clinical applications of radiosurgery was no accident; from its very inception, the desire to provide the precision of SRS to extracranial targets had been at the heart of the development of image-guided therapeutic radiation.

A. Golby (Ed): Image-Guided Neurosurgery
DOI: http://dx.doi.org/10.1016/B978-0-12-800870-6.00015-7

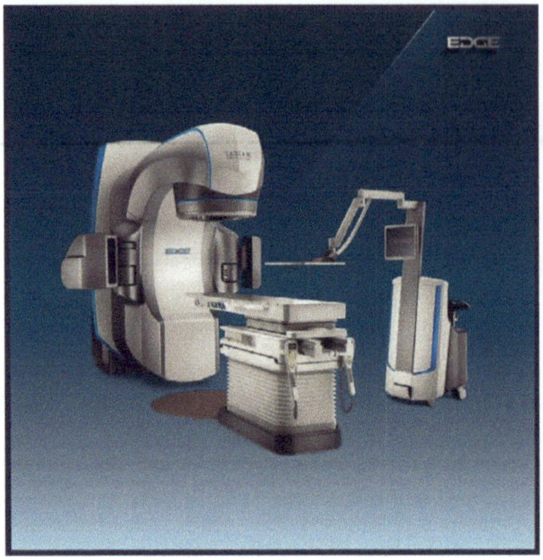

Figure 15.1 A Linac-based radiosurgery system.

In addition to providing stereotactic accuracy for all regions of anatomy, an important benefit of image guidance is the notion of fractionating radiosurgery. Traditionally, radiosurgery, as taught by Lars Leksell, was done in a single fraction, in large part because this was all that stereotactic technology was capable of supporting. The challenge of keeping a stereotactic frame in place or repeatedly replacing it over the course of multiple fractions proved to be an insurmountable barrier to the adoption of multifraction SRS, despite potential theoretical biological benefits. It was only after the development of accurate and reliable frameless (image-guided) technologies that it became possible to precisely treat with radiosurgery over the course of multiple fractions; and, after subsequent clinical experience, the associated benefits of this approach have been gradually recognized.[4,5]

Throughout the past two decades, numerous advances in medical imaging and supporting technology have made possible several new, primarily Linear accelerator (Linac) based, devices capable of performing image-guided radiosurgery. An example of one such modern radiosurgery system that is flexible enough to accommodate both frameless and frame-based treatments is the Edge Radiosurgery System (Varian Medical Systems, Palo Alto, CA, USA) (Figure 15.1). Although noninvasive immobilization is commonly used with all image-guided Linacs today (such as Edge), it is important to note that these devices are used only for restricting (i.e., not eliminating) patient motion, and, unlike stereotactic frames, have no role in actual target localization. This is an important differentiator between image-guided radiosurgery and those procedures that entail more traditional stereotactic frame-based targeting in which the frame explicitly immobilizes the head during treatment.

When performing conventional radiosurgery, the process begins with the attachment of a stereotactic frame to the patient's head, which is followed by imaging studies with the reference frame in place, then treatment planning, and finally the delivery of the actual treatment. Throughout planning and treatment itself, target location is correlated with respect to the stereotactic coordinates of the frame, which in turn is referenced to the ultimate machine isocenter. The conventional delivery of radiosurgery in a single session was predicated by this target-frame-machine linkage. The use of alternative immobilization devices (perhaps more aptly referred to as motion minimization schemes), as is true with modern radiosurgical Linacs and their use of the skull as a reference frame, breaks this linkage and introduces significant clinical flexibility.

Most modern Linac-based radiosurgical systems were developed with a view toward accurately treating both intracranial targets and extracranial targets, which by their nature can be static or move as a result of normal bodily functions. Besides adapting to such small movements, the flexibility of targeting enabled by modern Linacs makes single or multifraction irradiation a trivial process whether the lesion is situated in the brain, spine, prostate, lung, or liver. When performed on an image-guidance equipped Linac, this variety of treatments typically utilizes a combination of beam shaping, X-ray imaging, electromagnetic transponder-based target tracking, and/or patient surface mapping to achieve the requisite level of accuracy.

OVERVIEW OF IMAGE-GUIDED RADIOSURGICAL TECHNOLOGY

Modern, image-guided radiosurgery systems typically combine a linear accelerator with an X-ray based target tracking system, by means of which submillimeter treatment can be delivered. Ultimately, this combination of technical elements makes it possible to modulate radiation output in real time, treat both isocentrically and nonisocentrically, readily fractionate, and even do all this for moving targets.

Different modalities for locating and tracking target motion can be utilized depending on the anatomic location and characteristics of the lesion. While X-ray imaging using either or both projection X-ray and cone-beam CT (CBCT) is the primary tool for intermittent localization, real-time target positioning is also possible for some anatomic locations via electromagnetic transponders or real-time patient surface monitoring. Although ultrasound has also been explored for patient positioning over almost two decades, it is not commonly used.[6]

PROCEDURAL STEPS

Treatment planning

Treatment planning begins with the acquisition of one or more high-resolution 3D, or in those situations where respiratory motion is a concern, time-resolved 4D image

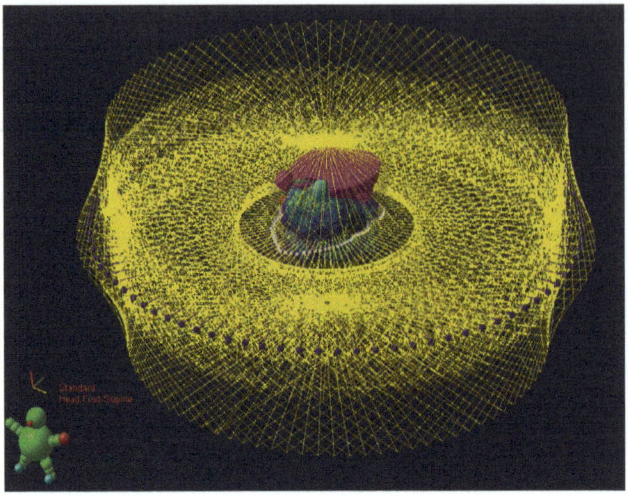

Figure 15.2 Beam distributions in a RapidArc approach.

sets. These images allow the visualization and identification of the target to be treated as well as nearby critical anatomy, often referred to as organs at risk (OAR). Once acquired, these image sets are transferred to the treatment planning system. Within the contemporary treatment planning systems, the OARs can now be auto-contoured using a combination of template, anatomy, or disease-based segmentation tools. After this process is complete, dosimetric planning is performed.

An important new technique for administering modern radiosurgery is volumetric arc therapy (VMAT), a commercial implementation of which is RapidArc® (Varian Medical Systems, Palo Alto, CA) (Figure 15.2). Using the now ubiquitous multileaf collimators (MLC) that are found on all modern Linacs, and a sophisticated set of optimization algorithms, VMAT makes it possible to deliver very large and highly conformal doses with amazing efficiency. This is principally achieved by delivering dose from a full 360° set of planar directions around the patient using beamlets (via the MLC) delivered at 2° intervals. It should be stated that the beam distribution is optimized algorithmically based on the size, location, and proximity of OARs to the target.

Once optimization is complete, a final dose calculation is performed. Recent advances in computing technology have allowed the mainstream availability of advanced dose calculation techniques, such as Monte Carlo,[7] that not only provide very good accuracy in heterogeneous media such as the nasopharyngeal cavity where air-tissue interfaces are the norm, but may also be faster than older, simpler algorithms. The net result of volumetric dose delivery is analogous to Leksell's original premise of subdividing a delivered dose into a multitude of cross-fired beams. When combined with multiple patient orientations using relative treatment couch positioning, the number of dose entrance angles can be increased significantly to even further spread out entrance dose. The use of volumetric techniques in combination with multiple couch orientations has been well studied and is utilized on a routine basis in clinics for both SRS and stereotactic body (SBRT) treatments.[8–10]

Treatment delivery

Patient positioning

The treatment delivery process begins with accurate patient positioning using any number of the previously mentioned image-guidance techniques; the simple goal is to align the patient isocenter with the Linac isocenter. The most commonly used sequence starts with a gross visual patient alignment that uses external markers and a system of calibrated lasers. Fine positioning is then done by acquiring a CBCT and adjusting the patient position within a 6 degree of freedom (6DoF) dimensional space. For a growing number of image-guided Linacs, an integrated 6DoF couch allows the machine software to automatically reposition the patient based on the output of the CBCT and registration software. In contrast, such repositioning is done with the CyberKnife system by adjusting the position of the Linac via the robotic manipulator.

X-ray based target tracking

During patient set-up and repeatedly during treatment, two independent X-ray imaging systems, positioned both orthogonally and in line with the treatment beam, may be utilized to acquire real-time images of the patient's anatomy. The latest radiosurgery systems utilize a graphics processing unit (GPU)-based computational process to instantaneously generate digitally reconstructed radiographs (DRRs) corresponding to the imaging system orientation; this approach is also much akin to the original CyberKnife technology. Live X-ray images are automatically registered to the DRRs, thereby enabling shifts from the reference position to be calculated. Patient position is adjusted via the couch to assure proper initial anatomic set-up with respect to the treatment isocenter. During treatment, live images may be acquired based on a multitude of treatment parameters including the time elapsed since last image acquisition, number of radiation monitor units delivered since last image acquisition, Linac rotation in degrees, or on an *ad hoc* basis. The best treatment delivery systems integrate information from treatment planning into the treatment delivery sequence allowing objects and structures from treatment planning to be displayed on live images. This facilitates a quasi real-time assessment of the location of critical structures at the time of beam delivery and can minimize, or even eliminate, the need for treatment margin.

Intracranial target tracking

The advent of real-time surface imaging has introduced a new way to deliver image-guided radiosurgery. This technique employs an open face mask and real-time tracking of a predefined region of interest of the patient's face, which in turn can be correlated with either the Linac isocenter or the target location.

When tracking with surface imaging, preliminary setup is achieved by matching live 3D surface data to the extracted skin contour of the CT planning volume. Final checks and corrections to the target location are applied by means of CBCT and image correlation, after which real-time tracking of the visible surface anatomy is utilized to detect movement during treatment delivery (intrafraction motion). Should the patient move outside a predefined tolerance, the radiation beam is automatically turned off until patient position is corrected.

Despite the deformable nature of skin, a series of independent studies confirm that the cranium can be tracked using surface (skin) monitoring with submillimeter accuracy.[11–13] Cerviño et al.[14] present their initial clinical experience using AlignRT® (Vision RT, London, UK) for frameless and maskless SRS treatments on their first 23 patients treated with this technique. Meanwhile, in a prospective series of 44 consecutive patients with a total of 115 intracranial metastases, Pan et al.[15] showed that this open-face mask technique (Figure 15.3) for treating intracranial metastases produced clinical outcomes comparable to those obtained with conventional frame-based radiosurgery.

Figure 15.3 Open-face mask immobilization being used in conjunction with surface imaging and tracking.

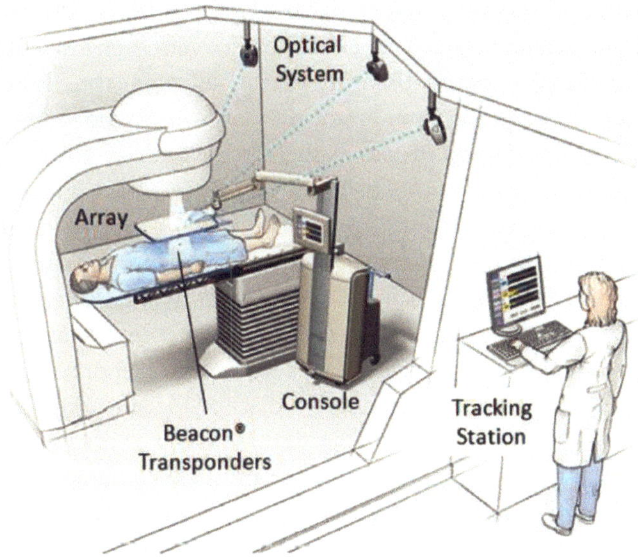

Figure 15.4 Electromagnetic target tracking of embedded beacon transponder.

Extracranial target tracking

Targets outside the head exhibit motion characteristics that are highly dependent on their anatomical location. While movement of lesions within the lung, liver, pancreas, and kidney is well correlated with respiration, the prostate is most affected by functions such as bladder fill and rectal gas. A novel technique developed specifically for prostate SBRT utilizes electromagnetic tracking for target tracking. Commercially known as Calypso® (Varian Medical Systems, Palo Alto, CA), this technology utilizes "passive" implanted electromagnetic markers (Figure 15.4). An antennae array positioned over the patient generates an electromagnetic signal that transiently excites previously implanted transponders. The system can detect in 3D the brief response signal emitted by transponders and in doing so determine the target location relative to the machine isocenter. By virtue of its 25 Hz frequency, this device provides near real-time target localization.

BROAD CLINICAL ADVANTAGES OF IMAGE-GUIDED RADIOSURGERY

Intracranial lesions

Rigid frames can exclude some skull base locations from radiosurgery treatment. In contrast, image-guided methods make it possible to treat some brain targets with a wider array of beam angle trajectories. Importantly, this capability means that image-guided methods can exploit beams penetrating through the splanchnocranium (portion of the skull arising from the first three branchial arches and forming the supporting structure of the jaw), which means that the brain can be largely spared entrance dose from beams targeted at extra-axial targets such as vestibular schwannomas or the trigeminal nerve. Moreover, when indicated, the ability to hypofractionate radiosurgery may permit safer irradiation of tumors and other lesions located close to exquisitely radiosensitive structures such as the optic pathways and cochlea.

One very important application of radiosurgical (hypo) fractionation is for treating lesions that are in close proximity to the anterior visual pathways, such as peri-optic meningioma, pituitary adenoma, or, less commonly, metastasis; when ablating lesions in this location, single fraction SRS must frequently irradiate the optic nerves and/or chiasm to maximal tolerance, which not infrequently is less than the optimal dose required for tumor control. Multiple studies have now reported that multisession radiosurgery for lesions located immediately adjacent to the optic apparatus is both safe and effective.[16–18]

Despite the proximity of the optic apparatus, these studies demonstrate that fractionated (2–5 sessions) radiosurgery has minimal toxicity; and, after many years of follow-up, tumor control rates appear high ($>92\%$).

Spinal lesions

Image-guided radiosurgery provides an important tool for treating a broad spectrum of spinal lesions with the same measure of accuracy and conformality that characterizes intracranial radiosurgery. Two significant challenges of spinal radiosurgery are the tracking of lesions without rigid fixation and mitigating the relative radiation sensitivity of the spinal cord, the consequences of injury being devastating in most patients. Different image-guided radiosurgery systems are able to address the challenge of targeting through a combination of various imaging modalities, the vast majority of which involve X-ray based intrafraction imaging of skeletal anatomy or implanted metallic fiducial markers. Meanwhile, volumetric delivery and robotic flexibility enable exquisitely steep dose gradients, the strategic placement of which can significantly limit the exposure of the spinal cord to radiation. However, when in doubt—i.e., larger lesions displacing the spinal cord or a history of prior radiation therapy—hypofractionated radiosurgery is easily (and very efficiently) utilized in the new world of image-guided targeting.[19]

Extracranial radiosurgery

The combination of real-time target tracking and volumetric dose delivery has been demonstrated in many published reports to be a safe and effective minimally or noninvasive tool to treat thoracic, abdominal, and pelvic lesions. The advantages of using image-guided radiosurgery to ablate nonneurological extracranial lesions (as compared to conventional fractionated radiation therapy) are the ability to administer much more conformal radiation and do so in a manner that maximally spares the surrounding normal tissue; once again treatment can be administered in a single or limited number of sessions as needed from a radiobiological perspective. Many ongoing clinical trials are being performed worldwide to evaluate the role of extracranial radiosurgery for a wide variety of benign and malignant lesions. Given the ongoing nature of these patient studies, the full impact of image-guided radiosurgery on the gamut of human pathology is hard to estimate; but it is almost certain to grow significantly. Based merely on the explosive growth in extracranial radiosurgery over recent years, it seems likely that image-guided radiosurgery will at a minimum assume a substantial role in the setting of many/most multimodality cancer therapy protocols.

CONCLUSIONS

For almost two generations, pioneering work in radiosurgery by neurosurgeons has placed neurosurgery at the forefront of this field of medicine. Neurosurgeons' earliest involvement in this field is not that surprising given the important role stereotactic frames had in targeting and immobilization of the skull and brain. However, recent developments in image processing and real-time machine control systems helped to extrapolate Lars Leksell's original concepts into a realm of frameless radiosurgery, thereby enabling new and increasingly ubiquitous techniques that involve extracranial diseases and/or a multisession approach. Image-guided radiosurgical innovations such as the CyberKnife and the Edge, the technical specifications for which have been detailed within this chapter, continue to advance the frontiers of radiosurgery and ultimately the world of noninvasive interventions.

REFERENCES

1. Leksell L. The stereotaxic method and radiosurgery of the brain. *Acta Chir Scand.* 1951;102:316–319.
2. Adler J, Mould R. Historical vignette on radiation cross-fire: Paris 1905 to Stanford University 1994. *Robotic Radiosurgery.* Vol. 1. Sunnyvale, CA: The CyberKnife Society; 2005;3–12.
3. Adler Jr. JR, Murphy MJ, Chang SD, et al. Image-guided robotic radiosurgery. *Neurosurgery.* 1999; 44:1299–1306.
4. Hansasuta A, Choi CY, Gibbs IC, et al. Multisession stereotactic radiosurgery for vestibular schwannomas: single-institution experience with 383 cases. *Neurosurgery.* 2011;69(6):1200–1209.
5. Choi CY, Chang SD, Gibbs IC, et al. What is the optimal treatment of large brain metastases? an argument for a multidisciplinary approach. *Int J Radiat Oncol Biol Phys.* 2012;84(3):688–693.
6. Chandra A, Dong L, Huang E, et al. Experience of ultrasound-based daily prostate localization. *Int J Radiat Oncol Biol Phys.* 2003;56(2):436–447.
7. Jabbari K. Review of fast Monte Carlo codes for dose calculation in radiation therapy treatment planning. *J Med Signals Sens.* 2011;1(1):73–86.
8. Clark GM, Popple RA, Prendergast BM, et al. Plan quality and treatment planning technique for single isocenter cranial radiosurgery with volumetric modulated arc therapy. *Pract Radiat Oncol.* 2012;2 (4):306–313.
9. Audet C, Poffenbarger BA, Chang P, et al. *Med Phys.* 2011;38(11):5863–5872.

10. Clark GM, Popple RA, Young PE, Fiveash JB. Feasibility of single-isocenter volumetric modulated arc radiosurgery for treatment of multiple brain metastases. *Int J Radiat Oncol Biol Phys*. 2010;76(1): 296–302.
11. Cerviño L, Pawlicki T, Lawson J, Jiang S. Frame-less and mask-less cranial stereotactic radiosurgery: a feasibility study. *Phys Med Biol*. 2010;55:1863–1873.
12. Peng JL, Kahler D, Li JG, et al. Characterization of a real-time surface image-guided stereotactic positioning system. *Med Phys*. 2010;37(10):5421–5433.
13. Li G, Ballangrud A, Kuo LC, et al. Motion monitoring for cranial frameless stereotactic radiosurgery using video-based three-dimensional optical surface imaging. *Med Phys*. 2011;38(7).
14. Cerviño L, Detorie N, Taylor M, et al. Initial clinical experience with a frameless and maskless stereotactic radiosurgery treatment. *Pract Radiat Oncol*. 2012;2(1):54–62.
15. Pan H, Cerviño L, Pawlicki T, et al. Frameless, real-time, surface imaging-guided radiosurgery: clinical outcomes for brain metastases. *Neurosurgery*. 2012;71(4):844–852.
16. Adler Jr. JR, Gibbs IC, Puataweepong P, et al. Visual field preservation after multisession CyberKnife radiosurgery for perioptic lesions. *Neurosurgery*. 2006;59:244–254.
17. Iwata H, Tatewaki K, Inoue M, et al. Single and hypofractionated stereotactic radiotherapy with CyberKnife for craniopharyngioma. *J Neurooncol*. 2012;106(3):571–577.
18. Romanelli P, Bianchi L, Muacevic A, Beltramo G. Staged image guided robotic radiosurgery for optic nerve sheath meningiomas. *Comput Aided Surg*. 2011;16(6):257–266.
19. Choi CY, Adler JR, Gibbs IC, et al. Stereotactic radiosurgery for treatment of spinal metastases recurring in close proximity to previously irradiated spinal cord. *Int J Radiat Oncol Biol Phys*. 2010; 78(2):499–506.

CHAPTER 16

MRI-Guided Stereotactic Laser Ablation

Jon T. Willie[1,2,3], Jack K. Tung[4], and Robert E. Gross[1,2,3,4]
[1]Department of Neurosurgery, Emory University School of Medicine, Atlanta, GA, USA
[2]Department of Neurology, Emory University School of Medicine, Atlanta, GA, USA
[3]Interventional MRI Program, Emory University Hospital, Atlanta, GA, USA
[4]Coulter Department of Biomedical Engineering, Georgia Institute of Technology, Atlanta, GA, USA

INTRODUCTION

MR-guided laser interstitial thermal therapy (MRgLITT) is a minimally invasive method for thermal destruction of benign or malignant tissue. MRgLITT entails (1) the use of an interstitial laser probe (optical fiber within an internally cooled cannula) to heat tissue in a controlled fashion, and (2) near real-time MR thermal imaging, or MR thermography, to monitor tissue temperatures and treatment volumes. When MRgLITT is utilized for neurosurgery, various stereotactic techniques are used to accurately deliver the laser probe within the intracranial space, hence the alternative name stereotactic laser ablation (SLA) to describe the composite procedure.

SLA facilitates spatially and temporally precise destruction of intracranial tissue via a minimally invasive approach. For focal lesions associated with epilepsy and metastatic disease, ablation may be an effective alternative to open resection that reduces risks of exposure and collateral injury to off-target structures. As such, SLA may offer improved neurocognitive outcomes and procedure tolerability. SLA may also be used for primary or salvage cytoreduction of diffuse, difficult to access, high risk, or otherwise "unresectable" neoplastic lesions (such as gliomas) precluded from resective surgery. Compared with stereotactic-focused radiation ("radiosurgery") approaches, SLA offers immediate and definitive results without dose limitations of ionizing radiation and without the prolonged symptomatic edema associated with radiation necrosis. In this chapter, we will discuss the history, technology, methods, and clinical utility of MRgLITT/SLA in the brain for treatment of selected functional, neoplastic, and cerebrovascular disorders.

HISTORY

Laser, an acronym for *light amplification by stimulated emission of radiation*, involves the controlled release of photons from energized atoms. Laser energy was first applied to medicine shortly after its description in the early 1960s by Maiman.[1] Following several technological developments [2–5] and pioneering animal studies,[6–12] the neurosurgical use of lasers in patients was first described in 1966 by Rosamoff.[13] Ergonomic and practical considerations, however, limited the early proliferation of lasers as neurosurgical tools.

A. Golby (Ed): Image-Guided Neurosurgery
DOI: http://dx.doi.org/10.1016/B978-0-12-800870-6.00016-9

With the development of new types of lasers (e.g., argon and neodimium-doped yttrium aluminum garnet [Nd:YAG] diode lasers) and improved practicality of use in the operating room, various studies emerged reporting the utility of lasers for smaller surgical corridors as a minimally invasive neurosurgical tool.[14–16]

The concept of laser interstitial thermal therapy (LITT), initially termed "phototherapy," to induce thermal tissue injury to tumors was first described in 1983 by Bown.[17] Laser energy was delivered via implanted bare optical fibers *within* the target tissue, where light scattering induces local heating. Thermal damage is a time- and temperature-dependent process: below 43°C tissue is not damaged regardless of exposure time, while tissue heated to ~44–59°C undergoes thermal damage in a predictable time-dependent manner. Tissue heated to greater than 60°C undergoes rapid thermocoagulation as proteins are instantaneously denatured, and above 100°C tissue is vaporized. The therapeutic goal of LITT is to produce controlled hyperthermia-induced necrosis at 50–90°C within target tissue. Bown and Sugiyama reported their initial results in humans,[17,18] and other groups likewise demonstrated that LITT could be a safe and effective means for treating deep-seated tumors.[19–22]

Historically, widespread adoption of LITT was impeded by available technology. In particular, early LITT studies were unable to monitor and control therapeutic "dosages" and related tissue temperature changes in real time. Postoperative images could be used to estimate thermal damage zones, but provided no feedback during the actual procedure. A critical advance, therefore, was the coupling of LITT to continuous noninvasive MR thermography, based upon temperature-sensitive MR parameters (e.g., proton resonance frequency, the diffusion coefficient, and T1 or T2 relaxation times) to allow rapid monitoring of tissue heating of both the intended ablation target as well as nearby off-target tissue at risk of unwanted injury. MR thermography was first used to monitor temperature changes during MRgLITT by Tracz et al. in 1992.[23] This technique was further refined with proton resonance frequency methods established by De Poorter in 1995 to an accuracy of ±0.2°C.[24] In addition to continuous monitoring of thermal dosing, algorithms to predict thermal damage zones have improved the utility of MRgLITT for thermal ablation.[25]

Another technological development now incorporated into commercially available MRgLITT probes is that of housing optical fibers within cooling sheaths. Sheaths that transmit light energy to tissue, but keep the fiber irrigated with fluid or gas have two technical advantages. First, internal cooling prevents overheating and damage to the fiber itself, prolonging service life. Second, cooling sheaths prevent charring of immediately adjacent tissue, maintaining optimal conditions for deeper light penetration.

The increased availability and accuracy of neuronavigation devices, improved design of laser delivery probes, and integration with intraoperative/interventional MRI suites, have converged in commercialization of MRgLITT and mainstream adoption for neurosurgical applications. Modern MRgLITT devices and accurate delivery platforms now allow precise ablations with reliable real-time dose monitoring for neurosurgery.

PHYSICAL–ANATOMICAL PRINCIPLES OF MRgLITT

MRgLITT is based upon thermal coagulation of tissue by the delivery of laser energy. Implanted optical fibers emit photons that are absorbed by tissue, generating and conducting heat. Heat generation depends upon laser power, wavelength, beam density, duration of exposure, and tissue-related differences in absorbance and perfusion. The two currently available diode semiconductor laser systems are available for wavelengths of 980 nm (Visualase® Thermal Therapy System, Medtronic, Inc., Minneapolis, MN) and 1064 nm (NeuroBlate® System, Monteris Medical, Inc., Plymouth, MN). The shorter wavelength (Visualase) has the theoretical advantage of matching the wavelength absorption of water (800–980 nm), which should create faster lesions with sharper thermal gradients and smaller transition zones. The longer wavelength laser (NeuroBlate) has the theoretical advantage of matching the infrared spectrum, which should lessen tissue absorption and extend tissue penetration, possibly at the expense of sharp demarcation of the ablation zone. In clinical practice, however, it should be emphasized that any significant relative advantages of different wavelengths used by the Visualase and NeuroBlate systems are unproven.

Generally, tissue nearest a laser energy source will heat first, but the commercially available systems utilize (1) diffusion sheaths to improve light penetration and yield a wider ablation, and (2) active internal cooling with either saline (Visualase) or CO_2 gas (NeuroBlate) to keep the probe and immediately adjacent tissue from overheating. As the laser heats tissue, thermal energy spreads circumferentially by conduction and convection, potentially confounding directionality of the laser energy with prolonged use. Maintaining precise control over the ablation process therefore necessitates a system of feedback control based upon thermal imaging. Another stereotactic thermal ablation technique, radio-frequency lesioning, utilizes a probe thermocouple to directly monitor tissue temperature and has been widely used for generating small focal lesions (e.g., pallidotomy for Parkinson's disease and thalamotomy for tremor). However, this approach may be less facile for complex or larger targets such as tumors and epileptic foci, in part because it cannot monitor tissue temperature at a distance from the probe tip.

The commercial laser ablation systems are engineered upon a predictable relationship between induced tissue temperature and resulting cellular viability (Arrhenius theory of time and temperature[26]). Above 43°C, progression to cell death increases with tissue temperature and time of exposure, resulting in partial injury and edema. Above 60°C, however, tissue undergoes immediate irreversible coagulative necrosis. Above 100°C, charring/carbonization, vaporization (and gas expansion), and unpredictable heat spread to surrounding tissues results. Charring reduces light penetration (impeding further ablation), causes tissue adherence (promoting tissue shear and hemorrhage with probe movement), or directly damages the probe. Vaporization and gas expansion potentially increase local pressure within tissue, engender unpredictable heat spread, and threaten injury to off-target structures. Thus, the goal of MRgLITT is to bring all target tissue to 50–90°C while maintaining spatial "safety margins" in which off-target tissues remain <43°C. Thus, spatial and temporal resolution of MR thermography is critical to the safe and precise ablation required of neurosurgical SLA.

In the brain, ablation diameters along a trajectory path may reach 2–4 cm with available devices, but can depend upon local structural and thermodynamic constraints. For SLA, different options exist with respect to device diameter, direction of light emission (e.g., side-firing tip vs. diffusion tip) and tip length. The surgeon may match these options to the volume and geometry of the intended target. For instance, we have routinely used a 10-mm length diffusion tip (Visualase) for long-trajectory ablations (e.g., amygdalohippocampotomy), whereas a 3-mm diffusion tip (Visualase) may be used for smaller ablations such as hypothalamic hamartomas, cavernous malformations, pallidotomies, and metastases. Where a larger volume or complex geometry is present, multiple trajectories, staged procedures, or use of directional probe designs (NeuroBlate) may be considered. Multiple trajectories can be performed with multiple laser assemblies, or can be performed serially with reuse of a single probe over the same surgical session. The latter is facilitated in our hands by using an MRI-guidance miniframe stereotactic platform (MRI Interventions ClearPoint SmartFrame®) and performing the entire procedure in the MRI environment (discussed further below).

When performing SLA, specific local thermodynamic factors may limit the extent of ablation of a given structure. Brain tissue borders or interfaces with differing reflectance (e.g., gray/white matter boundaries, pial/sulcal boundaries, or local blood products), or nearby heat sinks (e.g., vessels, and CSF ventricles/cisterns) may confine an ablation to an intended target but also serve to avoid vascular injury or other off-target complications. Multiple trajectories or prolonged treatment times may be necessary to ablate across anatomical boundaries such as pia, when desired. Anticipating these local thermodynamic factors will obviously influence stereotactic planning.

In our experience, tissue injury and resulting edema from laser ablations <10 cm^3 in appropriately selected patients is well tolerated, but larger ablations of tumors (>50 cm^3) can be performed. Nevertheless, MRgLITT can temporarily exacerbate mass effect with peak postoperative edema within 1–2 weeks. The significance of postoperative edema may depend in part upon lesion type and location. Greater caution may be required for lesions with preexisting mass effect (e.g., tumors with cytotoxic edema) relative to targets without mass effect (e.g., mesial temporal sclerosis). For stereotactic laser amygdalohippocampotomy (SLAH), which has been performed at multiple clinical centers[27,28] and is well tolerated by patients, surrounding CSF cisterns and the lateral ventricle provide not only physical impediments against injury to off-target structures (e.g., temporal stem, thalamus, cranial nerves, etc.), but also buffer zones into which the ablated mesial temporal structures may possibly expand. Notably, despite routine ablation of the contents of the uncus during SLAH, at no time have we or other clinical centers reported any radiological or clinical evidence of uncal herniation (e.g., third cranial nerve palsy). Studies of clinical outcomes and complication rates with ablations of various sizes, locations, and underlying pathologies are still needed to guide users and provide patient counseling.

TECHNICAL CONSIDERATIONS: DEVICES AND WORKFLOW

At present there are two device systems marketed and approved for MRgLITT/SLA in the brain in the United States (Table 16.1): The Visualase® Thermal Therapy System (recently acquired by Medtronic, Inc., Minneapolis, MN) and the Monteris Medical NeuroBlate® System (previously called The Monteris Medical AutoLITT® System, Monteris Medical. Inc., Plymouth, MN). The Food and Drug Administration (FDA)

Table 16.1 Technical specifications of commercially available systems for SLA

	Visualase thermal therapy system (Medtronic, Inc.)	NeuroBlate system/AutoLITT (Monteris Medical, Inc.)
Probe diameters	1.65 mm	3.3 mm, 2.2 mm
Tip options	Diffusing/circumferential (10 or 3 mm lengths)	Diffusing/circumferential (6 mm focal length) Directional ("side-fire")
Predominant method for device guidance*	Stereotactic anchor bolt, MRI guidance miniframe	Stereotactic anchor bolt, AXiiiS stereotactic miniframe, MRI guidance miniframe
Coolant	Liquid (saline)	Compressed gas (CO_2)
Laser source	980 nm 15 W continuous diode laser	1064 nm 12 W pulsed diode laser
Workstation Software	Proprietary: image fusion, multiplanar ablation monitoring, semi-automated laser control based on safety markers	Proprietary (M-Vision): image fusion, multiplanar contoured planning, volumetric ablation monitoring, and robotic probe driving applications
Temperature monitoring	Real-time MR thermography	Real-time MR thermography Probe internal thermocouple
Typical ablation diameter	2–3 cm	3–4 cm

*Both systems currently offer compatibility with multiple stereotactic delivery methods/devices. See text and Table 16.2 for details.

has cleared both systems for marketing under the Section 510(k) mechanism to "necrotize or coagulate soft tissue through interstitial irradiation or thermal therapy under magnetic resonance imaging (MRI) guidance in medicine and surgery... for wavelengths 800 nm through 1064 nm."[29] At the time of chapter submission, more than 1000 Visualase and more than 300 NeuroBlate neurosurgical procedures had been reportedly performed in the United States.

Visualase system and workflow

The Visualase Thermal Therapy System (Table 16.1) is comprised of a computer workstation, a 15 W 980-nm wavelength diode laser, a liquid cooling pump, and a disposable laser applicator set composed of a 400 μm core silica optical fiber with a cylindrical diffusing tip (optional 10-mm or 3-mm lengths) housed within a 1.65-mm diameter saline-cooled polycarbonate cooling cannula (Figure 16.1a). The workstation connects via Ethernet to the clinical MR scanner and retrieves images in user-defined planes as they

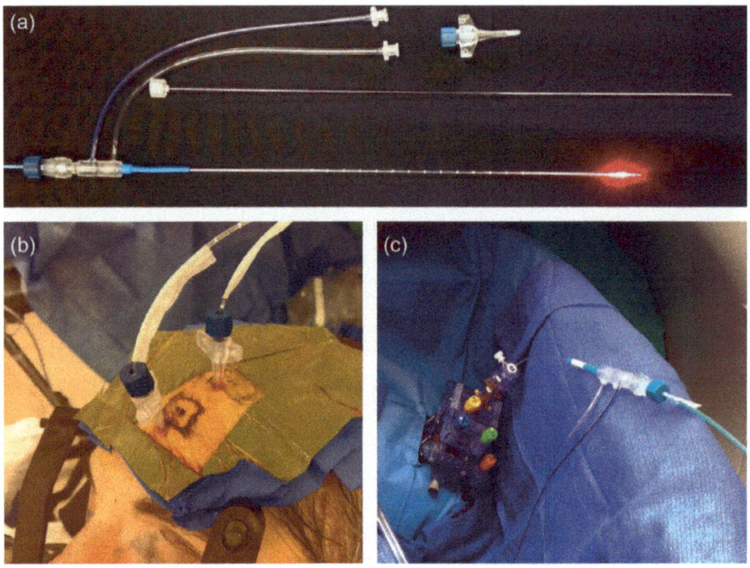

Figure 16.1 Hardware components for SLA using Visualase laser thermal therapy system using different stereotactic methods. (a) 15 W 980-nm diode laser energy is directed along a 400-μm core silica optical fiber that terminates in a circumferential diffusing tip (red). This fiber optic is housed within a 1.65-mm diameter saline-cooled polycarbonate cooling cannula (bottom). A threaded plastic bone anchor (top right) and stiffening stylet (middle) are used to stereotactically deliver the device to brain structures. (b) Stab incisions and 3.2-mm twist drill holes are made using a stereotactic headframe. Anchor bolts are threaded into twist holes under stereotactic control; Visualase laser applicators are passed through bolts, secured, and flagged with Steristrips. (c) Alternative direct real-time MRI-guided placement of Visualase laser applicator via an MRI guidance miniframe (SmartFrame, MRI Interventions, Inc.) within an MRI suite.

become available. Extracted thermal data generate color-coded "thermal" and "damage" images based on an Arrhenius rate process model,[26] which are displayed on the workstation (Figure 16.2b and 16.2c). The damage image accounts for the cumulative effects of the time–temperature history of each image voxel. In addition to this visualization, the interface allows the user to define safety temperature limits to specific marker points on

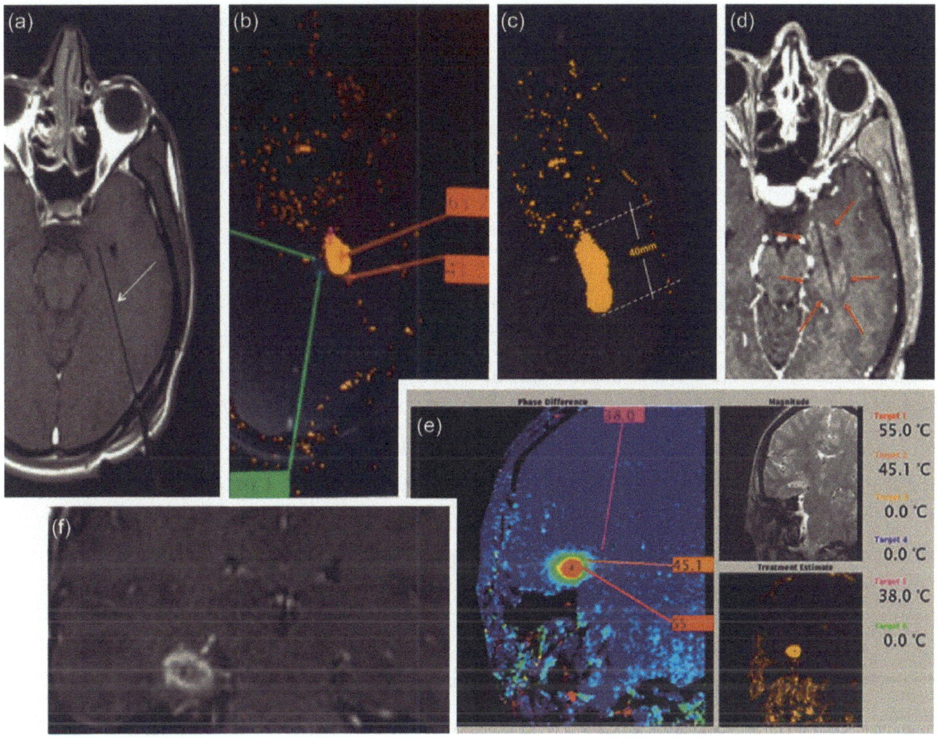

Figure 16.2 MR images associated with stereotactic laser amygdalohippocampotomy (SLAH) for epilepsy using the Visualase system. (a) Pre-ablation T1 MRI (axial) showing placement of laser optical fiber (white arrow) in mesial temporal structures. (b) Axial MR thermography screenshot from Visualase workstation illustrating irreversible damage estimate (orange) of initial anterior ablation and concurrent temperatures (degrees Celcius) at user-defined margins. Green indicates safe (i.e., midbrain), and red damaging temperatures. (c) Combined irreversible damage estimate (orange) of entire ablation (area approximately 40 mm × 17 mm in axial cross-section) from Visualase workstation. (d) Immediate post-ablation gadolinium-contrasted axial T1 MRI demonstrating ablation zone (red arrows demarcate borders) which is comparable to the damage estimate from the Visualase workstation (panel c). (e) Coronal MR thermography screenshot from Visualase workstation of a different patient illustrating real-time MR thermography (left subpanel), baseline correlating T2 anatomic image (right superior subpanel), and concurrent irreversible damage estimate (right inferior subpanel) during ablation in the hippocampus. Magenta indicates safe temperature (basal ganglia), orange and red indicate damaging temperatures (hippocampus). (f) Post-ablation contrasted coronal T1 MRI demonstrating ablation centered in hippocampus (comparable to damage estimate in panel e) with relative sparing of parahippocampal region.

the images. If during treatment, the computed temperature at any marker exceeds the specified temperature limit, the laser is automatically deactivated as a safety measure. The user may relocate these safety markers in real time during different stages of therapy.

Standard workflows using Visualase for SLA (Figure 16.3) may vary based upon the stereotactic delivery method (discussed further below). The procedure may be performed under general or local anesthesia. Stereotactic trajectory planning is performed from gadolinium-enhanced volume images (to visualize blood vessels and the target structure/lesion) using third-party planning software (e.g., Medtronic Stealth®, BrainLAB®, MRI Interventions ClearPoint®, etc.). One or more 3.2-mm twist-drill craniostomy/durotomy holes are drilled under stereotactic control. A threaded polycarbonate bone anchor bolt (Figure 16.1b) is screwed into each twist-hole using stereotactic control. Through each anchor bolt, the laser applicator is placed to the prescribed depth over a stiffening stylet (Figure 16.1a); the stylet is then exchanged for the fiber-optic and secured by a Tuouy-Borst adapter on the anchor bolt.

At this point, workflow dictates transfer of the patient from the OR to the MRI suite (Figure 16.3) unless the entire procedure is being performed in an intraoperative or diagnostic ("interventional") MRI suite with an MRI guidance frame (Figure 16.4) or in an interventional suite equipped with a moveable MR instrument. With the applicator and laser probe secure, the patient is placed in the MRI bore where a third-party flexible transmit–receive MR head coil is utilized. Applicator cooling lines and laser optical fiber are connected through a waveguide to the system workstation in the control room. Pretreatment anatomic (T1 or T2), and T1-weighted fast spoiled gradient-recalled echo

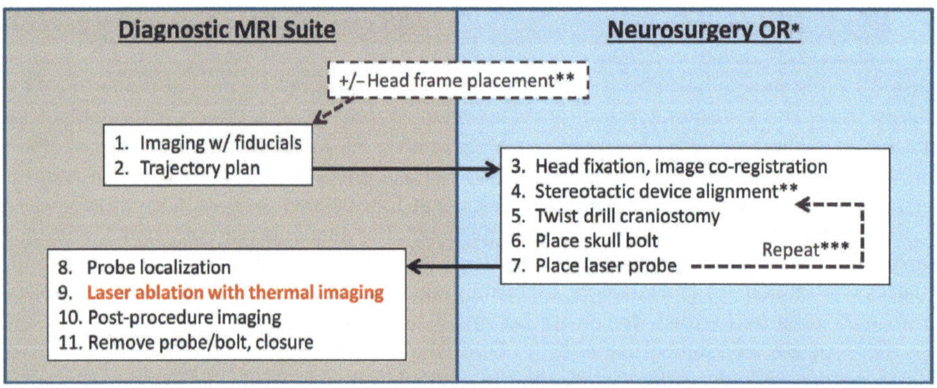

Figure 16.3 Typical workflow for SLA with standard stereotactic approaches and Visualase. Footnotes: *Patient transport between OR and diagnostic MRI suite eliminated if using intraoperative MRI suite. **The Visualase system is compatible with a variety of stereotactic platforms, of which traditional rigid head frame stereotaxis and anchor bolt is a common approach. ***Repeat as needed for multiple trajectories: while in OR, additional bolts/probes may be placed prior to therapy in MRI suite.

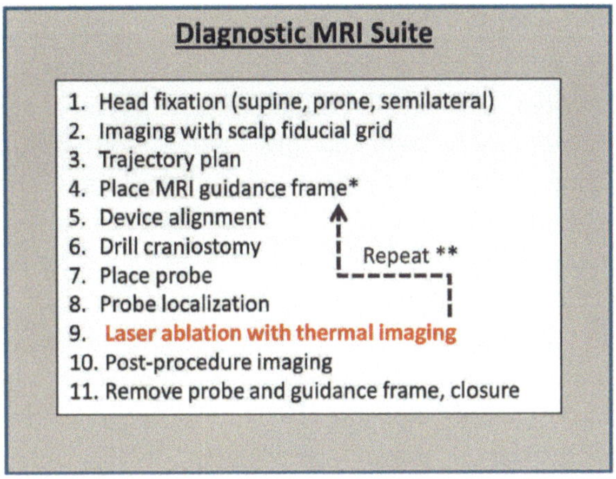

Figure 16.4 Workflow for SLA combining Visualase with direct MRI guidance. Footnotes: *ScalpMount SmartFrame (MRI Interventions, Inc.). **Repeat as needed for multiple trajectories by removing probe followed by moving/realigning guidance frame.

(FSPGR) volume images are acquired (Figure 16.2a). For treatment monitoring, MR thermal images are continuously acquired during laser delivery using the FSPGR sequence which require approximately 4 sec acquisition time per monitoring plane and are reconstructed into complex images (treatment overlaid upon real images) (Figure 16.2b and 16.2c).

A low power test pulse (e.g., 4 W, 30 sec) is used to visualize thermal change below threshold for damage, verifying laser tip position. Pulsed treatment doses (10–15 W, 30–180 sec) are then applied. Brief periods without laser energy are provided between doses to allow tissue cooling and to avoid unwanted heat spread to off-target structures. Treatment is completed at the surgeon's discretion when a predicted thermal damage zone sufficiently covers the desired target or when heat encroaches upon off-target structures, as indicated by safety markers. Contiguous overlapping ablation zones along a probe trajectory arc created by manually translating the optical fiber within the cannula. Multiple trajectories may be treated sequentially if appropriate. Following treatment, T2/FLAIR, diffusion-weight, and gadolinium-enhanced T1-weighted images all provide verification of the ablation zone, which is T2/FLAIR-intense, diffusion-restricting, and rim-enhancing with a hypointense coagulated center, respectively (Figure 16.2d and 16.2f). Applicators, bone anchors, etc. are removed and the entry site closed with a suture.

Following the procedure, patients are typically treated with a brief dexamethasone taper and discharged on postprocedure day 1 without need for intensive care unit monitoring. Contrast enhancement upon repeat imaging varies by target, but typically dissipates over subsequent months.

NeuroBlate system and workflow

The Monteris Medical NeuroBlate System (previously branded as "The Monteris Medical AutoLITT System" prior to recent addition of a remote robotic probe driver) is comprised of a computer workstation (M-Vision® software), a 12 W 1064-nm wavelength pulsed diode laser, CO_2 gas coolant tank and lines, and an assortment of disposable laser applicators (3.3-mm and 2.2-mm diameter, variable length options, diffusion tip or directional side-firing options, Table 16.1). All probes consist of a clear (light transmitting) cylindrical sapphire tip and rigid polymer shaft containing the laser optical fiber, cooling tube, and internal thermocouple (Figure 16.5). A probe is stereotactically inserted and stabilized either through a proprietary adjustable stereotactic miniature tripod frame (AXiiiS®, Figure 16.6), or more recently, through a cranial anchor bolt affixed through a 4.5-mm stereotactic twist drill craniostomy. When using the AXiiiS miniframe, the device mounts to the skull percutaneously via adjustable pedestals connected to a locking central ball around a directing cannula. The device is aligned using a stereotactic pointer from a third-party image-guided neuronavigation system (e.g., Medtronic Stealth, BrainLAB). The NeuroBlate system also utilizes the proprietary AtamA® Stabilization System, which combines a patient transfer

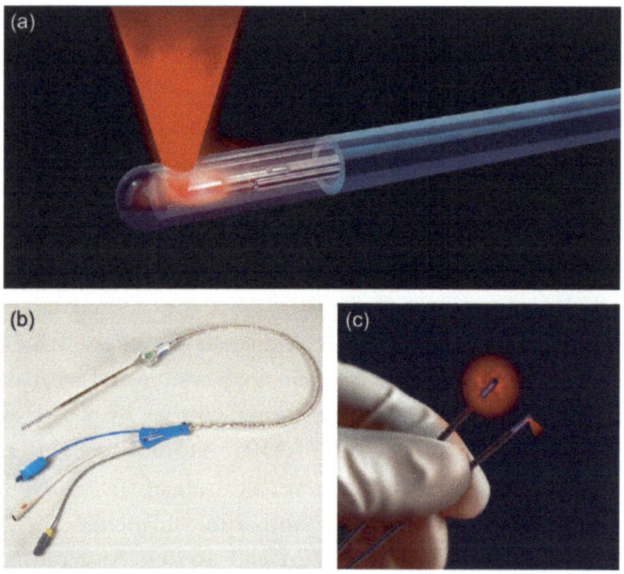

Figure 16.5 Laser delivery probes for the Monteris NeuroBlate System. (a) Illustration of the Monteris NeuroBlate SideFire™ laser probe tip depicting the laser fiber, thermocouple, and the CO_2 cooling tube. (b) The Monteris NeuroBlate laser probe depicted within its protective cover, which also serves as the ruler for setting the depth stop prior to probe insertion. (c) Recently released reduced diameter tip options for the NeuroBlate probe include a diffusing tip (left) and directional SideFire tip (right) with laser light digitally illustrated. *Images courtesy of Monteris Medical, Corp.*

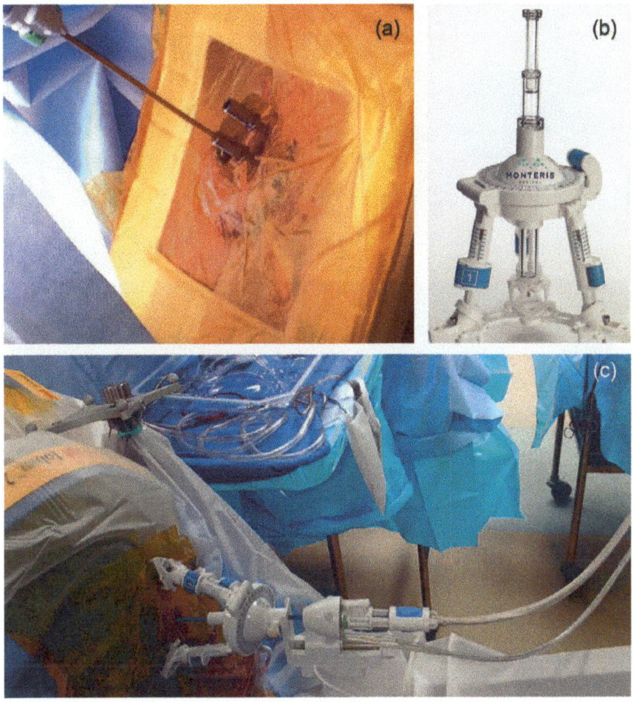

Figure 16.6 Monteris stereotactic platform for SLA with NeuroBlate. (a) The Monteris disposable titanium bone anchor bolt, shown as the NeuroBlate laser probe is inserted for one of two planned trajectories. (b) The Monteris AXiiiS stereotactic miniframe, a disposable, MRI compatible, trajectory alignment device that consists of three linearly translating legs, ball-socket and a 360-degree directional interface. (c) The AXiiiS miniframe, mounted percutaneously to a patient's skull, supports using standard indirect image-guided stereotaxis for placement of the NeuroBlate laser probe. *Images courtesy of Monteris Medical, Corp.*

board and head fixation ring that provides access for treatment (Figure 16.7). The AtamA protrudes from the MR coil, accommodates third-party flexible transmit–receive head coils, and also holds the robotic probe driver. Probes lock into an interface that in turn mates with an MRI-compatible piezomotor robotic probe driver to remotely translate and rotate the entire probe within the brain to direct laser energy. The optical fiber, CO_2 lines, and electrical cords run out to the laser source, gas delivery controller, and computer workstation in the control room.

The workstation for NeuroBlate features proprietary software (M-Vision) for multiplanar planning and executing SLA (Figure 16.8). Planning software has tools for defining an ablative area of interest, aligning the actual and planned treatment trajectories, arming the laser, and monitoring and mapping temperature changes in brain tissue as treatment progresses. Delivery of therapy along a trajectory is controlled remotely from the workstation, which drives the robot to rotate or translate the probe through tissue

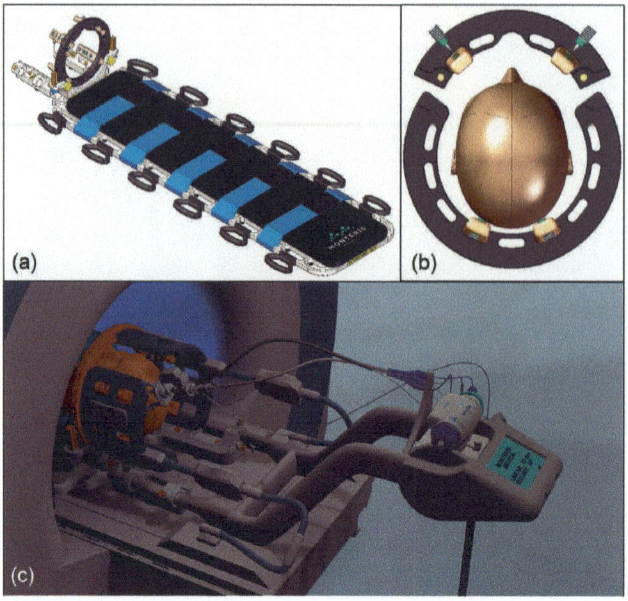

Figure 16.7 Patient stabilization/transport system and robotic probe driver for the Monteris NeuroBlate® System. (a) Illustration of the Monteris AtamA® System which provides integrated head stabilization with a patient transfer board for transport from OR to MRI. (b) Illustration of the AtamA two-piece head fixation ring. (c) Illustration of the NeuroBlate Robotic Probe Driver hardware as affixed to the AtamA patient board and attached to the AXiiiS stereotactic miniframe with the NeuroBlate laser probe deployed depicted at the backside of the MRI scanner. *Images courtesy of Monteris Medical, Corp.*

during ongoing ablation. Multiple targets/trajectories are treated serially by withdrawing the probe, redirecting or remounting the AXiiiS miniframe, creating a new scalp/skull opening, and inserting the device along a new trajectory.

Workflow using NeuroBlate for SLA (Figure 16.9) has historically utilized the AXiiiS miniframe, but use with a cranial anchor bolt or an MRI guidance frame (Figure 16.4) is now also feasible. The procedure is performed under general anesthesia. Stereotactic trajectory planning is performed from gadolinium-enhanced volume images (to visualize blood vessels and the target structure/lesion) using third-party planning software (Medtronic Stealth, BrainLAB, MRI Interventions ClearPoint, etc.). Either a 14-mm burr hole or a 4.5-mm stereotactic twist-drill craniostomy/durotomy is created. The laser applicator probe is inserted intraparenchymally to the prescribed depth.

The patient is then transported via the AtamA transfer board and head holder (Figure 16.7) from the neurosurgical OR to the MRI suite (unless utilizing an intraoperative MRI suite). Additional imaging and the M-Vision software are used to verify probe location, select monitoring planes, and generate a multiplanar ablation plan (Figure 16.8). MR thermography (FSPGR sequence) is continuously acquired during

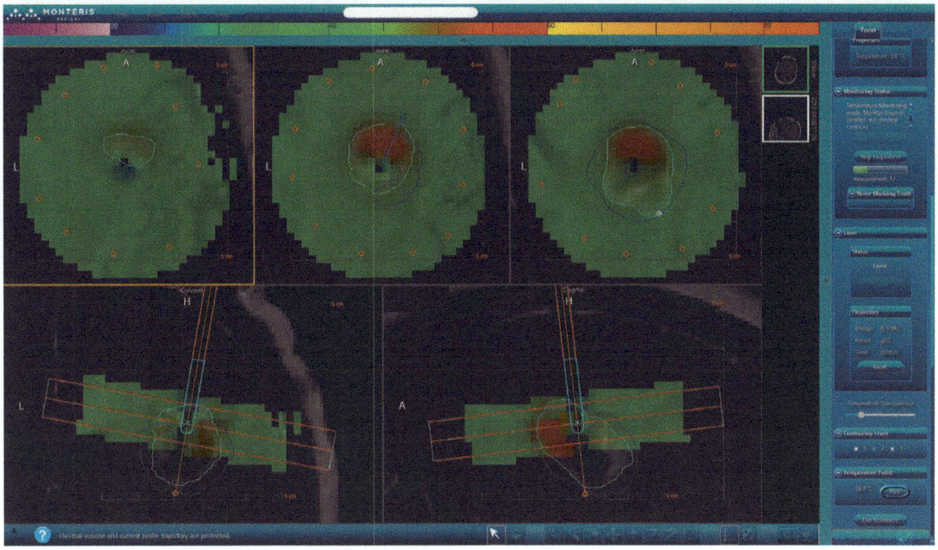

Figure 16.8 Workstation view of Monteris M-Vision software. A screenshot showing three paraaxial treatment slices (above, at, and below laser firing) and reconstructed coronal and sagittal images. The light blue contour line is created during planning to indicate the intended ablation margins. The green zone indicates the zone of real-time temperature mapping. Positions of the three thermal monitoring planes (indicated by red lines) and laser probe in the brain are graphically rendered on the reconstructed images. The arrow displayed over the middle paraxial image indicates the orientation of the SideFire Probe. These displays serve as a graphic user interface allowing control of the linear and rotational movements of the Robotic Probe Driver from the software while delivering laser energy. The cumulative treatment effect (dark blue contour line above) is displayed by thermal dose threshold (TDT) lines throughout the ablation procedure. The user can select to view one of three different dose thresholds (representing increasing likelihood of irreversible cellular injury) at any given time during thermal delivery. *Images courtesy of Monteris Medical, Corp.*

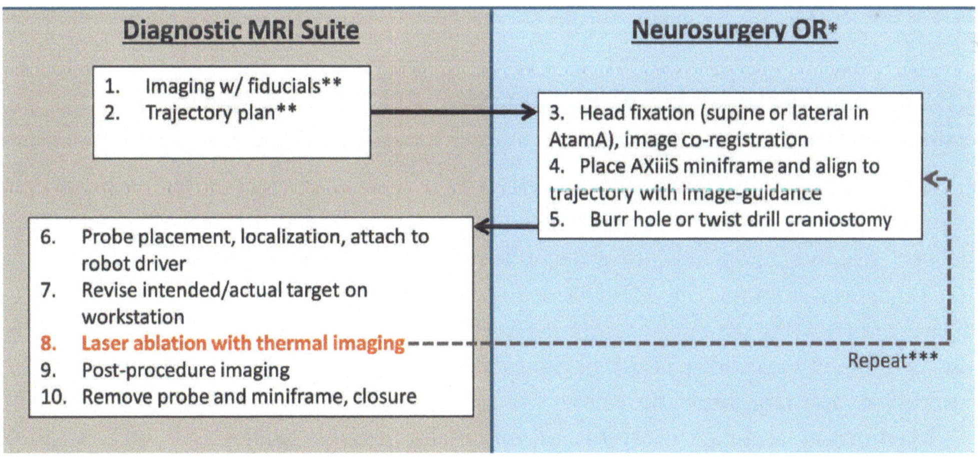

Figure 16.9 Typical workflow for SLA with the NeuroBlate system. Footnotes: *Patient transport between OR and diagnostic MRI eliminated if using intraoperative MRI. **Steps may be carried out prior to day of surgery. ***Repeat as needed for multiple trajectories: may require return to OR for probe removal, miniframe relocation/realignment, and new craniostomy.

laser delivery to construct thermal damage threshold (TDT) lines, which are overlaid upon previous real images (e.g., gadolinium-contrasted T1) and expand outward as therapy continues. The user may alter the direction of the laser emission (when using the side-firing option) or depth of the probe without discontinuing therapy. Laser therapy time varies widely by location, lesion identity and volume, number of trajectories required, and probe configuration. Using side-firing probes may have advantages in contouring the ablation, but directionality will be constrained in part by thermodynamic factors discussed above. Diffusional probes may produce rapid circumferential ablations within the constraints discussed. Treatment doses and times are not standardized but vary widely by target pathology, location, size, and probe configuration. Laser irradiation is stopped at the surgeon's discretion when the predicted thermal ablation zone (guided by TDT lines in different monitoring planes) sufficiently covers the volumetrically constructed plan.

If an intraoperative MRI suite or MRI guidance frame is unavailable, probe placement in additional trajectories will likely mandate return to the neurosurgical operating room or performance of staged procedures (Figure 16.9). Recent introduction of a cranial bolt will also facilitate treatment of multiple targets. Following treatment, T2/FLAIR, diffusion-weight, and gadolinium-enhanced T1-weighted images all provide verification of the ablation zone(s) with ablations being T2/FLAIR-intense, diffusion-restricting, and rim-enhancing with a hypointense coagulated center, respectively. Devices are removed and wounds are closed in standard fashion.

Following the procedure, patients may be treated with a brief dexamethasone taper and are typically discharged on postprocedure day 1–2. Contrast enhancement upon repeat imaging varies by target, but typically dissipates over subsequent months (Figure 16.10).

CHOICE OF STEREOTACTIC PLATFORMS AND TARGETING VERIFICATION

Increasingly, both Visualase and NeuroBlate laser probes are compatible with delivery via multiple stereotactic platforms (Table 16.2). Choice of a platform for stereotactic delivery depends upon institution-dependent factors and physician preference, including limitations imposed by MRI bore size/working space, accuracy/precision tolerances, and requisite patient position or target. Procedure times may also be affected by use of different platforms and workflows, and thus are variable and expected to be institution- and target-specific.

Maximizing accuracy/precision of minimally invasive procedures such as SLA entails a solid understanding of the principles and limitations of stereotactic platforms. The ultimate goal of all surgical navigation is to achieve accurate action (e.g., probe implantation) within the physical reference frame of the patient, and in particular

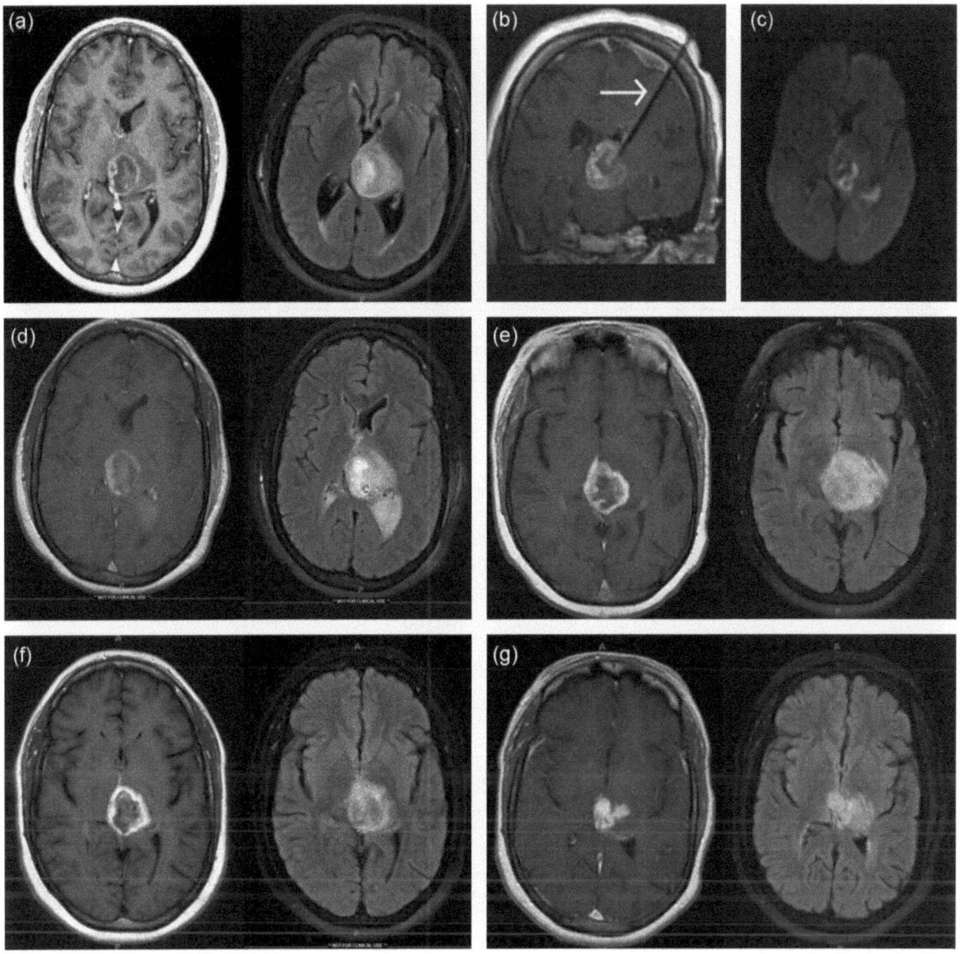

Figure 16.10 MR images showing SLA of a thalamic glioblastoma. (a) Preoperative axial gadolinium-enhanced T1-weighted (left) and T2 FLAIR (right) images showing a 3.5 × 4.4-cm enhancing lesion with surrounding edema involving the left thalamus in a 34-year-old woman presenting with headache and syncope. She underwent placement of a ventriculoperitoneal shunt and stereotactic needle biopsy confirmed pathological diagnosis of glioblastoma. (b) Intraoperative coronal-enhanced T1-weighted image demonstrating placement of the NeuroBlate probe (arrow). On postoperative day 1, diffusion-weighted (c) enhanced T1, and T2 FLAIR images were obtained (d) She underwent standard postoperative fractionated radiation and temozolamide chemotherapy. Similar imaging was obtained at 3 (e), 7 (f), and 15 (g) months postoperatively. *Reprinted with permission.*[30]

within the specific tissue being targeted (e.g., brain); we will call this *tissue space*. Since in closed procedures tissue space is not directly visualized (i.e., by eye), clinical imaging defines an *image space*, which is accepted as a proxy for tissue space. Stereotactic methodology constructs yet another spatial reference (*stereotactic space*) around part of

Table 16.2 Comparison of potential stereotactic platforms for SLA

Method	Target visualization during probe placement	Demonstrated compatibility (Visualase, NeuroBlate)	Workflow*	Accuracy**	Advantages	Disadvantages
Rigid head frame (CRW, Leksell, Compass) with anchor bolt	Indirect (image co-registration)	Both	OR → MRI vs. intraoperative MRI suite	<1 mm[25,31]	Gold standard accuracy, precision, and trajectory control.	Patient transport Inconvenience
Frameless navigated articulating arms and drills (Medtronic Stealth, BrainLAB) with anchor bolt	Indirect (image co-registration)	Both	OR → MRI vs. intraoperative MRI suite	<6 mm[31]	Availability. Convenience.	Reduced accuracy/ precision (best with bone fiducial insertion) Patient transport
Stereotactic robot (Rosa) with anchor bolt	Indirect (image co-registration)	Visualase	OR → MRI vs. intraoperative MRI suite	<2 mm [31,32]	Rapid placement of multiple probes.	Reduced availability Accuracy best with bone fiducials Patient transport
Customized (3D printed) miniframe (The FHC microTargeting Platform a.k.a. STarFix)	Indirect (image co-registration)	Both	OR → MRI vs. intraoperative MRI suite	<2 mm[33]	Reduced OR time.	Advance implantation of bone fiducial fixtures Advance trajectory planning required for device manufacture Patient transport

Image-guided scalp-mounted miniframe (Monteris AXiiiS)	Indirect (image co-registration)	NeuroBlate	OR → MRI vs. intraoperative MRI suite	Not reported[30,34–38]	Facilitates robotic driving of NeuroBlate probe depth and direction of light emission ("side-fire") during ongoing treatment.	Patient transport. Restricted patient positioning or trajectory due to device profile.
MRI guidance scalp-mounted miniframe (ClearPoint/SmartFrame, MRI Interventions, Inc.)	Direct (real time)	Both	Diagnostic MRI vs. intraoperative MRI suite	<0.5 mm (Willie JT and Gross RE, unpublished findings)	Accurate. All steps carried out in MRI (no patient transport). Facilitates sequential treatment of multiple targets/trajectories with reuse of a single probe.	Additional scanner time for frame alignment. MR distortion fields influence selection of device orientation in the magnet bore, which may affect patient positioning (see text).

*Additional compatibilities of Monteris NeuroBlate system with multiple stereotactic platforms are pending or anecdotally reported.
**Methodological accuracies inferred in part from literature regarding stereotactic depth electrode and DBS electrode placements (see references).

the patient, based on the particular stereotactic instrument (e.g., a stereotactic frame), and navigation within the stereotactic space is taken as a proxy for navigation within tissue space to which it is linked via the image space. Imaging control provides a means for verification of stereotactic navigation by ascertaining accuracy within image space, since ascertaining accuracy within tissue space is not possible without direct visualization of tissue. In fact, imaging control can be further subdivided into those radiological techniques that provide accurate images of the tissue target per se (i.e., MRI) and those that only provide images of surrounding structure (i.e., X-ray and computerized tomography, CT). The latter rely on sustained co-registration of the physical reference frame of the brain tissue to that of the bone (which X-ray and CT image best), which under surgical conditions are indeed prone to shift relative to one another. Therefore, we consider only MR imaging control to provide *direct* verification of targeting because of the highly accurate co-registration of MR image space to tissue space (see exception below). In contrast, targeting verification within stereotactic space and X-ray/CT-defined imaging space is considered *indirect*, since co-registration between these two spatial reference frames to tissue space is prone to various sources of error, or "shifts."

Procedural accuracy (i.e., accuracy within tissue space) must rely upon direct (MRI) and/or indirect (stereotactic, CT, X-ray) control depending upon available technology (Table 16.2). Stereotactic navigation begins with co-registering the stereotactic space (the space of the instrument) to brain tissue image space. This is accomplished by capturing an image set that contains both brain tissue and references to stereotactic space (i.e., fiducials) affixed to a stereotactic head frame or the patient's scalp or skull. With a stereotactic frame system, the delivery instrument (e.g., arc) mounts on the same base that surrounds the head and holds the fiducial box, thus physically co-registering the instrument space, fiducial space, and brain tissue image space. Navigation within stereotactic space is then considered an accurate proxy for navigation within image space, and thereby also within tissue space. With so-called "frameless" approaches, an instrument not attached to the patient is co-registered to the fiducial/patient image space by tracking an instrument in optical space with an infrared camera (e.g. Medtronic Stealth or BrainLAB workstations), or in physical space with a robot arm (e.g., ROSA®, Medtech Surgical, Inc.), as it is touched to the fiducials.

Typically, stereotactic navigation is used for probe placement procedures in a neurosurgical OR. Implantation accuracy is assumed when the stereotactic trajectory is attained, but this does not account for significant potential sources of inaccuracy such as deviation or shifts. The first order of verification may be indirect with intraoperative radiological imaging such as X-ray (2D, or when available, 3D-fluoroscopy). This may confirm stereotactic accuracy (i.e., of the probe with respect to the stereotactic frame) but does not account for brain shift within tissue space, which is not imaged with these techniques. If intraoperative MRI is not available then the patient is transferred to a diagnostic MRI suite for brain target confirmation and subsequent therapy. If upon confirmatory imaging the placement is found to be inaccurate, or subsequent therapy is deemed inadequate, it may be necessary to return the patient to the OR for new trajectory planning and probe placement. Use of an intraoperative MRI suite, if available, obviates patient relocation and facilitates multitrajectory treatments.

Accuracy and precision of indirect targeting verification methods depends upon the rigidity of the stereotactic device and/or co-registration error tolerances. Of such devices, traditional rigid head frames (historically most accurate/precise[39–41]), frameless neuronavigation systems (historically least accurate/precise[40,41]), stereotactic robots, patient-specific (customized) 3D-printed platforms,[33,41,42] and certain mini-frame devices (e.g., Monteris AXiiiS) are technically compatible with both commercially available laser probes (Table 16.2). To date, however, the Visualase system has been used most extensively with head frames, frameless systems, and an MRI guidance frame (discussed below), while NeuroBlate has predominantly been used with the Monteris AXiiiS device, which enables remote robotic control of probe depth and direction of light emission (side-firing) during therapy. All such devices except the MRI guidance frame depend upon indirect targeting verification at the time of probe placement. Additionally, as some of the stereotactic devices (e.g., head frames and frameless neuronavigation) themselves either require wide working spaces (standard MR bores are only 60–70 cm in diameter) or have MRI-incompatible components, MRI-compatible anchor bolts may be required to maintain probes in position for patient transport and subsequent therapy. Both Visualase and NeuroBlate now offer anchor bolt options for such workflows (Figures 16.1b and 16.6a).

In contrast to these options, which depend upon indirect targeting verification, MRI-based direct targeting verification within the brain (i.e., in tissue space) is relatively novel in practice, and has several major advantages. First, it allows all steps of the SLA/MRgLITT procedure (planning, probe placement, laser therapy, and ablation confirmation) to be carried out with the patient in a single location (i.e., within a diagnostic or intraoperative MRI bore) (Figure 16.1c, Figure 16.4, and Table 16.2). Second, MRI-based direct targeting verification allows confirmation of probe placement accuracy in MRI image space, in real time *during* the stereotactic procedure. As discussed above, all other techniques only provide real-time indirect targeting verification in stereotactic and/or X-ray/CT image space, which remains prone to error associated with successive co-registration steps from preoperative MRI image space to stereotactic space and/or optical space. In the latter, direct verification does not occur until transfer to the MRI suite, when corrections cannot be made without returning to the OR. The third advantage is that MRI-based direct targeting verification facilitates sequential multitrajectory treatments with a single expendable probe for complex or multiple targets. Finally, performing the procedure in an MR environment eliminates time spent in the neurosurgical OR at the expense of significantly increasing scanner time. This may be associated with improved institutional resource utilization, although there may be losses to opportunity costs when performing the entire procedure within a dedicated diagnostic MR scanner.

MRI-based direct targeting verification is based on the tight co-registration of the MR image space with tissue (i.e., anatomical) space. However, even MR images—and the stereotaxis based on them—are susceptible to predictable MR distortion fields that must be accounted for with correction algorithms. We have found that MR distortion errors when using MRI-based stereotactic methods and direct targeting verification are greatly minimized by positioning patients with the goals of (1) centering targets/trajectories within the magnet bore and (2) orienting anticipated trajectories either parallel or perpendicular to the bore axis (unpublished observations).

To perform MRI-based direct targeting, the authors have made extensive use of an expendable percutaneously mounted MRI guidance miniframe (ScalpMount SmartFrame®, MRI Interventions, Inc.) for stereotactic twist drilling, probe insertion, and securing the probe throughout therapy. The associated workstation/software (ClearPoint®, MRI Interventions, Inc.) facilitates stereotactic planning, automates image distortion correction algorithms, guides frame alignment, predicts trajectory and target accuracies prior to probe insertion, and calculates final accuracy data. Originally designed for direct targeting verification of deep brain stimulation electrode placement via a supine transfrontal approach (trajectory parallel to the axis of the magnet bore), we have used this system extensively for laser applications from a variety of approaches, most notably a prone transoccipital approach (trajectory perpendicular to the axis of the magnet bore) to ablate mesial temporal lobe structures for epilepsy.

In particular, SLAH[28] from a transoccipital approach, and stereotactic laser corpus callosotomy from combined transfrontal and transparietal approaches, are relatively demanding in terms of accuracy given long trajectory corridors (sclerotic hippocampus or corpus callosum), which are tubular, curved, and/or narrow. Passing accurately through multiple elements of the target(s) (i.e., hippocampal body, pes hippocampus, and amygdala) while also avoiding vasculature, choroid plexus, and the brainstem places high demands upon trajectory tolerances. By contrast, other applications targeting a globular lesion (e.g., tumor, hamartoma, or cavernous malformation) may be less demanding with respect to choice of trajectory or corridor, as long as the probe ends up in the target while avoiding large vessels (Figure 16.11). While a standard stereotactic head frame and anchor bolt placement provides predictable and facile trajectory control for SLAH or laser corpus callosotomy, we have found direct MRI guidance to provide the best combination of accuracy/precision, trajectory control, and elimination of patient transport following probe placement. Moreover, direct MRI guidance facilitates sequential treatment of multiple trajectories/targets with a single expendable probe and guidance frame during one surgical session.

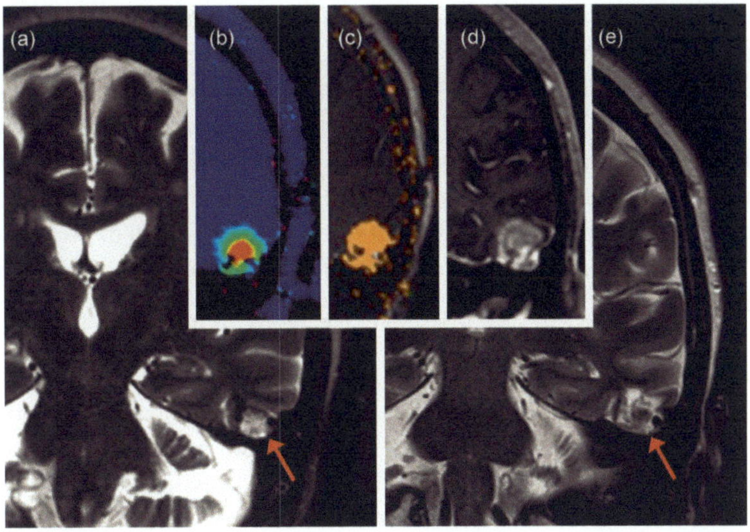

Figure 16.11 Ablation of epileptogenic tissue associated with a cavernous malformation while sparing an associated vein. (a) Preoperative coronal T2-weighted image exhibiting cavernous malformation (red arrow) and adjacent vein of Labbé. (b) Intraoperative coronal MR thermography screenshot (with red = heat and blue = cold) as shown on the Visualase workstation during laser interstitial thermal therapy. (c) Intraoperative coronal estimate of the total zone of irreversible laser ablation as shown on the Visualase workstation during therapy. (d) Immediate postprocedure coronal gadolinium-contrasted T1 image confirming the ablation zone (area of enhancement). (e) 6–month postoperative coronal T2-weighted image demonstrates encephalomalacia surrounding the lesion and an intact vein of Labbé, illustrating the relative safety of appropriately administered therapy around blood vessels. The patient remained free of seizures impairing awareness at 12 months follow up.

INDICATIONS AND OUTCOMES

SLA for epilepsy

The goal of epilepsy surgery is to eliminate or reduce seizures while minimizing neurological and cognitive deficits. SLA is an emerging minimally invasive approach to epilepsy surgery under evaluation for safety and effectiveness. At the authors' institution, patients with medically refractory epilepsy considered for SLA undergo the same rigorous preoperative workup prior to considering any resective approach. "Phase I" evaluation includes at a minimum noninvasive studies such as 3-Tesla MRI, 18-fluorodeoxyglucose positron emission tomography (18-FDG-PET), neuropsychological testing, and inpatient video-EEG monitoring. Depending upon the circumstances, magnetoencephalography (MEG), single-photon emission CT, functional MRI, diffusion tensor imaging tractography, Wada test, and invasive intracranial electrode monitoring studies may also be appropriate. A multidisciplinary committee consisting of epileptologists, neuropsychologists, neuroradiologists, and neurosurgeons review all preoperative data for consensus recommendations regarding the need for surgery.

The details of surgical decision-making with respect to patient selection are beyond the scope of this review; for these the reader is referred to dedicated texts[43] (see also Chapter 11 of this volume). In general, with well-defined epilepsy syndromes (i.e., mesial temporal lobe epilepsy) and where multiple noninvasive studies provide sufficient localizing information, proceeding to ablation may be pursued. In contrast, where results are nonfocal, multifocal, discordant, or incomplete, invasive "phase II" intracranial electrode monitoring of seizures (e.g., subdural and/or stereotactic depth electrode arrays) may be required to sufficiently localize epilepsy before proceeding with ablation.

Epilepsy indications

To date, the greatest experience with using SLA for focal epilepsy has been for mesial temporal lobe epilepsy (MTLE),[27,28,44] hypothalamic hamartomas (HH),[27,28,44,45] cerebral cavernous malformations (J.T. Willie, R.E. Gross, unpublished series), cortical dysplasias, cortical tubers, gray matter heterotopias, and low-grade neoplastic lesions (e.g., dysembryoplastic neuroepithelial tumors[46]). Disconnections (i.e., corpus callosotomy) are also safe, feasible, and effective (D. Curry, University of Texas, Dallas, personal communication, and J.T. Willie, unpublished experience). With a handful of exceptions, almost all of these epilepsy procedures were performed using the Visualase system.

Epilepsy outcomes and complications

For MTLE associated with mesial temporal sclerosis, the authors' early studies showed that 75% (9/12) of patients became free of disabling seizures (Engel class I) at one year following SLA alone.[27,28] Likewise, early experience with HH from a single center suggests that 86% (12/14) maintain freedom from gelastic seizures at 9 months.[44,45] Our preliminary experience with SLA of epileptogenic cortical cavernous malformations without prior intracranial monitoring achieved 4 of 5 patients (80%) becoming free of disabling seizures at 12–28 months (J.T. Willie and R.E. Gross, unpublished observations). While each of these series is small, such preliminary results are consistent with previous studies suggesting that, where focal epilepsy can be localized and completely resected, seizure freedom on the order of 70–80% may be expected (reviewed in Okonma et al.[47]).

Minimally invasive SLA procedures for epilepsy have the added potential to minimize neurocognitive complications. For instance, MTLE patients following standard open surgical resections (including both selective amygdalohippocampectomy as well as anterior temporal lobectomy/amygdalohippocampectomy procedures) are known to experience significant deficits in category-related object recognition and naming. Such deficits likely result from collateral damage to regions outside the hippocampus (such as temporal stem) transected or retracted by standard surgical approaches.[48]

Recently, we demonstrated that neurocognitive decline was significantly greater following both dominant and nondominant open temporal lobe procedures than for corresponding SLA procedures, despite comparable efficacy with respect to seizure control.[49] Naming and recognition functions in particular were spared by SLA. Thus, SLA may be a particularly attractive alternative to open surgery where the epileptogenic zone is deep or imbedded in functional structures and networks.

Large clinical studies are still needed to accurately judge complication rates for SLA procedures, which may depend upon indication, target location, pathology, and trajectory. However, early results suggest that complication rates may be similar to analogous stereotactic ablative procedures such as radio-frequency lesioning. Specifically, recent review of published case reports and case series including the cumulative 66 earliest patients who underwent SLA for epileptogenic lesions using Visualase reported 5 neurological deficits, 2 clinically significant acute hemorrhages,[28,45] 2 initially misplaced probes,[28,45] and no perioperative deaths or infections. Of the reported neurological deficits, transient visual field defects (with mesial temporal ablations[27,28]) and transient hypothalamic–pituitary dysfunction (with hypothalamic hamartoma ablations[27,28,44,45]) were notable. No hemorrhagic complications have been observed in our unpublished experience ablating a small number of epileptogenic cavernous malformations (personal observations). In our unpublished experience of over 60 SLA cases using Visualase for a variety of epileptogenic lesions, the most common postoperative complaint was transient headaches, which were generally tolerable and responsive to a brief dexamethasone taper.

Contraindications in SLA for epilepsy

As ablated tissue volume increases slightly for days to weeks following a typical procedure, due to temperature–time dependent necrosis, and also as a result of perilesional edema, possible brain shift or mass effect must be considered. Thus, ablation of particularly large, complex, or neoplastic lesions (e.g., large cortical dysplasias, low-grade gliomas) may be contraindicated unless carefully staged. Depending upon location, ablation volumes of approximately 5–10 cm^3 (e.g., amygdalohippocampal complex) are well tolerated. Other contraindications may be similar to that of other epilepsy resections, such as insufficient localization of seizure onset zone or insufficient differentiation of target from eloquently functional structures. Where medical comorbidities preclude general anesthesia, awake procedures on patients without claustrophobia are technically feasible. Finally, MRI-related contraindications, such as implanted metal devices (e.g. cardiac pacemaker, vagus nerve stimulator) must be considered. In the presence of such devices, an appropriate transmit–receive head coil and an awake procedure may address safety concerns. Consultation and protocols involving the institutional MR physicist are mandatory.

SLA for malignant tumors and post-radiation necrosis

Tumor patient selection

The goal of tumor surgery is to enhance progression-free survival (PFS) by maximal safe cytoreduction. However, patients presenting with difficult to access tumors may not be considered for open resection due to risk of collateral injury. SLA offers a treatment option for tumors in deep or eloquent functional structures, and may reduce collateral injury or cosmetic disfigurement compared with open resection. Clinical considerations depend upon a wide array of issues such as suspected diagnosis (e.g., infiltrative high-grade glioma vs. noninfiltrative metastatic tumor), expected sensitivity to medical or radiation therapy, overall prognosis, specific tumor location, symptoms from infiltration versus mass effect, and other patient factors (age, comorbidity, and preoperative functional status), the discussion of which is beyond the scope of the chapter.

Tumor indications, outcomes, and complications

Two emerging indications for SLA of tumors include (1) difficult to access or recurrent glioblastoma multiforme (GBM) and (2) metastatic tumors. Maximal open surgical resection of high-grade gliomas (WHO grades 3 and 4) is associated with improved survival, but currently fewer than half of patients are candidates for open resection at presentation, often due to lesions being difficult to safely access. Safety and feasibility of SLA has been demonstrated in select patients with recurrent GBM who were not considered good candidates for open surgery (as illustrated in Figure 16.10).[34] Furthermore, in a retrospective multicenter study of patients receiving SLA with NeuroBlate as upfront or salvage therapy, overall median PFS was ~5 months.[35] More complete ablation of tumor volume was associated with longer PFS (~10 months), analogous to the extent of resection concept for open surgery. Case series with SLA using the Visualase system similarly suggest safety and feasibility of this approach to difficult to access or recurrent GBM.[50] MRgLITT-based therapies are likely to prove superior to biopsy alone in patients who are not otherwise candidates for open surgery. The predominant limitation, like surgical resection for high-grade gliomas, is that MRgLITT remains another focal treatment for an infiltrative disease that cannot be targeted exclusively with focal treatments. Nevertheless, aggressive cytoreduction may reduce mass effect and improve responses to subsequent adjuvant chemotherapy and radiation in a patient population that would not otherwise be eligible for surgery.[51]

With respect to treating metastatic brain tumors, SLA may have advantages compared to both open surgery and stereotactic radiosurgery (SRS). Minimally invasive access improves patient healing relative to open surgery, and chemotherapy may therefore not require interruption. Since MRgLITT is based on thermal destruction of the target, it is not constrained by a maximum dose limit and, in contrast to SRS, may be used opportunistically multiple times when needed. Enhancing lesions that progress after

SRS may result from tumor recurrence, postradiation necrosis, or both. Nonspecific imaging and mixed pathological features make these distinctions practically difficult, and either recurrence or symptomatic postradiation necrosis may require surgical management. With respect to this particular problem, SLA improves local control for postradiosurgery recurrence and/or radiation necrosis.[36,52,53] While outcome studies are preliminary, the largest case series by Rao et al. demonstrated local control of 76% at 24 weeks, and a median PFS >9 months.[53]

In general, "noninvasive" SRS remains a first line therapy for solitary or multiple, small (<2 cm), radiation-responsive metastatic tumors (e.g., metastatic non small-cell lung cancer, adenocarcinoma) in neurologically stable patients in which a pathological diagnosis has been established. Alternatively, SLA may be an approach to solitary or multiple, larger (1−4 cm), relatively radiation-resistant metastatic tumors (e.g., small-cell lung cancer, melanoma), postradiation failures (progression), or symptomatic postradiation necrosis. Additionally, where primary resection of symptomatic metastatic tumors would be preferred, but access is difficult or unacceptably morbid (e.g., periventricular, thalamic, brainstem, or skull-base lesions), SLA may be a reasonable primary approach, particularly since SLA is likely associated with a more rapid resolution of mass effect than with SRS. Additional studies of SLA need to address clinical algorithms and outcomes for difficult to access gliomas and metastatic tumors.

Both Visualase and NeuroBlate have been widely utilized for indications associated with malignant tumors including postradiation necrosis. Review of 54 published cases of SLA using Visualase for these indications reported 5 new or worsening neurological deficits, 2 clinically significant acute hemorrhages, and no new-onset seizures, infections, or perioperative deaths.[25,50,52,53] By comparison, review of 69 published cases of SLA using NeuroBlate for malignant tumors and postradiation necrosis reported 34 new or worsening neurological deficits, 3 clinically significant acute hemorrhages (one of which resulted in death from thalamic hemorrhage), 1 delayed hemorrhage associated with pseudoaneurysm, 2 new-onset seizures, and 5 infections (including 1 death from meningitis), for a total of 2 perioperative fatalities.[30,34,35,37,38] Likely factors contributing to these higher rates of reported complications in NeuroBlate case series include an early selection bias toward treating deep-seated glioblastomas (a particularly ill patient population) with NeuroBlate, and the larger thermal doses, longer treatment times, and larger ablation volumes that have been reported with NeuroBlate cases. However, the case of a delayed pseudoaneurysmal hemorrhage and the higher apparent infection rate with NeuroBlate also raise a question of other technical aspects of therapy specific to NeuroBlate, possibly including (1) the wider diameters of cranial access and the original NeuroBlate laser probe, (2) the implications of the higher laser wavelength and prolonged thermal dosing of this therapy, and/or (3) possible brain shear associated with movement of the probe during NeuroBlate therapy. Greater clinical experience with each laser system in comparable lesions is necessary.

Contraindications in SLA for tumors

Possible contraindications for ablation of a tumor may be lack of a definitive pathological diagnosis (however, biopsy for preliminary diagnosis can be performed in the same surgical session as ablation), lesions in which gold-standard open surgery with definitive gross total resection may be achieved with a favorable risk-benefit profile, or where the size or mass effect of the lesion requires a more surgically definitive or acute reduction in mass. As discussed above, in the case of small and/or multiple metastatic lesions with a reasonable expectation of control by SRS, up front SLA may be initially less favored (although not necessarily contraindicated) with current levels of evidence.

Emerging indications

Cavernous malformations

Cavernous malformations of the brain may be associated with seizures, headaches, or focal neurological deficits referable to mass effect or locally irritating blood products; any such persistent or severe symptomology may be sufficient to indicate surgery. Preliminary experience with a small number of lesions causing epilepsy in the temporal lobe suggests that SLA of cavernous malformations in eloquent temporal structures may be a safe and effective means for treating focal epilepsy (J.T. Willie, R.E. Gross; unpublished series). Notably, no hemorrhagic complications of probe insertion or ablation have been encountered with this approach. The safety and efficacy of SLA for reducing mass effect and/or recurrent hemorrhage risk in otherwise inaccessible brainstem cavernous malformations is a matter of clinical importance that remains to be investigated.

Functional disorders

Our group has utilized SLA to perform pallidotomies for movement disorders and amygdalotomies for self-injurious behavior, demonstrating its feasibility (R.E. Gross, unpublished observations). Cingulotomy for medically refractory cancer pain and obsessive-compulsive disorder also appears feasible (S. Danish, personal communication). At this time, the advantages and disadvantages of utilizing laser ablation over more traditional radio-frequency ablation, focused radiation, or focused ultrasound remain to be defined.

CONCLUSION

Modern SLA/MRgLITT using commercially available systems is the culmination of technological advancements providing the ability to precisely, safely, and effectively perform thermal tissue ablations in the brain under the direction of real-time MR thermometry. Available laser systems may be placed accurately using a variety of stereotactic platforms, of which the authors favor direct MRI guidance. Early clinical results targeting a wide assortment of lesions in the brain with both commercially available laser systems (Visualase and NeuroBlate) suggest significant potential of the minimally invasive SLA/MRgLITT approach for treating focal epilepsy and malignant brain tumors (primary or metastatic), especially those judged difficult to access, as well as symptomatic postradiation necrosis. In appropriately selected epilepsy patients, ablation with the Visualase system appears as effective as open surgery while minimizing the detrimental cognitive impact caused by collateral injury associated with open approaches. Novel applications to cavernous malformations and functional neurological disorders are also emerging. Further clinical investigations regarding outcomes with respect to efficacy, complications, and cost are needed for this therapeutic approach to best be utilized in the context of MRI-guided neurosurgical procedures.

REFERENCES

1. Maiman T. Stimulated optical radiation in ruby. *Nature.* 1960;187:493–494.
2. Patel C. Continuous-wave laser action on vibrational-rotational transitions of CO_2. *Phys Rev.* 1964; 136A:1187–1193.
3. Hall RR, Baker E, Morison PC. Incision of tissue by carbon dioxide laser. *Nature.* 1971;232: 131–132.
4. Stellar S, Bredemeier HC. Experimental studies with the carbon dioxide laser as a neurosurgical instrument. *Med Biol Eng.* 1970;8:549–558.
5. Ascher P. Newest ultrastructural findings after the use of a CO_2-laser on CNS tissue. *Acta Neurochir Suppl.* 1979;28:572–581.
6. Earle KM, Carpender S, Roessmann U, Ross MA, Hayes JR, Zeitler E. Central nervous system effects of laser radiation. *Fed Proc.* 1965;24:129–139.
7. Fine S, Klein E, Nowak W. Interaction of laser radiation with biologic systems. I. Studies on interaction with tissues. *Fed Proc.* 1965;14:S35–S45.
8. Fox JL, Hayes J, Stein MN, Green RC, Paananen R. Experimental cranial and vascular studies of the effects of pulsed and continuous wave laser radiation. *J Neurosurg.* 1967;27:126–137.
9. Rosomoff HL. Effect of laser on brain and neoplasm. *Surg Forum.* 1965;16:431–433.
10. Brown TE, True C, Mclaurin RL, Hornby P, Rockwell RJ. Craniocerebral trauma induced by laser radiation. A method for experimental production of subdural and subarachnoid hematoma. *Life Sci.* 1966; 5:81–87.
11. McGuff PE, Gottlieb LS, Fahimi HD, Bushnell D. Surgical applications of laser. *Ann Surg.* 1964;160: 765–777.
12. Minton JP, Ketcham AS. The effect of ruby laser radiation on the Cloudman S-91 melanoma in the CDBA-2F hybrid mouse. *Cancer.* 1964;17:1305–1309.
13. Rosomoff HL. Reaction of neoplasm and brain to laser. *Arch Neurol.* 1966;14:143–148.
14. Ascher PW, Heppner F. CO_2-Laser in neurosurgery. *Neurosurg Rev.* 1984;7:123–133.
15. Beck OJ. The use of the Nd-YAG and the CO_2 laser in neurosurgery. *Neurosurg Rev.* 1980;3: 261–266.
16. Takizawa T. The carbon dioxide laser surgical unit as an instrument for surgery of brain tumours—its advantages and disadvantages. *Neurosurg Rev.* 1984;7:135–144.
17. Bown SG. Phototherapy of tumors. *World J Surg.* 1983;7:700–709.
18. Sugiyama K, Sakai T, Fujishima I, Ryu II H, Uemura K, Yokoyama T. Stereotactic interstitial laser-hyperthermia using Nd:YAG laser. *Stereotact Funct Neurosurg.* 1990;54–55:501–505.
19. Ascher PW, Justich E, Schröttner O. A new surgical but less invasive treatment of central brain tumours: preliminary report. *Acta Neurochir Suppl (Wien).* 1991;52:78–80.
20. Kahn T, Bettag M, Ulrich F, et al. MR-imaging guided laser-induced interstitial thermotherapy in cerebral neoplasm. *J Comput Assist Tomogr.* 1994;18:519–532.

21. Reimer P, Bremer C, Horch C, Morgenroth C, Allkemper T, Schuierer G. MR-monitored LITT as a palliative concept in patients with high grade gliomas: preliminary clinical experience. *J Magn Reson Imaging*. 1998;8:240–244.
22. Schwarzmaier HJ, Yaroslavsky IV, Yaroslavsky AN, Fiedler V, Ulrich F, Kahn T. Treatment planning for MRI-guided laser-induced interstitial thermotherapy of brain tumors—the role of blood perfusion. *J Magn Reson Imaging*. 1998;8:121–127.
23. Tracz RA, Wyman DR, Little PB, et al. Magnetic resonance imaging of interstitial laser photocoagulation in brain. *Lasers Surg Med*. 1992;12:165–173.
24. Poorter JD. Noninvasive MRI thermometry with the proton resonance frequency method: study of susceptibility effects. *Magn Reson Med*. 1995;34:359–367.
25. Carpentier A, McNichols RJ, Stafford RJ, et al. Real-time magnetic resonance-guided laser thermal therapy for focal metastatic brain tumors. *Neurosurgery*. 2008;63:ONS21–ONS29.
26. Svaasand L, Welsch A, van Gemert ME. *Optical-Thermal Response of Laser-Irradiated Tissue*. New York: Plenum Press; 1995.
27 Gross R, Willie JT, Sharan AD, et al. 103 Stereotactic laser amygdalohippocampotomy for mesial temporal lobe epilepsy: collective experience from 7 single-center, prospective, investigator-initiated studies. *Neurosurgery*. 2014;61(Suppl 1):192.
28 Willie JT, Laxpati NG, Drane DL, et al. Real-time magnetic resonance-guided stereotactic laser amygdalohippocampotomy for mesial temporal lobe epilepsy. *Neurosurgery*. 2014;74(6):569–584.
29. U.S. Food and Drug Administration (FDA) website. Accessed 1/24/2015. Available from: http://www.accessdata.fda.gov/cdrh_docs/pdf8/K081509.pdf; 2015.
30. Hawasli AH, Bagade S, Shimony JS, Miller–Thomas M, Leuthardt EC. Magnetic resonance imaging-guided focused laser interstitial thermal therapy for intracranial lesions: single-institution series. *Neurosurgery*. 2013;73:1007–1017.
31. Cardinale F, Cossu M, Castana L, et al. Stereoelectroencephalography: surgical methodology, safety, and stereotactic application accuracy in 500 procedures. *Neurosurgery*. 2013;72:353–366.
32. Gonzalez-Martinez J, Vadera S, Mullin J, et al. Robot-assisted stereotactic laser ablation in medically intractable epilepsy: operative technique. *Neurosurgery*. 2014;10(Suppl 2):167–172.
33. Konrad PE, Neimat JS, Yu H, et al. Customized, miniature rapid-prototype stereotactic frames for use in deep brain stimulator surgery: initial clinical methodology and experience from 263 patients from 2002 to 2008. *Stereotact Funct Neurosurg*. 2011;89:34–41.
34. Sloan AE, Ahluwalia MS, Valerio-Pascua J, et al. Results of the NeuroBlate System first-in-humans Phase I clinical trial for recurrent glioblastoma: clinical article. *J Neurosurg*. 2013;118:1202–1219.
35. Mohammadi AM, Schroeder JL. Laser interstitial thermal therapy in treatment of brain tumors—the NeuroBlate System. *Expert Rev Med Devices*. 2014;11:109–119.
36. Rahmathulla G, Recinos PF, Valerio JE, Chao S, Barnett GH. Laser interstitial thermal therapy for focal cerebral radiation necrosis: a case report and literature review. *Stereotact Funct Neurosurg*. 2012;90:192–200.
37. Hawasli AH, Ray WZ, Murphy RKJ, Dacey RG, Leuthardt EC. Magnetic resonance imaging-guided focused laser interstitial thermal therapy for subinsular metastatic adenocarcinoma: technical case report. *Neurosurgery*. 2012;70:332–337.
38. Mohammadi AM, Hawasli AH, Rodriguez A, et al. The role of laser interstitial thermal therapy in enhancing progression-free survival of difficult-to-access high-grade gliomas: a multicenter study. *Cancer Med*. 2014;3:971–979.
39. Maciunas RJ, Galloway RLJ, Latimer JW. Application accuracy of stereotactic frames. *Neurosurgery*. 1994;35:682–695.
40. Maciunas RJ, Galloway Jr RL, Latimer J, et al. An independent application accuracy evaluation of stereotactic frame systems. *Stereotact Funct Neurosurg*. 1992;58:103–107.
41. Bjartmarz H, Rehncrona S. Comparison of accuracy and precision between frame-based and frameless stereotactic navigation for deep brain stimulation electrode implantation. *Stereotact Funct Neurosurg*. 2007;85:235–242.

42. Quiñones-Hinojosa A, Ware ML, Sanai N, et al. Assessment of image guided accuracy in a skull model: comparison of frameless stereotaxy techniques vs. frame-based localization. *J Neurooncol.* 2006;76: 65–70.
43. Lüders H, ed. *Textbook of Epilepsy Surgery.* New York: Informa Healthcare; 2008.
44. Curry DJ, Gowda A, McNichols RJ, Wilfong AA. MR-guided stereotactic laser ablation of epileptogenic foci in children. *Epilepsy Behav.* 2012;24:408–414.
45. Wilfong A, Curry D. Hypothalamic hamartomas: optimal approach to clinical evaluation and diagnosis. *Epilepsia.* 2013;54(Suppl 9):109–114.
46. Jethwa PR, Lee JH, Assina R, Keller IA, Danish SF. Treatment of a supratentorial primitive neuroectodermal tumor using magnetic resonance-guided laser-induced thermal therapy. *J Neurosurg Pediatr.* 2011;8(5):468–475.
47. Okonma S, Blount J, Gross R. Planning extent of resection in epilepsy: limited versus large resections. *Epilepsy Behav.* 2011;21:494.
48. Helmstaedter C, Van Roost D, Clusmann H, Urbach H, Elger CE, Schramm J. Collateral brain damage, a potential source of cognitive impairment after selective surgery for control of mesial temporal lobe epilepsy. *J Neurol Neurosurg Psychiatry.* 2004;75:323–326.
49. Drane DL, Loring DW, Voets NL, et al. Better object recognition and naming outcome with MRI-guided stereotactic laser amygdalohippocampotomy for temporal lobe epilepsy. *Epilepsia.* 2015;56(1):101–113.
50. Carpentier A, Chauvet D, Reina V, et al. MR-guided laser-induced thermal therapy (LITT) for recurrent glioblastomas. *Lasers Surg Med.* 2012;44(5):361–368.
51. Elder JB, Chiocca EA. Editorial: glioblastoma multiforme and laser interstitial thermal therapy. *J Neurosurg.* 2013;118(6):1199–1200.
52. Carpentier A, McNichols R, Stafford RJ, et al. Laser thermal therapy: real-time MRI-guided and computer-controlled procedures for metastatic brain tumors. *Lasers Surg Med.* 2011;43:943–950.
53. Rao MS, Hargreaves EL, Khan AJ, Haffty BG, Danish SF. Magnetic resonance-guided laser ablation improves local control for postradiosurgery recurrence and/or radiation necrosis. *Neurosurgery.* 2014; 74:658–667.

SECTION III

New Directions

CHAPTER 17

Advances in Molecular Imaging for Surgery

Olutayo I. Olubiyi, Fa-Ke Lu, David Calligaris, Ferenc A. Jolesz, and Nathalie Y. Agar
Department of Neurosurgery and Department of Radiology, Harvard Medical School, Brigham and Women's Hospital, Boston, MA, USA

INTRODUCTION

The development and implementation of molecular imaging approaches for neurosurgery promise to increase the precision of neurosurgical interventions through improved surgical planning and real-time characterization of tissue. Positron emission tomography (PET) and magnetic resonance spectroscopy (MRS) are established in vivo imaging modalities that can inform surgical planning by providing increased diagnostic value to the more conventional magnetic resonance imaging (MRI) protocols, but also be used intraoperatively in some centers to detect residual tumor during surgery. The development and validation of new molecular probes or markers should contribute to their more widespread use in image-guided surgery. Molecular imaging based on optical modalities is promising to offer valuable tools for intraoperative inspection of the surgical cavity with a few millimeters penetration depth and a broad range of molecular specificity. The most developed and specific approaches to date are based on fluorescence imaging of either biomolecules such as hemoglobin providing contrast for the visualization of tissue architecture or administered molecular probes targeting tumor cells. Mass spectrometry is a newer modality that allows the sensitive detection, quantitation, and characterization of biomolecules directly from tissue without the need for molecular probes. Mass spectrometry techniques provide information on hundreds of molecules found in the tissue, which may be used as specific biomarkers and chemical signatures for the real-time characterization of surgical tissue. The continuously developing arsenal of molecular imaging modalities compatible with operating room workflows and requirements is beginning to offer a variety of tools that may be matched to the wide range of surgical approaches.

IN VIVO IMAGING MODALITIES

Positron emission tomography

PET is based on the detection of particles emitted by the nucleus of radioactive atoms and produces three-dimensional (3D) images of functional processes in the body.[1] PET

A. Golby (Ed): Image-Guided Neurosurgery
DOI: http://dx.doi.org/10.1016/B978-0-12-800870-6.00017-0

is used to measure important bodily functions, such as blood flow, oxygen use, and glucose metabolism to help evaluate how well organs and tissues are functioning. It involves the introduction of a radioactive tracer—a biological compound of interest labeled with a positron-emitting isotope, such as ^{11}C, ^{18}F, and ^{15}O—into the human body, usually via an intravenous injection. Common PET tracers are listed in Table 17.1.[2,3] The isotopes used have relatively short half-lives (minutes to less than 2 hours), to avoid exposing the subjects to prolonged periods of radiation and to allow maximal signal within relatively short imaging acquisitions.[4] Its use in combination with other imaging methods, such as MRI and MRS, has been known to greatly improve the efficiency of tissue visualization in many disease conditions.

Briefly, radioisotopes are produced in a cyclotron, which accelerates a beam of protons using two high-voltage electrodes and directs it toward target nuclei, thereby incorporating an extra proton into them. These newly formed isotopes with a neutron-to-proton ratio that is energetically unstable are then coupled to a molecular probe of interest, producing radioactive tracers of bodily functions. Most of these unstable isotopes undergo radioactive decay whereby the excess proton is converted into a neutron, a positron (similar to an electron but positively charged), and a neutrino. The emitted positron travels up to a range of a few millimeters in tissue before being annihilated along with an electron from the surroundings, producing two photons of equal energy (511 keV) traveling in opposite directions. Almost all PET scans are performed on instruments that have combined PET and computed tomography (CT) scanners. These combined PET/CT images pinpoint the anatomic location of metabolic activity within the body and have been shown to have higher diagnostic accuracy than the two scans performed separately. Diagnostic accuracy of

Table 17.1 Common PET tracers used to study neurological disorders

Application	Tracer
Cerebral blood flow	$H_2{}^{15}O$
Oxygen metabolism	$^{15}O_2$
Glucose metabolism	^{18}F-2-deoxyglucose (^{18}FDG)
Dopamine storage	^{18}F-6-fluorodopa (^{18}F-dopa)
Carrier-mediated active transport	O-(2-[18F]fluoroethyl)-L-tyrosine (FET)
DNA synthesis	3-Deoxy-3-[18F] flurothymidine (FLT)
Cellular amino acid uptake	^{11}C-methionine
Dopamine D_1 receptors	^{11}C-SCH23390
Activated microglia	^{11}C(R)-PK11195
Dopamine D_2 receptors	^{11}C-raclopride
Central benzodiazepine binding	^{11}C-flumazenil
Opioid binding	^{11}C-diprenorphine

Source: S. R. Cherry, M. E. Phelps.[2]
Source: D'Souza et al.[3]

PET is 8–43% higher than conventional procedures and changes treatment in 20–40% of the patients in the management of lung and colorectal cancers, melanoma, and lymphoma, with similar findings in breast, ovarian, head and neck, and renal cancers.[5]

PET applications for neurosurgery

Brain tumors

PET identifies tumor tissues, especially of higher histological grades, because they typically have increased metabolic and mitotic rates compared to normal brain tissue.[2] It can also detect transformation of low-grade to high-grade glioma.[6] ^{18}F-fluorodeoxyglucose (FDG) PET can provide an important prognostic information as increased glucose metabolism of gliomas correlates with higher histological grades (III and IV) and shorter survival period.[7,8] Similarly, increased uptake of ^{11}C-methionine, which reflects cellular amino acid uptake, is indicative of high-grade glioma and poorer survival.[9] The information obtained may influence the therapeutic approach.

Integrating PET images into image-guided resection of high-grade gliomas has shown that metabolic information on tumor heterogeneity and distribution is useful for planning surgery, improves tumor delineation, and provides a final target contour different from that obtained with MRI alone in about 80% of procedures.[10] FDG PET images have been shown to provide clinically significant information by appropriately directing biopsy efforts when CT and MRI were inconclusive or misleading, thereby increasing the diagnostic yield of stereotactic biopsies.[11]

Functional brain mapping using PET scans and coregistered MR images provided the neurosurgeon with precise definitions of structural and functional cortical areas; this altered surgical management in some cases and/or was used to predict the outcome. In a 1999 study of pediatric brain tumors, PET imaging with FDG and/or CMET was used for the characterization and grading of the tumor and imaging of ^{15}O-water uptake for the definition of structural and functional cortical areas upon activation (Figure 17.1). The combination of PET FDG, CMET, ^{15}O-water, and MRI was then used to characterize the relationship between potentially resectable tumors and functionally eloquent brain areas, which provided the neurosurgeon with precise definitions of structural and functional cortical areas.[13]

Response to chemotherapy and radiotherapy is associated with a significant reduction in tumor glucose metabolism.[4,6,14] Therefore, ^{18}FDG PET may also provide objective imaging evidence of response to treatment. A reduction of 15–25% in glucose uptake on PET scan after one cycle of chemotherapy would be classified as partial metabolic response, and complete resolution of glucose uptake within the tumor volume so that it is indistinguishable from the surrounding normal tissue is classified as complete metabolic response (European Organization for Research and Treatment of Cancer PET Study Group recommendation). The assessment may be challenging to make considering factors such as brain and tumor heterogeneity, tumor volume

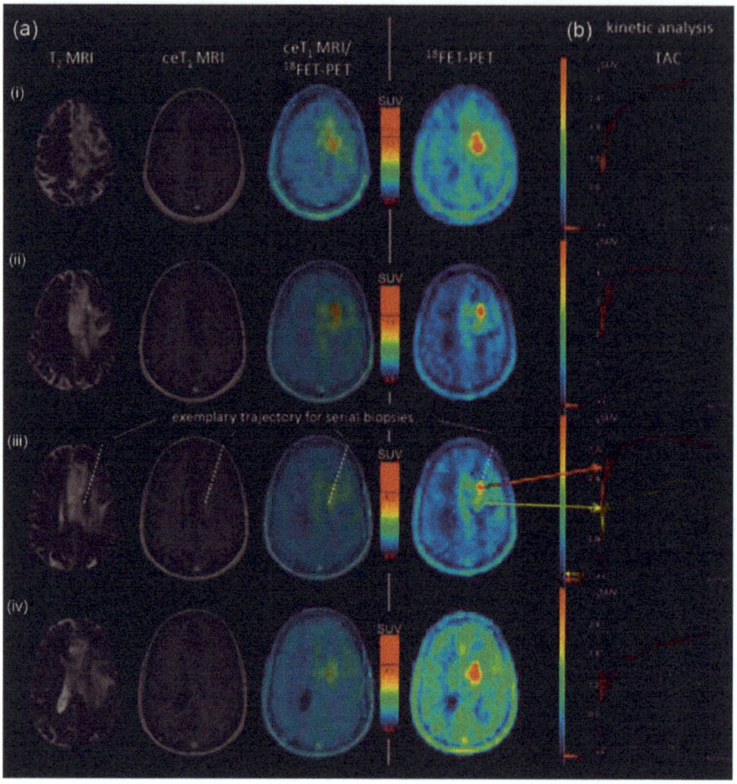

Figure 17.1 Representative MRI and PET images for a tumor exhibiting a focal decreasing time-activity curve (TAC) pattern. (a) (i)–(iv), consecutive axial planes of T_2-weighted sequences (first column); ceT_1-weighted sequences (second column); superimposed ceT_1-weighted MRI, corresponding ^{18}F-FET PET images (third column); and ^{18}F-FET PET images alone (fourth column). (b) Intratumoral analysis of dynamic ^{18}F-FET uptake. Decreasing kinetics were seen in rostral tumor parts [(i) and the upper curve in (iii)]. All other areas exhibited an increasing TAC pattern. A representative trajectory for serial biopsies is depicted in (iii) (dashed lines). ^{18}F-FET PET, O-(2-[^{18}F]fluoroethyl)-L-tyrosine positron emission tomography; MRI, magnetic resonance imaging; SUV, standard uptake value; TAC, time–activity curve.[12]

and spatial resolution, and in the case of hypometabolic tumors.[2,15] Conversely, increased glucose metabolism in recurrent cerebral glioma can help differentiate progression from postradiotherapy changes (pseudoprogression) that may be indistinguishable with conventional imaging techniques such as CT or MRI.[4,16,17] Similarly, ^{11}C-methionine is useful for discriminating between recurrences of local or metastatic tumors and radiation-induced changes. In a recent study, ^{11}C-methionine PET had a sensitivity of 77.8% and a specificity of 100% for differentiating recurrence of metastatic brain tumors from postradiotherapy changes.[4,18] However, disruption of the blood–brain barrier (BBB), such as cerebral hematoma or even necrotic areas caused by radiotherapy, reduces its specificity,[4,19] whereas glucose metabolism may be normal or low in lower grade tumors compared with the surrounding cortex. Combined use of ^{11}C-methionine and ^{18}FDG PET enhances the accuracy of discrimination between recurrent tumor and postradiotherapy changes.[4,20]

A recent study reported the use of dynamic O-(2-[^{18}F]fluoroethyl)-L-tyrosine (^{18}F-FET) PET to subtype low-grade gliomas. The ^{18}F-FET PET-guided biopsies were histopathologically evaluated with the inclusion of molecular genetic markers such as O^6-methylguanine-DNA methyltransferase (MGMT) promoter methylation, isocitrate dehydrogenase (IDH1/2) mutational, and 1p/19q codeletion status. Homogeneous increasing, focal decreasing, and homogeneous decreasing time–activity curves (TACs) were correlated with progression-free survival (PFS) and histopathology results. Overall, the results indicated that dynamic ^{18}F-FET PET could be an important imaging biomarker for the stratification of patients with suspected World Health Organization (WHO) grade II gliomas to further inform therapeutic strategies.[12]

Epilepsy

Surgical removal of epileptogenic foci in partial seizures such as intractable temporal lobe epilepsy results in a significant improvement in control of the seizures and quality of life[4,21]; 20–30% of potential surgical candidates with focal epilepsy have normal MRI.[4,22] Microscopic structural abnormalities, which may only be evident on histological examination, may not be detected on MRI.[4,23] The main clinical uses of PET in epilepsy are localization of epileptogenic foci in potential surgical candidates with partial seizures and corroborating findings from other investigational modalities such as electroencephalography (EEG).[2] Overall, for epileptogenic foci ^{18}FDG PET may be more suitable as a lateralizing rather than a localizing tool. However, ^{11}C-flumazenil (FMZ) PET provides an in vivo marker of γ-aminobutyric acid ($GABA_A$) receptor binding.[2] GABA is the principal inhibitory neurotransmitter in the brain, acting at up to 40% of synapses, and FMZ is a specific reversible antagonist that binds to the benzodiazepine binding site of the $GABA_A$-central benzodiazepine receptor complex.[2] A study that examined the localizing abilities of ^{11}C-FMZ and ^{18}FDG PET, using extra- and intracranial EEG recordings as reference, found the former to be more sensitive and accurate.[4,24] PET is likely to be most useful in situations where MRI is equivocal or normal. Studies have also demonstrated that the combination of PET imaging and MRI was superior in recognition of mesial temporal lobe epilepsy by visual inspection.[25]

Future directions

Potential applications of PET in clinical practice include the use as an adjunct to clinical diagnosis in atypical cases of Parkinsonian syndromes and dementia, and in early and presymptomatic diagnosis of individuals at risk for neurodegenerative disorders such as Alzheimer's disease and Parkinson's disease if an effective neuroprotective agent becomes available.[2] Another area of relevance of PET imaging with regard to tumor management is tumor hypoxia measurement. Hypoxia frequently occurs in solid cancers and can lead to therapeutic resistance[26] and enhanced tumor progression.[27] PET imaging has demonstrated efficacy in tumor hypoxia imaging with both fluoromisonidazole (FMISO) and copper(II)-diacetyl-bis(*N*(4)-methylthiosemicarbazone) (Cu-ATSM).[26] These two PET methods and BOLD-MRI are the lead contenders for noninvasive tumor hypoxia mapping, but there needs to be standardization of imaging procedures and analysis methods to complete validation.[27]

Magnetic resonance spectroscopy

MRS offers a unique opportunity for noninvasive measurement of brain metabolism in vivo.[28] It typically uses a conventional MRI instrument to analyze molecules such as metabolites, amino acids, lipids, lactate, and creatine in a tissue, and it is recorded on a graph as peaks of varying height (Figure 17.2). The first MRI medical scanners were introduced for clinical use in the 1980s, and since then improvements have been made, especially in relation to higher field strengths.[29] Both MRS and MRI have their origin in nuclear magnetic resonance (NMR), first described in 1946 simultaneously by the Nobel Prize winners Edward Purcell and Felix Bloch, from Harvard and Stanford Universities, respectively.[29] Lauterbur, Mansfield, and Grannell introduced gradient into the magnetic field, enabling them to determine the location of the emitted signal and to reproduce it on an image in the 1970s. NMR spectroscopy used in vivo is now named MRS because the term "nuclear" had been erroneously associated with nuclear medicine.[29] Years of considerable effort have been devoted to applying pattern recognition techniques to the complex task of data analysis in

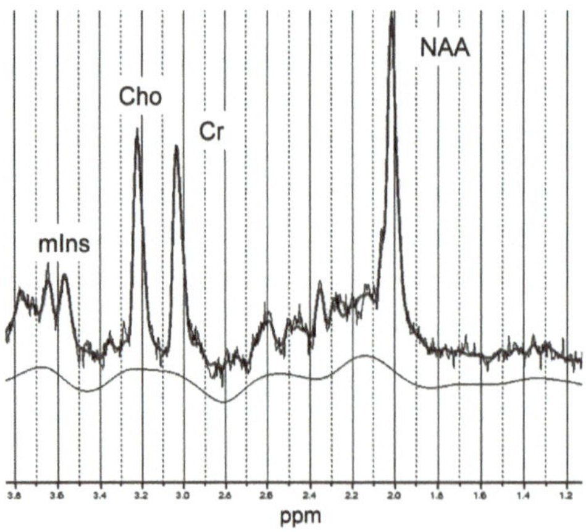

Figure 17.2 Proton spin-echo spectrum from normal human frontal lobe in vivo at 3 T, with a short (35 ms) echo time, analyzed using the software package LCModel (http://s-provencher.com/pages/lcmodel.shtml).

MRS.[30] MRS is now most widely used for the evaluation of central nervous system (CNS) disorders,[29] including common dementias, epilepsy, metabolic disorders, hypoxic–ischemic injuries, multiple sclerosis, brain infections, and pediatric and adult brain neoplasms.[31] Some of the main advantages of MRS are that it offers a noninvasive chemical mapping of tissue, without exposure to ionizing radiation.

Basic principles of MRS

When magnetic nuclei (such as ^{1}H, ^{31}P, ^{13}C, or ^{19}F) are placed in a magnetic field, they absorb and reemit electromagnetic radiation (resonate) at specific frequencies according to the magnetic field strength, e.g., hydrogen (^{1}H) at about 128 MHz in 3 Tesla (T) and 64 MHz in 1.5 T magnetic fields. In practice, ^{1}H (often called proton) is by far the most widely used nucleus due to its natural abundance and strong MR signal, but also because it is detected using the same coils and other hardware as MRI. As electrons shield nuclei from the applied magnetic field to varying extents, depending on the electronegativity of the chemical bonds in which they take part, hydrogen atoms contained in different molecules resonate at slightly different frequencies. The very small resulting frequency shift, generally reported in parts per million (ppm) instead of hertz, is what allows different metabolites to be distinguished on an MR spectrum. Water resonates at approximately 4.7 ppm at 37.8°C, and most metabolites of interest lie in the narrow range of frequencies between 0 and 4.7 ppm. The MRS experiment involves transmitting a pulse to excite all the nuclei in a given volume and then receiving the signal that returns. The resultant spectrum (Figure 17.2) is a plot of signal intensity vs. frequency over the excited range. A given metabolite has a specific spectral fingerprint: the number, location, and relative sizes of peaks are often constant, allowing for proper identification of the chemicals that are present and their concentrations if the peak areas are carefully calibrated relative to a standard reference.[28]

MRS in neurosurgery

MRS is of great relevance to the field of neurosurgery as most of its accepted applications relate to lesions of the CNS. MRS measurements are common practice in the preoperative diagnostic regimen,[32] and a handful of demonstrations of its intraoperative application have also been described.

Brain tumors

The diagnostic and prognostic values of proton MRS (Table 17.2) in the characterization of infiltrating gliomas have been well illustrated to differentiate gliomatosis cerebri from low-grade gliomas.[34] Its sensitivity and specificity for tumor detection were shown to be 100% and 81%, respectively (Figure 17.3). The accuracy of distinguishing neoplastic from nonneoplastic tissue could be from 68% with the use of MRI alone to 97% with MRI in conjunction with MRS and FET PET (see previous section on

Table 17.2 Metabolites displayed in proton MRS[33]

Metabolites	Location on spectrum	Physiologic significance
N-Acetyl aspartate (NAA)	2.02 ppm	Seen only in neural tissue. Marker of neuronal integrity. Reduced in most types of brain insults. Increases in Canavan's disease.
Choline and other choline-containing compounds (Cho)	3.2 ppm	Linked to cell membrane turnover, as in rapid cell division or breakdown. Tumors or demyelination can increase levels.
Creatine and phosphocreatine (Cr)	3.03 and 3.94 ppm	Related to compounds of energy storage. Often used as internal reference because it is relatively stable in metabolic disease.
Lipids	0.9–1.5 ppm	Not seen in normal brain. Represents membrane breakdown products. Increase noted in necrotic tumors and acute inflammation.
Lactate (Lac)	1.32 ppm–doublet	Not detected in normal brain. Presence indicates anaerobic metabolism or oxidative phosphorylation failure as in mitochondrial diseases, ischemia, inflammation, and tumors.
Myoinositol (mI)	3.56 ppm	Glial marker. Increased in some forms of dementia and human immunodeficiency virus encephalopathy. High in infant brain.
	2.1 and 2.4 ppm	Increased in hepatic encephalopathy/ hyperammonemias.

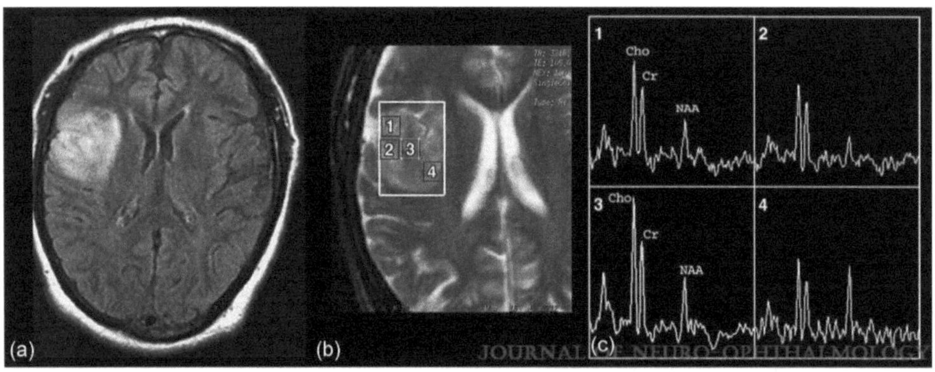

Figure 17.3 MRS: Astrocytoma. (a) Axial FLAIR image shows a right peri-insular mass. (b) MRS scout image shows the measurement area (large rectangle) and multiple voxels (small numbered rectangles) corresponding to the spectra in (c). (c) MRS spectra show an increase in Cho and a reduction in NAA in different areas of the lesion.[33]

PET).[35] Similarly, MRS is a useful diagnostic method for determining the proliferative or malignant potential of meningiomas according to the Cho/Cr ratio, and a lactate and/or methylene signal suggests a high-grade tumor.[36]

MRS can also be useful in differentiating tumor histological types. Hemangiopericytoma is a rare type of brain tumor that is very similar to meningioma in appearance and symptoms but requires different treatment such as the recommendation for preoperative embolization and postoperative irradiation. It is not normally possible to distinguish them by MRI and CT studies, but it has been shown that MRS could clearly distinguish hemangiopericytomas from meningiomas by a larger peak at 3.56 ppm. Measurements of extracts of the tumors and comparison of spectra acquired with MRS at long (135 ms) and short (20 ms) echo times established that this was due to the much higher levels of myoinositol in the hemangiopericytomas.[37] Likewise, in vivo ^{1}H-MRS is a reliable technique for presurgical grading of neuroepithelial brain tumors.[38] In treatment planning and monitoring, especially in radiosurgery, MRS has been indicated to add significant benefits by avoiding overtreatment of enhancing but nonmetabolically active regions and undertreatment of nonenhancing but metabolically active regions of the tumors. Evidence has suggested that patients undergoing gamma knife surgery may benefit from the inclusion of ^{1}H-MRS imaging in the treatment planning process.[39]

2-Hydroxyglutarate

Mutations in the gene isocitrate dehydrogenase 1 (IDH1) are present in up to 86% of grade II and III gliomas and secondary glioblastomas.[40] Regardless of the specific molecular consequences of *IDH1* alterations, detection of mutations in *IDH1* is likely to be clinically useful.[41] The IDH1(R132) mutation results in 2-HG resonance at 2.25 ppm and a reduction of glutamate levels as determined by MRS.[42] In addition to 2-HG, alterations in several other metabolites measured by MRS correlate with *IDH1* mutation status,[43] but 2-HG is best characterized and most widely utilized. The feasibility, specificity, and selectivity of using MR detection and quantification of 2-HG for the diagnosis and classification of *IDH1/2* mutation-positive brain tumors have been demonstrated.[44] 2-HG detection by in vivo or ex vivo MRS enabled detailed molecular characterization of a clinically important subset of human gliomas (Figure 17.4).[40] Such information may augment the ability of clinicians to monitor therapeutic response and provide criteria for stratifying patients to specific treatment regimens.[45] It is conceivable that new treatments could be designed to take advantage of IDH1 alterations in these patients, because inhibition of a different IDH enzyme (IDH2) has recently been shown to result in increased sensitivity of tumor cells to a variety of chemotherapeutic agents.[41,46] MRS provides a noninvasive measure of 2-HG in gliomas and may serve as a potential biomarker for patients with *IDH1* mutant brain tumors.[43,47] Caution must be exercised, however, when acquiring and

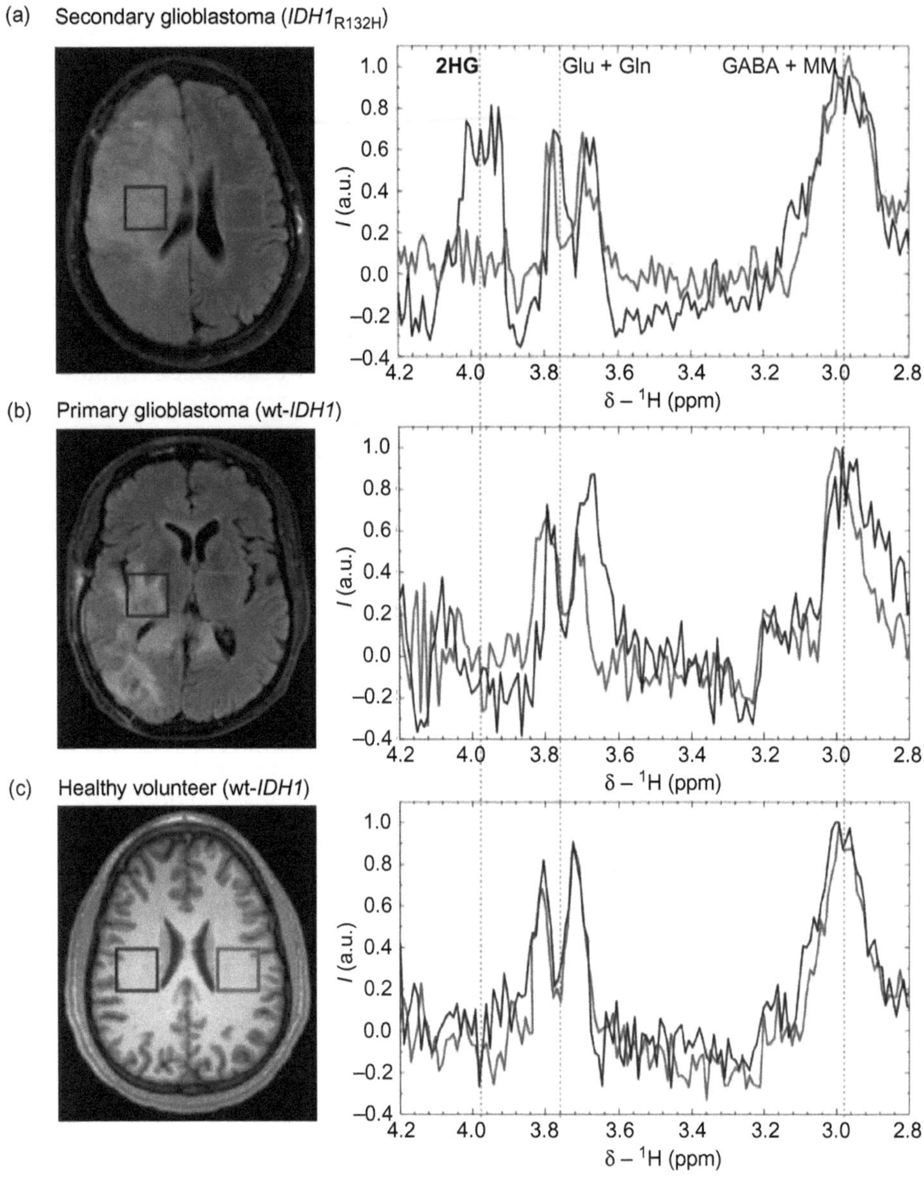

Figure 17.4 2-HG MRS in human glioma. One-dimensional MEGA-LASER spectra in human subjects at 3 T. In all subjects, two voxels ($3 \times 3 \times 3$ cm^3 each) were placed in both brain hemispheres, symmetrically from the middle line. (a) A secondary glioblastoma patient with *IDH1*$_{R132H}$ mutation. The spectra from subjects with wt-*IDH1*—primary glioblastoma (b) and healthy volunteer (wt-*IDH1*) (c). 2-HG is present only in the tumor voxel of *IDH1*$_{R132H}$ patient. (MM denotes contamination of GABA signal with macromolecule signal.)[40]

analyzing in vivo MRS data, as reliable measurements of 2-HG require customized in vivo MRS sequences to avoid false-positive results.[47]

Inflammation

MRS has been demonstrated to be very powerful in differentiating brain abscesses from other brain pathologies, especially when combined with MRI. Amino acids, probably originating from extracellular proteolysis, and other compounds, such as acetate, arising from bacterial metabolism, were visible in the MRS spectra of the abscess, but were undetected in spectra of neoplasms. In this sense, MRS complemented the information provided by MRI to achieve a correct diagnosis of brain abscesses,[48] though other MR sequences are more routinely used.

Epilepsy

Proton MRS imaging of the ratio of *N*-acetylaspartate (NAA) to creatine (Cr) has proved efficacious as a localizing tool in demonstrating the metabolic changes associated with temporal lobe epilepsy. The significant association of the NAA/Cr ratio with the GFAP immunoreactivity of most *Cornu Ammonis* sectors indicates that the NAA/Cr ratio may provide a more accurate measurement of recent neuronal injury caused by epileptic activity.[49]

Future directions

Growing evidence supports the relevance of MRS in surgical planning and guidance. It has been shown that intraoperative MRS may allow better intraoperative identification of tumor border zones based on metabolic changes due to tumor infiltration using maps of Cho/NAA ratio and automatic segmentation of the tumors.[50] MRS in an intraoperative MRI setting is feasible, facilitating preoperative glioma staging as well as satisfactory characterization of suspected tumor remnants. Thus, it may be a helpful tool for an extended tumor resection.[32] Furthermore, MRS may also be useful for prognostication of patient outcomes in glial brain tumors. The hypothesis that mobile lipids observed in vivo in ^{1}H-MR spectroscopy may correlate independently with prognosis of individual patients was borne out of the finding that mobile lipids occurred in 41% of high-grade tumors with higher mean amounts found in glioblastomas.[51] MRS has evolved to be a useful clinical tool over the past four decades especially in the care of CNS diseases. Using regular MRI instruments to measure metabolic spectra, it offers tremendous benefit for noninvasive distinction of neoplastic and nonneoplastic lesions.

Optical modalities

In optical imaging, the phenomena during the light–matter interaction, such as reflection, absorption (nonemission or emission), elastic scattering (linear), and inelastic

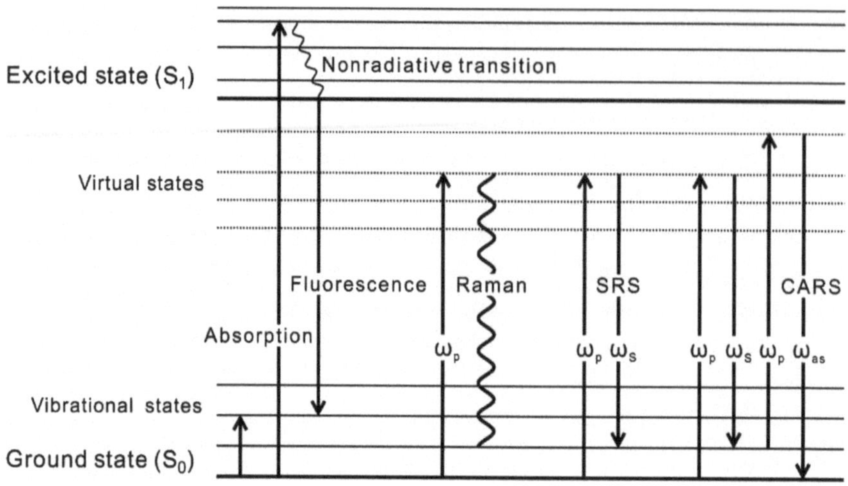

Figure 17.5 Energy diagrams of optical absorption, fluorescence emission, spontaneous Raman scattering, stimulated Raman scattering (SRS), and coherent anti-Stokes Raman scattering (CARS). By absorbing photons, the molecule is excited from the ground state (S_0) to higher energy levels (including rotational state, vibrational state, and excited state S_1). Then the excited molecule will dissipate its energy through both nonradiative transition (vibrational relaxation) and radiative transition by emitting a photon with a longer wavelength (fluorescence). In spontaneous Raman scattering, the pump photons (ω_p) are inelastically scattered off molecular vibrations, generating new photons at lower frequencies. In SRS, both the pump (ω_p) and Stokes (ω_S) beams are incident on the molecules. If the frequency difference of the pump and Stokes beams ($\omega_p - \omega_S$) matches a vibrational level of the molecule, stimulated excitation of vibration transitions occurs. Similar to SRS, CARS is a four-wave mixing process generating new photons at the anti-Stokes frequency ($\omega_{as} = 2\omega_p - \omega_S$). When the frequency difference matches a vibrational level of the molecule, the scattering process is resonantly enhanced.

scattering (nonlinear), are detected to reveal the chemical, morphological, and functional characteristics of the biological system.[52] Optical imaging offers high spatial resolution with molecular contrast from either intrinsic molecules or exogenous probes holding great potential for neurosurgical guidance. Energy diagrams of optical absorption, fluorescence emission, spontaneous Raman scattering, stimulated Raman scattering (SRS), and coherent anti-Stokes Raman scattering (CARS) are presented in Figure 17.5 to illustrate the different underlying optical phenomena between the presented optical modalities.

Fluorescence image-guided neurosurgery

Fluorescence image-guided surgery (FIGS) is a medical imaging technique that uses fluorescent contrast agents to detect labeled cancer tissue during surgery. A few review articles about this approach and its clinical applications provide the interested reader a more exhaustive coverage of the approach.[53–57] Although many fluorescent contrast agents are being developed for intraoperative neurosurgical oncology, very few have been subjected to clinical trials. We will focus on the most widely used fluorescence agents: indocyanine green (ICG), fluorescein sodium (FS), and 5-aminolevulinic acid (5-ALA).

Indocyanine green

ICG is a near-infrared (NIR) passively targeted fluorescent cyanine dye. It is commonly used for determining cardiac output, hepatic function, and liver blood flow, and ophthalmic angiography for imaging retinal blood vessels (FDA approved).[58] The absorption spectrum of ICG depends on the nature of the solvent, as well as the dye concentration. Binding of the dye to serum proteins shifts the absorption spectrum toward a maximum at 805 or 810 nm, which is the excitation wavelength range used for medical applications. Its fluorescence emission is from 800 to 850 nm with a major peak at 830 nm. ICG is administered intravenously, and the toxicity of ICG is thought to be low, as it is not absorbed by the intestinal mucous membrane. It can be metabolized and excreted via the liver and the hepatobiliary system (bile ducts) with a half-life of around 3–4 minutes.[59] Some side effects have been reported, including sore throat, hot flashes, and occasional anaphylaxis.

The ability of ICG to delineate brain tumor margins has been demonstrated in rodent models.[60] Haglund et al. reported a clinical trial with nine patients who underwent brain tumor resection with ICG fluorescence[61] and showed differences in the dynamic optical signals between normal brain, low-grade astrocytoma, and malignant astrocytoma. Optical imaging of the resection margins in malignant tumors showed differences between adjacent normal tissue and residual tumor tissue. Raabe et al. reported their initial clinical experience and evaluation of ICG videoangiography (ICGA) for intraoperative blood flow assessment in neurovascular cases in 14 patients. The postoperative angiographic results corresponded to the intraoperative ICG video-angiographic findings, and in three cases, ICGA significantly changed the surgical approach by providing real-time information on the patency of arterial and venous vessels.[62]

Zaidi et al. applied ICG angiography to map the vascular configuration of cerebral arteriovenous malformations (AVMs) during surgical resection in 130 consecutive cases supporting ICGA as a rapid and safe method for intraoperative mapping of the angioarchitecture of superficial AVMs. The approach proved less helpful for deep-seated lesions and alone did not improve the identification of residual disease or prediction of patient outcomes.[63] Sharma et al. performed a retrospective study including 112 patients with 126 aneurysms who underwent microsurgical clipping with ICGA during aneurysm surgery. Their results showed that ICGA could be useful for intraoperative assessment of the adequacy of aneurysm obliteration and the patency of parent and perforating vessels, but showed increased reliability when used as a complementary modality to intraoperative digital subtraction angiography (DSA).[64] (Further discussion on the use of ICG is presented in Chapter 12.)

Fluorescein sodium

Similar to ICG, FS is a synthetic organic fluorescent dye. The FDA has approved its use for diagnostic fluorescein angiography or angioscopy of the retina and iris vasculature. Its major excitation peak is between 465 and 490 nm and the fluorescence emission band is from 510 to 530 nm. FS has been used for navigation and resection of intracranial tumors as early as 1948.[54] Some recent studies have investigated the use of ICG and FS to guide brain tumor resection using both wide-field imaging[65–67] and microscopic imaging.[68–71]

Koc et al. has evaluated FS-guided surgery in glioblastoma multiforme (GBM) in 47 patients with a control group of 33 patients. The data showed that the use of FS as a simple procedure led to a significant increase in the number of patients having gross total resection (GTR) (83% versus 55%).[72] Okuda et al. applied high-dose FS (20 mg/kg) to guide GBM surgery in 10 patients using filters to distinguish tumor from brain surface and tumor vessels from surrounding normal vessels. No additional complications or permanent side effects were found in their study.[65] Low-dose (3–4 mg/kg bodyweight) fluorescein imaging has also shown to be partly beneficial in a study of 35 malignant brain tumor patients. In 28 patients, FS imaging appeared to improve visualization of the tumor margin. However, they found that accumulation of FS in regions with a disrupted BBB seemed to depend on histopathology and/or pretreatment with surgery, radiation, and chemotherapy.[73]

5-Aminolevulinic acid

5-ALA is the first compound in the porphyrin synthesis pathway that leads to heme synthesis. Oral administration of nonfluorescent 5-ALA functions as a prodrug, which is metabolized to promote the accumulation of fluorescent protoporphyrin IX (PpIX) in malignant gliomas (Figure 17.6). However, its use is presently FDA approved solely for topical application. Stummer et al. demonstrated intraoperative detection of malignant glioma by 5-ALA-induced porphyrin fluorescence in 10 patients. Tumor tissue was distinguished by bright red fluorescence with emission peaks at 635 and 704 nm, whereas normal brain tissue revealed no porphyrin fluorescence, suggesting the specific labeling of malignant gliomas.[75,76] In 2006, Stummer et al. conducted a randomized controlled multicenter phase III trial of fluorescence-guided surgery with 5-ALA for resection of malignant glioma. About 322 patients with suspected malignant glioma amenable to complete resection of contrast-enhancing tumor were randomly assigned to 5-ALA fluorescence-guided resection ($n = 161$) or conventional microsurgery with white light ($n = 161$). The study resulted in complete resection of contrast-enhancing tumor in 90 of 139 patients (65%) from the 5-ALA-assisted arm in comparison with only 47 of 131 (36%) from the white light arm. Moreover, the patients from the 5-ALA arm of the study showed a higher 6-month PFS than the white light arm.[77]

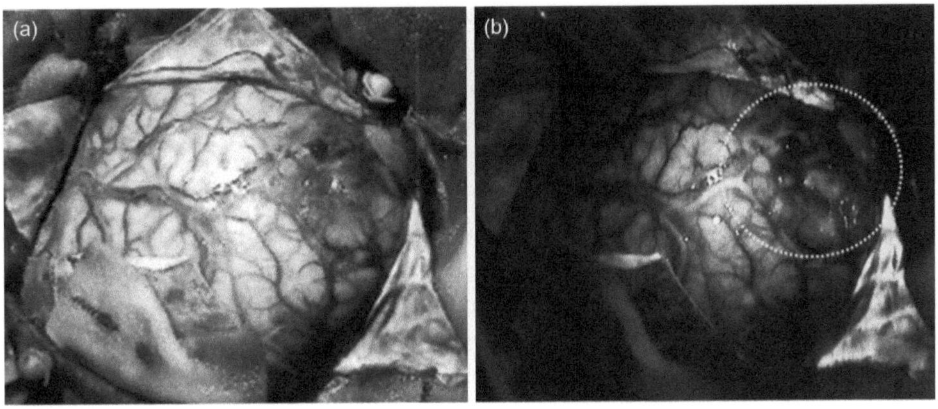

Figure 17.6 Intraoperative fluorescence imaging of malignant glioma: (a) Image captured under white light illumination; (b) red fluorescence image (5-ALA-induced PpIX) excited by blue light. *Adapted from*[74] *with permission.*

Hefti et al. reported their 1-year experience on using 5-ALA-induced porphyrin fluorescence in high-grade glioma surgery in 74 cases undergoing GTR, partial resection, and biopsy. They reported that a solid fluorescence signal defines tumor with high sensitivity and specificity of 0.98 and 1.0, respectively, but vague fluorescence reduced sensitivity and specificity to 0.76 and 0.85, respectively. Heterogeneous tumors with lower grade elements and satellite lesions could not be reliably resected using this modality alone.[78] Nabavi et al. conducted a multicenter prospective phase II study using 5-ALA to guide resection of recurrent malignant gliomas in 36 patients and reported 5-ALA as an effective molecular imaging modality providing relevant information for surgical decision making in recurrent malignant gliomas.[79]

Other techniques and data analysis have been introduced to further improve the sensitivity and specificity for 5-ALA-induced fluorescence surgery. Sanai et al. used intraoperative confocal microscopy to visualize 5-ALA tumor fluorescence in low-grade gliomas during microsurgical resection in 10 patients. Although macroscopic tumor fluorescence was not evident in any of the patients, in each case, intraoperative confocal microscopy identified tumor fluorescence at a cellular level at the brain–tumor interface, which corresponded to tumor infiltration on matched histological analysis.[80] Rapp et al. also investigated the combination of an endoscope with 5-ALA-induced fluorescent contrast in 9 patients. In all cases, this new modality identified 5-ALA-positive tissue not sufficiently exposed by conventional microscopic visualization. In eight patients, subsequent histopathology examination of tumor tissue confirmed residual tumor, whereas in one patient the endoscopic visualization gave falsely ALA-positive signals and the tissue was finally confirmed to be radiation necrosis (not tumor) through histopathologic examination. This case indicated that endoscopic visualization may overestimate the amount of tumor.[81] Valdés et al. performed quantitative analysis of the 5-ALA-induced fluorescence in 14 patients with diagnoses of low-grade glioma, high-grade glioma, meningioma, and brain metastasis. Receiver operating characteristic curve analysis of porphyrin PpIX concentration as a diagnostic variable for the detection of neoplastic tissue yielded a classification efficiency of 87% compared with 66% for nonquantitative conventional fluorescence imaging.[82] 5-ALA has been widely tested, including class I evidence, for fluorescence-guided brain tumor resection,[77] and its use for the resection of primary malignant brain tumors is considered to be a milestone in neurosurgery.

Tumor Paint

Chlorotoxin is a small 36-amino acid peptide initially isolated from the venom of the deathstalker scorpion (*Leiurus quinquestriatus*) and shown to inhibit reconstituted small conductance chloride ion channels.[83,84] Targeting molecules such as antibodies and peptides are being investigated as tumor-specific ligands, and further studies supported chlorotoxin as a specific target of human glioma cells in a xenograft mouse model.[85] Moreover, an immunochemical screening of biopsy samples from more than 250 patients showed chlorotoxin to be a highly specific marker for four grades of gliomas (WHO I–IV) and tumors of neuroectodermal origin, such as medulloblastomas, neuroblastomas, melanomas, primitive neuroectodermal tumors, and small cell lung carcinoma. By contrast, normal tissue from the brain, skin, kidney, and lung showed consistently negative binding.[86] Due to its high specificity to cancer cells, a radiolabeled form of chlorotoxin (^{131}I-TM-601) was developed and used as a SPECT imaging agent to evaluate tumor extent in phase I/II clinical trials.[87]

In 2007, Olson et al. developed the chlorotoxin:Cy5.5 bioconjugate (Tumor Paint), which emits in the NIR fluorescence range. This probe delineated glioma and medulloblastoma from adjacent non-neoplastic tissue in mouse models, and metastatic cancer foci with a few hundred cells were detected in lymph channels.[88] A new chemical entity that only bound a single NIR fluorescent molecule was developed by the Olson lab in 2011 and shown to be functionally equivalent to native chlorotoxin:Cy5.5.[89,90] A variant of Tumor Paint conjugated to ICG (BLZ-100, Blaze Bioscience, FDA approved) was developed as the first Tumor Paint product candidate. This molecular probe has been demonstrated to effectively target cancer cells in a wide range of cancer types, including brain, prostate, breast, prostate, lung, colorectal, skin, and sarcomas as preclinical studies,[91,92] and is currently in phase I clinical trials for brain and skin cancers (Blaze Bioscience). In contrast to 5-ALA, which is administered topically, Tumor Paint is administered intravenously and circulates within the body prior to surgery.

Optical coherence tomography

The use of lower energy light such as in the NIR part of the optical spectrum allows for increased depth of penetration in tissue compared to higher energy light such as in the visible and ultraviolet parts of the optical spectrum. Optical coherence tomography (OCT) takes advantage of the NIR depth penetrance such as Tumor Paint fluorescence, but is not a molecular imaging modality. Due to its wide acceptance as a tool for tissue characterization and associated potential for intraoperative decision making, we opted to include a brief discussion of this method here. OCT is an established biomedical imaging technique based on low-coherence optical interferometry for noninvasive 3D imaging of the internal microstructures by measuring light scattering from the sample.[93] Time-domain OCT requires rapid scanning of the reference arm to obtain the depth profiles (A-scan) of the sample. Next-generation techniques, such as spectral domain and swept source systems, allow for much higher line acquisition rate with improved sensitivity by several orders of magnitude compared to the time-domain system.[94–96] OCT has been used for several clinical applications, including ophthalmology, cardiology, oncologic gastroenterology, and dermatology.[97]

Böhringer et al. analyzed biopsy specimens of human brain tumor with OCT. As shown in Figure 17.7, human brain cortex shows a characteristic image with a fine line of an intense surface, a homogeneous decrease in the signal with increasing tissue depth, and no prominent structures are identified within the tissue. By contrast, glioblastoma

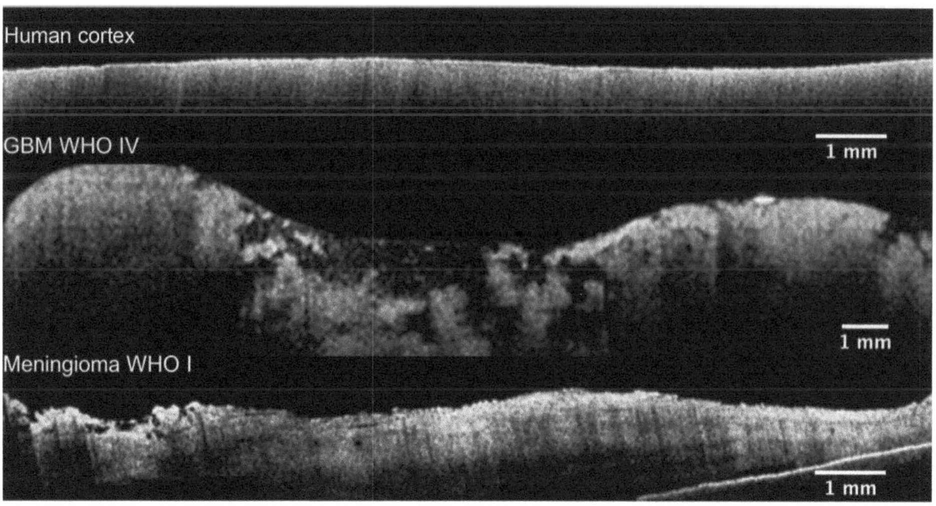

Figure 17.7 Time-domain optical coherence tomography (OCT) imaging of native tissue obtained from surgical specimens of human cortex, a glioblastoma (WHO grade IV) and a meningioma (WHO I). *Modified and adapted from Bohringer et al.*[98]

specimens showed a heterogeneous picture with a prominent microstructure of the tumor parenchyma clearly distinct from the signal of human cortex. Imaging of a meningioma specimen and its dural insertion showed layering of the dura, thickening of the dura, and loss of layering closer to the tumor mass. These observations confirmed different microstructures and characteristic signal attenuation profiles of tumor versus normal brain, demonstrating that OCT could be used to discriminate normal brain, areas of tumor infiltrated brain, and solid tumor.[98] To expand the clinical use of OCT, the ongoing research is focusing on improving image quality, increasing acquisition speed, and adding molecular sensitivity. OCT may become a useful tool for neurosurgical guidance in the near future.

Functional near-infrared spectroscopy

Functional NIR spectroscopy (fNIRS) is a neuroimaging and brain mapping technique for the noninvasive monitoring of human brain neural activity and regional cerebral hemodynamics by detecting light absorption and scattering mainly associated with detectable chemicals such as hemoglobin and cytochromes through the skull. The basic concept of fNIRS for monitoring the hemodynamics of human tissue was presented by Jobsis.[99] Later, the NIR technique was rapidly applied to measure hemodynamics associated with functional cortical activity.[100–102] Watanabe et al. performed noninvasive assessment of language dominance with fNIRS.[103] As shown in Figure 17.8, cerebral activation, such as hemoglobin dynamics, was observed when the volunteers performing a task (a 17-second word generation). This work demonstrated that fNIRS could be a feasible noninvasive alternative to the Wada test. Over the past two decades, this technique has been used for neurology, including Alzheimer's disease, dementia, depression, epilepsy, Parkinson's disease, postneurosurgical deficits, rehabilitation, psychiatry, and psychology/education.[104–106] Sato et al. investigated the expressive and receptive language functions separately in 10 presurgical glioma patients using NIRS. The hemodynamic changes during a verb generation task or story listening task were measured in the cerebral hemisphere on either side covering the language areas. Their results demonstrated that this technique could be used for preoperative assessment of language function and exemplifies how a preoperative NIRS study can allow detection of unforeseen language lateralization.[107]

Raman spectroscopy

Raman spectroscopy is a spectroscopic technique used to detect vibrational, rotational, and other states in a molecular system, capable of probing the chemical composition of materials. Raman spectra provide a fingerprint of the molecules and can be used to identify the chemical species and composition. It can be used for analyzing many kinds of materials such as gases, vapors, aerosols, liquids, and solids including

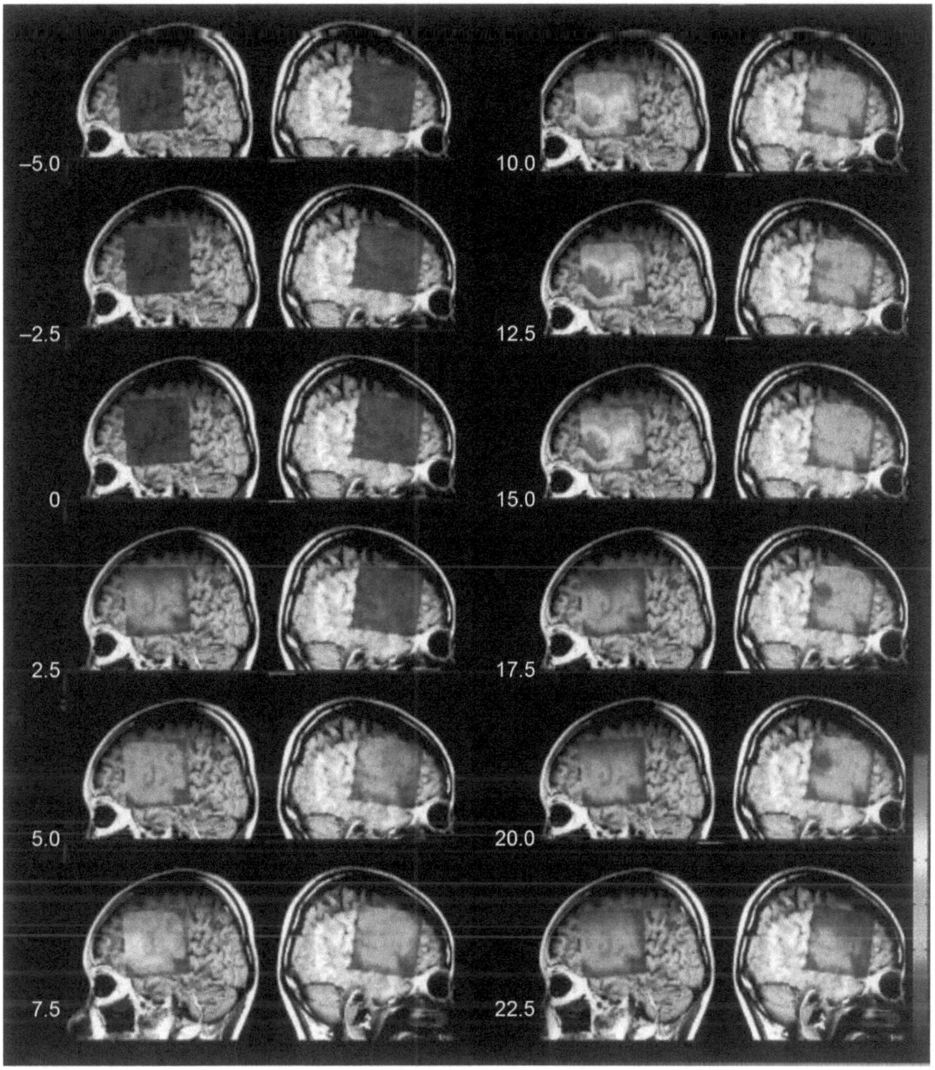

Figure 17.8 Serial fNIR pseudocolor maps (12 channels on each hemi) of total hemoglobin changes overlaid on the lateral MRI surface image of the subject. Total hemoglobin increased after the beginning of the task (a 17-second word generation) and then gradually decreased. The peak was observed in the left inferior frontal area, which is in the vicinity of the Broca's area. The numbers indicate time in seconds. *Adapted from Watanabe et al.*[103]

biological tissues. Biomedical applications of Raman spectroscopy and microscopy for disease diagnosis, such as cancers, have also been widely studied for many tissues and organs in vivo.[108] In this section, we will review the development of Raman microspectroscopy for brain tumor diagnosis and its potential for neurosurgical guidance.

Table 17.3 Raman peaks and assignments for raman bands of native brain tissue and intracranial tumors

Raman shift (cm^{-1})	Assignments	Raman shift (cm^{-1})	Assignments
670	Hemoglobin	1267	Lipid $\delta(C{=}CH_2)$, protein (amide III)
700	Cholesterol	1298	Lipid $\delta(CH_2)$
717	Choline $N^+(CH_3)_3$	1440	Lipid $\delta(CH_2)$
757	Protein (Trp[a]), hemoglobin	1450	Protein $\delta(CH_2/CH_3)$
857	Protein (Tyr), collagen	1566	Hemoglobin
940	Protein, collagen	1623	Hemoglobin
1005	Protein (Phe)	1635	Water $\delta(OH)$
1064	Lipid $\nu(C{-}C)$	1660	Protein (amide I), lipid $\nu(C{=}C)$
1129	Lipid $\nu(C{-}C)$	2888	$\nu(CH_2)$
1210	Protein (Phe, Tyr)	2937	$\nu(CH_3)$
1225	Hemoglobin	3230	Water $\nu(OH)$
1247	Collagen, protein (amide III)	3400	Water $\nu(OH)$

[a]Standard notation for amino acids. δ, deformation vibrations; ν, stretching vibrations.
Source: Adapted from Krafft et al.[109] with permission.

Krafft et al. evaluated the ability of Raman spectroscopy and Raman mapping to distinguish between normal tissue and human intracranial tumors (gliomas and meningiomas). Through cluster analysis, they reported that normal brain tissue contains higher levels of lipids, intracranial tumors have more hemoglobin and lower lipid-to-protein ratios, and meningiomas contain more collagen with maximum collagen content in normal meninges.[109] Amharref et al. investigated molecular changes associated with glioma tissue of rat brains and reported that using hierarchical cluster analysis on the only cluster averaged spectra showed a clear distinction between normal, tumoral, necrotic, and edematous tissues.[110] Kast et al. demonstrated the proof of principle of Raman mapping for identification of the regions of normal brain, necrosis, diffusely infiltrating glioma, and solid glioblastoma, as well as the boundaries among them. Featured Raman peaks at 1003, 1313, 1441, 1585, and 1659 cm^{-1} were analyzed (Table 17.3). For their study, the spatial resolution was limited to 20–300 μm due to the slow speed of Raman mapping as one of the major limitations.[111] Kalkanis et al. analyzed 40 frozen brain tumor tissue sections and reported high accuracy (97.8%) in distinguishing tissue types of gray matter, necrosis, and GBM in the validation dataset.[112]

Several molecular biomarkers have been investigated with Raman microspectroscopy for brain tumor diagnosis, including lipids, proteins, collagen, and water. Nygren et al. reported increased levels of cholesterol esters in glioma tissue and surrounding areas of human brain.[113] Koljenović et al. studied the biochemical difference between vital and necrotic glioblastoma tissues and found that necrotic tissue consistently contains higher levels of cholesterol (-esters).[114] Beleites et al. used unsaturated lipids to differentiate between normal and tumor tissues, and used total lipid content to determine the tumor grade of astrocytoma tissues with a fiber-optic Raman probe and soft classification method.[115] Krafft et al. analyzed the correlation of malignancy and cell density of primary brain tumors using Raman microscopic imaging and conclude that Raman spectra of high-grade tumor cells were found to contain more intense spectral contribution of nucleic acids than those of low-grade tumor cells using spectral unmixing algorithm vertex component analysis.[116] In 20 patients, Koljenović et al. reported large difference between the Raman spectra of dura and meningioma, because of the high collagen content and the increased lipid content of tumors, indicating its potential for neurosurgical resection of meningioma.[117] Wolthuis et al. used Raman spectroscopy to assess brain edema, by measuring the water concentration in the tissue directly.[118] Karabeber et al. evaluated the ability of a handheld Raman scanner, guided by surface-enhanced Raman scattering (SERS) nanoparticles, to identify the microscopic tumor extent in a mouse model with brain tumors. In a simulated intraoperative scenario, they found SERS nanoparticles accurately outlined the extent of the tumors and yielded more accurate tumor resection.[119]

Many studies have been done for brain tumor diagnosis using Raman microspectroscopy. However, a study reported very similar spectra for glioma grades II and III, two acoustic neuromas and central neurocytoma, as normal gray matter spectra.[120] Technical advances in detection sensitivity and data acquisition speed are highly desirable before Raman microspectroscopy can be adapted for clinical image-guided neurosurgical resection.

Coherent Raman microscopy

Coherent Raman microscopy (CRM), in the form of CARS and SRS, has been developed into a valuable tool for label-free biomolecular imaging. Being a nonlinear optical microscopy, CRM is several orders of magnitude faster in image acquisition than spontaneous Raman microscopy, enabling rapid chemical mapping up to video rate, with 3D sectioning capability. CRM has been applied for many biological and biomedical applications through the label-free imaging of lipids, proteins, DNA, water, other metabolites, and drugs. More technical and instrumentation details for CARS and SRS microscopy are presented in the literature.[121–125]

CARS microscopy

Evans et al. demonstrated the use of CARS microscopy to image brain structure and pathology using fresh mouse brain tissue. White matter and gray matter could be clearly distinguished based on lipid concentration.[126] Uckermann et al. further evaluated the feasibility of CARS microscopy for brain tumor identification and delineation using human brain tumor tissue. They reported that the morphochemical contrast of CARS images enabled the delineation of brain tumors with cellular resolution, irrespective of the tumor type and properties, suggesting that CARS imaging may be a potential tool for diagnostic neuropathology as for tumor margin identification.[127]

SRS microscopy

SRS microscopy has a major advantage over CARS as it offers background-free and readily interpretable chemical contrast. Ji et al. used SRS imaging to differentiate tumor from normal tissue in an infiltrative human glioblastoma xenograft mouse model based on lipid and protein mapping. In vivo experiments in mice demonstrated that SRS could reveal tumor margins that were undetectable under conventional white light illumination (Figure 17.9) and correlated with histopathology.[128,129]

The essential diagnostic features for conventional histopathology critically rely on the evaluation of nuclear content and morphology. Our recent work on multispectral SRS imaging is now allowing the visualization of DNA, as shown in Figure 17.10. Technical advances in the last decade have made CRM microscopy a state-of-the-art technique for biological and biomedical applications. It has significantly contributed to biology and biomedicine in many fields, and its use for brain tumor diagnosis and potentially for neurosurgical resection guidance is under investigation.

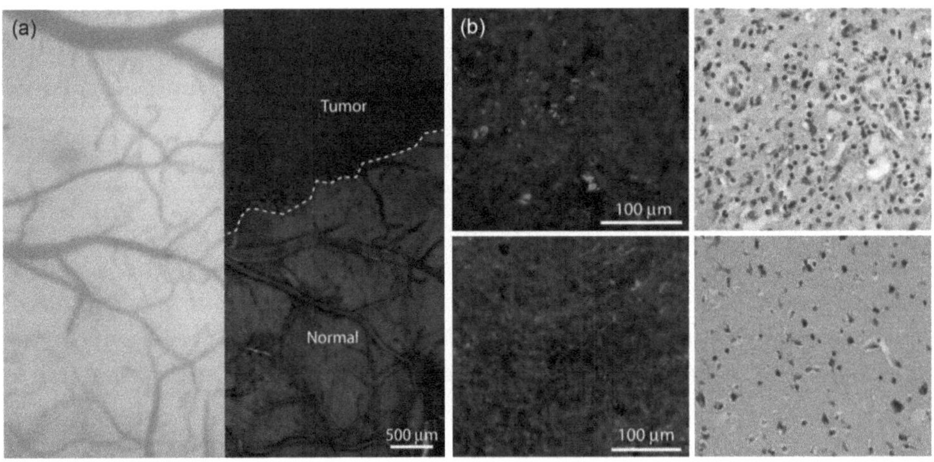

Figure 17.9 (a) In vivo SRS imaging of human GBM in a xenograft mouse model. Bright-field microscopy (left) appears grossly normal, whereas the SRS image (right) of the same field of view demonstrates distinctions between tumor-infiltrated areas and noninfiltrated brain with a clear border. (b) SRS (left) and hematoxylin and eosin (H&E) (right) microscopy of freshly excised tissue from a human brain tumor. The hypercellularity of viable tumor (upper) contrasts with normocellular regions of adjacent brain with minimal tumor infiltration (below). Green: Lipid at 2845 cm^{-1} (CH_2); blue: protein at 2930 cm^{-1} (CH_3) with SRS. *Adapted from*[128] *with permission.*

Figure 17.10 SRS imaging of a 1-mm fresh mouse brain tissue slice. Multispectral SRS with linear decomposition facilitates visualization of the cell nuclei based on retrieved nucleic acid signals (magenta). Lipid imaging maps the tissue morphology, as well as the myelin sheath distribution and orientation (green).

Chemical modalities

Despite the need for tissue characterization during surgery, there is still no available FDA-approved technique to assess the molecular content of surgical samples in real time. The most information provided to the neurosurgeon during surgery comes from frozen section analysis, which takes in general 30 minutes and provides limited diagnostic information. Mass spectrometry (MS) is an analytical tool routinely used for the detailed chemical analysis of a wide range of samples offering the required sensitivity and specificity for biomedical applications. For a molecule to be analyzed by MS, it must be introduced into the instrument and ionized, and further separated from other molecules according to its mass-to-charge ratio (m/z). The accurate mass and fragmentation pattern of a molecule can provide information on its elemental composition and structure.

The use of MS for the visualization of the spatial distribution of biomolecules in a tissue section has been introduced as MS imaging (MSI). Direct tissue analysis by MSI is performed using different desorption/ionization techniques such as matrix-assisted laser desorption ionization (MALDI)[130–133] and desorption electrospray ionization (DESI).[134–136] One important technical parameter to be considered for the analysis of surgical samples by MS in a near-real-time fashion during surgery is that it should

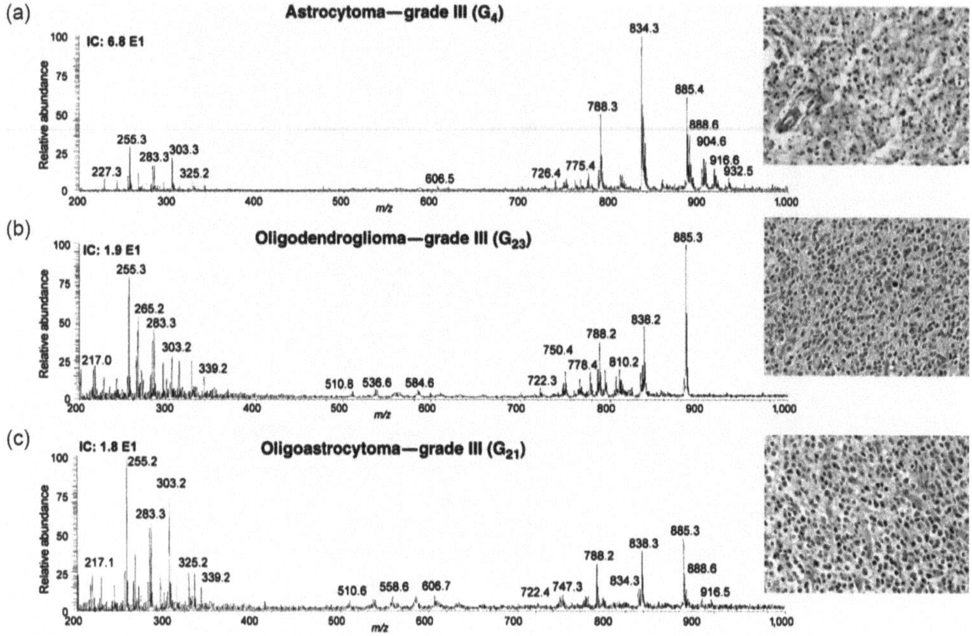

Figure 17.11 Glioma subtypes by MS. Discrimination between subtypes of grade III gliomas is achieved through lipid profiles detected by DESI MS. Representative negative-ion mode mass spectra from different grade III subtypes of human gliomas: (a) anaplastic astrocytoma, (b) anaplastic oligodendroglioma, and (c) anaplastic oligoastrocytoma. Insets show bright-field optical images (400 ×) of the corresponding diagnostic permanent H&E section. *Reprinted from Eberlin et al.*[141] *with permission from the American Association for Cancer Research.*

require minimal-to-no sample preparation.[137–139] MS offers the possibility of directly imaging multiple molecules simultaneously and with a high specificity from biological tissues without the application or systemic administration of molecular probes. Our group has undertaken the validation and implementation of real-time and near-real-time MS approaches to support surgical decision making. The validation of our work benefits from the Advanced Multimodal Image Guided Operating suite at Brigham and Women's Hospital Boston, Massachusetts, where imaging modalities can be correlated and cross-validated. Using neuronavigation, sampling positions are digitally registered and the coordinates later used to overlay the molecular characterization results derived from MS with 3D medical images such as MRI and fMRI using in-house developed and freely available image analysis software (3D Slicer).[140]

Over the past few years, our validation of MS for intraoperative characterization of brain tumors has been based on the use of DESI MS. Using this technique and a multivariate statistical approach, we have established correlations between lipidomic data and histopathologic information.[139,141] The approach allows the classification of human gliomas and meningiomas according to their subtype (Figure 17.11), grade

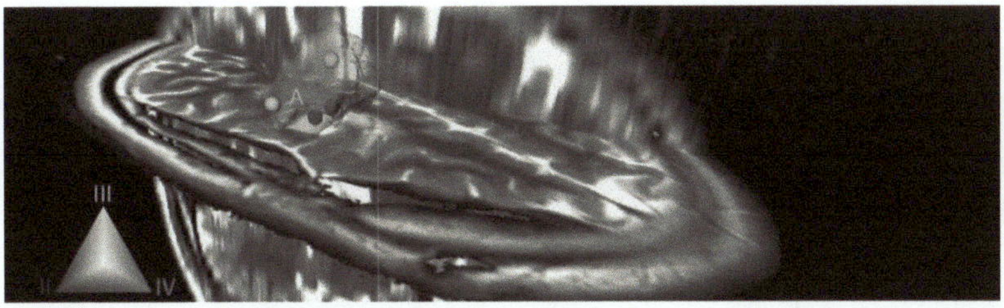

Figure 17.12 Mass spectrometry astrocytoma grading/heterogeneity. Glioma tumor grade classification results for surgical case 2 visualized in segmented preoperative 3D MRI volume reconstruction of the tumor and surrounding regions. Tumor volume is represented in light purple. *Reprinted from Eberlin et al.*[139] *with permission from the National Academy of Sciences.*

(Figure 17.12), and tumor cell concentration as a marker of tumor margins. MS analyses of tissue also provided distinction between viable glioblastoma and necrotic tissue from 12 surgical samples.[142]

Our most recent work has focused on the detection of tumor metabolites such as 2-HG found in tumors harboring a mutation in the Krebs cycle enzyme isocitrate dehydrogenases (IDH1 and IDH2).[143] Providing general 2-HG analysis during surgery could give the surgeon diagnostic and prognostic information, and the detailed stereotactic analysis of the metabolite can be used to delineate the tumor (Figure 17.13).

The required sample preparation for MALDI MSI has long been considered to be a limiting factor for its application to surgical decision making, but the approach has proven to be valuable for the rapid profiling of biomolecules from tissue.[131,132,144,145] The possibility of analyzing a wider mass range of biomolecules, including validated biomarkers such as peptides and proteins, makes MALDI MSI a coveted approach to support rapid diagnosis during surgery.[144] Our group is currently validating the approach to support surgical decision making for pituitary surgery through the monitoring of peptide and protein hormones in near real time. Newer interfaces for the introduction of surgical material into the mass spectrometer either from a stereotactic sample or a direct and continuous sampling from within the surgical cavity should further facilitate the adoption of MS as a tool for real-time tissue characterization given adequate validation of the analyses with the current gold standard of histopathology.

FUTURE DIRECTIONS

Validation is critical to translate new imaging modalities to relevant image-guided therapy tools. The integration of the clinical and research teams is also a determinant

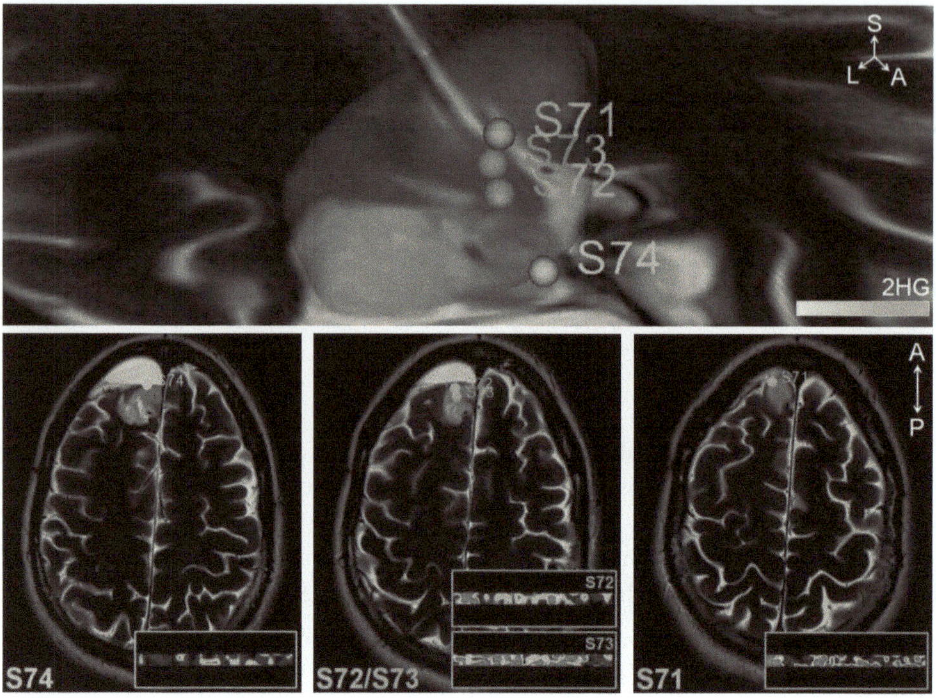

Figure 17.13 2-HG tumor distribution by MS. Normalized 2-HG signal is represented with a warm color scale as indicated by the scale bar, set from the lowest (yellow) to highest (orange) levels detected from this individual case. Stereotactic positions were digitally registered to the preoperative MRI using neuronavigation in a standard operating room. The 3D tumor volume is shown (upper). Classification results of samples S74, S72, S73, and S71 are further visualized on axial sections (lower). *Reprinted from Santagata et al.*[143] *with permission from the National Academy of Sciences.*

factor in that the resulting tool or approach is more likely to be relevant if it is iteratively based on the needs and the understanding of the needs of the practitioners. In this era of valuing innovation, the field is likely to see an increasing number of molecular imaging tools and methods coming its way to improve the care of neurosurgical patients.

REFERENCES

1. Bailey DL, Townsend DW, Valk PE, Maisey MN, eds. *Positron Emission Tomography*. New York: Springer; 2005.
2. Cherry SR, Phelps ME. Imaging brain function with positron emission tomography. In: Toga A, Mazziota J, eds. *Brain Mapping. The Methods*. San Diego, CA: Academic Press; 1996:191–221.
3. D'Souza MM, Sharma R, Tripathi M, Panwar P, Jaimini A, Mondal A. Novel positron emission tomography radiotracers in brain tumor imaging. *Indian J Radiol Imaging*. 2011;21(3):202.

4. Tai Y, Piccini P. Applications of positron emission tomography (PET) in neurology. *J Neurol Neurosurg Psychiatr.* 2004;75(5):669–676.
5. Phelps ME. Positron emission tomography provides molecular imaging of biological processes. *Proc Natl Acad Sci USA.* 2000;97(16):9226–9233.
6. Brock C, Young H, O'Reilly S, et al. Early evaluation of tumour metabolic response using [18F] fluorodeoxyglucose and positron emission tomography: a pilot study following the phase II chemotherapy schedule for temozolomide in recurrent high-grade gliomas. *Br J Cancer.* 2000;82(3):608.
7. Di Chiro G, DeLaPaz RL, Brooks RA, et al. Glucose utilization of cerebral gliomas measured by [18F] fluorodeoxyglucose and positron emission tomography. *Neurology.* 1982;32(12):1323–1329.
8. Patronas NJ, Chiro GD, Kufta C, et al. Prediction of survival in glioma patients by means of positron emission tomography. *J Neurosurg.* 1985;62(6):816–822.
9. De Witte O, Goldberg I, Wikler D, et al. Positron emission tomography with injection of methionine as a prognostic factor in glioma. *J Neurosurg.* 2001;95(5):746–750.
10. Pirotte BJ, Levivier M, Goldman S, et al. Positron emission tomography-guided volumetric resection of supratentorial high-grade gliomas: a survival analysis in 66 consecutive patients. *Neurosurgery.* 2009;64(3):471–481.
11. Maciunas R, Kessler R, Maurer C, Mandava V, Watt G, Smith G. Positron emission tomography imaging-directed stereotactic neurosurgery. *Stereotact Funct Neurosurg.* 1992;58(1–4):134–140.
12. Thon N, Kunz M, Lemke L, et al. Dynamic 18F-FET PET in suspected WHO grade II gliomas defines distinct biological subgroups with different clinical courses. *Int J Cancer.* 2015;136(9):2132–2145.
13. Kaplan AM, Bandy DJ, Manwaring KH, et al. Functional brain mapping using positron emission tomography scanning in preoperative neurosurgical planning for pediatric brain tumors. *J Neurosurg.* 1999;91(5):797–803.
14. Rozental JM, Levine RL, Mehta MP, et al. Early changes in tumor metabolism after treatment: the effects of stereotactic radiotherapy. *Int J Radiat Oncol Biol Phys.* 1991;20(5):1053–1060.
15. Young H, Baum R, Cremerius U, et al. Measurement of clinical and subclinical tumour response using [18F]-fluorodeoxyglucose and positron emission tomography: review and 1999 EORTC recommendations. *Eur J Cancer.* 1999;35(13):1773–1782.
16. Valk PE, Budinger TF, Levin VA, Silver P, Gutin PH, Doyle WK. PET of malignant cerebral tumors after interstitial brachytherapy: demonstration of metabolic activity and correlation with clinical outcome. *J Neurosurg.* 1988;69(6):830–838.
17. Barrington S. Clinical uses of PET in neurology. *Nucl Med Commun.* 2000;21(3):237–240.
18. Tsuyuguchi N, Sunada I, Iwai Y, et al. Methionine positron emission tomography of recurrent metastatic brain tumor and radiation necrosis after stereotactic radiosurgery: is a differential diagnosis possible? *J Neurosurg.* 2003;98(5):1056–1064.
19. Dethy S, Goldman S, Blecic S, Luxen A, Levivier M, Hildebrand J. Carbon-11-methionine and fluorine-18-FDG PET study in brain hematoma. *J Nucl Med.* 1994;35(7):1162–1166.
20. Ogawa T, Kanno I, Shishido F, et al. Clinical value of PET with 18F-fluorodeoxyglucose and L-methyl-11C-methionine for diagnosis of recurrent brain tumor and radiation injury. *Acta Radiol.* 1991;32(3):197–202.
21. Wiebe S, Blume WT, Girvin JP, Eliasziw M. A randomized, controlled trial of surgery for temporal-lobe epilepsy. *N Engl J Med.* 2001;345(5):311–318.
22. Duncan JS. Imaging and epilepsy. *Brain.* 1997;120(2):339–377.
23. Kuzniecky R, Garcia JH, Faught E, Morawetz RB. Cortical dysplasia in temporal lobe epilepsy: magnetic resonance imaging correlations. *Ann Neurol.* 1991;29(3):293–298.
24. Savic I, Ingvar M, Stone-Elander S. Comparison of [11C] flumazenil and [18F] FDG as PET markers of epileptic foci. *J Neurol Neurosurg Psychiatr.* 1993;56(6):615–621.
25. Boling WW, Lancaster M, Kraszpulski M, Palade A, Marano G, Puce A. Fluorodeoxyglucose-Positron emission tomographic imaging for the diagnosis of mesial temporal lobe epilepsy. *Neurosurgery.* 2008;63(6):1130–1138.
26. Bourgeois M, Rajerison H, Guerard F, et al. Contribution of [64Cu]-ATSM PET in molecular imaging of tumour hypoxia compared to classical [18F]-MISO—a selected review. *Nucl Med Rev.* 2011;14(2):90–95.

27. Padhani AR, Krohn KA, Lewis JS, Alber M. Imaging oxygenation of human tumours. *Eur Radiol.* 2007;17(4):861–872.
28. Mclean MA, Cross JJ. Magnetic resonance spectroscopy: principles and applications in neurosurgery. *Br J Neurosurg.* 2009;23(1):5–13.
29. Bertholdo Db, Watcharakorn A, Castillo M. Brain proton magnetic resonance spectroscopy: introduction and overview. *Neuroimaging Clin N Am.* 2013;23(3):359–380.
30. El-Deredy W. Pattern recognition approaches in biomedical and clinical magnetic resonance spectroscopy: a review. *NMR Biomed.* 1997;10(3):99–124.
31. Brandao LA. MR spectroscopy of the brain. *Neuroimag Clin N Am.* 2013;23(3):359–562.
32. Roder C, Skardelly M, Ramina KF, et al. Spectroscopy imaging in intraoperative MR suite: tissue characterization and optimization of tumor resection. *Int J Comput Assist Radiol Surg.* 2014;9(4):551–559.
33. Gujar SK, Maheshwari S, Björkman-Burtscher I, Sundgren PC. Magnetic resonance spectroscopy. *J Neuroophthalmol.* 2005;25(3):217–226.
34. Galanaud D, Chinot O, Nicoli Fo, et al. Use of proton magnetic resonance spectroscopy of the brain to differentiate gliomatosis cerebri from low-grade glioma. *J Neurosurg.* 2003;98(2):269–276.
35. Floeth FW, Pauleit D, Wittsack Jr H, et al. Multimodal metabolic imaging of cerebral gliomas: positron emission tomography with [18F] fluoroethyl-L-tyrosine and magnetic resonance spectroscopy. *J Neurosurg.* 2005;102(2):318–327.
36. Shiino A, Nakasu S, Matsuda M, Handa J, Morikawa S, Inubushi T. Noninvasive evaluation of the malignant potential of intracranial meningiomas performed using proton magnetic resonance spectroscopy. *J Neurosurg.* 1999;91(6):928–934.
37. Barba I, Moreno Á, Martínez-Pérez I, et al. Magnetic resonance spectroscopy of brain hemangiopericytomas: high myoinositol concentrations and discrimination from meningiomas. *J Neurosurg.* 2001;94(1):55–60.
38. Herminghaus S, Dierks T, Pilatus U, et al. Determination of histopathological tumor grade in neuroepithelial brain tumors by using spectral pattern analysis of in vivo spectroscopic data. *J Neurosurg.* 2003;98(1):74–81.
39. Chan AA, Lau A, Pirzkall A, et al. Proton magnetic resonance spectroscopy imaging in the evaluation of patients undergoing gamma knife surgery for grade IV glioma. *J Neurosurg.* 2004;101(3):467–475.
40. Andronesi OC, Kim GS, Gerstner E, et al. Detection of 2-hydroxyglutarate in IDH-mutated glioma patients by in vivo spectral-editing and 2D correlation magnetic resonance spectroscopy. *Sci Transl Med.* 2012;4(116):116ra4.
41. Parsons DW, Jones S, Zhang X, et al. An integrated genomic analysis of human glioblastoma multiforme. *Science.* 2008;321(5897):1807–1812.
42. Lazovic J, Soto H, Piccioni D, et al. Detection of 2-hydroxyglutaric acid in vivo by proton magnetic resonance spectroscopy in U87 glioma cells overexpressing isocitrate dehydrogenase–1 mutation. *Neuro-oncology.* 2012;14(12):1465–1472.
43. Pope WB, Prins RM, Thomas MA, et al. Non-invasive detection of 2-hydroxyglutarate and other metabolites in IDH1 mutant glioma patients using magnetic resonance spectroscopy. *J Neurooncol.* 2012;107(1):197–205.
44. Kalinina J, Carroll A, Wang L, et al. Detection of "oncometabolite" 2-hydroxyglutarate by magnetic resonance analysis as a biomarker of IDH1/2 mutations in glioma. *J Mol Med.* 2012;90(10):1161–1171.
45. Elkhaled A, Jalbert LE, Phillips JJ, et al. Magnetic resonance of 2-hydroxyglutarate in IDH1-mutated low-grade gliomas. *Sci Transl Med.* 2012;4(116):116ra5.
46. Kil IS, Kim SY, Lee SJ, Park J-W. Small interfering RNA-mediated silencing of mitochondrial NADP+-dependent isocitrate dehydrogenase enhances the sensitivity of HeLa cells toward tumor necrosis factor-α and anticancer drugs. *Free Radic Biol Med.* 2007;43(8):1197–1207.
47. Andronesi OC, Rapalino O, Gerstner E, et al. Detection of oncogenic IDH1 mutations using magnetic resonance spectroscopy of 2-hydroxyglutarate. *J Clin Invest.* 2013;123(9):3659–3663.
48. Martínez-Pérez I, Moreno Á, Alonso J, et al. Diagnosis of brain abscess by magnetic resonance spectroscopy: report of two cases. *J Neurosurg.* 1997;86(4):708–713.

49. Cohen-Gadol AA, Pan JW, Kim JH, Spencer DD, Hetherington HH. Mesial temporal lobe epilepsy: a proton magnetic resonance spectroscopy study and a histopathological analysis. *J Neurosurg.* 2004;101(4):613–620.
50. Stadlbauer A, Moser E, Gruber S, Nimsky C, Fahlbusch R, Ganslandt O. Integration of biochemical images of a tumor into frameless stereotaxy achieved using a magnetic resonance imaging/magnetic resonance spectroscopy hybrid data set. *J Neurosurg.* 2004;101(2):287–294.
51. Negendank WG, Sauter R, Brown TR, et al. Proton magnetic resonance spectroscopy in patients with glial tumors: a multicenter study. *J Neurosurg.* 1996;84(3):449–458.
52. Grinvald A. Optical imaging of architecture and function in the living brain sheds new light on cortical mechanisms underlying visual perception. *Brain Topogr.* 1992;5(2):71–75.
53. Liu JT, Meza D, Sanai N. Trends in fluorescence image-guided surgery for gliomas. *Neurosurgery.* 2014;75(1):61–71.
54. Li Y, Rey-Dios R, Roberts DW, Valdes PA, Cohen-Gadol AA. Intraoperative fluorescence-guided resection of high-grade gliomas: a comparison of the present techniques and evolution of future strategies. *World Neurosurg.* 2014;82(1–2):175–185.
55. Behbahaninia M, Martirosyan NL, Georges J, et al. Intraoperative fluorescent imaging of intracranial tumors: a review. *Clin Neurol Neurosurg.* 2013;115(5):517–528.
56. Panciani PP, Fontanella M, Schatlo B, et al. Fluorescence and image guided resection in high grade glioma. *Clin Neurol Neurosurg.* 2012;114(1):37–41.
57. Eljamel MS. Fluorescence image-guided surgery of brain tumors: explained step-by-step. *Photodiagn Photodyn Ther.* 2008;5(4):260–263.
58. Alander JT, Kaartinen I, Laakso A, et al. A review of indocyanine green fluorescent imaging in surgery. *Int J Biomed Imaging.* 2012;2012:940585.
59. Stanga PE, Lim JI, Hamilton P. Indocyanine green angiography in chorioretinal diseases: indications and interpretation: an evidence-based update. *Ophthalmology.* 2003;110(1):15–21.
60. Martirosyan NL, Cavalcanti DD, Eschbacher JM, et al. Use of in vivo near-infrared laser confocal endomicroscopy with indocyanine green to detect the boundary of infiltrative tumor. *J Neurosurg.* 2011;115(6):1131–1138.
61. Haglund MM, Berger MS, Hochman DW. Enhanced optical imaging of human gliomas and tumor margins. *Neurosurgery.* 1996;38(2):308–317.
62. Raabe A, Beck J, Gerlach R, Zimmermann M, Seifert V. Near-infrared indocyanine green video angiography: a new method for intraoperative assessment of vascular flow. *Neurosurgery.* 2003;52(1):132–139.
63. Zaidi HA, Abla AA, Nakaji P, Chowdhry SA, Albuquerque FC, Spetzler RF. Indocyanine green angiography in the surgical management of cerebral arteriovenous malformations: lessons learned in 130 consecutive cases. *Neurosurgery.* 2014;10(Suppl 2):246–251:discussion 251.
64. Sharma M, Ambekar S, Ahmed O, et al. The utility and limitations of intraoperative near-infrared indocyanine green videoangiography in aneurysm surgery. *World Neurosurg.* 2014;82(5):e607–e613.
65. Okuda T, Yoshioka H, Kato A. Fluorescence-guided surgery for glioblastoma multiforme using high-dose fluorescein sodium with excitation and barrier filters. *J Clin Neurosci.* 2012;19(12):1719–1722.
66. Schebesch KM, Proescholdt M, Hohne J, et al. Sodium fluorescein-guided resection under the YELLOW 560 nm surgical microscope filter in malignant brain tumor surgery—a feasibility study. *Acta Neurochir (Wien).* 2013;155(4):693–699.
67. Kuroiwa T, Kajimoto Y, Ohta T. Development of a fluorescein operative microscope for use during malignant glioma surgery: a technical note and preliminary report. *Surg Neurol.* 1998;50(1):41–48.
68. Eschbacher J, Martirosyan NL, Nakaji P, et al. In vivo intraoperative confocal microscopy for real-time histopathological imaging of brain tumors. *J Neurosurg.* 2012;116(4):854–860.
69. Sanai N, Eschbacher J, Hattendorf G, et al. Intraoperative confocal microscopy for brain tumors: a feasibility analysis in humans. *Neurosurgery.* 2011;68(2 Suppl Operative):282–290.
70. Snuderl M, Wirth D, Sheth SA, et al. Dye-enhanced multimodal confocal imaging as a novel approach to intraoperative diagnosis of brain tumors. *Brain Pathol.* 2013;23(1):73–81.

71. Rey-Dios R, Cohen-Gadol AA. Technical principles and neurosurgical applications of fluorescein fluorescence using a microscope-integrated fluorescence module. *Acta Neurochir (Wien)*. 2013;155 (4):701–706.
72. Koc K, Anik I, Cabuk B, Ceylan S. Fluorescein sodium-guided surgery in glioblastoma multiforme: a prospective evaluation. *Br J Neurosurg*. 2008;22(1):99–103.
73. Diez Valle R, Tejada Solis S. Answer to: "Sodium fluorescein-guided resection under the YELLOW 560-nm surgical microscope filter in malignant brain tumor surgery—a feasibility study" (April 2013, volume 155, issue 4, pp 693–669). *Acta Neurochir (Wien)*. 2013; 155(7):1319–1320.
74. Liao H, Noguchi M, Maruyama T, et al. An integrated diagnosis and therapeutic system using intra-operative 5-aminolevulinic-acid-induced fluorescence guided robotic laser ablation for precision neurosurgery. *Med Image Anal*. 2012;16(3):754–766.
75. Stummer W, Stocker S, Wagner S, et al. Intraoperative detection of malignant gliomas by 5-aminolevulinic acid-induced porphyrin fluorescence. *Neurosurgery*. 1998;42(3):518–525.
76. Stummer W, Stepp H, Moller G, Ehrhardt A, Leonhard M, Reulen HJ. Technical principles for protoporphyrin-IX-fluorescence guided microsurgical resection of malignant glioma tissue. *Acta Neurochir (Wien)*. 1998;140(10):995–1000.
77. Stummer W, Pichlmeier U, Meinel T, et al. Fluorescence-guided surgery with 5-aminolevulinic acid for resection of malignant glioma: a randomised controlled multicentre phase III trial. *Lancet Oncol*. 2006;7(5):392–401.
78. Hefti M, von Campe G, Moschopulos M, Siegner A, Looser H, Landolt H. 5-Aminolevulinic acid induced protoporphyrin IX fluorescence in high-grade glioma surgery: a one-year experience at a single institutuion. *Swiss Med Wkly*. 2008;138(11–12):180–185.
79. Nabavi A, Thurm H, Zountsas B, et al. Five-aminolevulinic acid for fluorescence-guided resection of recurrent malignant gliomas: a phase II study. *Neurosurgery*. 2009;65(6):1070–1076.
80. Sanai N, Snyder LA, Honea NJ, et al. Intraoperative confocal microscopy in the visualization of 5-aminolevulinic acid fluorescence in low-grade gliomas. *J Neurosurg*. 2011;115(4):740–748.
81. Rapp M, Kamp M, Steiger HJ, Sabel M. Endoscopic-assisted visualization of 5-aminolevulinic acid-induced fluorescence in malignant glioma surgery: a technical note. *World Neurosurg*. 2014;82(1–2): E277–E279.
82. Valdes PA, Leblond F, Kim A, et al. Quantitative fluorescence in intracranial tumor: implications for ALA-induced PpIX as an intraoperative biomarker. *J Neurosurg*. 2011;115(1):11–17.
83. DeBin JA, Strichartz GR. Chloride channel inhibition by the venom of the scorpion *Leiurus quinquestriatus*. *Toxicon*. 1991;29(11):1403–1408.
84. DeBin JA, Maggio JE, Strichartz GR. Purification and characterization of chlorotoxin, a chloride channel ligand from the venom of the scorpion. *Am J Physiol*. 1993;264(2 Pt 1):C361–C369.
85. Soroceanu L, Gillespie Y, Khazaeli MB, Sontheimer H. Use of chlorotoxin for targeting of primary brain tumors. *Cancer Res*. 1998;58(21):4871–4879.
86. Lyons SA, O'Neal J, Sontheimer H. Chlorotoxin, a scorpion-derived peptide, specifically binds to gliomas and tumors of neuroectodermal origin. *Glia*. 2002;39(2):162–173.
87. Hockaday DC, Shen S, Fiveash J, et al. Imaging glioma extent with 131I-TM-601. *J Nucl Med*. 2005;46(4):580–586.
88. Veiseh M, Gabikian P, Bahrami SB, et al. Tumor paint: a chlorotoxin:Cy5.5 bioconjugate for intraoperative visualization of cancer foci. *Cancer Res*. 2007;67(14):6882–6888.
89. Akcan M, Stroud MR, Hansen SJ, et al. Chemical re-engineering of chlorotoxin improves bioconjugation properties for tumor imaging and targeted therapy. *J Med Chem*. 2011;54(3):782–787.
90. Stroud MR, Hansen SJ, Olson JM. In vivo bio-imaging using chlorotoxin-based conjugates. *Curr Pharm Des*. 2011;17(38):4362–4371.
91. Butte PV, Mamelak A, Parrish-Novak J, et al. Near-infrared imaging of brain tumors using the Tumor Paint BLZ-100 to achieve near-complete resection of brain tumors. *Neurosurg Focus*. 2014;36:2.
92. Cheng Y, Zhao J, Qiao W, Chen K. Recent advances in diagnosis and treatment of gliomas using chlorotoxin-based bioconjugates. *Am J Nucl Med Mol Imaging*. 2014;4(5):385–405.

93. Huang D, Swanson EA, Lin CP, et al. Optical coherence tomography. *Science*. 1991;254 (5035):1178–1181.
94. Leitgeb R, Hitzenberger C, Fercher A. Performance of Fourier domain vs. time domain optical coherence tomography. *Opt Express*. 2003;11(8):889–894.
95. de Boer JF, Cense B, Park BH, Pierce MC, Tearney GJ, Bouma BE. Improved signal-to-noise ratio in spectral-domain compared with time-domain optical coherence tomography. *Opt Lett*. 2003;28 (21):2067–2069.
96. Choma M, Sarunic M, Yang C, Izatt J. Sensitivity advantage of swept source and Fourier domain optical coherence tomography. *Opt Express*. 2003;11(18):2183–2189.
97. Zysk AM, Nguyen FT, Oldenburg AL, Marks DL, Boppart SA. Optical coherence tomography: a review of clinical development from bench to bedside. *J Biomed Opt*. 2007;12(5):051403.
98. Bohringer HJ, Lankenau E, Stellmacher F, Reusche E, Huttmann G, Giese A. Imaging of human brain tumor tissue by near-infrared laser coherence tomography. *Acta Neurochir (Wien)*. 2009;151 (5):507–517.
99. Jobsis FF. Noninvasive, infrared monitoring of cerebral and myocardial oxygen sufficiency and circulatory parameters. *Science*. 1977;198(4323):1264–1267.
100. Chance B, Zhuang Z, UnAh C, Alter C, Lipton L. Cognition-activated low-frequency modulation of light absorption in human brain. *Proc Natl Acad Sci USA*. 1993;90(8):3770–3774.
101. Kato T, Kamei A, Takashima S, Ozaki T. Human visual cortical function during photic stimulation monitoring by means of near-infrared spectroscopy. *J Cereb Blood Flow Metab*. 1993;13(3):516–520.
102. Villringer A, Planck J, Hock C, Schleinkofer L, Dirnagl U. Near infrared spectroscopy (NIRS): a new tool to study hemodynamic changes during activation of brain function in human adults. *Neurosci Lett*. 1993;154(1–2):101–104.
103. Watanabe E, Maki A, Kawaguchi F, et al. Non-invasive assessment of language dominance with near-infrared spectroscopic mapping. *Neurosci Lett*. 1998;256(1):49–52.
104. Hoshi Y. Functional near-infrared spectroscopy: current status and future prospects. *J Biomed Opt*. 2007;12(6):062106.
105. Gibson A, Dehghani H. Diffuse optical imaging. *Philos Trans A Math Phys Eng Sci*. 2009;367 (1900):3055–3072.
106. Ferrari M, Quaresima V. A brief review on the history of human functional near-infrared spectroscopy (fNIRS) development and fields of application. *Neuroimage*. 2012;63(2):921–935.
107. Sato Y, Uzuka T, Aoki H, et al. Near-infrared spectroscopic study and the Wada test for presurgical evaluation of expressive and receptive language functions in glioma patients: with a case report of dissociated language functions. *Neurosci Lett*. 2012;510(2):104–109.
108. Huang Z, McWilliams A, Lui H, McLean DI, Lam S, Zeng H. Near-infrared Raman spectroscopy for optical diagnosis of lung cancer. *Int J Cancer*. 2003;107(6):1047–1052.
109. Krafft C, Sobottka SB, Schackert G, Salzer R. Near infrared Raman spectroscopic mapping of native brain tissue and intracranial tumors. *Analyst*. 2005;130(7):1070–1077.
110. Amharref N, Beljebbar A, Dukic S, et al. Discriminating healthy from tumor and necrosis tissue in rat brain tissue samples by Raman spectral imaging. *Biochim Biophys Acta*. 2007;1768 (10):2605–2615.
111. Kast RE, Auner GW, Rosenblum ML, et al. Raman molecular imaging of brain frozen tissue sections. *J Neurooncol*. 2014;120(1):55–62.
112. Kalkanis SN, Kast RE, Rosenblum ML, et al. Raman spectroscopy to distinguish grey matter, necrosis, and glioblastoma multiforme in frozen tissue sections. *J Neurooncol*. 2014;116(3):477–485.
113. Nygren C, von Holst H, Mansson JE, Fredman P. Increased levels of cholesterol esters in glioma tissue and surrounding areas of human brain. *Br J Neurosurg*. 1997;11(3):216–220.
114. Koljenovic S, Choo-Smith LP, Bakker Schut TC, Kros JM, van den Berge HJ, Puppels GJ. Discriminating vital tumor from necrotic tissue in human glioblastoma tissue samples by Raman spectroscopy. *Lab Invest*. 2002;82(10):1265–1277.
115. Beleites C, Geiger K, Kirsch M, Sobottka SB, Schackert G, Salzer R. Raman spectroscopic grading of astrocytoma tissues: using soft reference information. *Anal Bioanal Chem*. 2011;400 (9):2801–2816.

116. Krafft C, Belay B, Bergner N, et al. Advances in optical biopsy—correlation of malignancy and cell density of primary brain tumors using Raman microspectroscopic imaging. *Analyst.* 2012;137(23):5533–5537.
117. Koljenovic S, Schut TB, Vincent A, Kros JM, Puppels GJ. Detection of meningioma in dura mater by Raman spectroscopy. *Anal Chem.* 2005;77(24):7958–7965.
118. Wolthuis R, van Aken M, Fountas K, Robinson Jr. JS, Bruining HA, Puppels GJ. Determination of water concentration in brain tissue by Raman spectroscopy. *Anal Chem.* 2001;73(16):3915–3920.
119. Karabeber H, Huang R, Iacono P, et al. Guiding brain tumor resection using surface-enhanced Raman scattering nanoparticles and a hand-held Raman scanner. *ACS Nano.* 2014;8(10):9755–9766.
120. Mizuno A, Kitajima H, Kawauchi K, Muraishi S, Ozaki Y. Near-infrared Fourier-transform Raman-spectroscopic study of human brain-tissues and tumors. *J Raman Spectrosc.* 1994;25(1):25–29.
121. Zumbusch A, Holtom GR, Xie XS. Three-dimensional vibrational imaging by coherent anti-Stokes Raman scattering. *Phys Rev Lett.* 1999;82(20):4142–4145.
122. Evans CL, Xie XS. Coherent anti-Stokes Raman scattering microscopy: chemical imaging for biology and medicine. *Annu Rev Anal Chem (Palo Alto Calif).* 2008;1:883–909.
123. Freudiger CW, Min W, Saar BG, et al. Label-free biomedical imaging with high sensitivity by stimulated Raman scattering microscopy. *Science.* 2008;322(5909):1857–1861.
124. Min W, Freudiger CW, Lu S, Xie XS. Coherent nonlinear optical imaging: beyond fluorescence microscopy. *Annu Rev Phys Chem.* 2011;62:507–530.
125. Cheng J-X, Xie XS. *Coherent Raman Scattering Microscopy.* Boca Raton, FL: CRC Press; 2013.
126. Evans CL, Xu X, Kesari S, Xie XS, Wong ST, Young GS. Chemically-selective imaging of brain structures with CARS microscopy. *Opt Express.* 2007;15(19):12076–12087.
127. Uckermann O, Galli R, Tamosaityte S, et al. Label-free delineation of brain tumors by coherent anti-Stokes Raman scattering microscopy in an orthotopic mouse model and human glioblastoma. *PLoS One.* 2014;9(9):e107115.
128. Ji M, Orringer DA, Freudiger CW, et al. Rapid, label-free detection of brain tumors with stimulated Raman scattering microscopy. *Sci Transl Med.* 2013;5(201):201ra119.
129. Bentley JN, Ji M, Xie XS, Orringer DA. Real-time image guidance for brain tumor surgery through stimulated Raman scattering microscopy. *Expert Rev Anticancer Ther.* 2014;14(4):359–361.
130. Ait-Belkacem R, Berenguer C, Villard C, et al. MALDI imaging and in-source decay for top-down characterization of glioblastoma. *Proteomics.* 2014;14(10):1290–1301.
131. Stoeckli M, Chaurand P, Hallahan DE, Caprioli RM. Imaging mass spectrometry: a new technology for the analysis of protein expression in mammalian tissues. *Nat Med.* 2001;7(4):493–496.
132. Chaurand P, Norris JL, Cornett DS, Mobley JA, Caprioli RM. New developments in profiling and imaging of proteins from tissue sections by MALDI mass spectrometry. *J Proteome Res.* 2006;5(11):2889–2900.
133. McDonnell LA, Corthals GL, Willems SM, van Remoortere A, van Zeijl RJ, Deelder AM. Peptide and protein imaging mass spectrometry in cancer research. *J Proteomics.* 2010;73(10):1921–1944.
134. Chen H, Talaty NN, Takats Z, Cooks RG. Desorption electrospray ionization mass spectrometry for high-throughput analysis of pharmaceutical samples in the ambient environment. *Anal Chem.* 2005;77(21):6915–6927.
135. Paglia G, Ifa DR, Wu C, Corso G, Cooks RG. Desorption electrospray ionization mass spectrometry analysis of lipids after two-dimensional high-performance thin-layer chromatography partial separation. *Anal Chem.* 2010;82(5):1744–1750.
136. Wiseman JM, Ifa DR, Song Q, Cooks RG. Tissue imaging at atmospheric pressure using desorption electrospray ionization (DESI) mass spectrometry. *Angew Chem Int Ed Engl.* 2006;45(43):7188–7192.
137. Balog J, Sasi-Szabo L, Kinross J, et al. Intraoperative tissue identification using rapid evaporative ionization mass spectrometry. *Sci Transl Med.* 2013;5(194):194ra193.

138. Cooks RG, Ouyang Z, Takats Z, Wiseman JM. Detection technologies. Ambient mass spectrometry. *Science.* 2006;311(5767):1566–1570.
139. Eberlin LS, Norton I, Orringer D, et al. Ambient mass spectrometry for the intraoperative molecular diagnosis of human brain tumors. *Proc Natl Acad Sci USA.* 2013;110(5):1611–1616.
140. Fedorov A, Beichel R, Kalpathy-Cramer J, et al. 3D Slicer as an image computing platform for the quantitative imaging network. *Magn Reson Imaging.* 2012;30(9):1323–1341.
141. Eberlin LS, Norton I, Dill AL, et al. Classifying human brain tumors by lipid imaging with mass spectrometry. *Cancer Res.* 2012;72(3):645–654.
142. Calligaris D, Norton I, Feldman DR, et al. Mass spectrometry imaging as a tool for surgical decision-making. *J Mass Spectrom.* 2013;48(11):1178–1187.
143. Santagata S, Eberlin LS, Norton I, et al. Intraoperative mass spectrometry mapping of an onco-metabolite to guide brain tumor surgery. *Proc Natl Acad Sci USA.* 2014;111(30):11121–11126.
144. Thomas A, Patterson NH, Marcinkiewicz MM, Lazaris A, Metrakos P, Chaurand P. Histology-driven data mining of lipid signatures from multiple imaging mass spectrometry analyses: application to human colorectal cancer liver metastasis biopsies. *Anal Chem.* 2013;85(5):2860–2866.
145. Caprioli RM, Farmer TB, Gile J. Molecular imaging of biological samples: localization of peptides and proteins using MALDI-TOF MS. *Anal Chem.* 1997;69(23):4751–4760.

CHAPTER 18

Drug Delivery to the Brain via Focused Ultrasound

Costas D. Arvanitis and Nathan McDannold
Department of Radiology, Brigham and Women's Hospital, Harvard Medical School, Boston, MA, USA

INTRODUCTION

The worldwide economic burden caused by diseases and disorders of the central nervous system (CNS) has been estimated to be as high as $2 trillion a year[1] and is expected to increase as the population ages. Three of the mainstays of medicine—drugs, surgery, and radiation—all have limitations when it comes to the brain. With the unique biology of the brain,[2] and the distinct requirement to preserve and protect neural function, the need for novel approaches and strategies to target CNS disorders is clear.

Ultrasound is a unique method for localizing energy deep into the body. Acoustic energy can be focused to small volumes, sparing the surrounding tissues of any effects. The mechanical energy deposited in the focal region can be utilized in a surprisingly large number of ways and has fascinated neuroscientists for more than 60 years.[3–8]

In the last 10–15 years significant technological developments have made the use of focused ultrasound (FUS) in the brain a reality. The creation of devices that can safely and precisely focus high-intensity ultrasound beams through the intact human skull,[9,10] and the integration of these devices to high-field MRI,[11–13] have made it realistic to expect that this technology will reach its potential in the clinic. Initial human trials with these devices are ongoing,[14–18] and the research community interested in FUS technology is growing exponentially.

This chapter will review the current status of MRgFUS in the brain, particularly how it can be used to "permeabilize" blood vessels in the brain and in brain tumors to enable the targeted delivery of therapeutics. It begins with an introduction to therapeutic ultrasound, and how systems were developed so that it can be applied noninvasively in the brain. We then provide an overview of ultrasound-induced bioeffects, and provide a detailed account of work performed to date on the use of FUS for drug delivery via disruption of the blood–brain barrier (BBB). Finally, we discuss other applications of FUS in the brain that are being investigated.

A. Golby (Ed): Image-Guided Neurosurgery
DOI: http://dx.doi.org/10.1016/B978-0-12-800870-6.00018-2

THERAPEUTIC ULTRASOUND

Various devices and techniques can be used to locally deliver energy to targeted areas in the body, for example, to induce local temperature changes. Devices for thermal therapy include ultrasound transducers, lasers, and antennas operating at microwave or radio frequencies. Other thermal therapy techniques include the use of heated fluids or the injection of magnetic fluids and the application of a time-varying magnetic field. All of these techniques aim to induce bioeffects in a target area (such as a tumor) while leaving the surrounding healthy tissues unharmed. These techniques are promising because they can be highly localized, they can be applied via minimally invasive or noninvasive procedures, and they can be repeated until the desired outcome is achieved (no ionizing radiation is used).

Of all these methods, ultrasound is unparalleled in flexibility and effectiveness. Ultrasound transducers can be constructed to produce a sharp focus (Figure 18.1). The fields can penetrate from an external source deep into tissues, allowing for completely noninvasive localization of acoustic energy. They can also be made small enough to fit on an interstitial probe,[19–21] an intracavitary device,[22,23] or inside a catheter.[24,25] The size, distribution, and temperature rise of the heated region can be controlled through the geometry of the transducer, the choice of ultrasound frequency, or with the use of acoustic lenses[26,27] or phased arrays[28] and electric power applied to the piezoelectric elements.[29–34] The energy delivery can also be enhanced by taking advantage of nonlinear ultrasound propagation,[35–37] the induction of cavitation, or the introduction of preformed gas bubbles (ultrasound contrast agents).

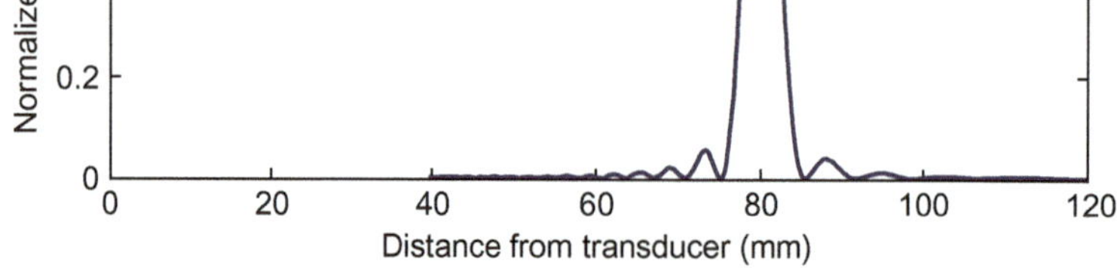

Figure 18.1 Precise deposition of acoustic energy deep in the brain. Acoustic intensity as a function of depth from a simple, geometrically focused transducer. Depending on the curvature, the F-number and frequency of the transducer, the focal depth and size can be altered, providing a localized region (a few millimeters) of energy deposition deep in the tissue (several centimeters) that can be used for different therapies.

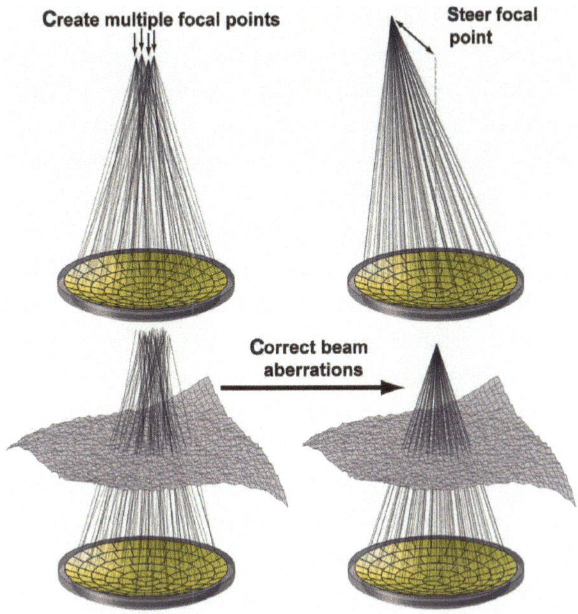

Figure 18.2 Multiphase arrays, composed of hundreds of piezoelectric elements, increase the treatment envelope and targeting precision of FUS in the brain. By manipulating the phase and amplitude of the signal at each element, phased array transducers can be used to focus the US beam, steer it to different targets, create multiple focal spots, and correct for aberrations caused by tissue structures. This ability demonstrates the flexibility of TcMRgFUS, which is essential for clinical translation.

The use of multichannel phased array transducers expands these abilities. Such arrays, which are generally controlled by individual FUS driving systems, increase the flexibility of the energy deposition. By manipulating the phase and amplitude of the signal at each element, phased array transducers can be used to focus the US beam, steer it to different targets, create multiple focal spots, and correct for aberrations caused by tissue structures such as differing skull thickness (Figure 18.2).

In addition to thermal effects, the mechanical effects produced in the body by high-intensity US can also be used therapeutically. The thermal and mechanical effects produced by a FUS beam have been investigated for numerous procedures for the treatment of disease, including tumor ablation,[38–51] vessel occlusion,[52–54] hemostasis,[55,56] thrombolysis,[57,58] drug delivery,[59] gene therapy,[60–66] focal opening of the BBB,[67,68] neuromodulation,[69–71] and direct mechanical damage.[67,72,73]

Most of these mechanical effects are produced via cavitation, which is the formation and subsequent interaction of microbubbles with an acoustic field. When microbubbles interact with an ultrasound beam, a range of biological effects has been observed.[74] Depending on their size, the bubbles can oscillate within the ultrasound field, and in some cases they may grow via rectified diffusion.[75] At high enough acoustic pressures, they can collapse during the positive pressure cycle, a phenomena known as inertial cavitation, producing shock waves and high-velocity jets,[76] free radicals,[77] and high local temperatures.[78,79] In addition, the medium surrounding the bubbles undergoes acoustic streaming,[80] which may be associated with large shear stresses. Further, a radiation force on the bubbles is produced along the direction of the ultrasound beam.[76]

Microbubbles can be formed from gas nuclei in tissues when sonicated at high intensities. One can also intravenously inject microbubble agents. These agents, which are available clinically as ultrasound imaging contrast agents, consist of semirigid lipid or albumin shells that encapsulate a gas, typically a perfluorocarbon. They range in size from about 1–10 μm and are constrained to the vasculature. The presence of the circulating preformed microbubbles concentrates the ultrasound effects to the microvasculature. They can also make cavitation-related therapies more predictable while greatly reducing the FUS exposure levels needed to produce bioeffects. Since they do not extravasate, their use is largely limited to applications related to blood vessels. For example, to vessels with restricted permeability or to tumors, which are hypervascular (see below). The preformed microbubbles used in ultrasound contrast agents presumably can exhibit all of the cavitation-related behaviors described above, either with their shells intact or after being broken apart by the ultrasound beam and their gas contents released.

This flexibility and promise of ultrasound as a therapeutic tool and the number of different target applications have led to intense interest in this technology for more than 60 years. Before being applied to imaging, the first use of ultrasound in the medical field was for therapy.[3,81–91] Since these fundamental initial works, numerous groups have made technical and theoretical advances in the field of therapeutic ultrasound, most notably the development of phased arrays, the integration with image guidance for closed-loop control, and the development of clinical devices. Based on these works, multiple clinical devices that use FUS for thermal ablation are now being tested around the world or are approved for thermal ablation of targets in practically every part of the body.

FOCUSED ULTRASOUND IN THE BRAIN

Until fairly recently, clinical tests of FUS in the brain have required the removal of a section of the skull to allow for ultrasound propagation into the brain[92–98] due to high ultrasound absorption and heating of the skull bone and beam aberration caused by the skull's irregular shape and large acoustic impedance.

In the past decade, FUS thermal ablation systems have been developed that overcome these obstacles posed by the skull making *transcranial* FUS treatment possible.[34,99,100] These approaches reduce skull heating through active cooling of the scalp and a transducer design with a large aperture to distribute the ultrasound energy over a large skull region. These FUS systems are also able to correct for beam aberrations using a phased array transducer design. When combined with methods that use acoustic simulation based on CT scans of the skull bone to determine the phase and

Figure 18.3 A clinical Transcranial MR guide FUS system. Transducer for the ExAblate Neuro TcMRgFUS system, which combines a 30-cm diameter, 1024-element phased array hemisphere transducer and a 3 T MRI system. The system is coupled to a patient's head with degassed and cooled water using a flexible membrane stretched around the head and attached to the outside of the transducer. The head is fixed with a standard MR-compatible stereotactic frame. Numerous multicenter clinical trials are underway with this system. *Image courtesy InSightec (Haifa, Israel).*

amplitude corrections for the phased array[9,10] and MR temperature imaging (MRTI) to monitor the heating,[101–103] a completely noninvasive alternative to surgical resection in the brain becomes possible. These systems have been tested in animals[11,104] and in initial human trials.[14–18] Pictures of a clinical FUS system developed for thermal ablation in the brain (ExAblate, InSightec Ltd., Haifa, Israel) are shown in Figure 18.3. The ability of the phased array to correct for skull-induced aberrations is demonstrated in Figure 18.2. Such systems may also be useful for ultrasound-based targeted drug delivery methods in the brain,[68] the treatment of stroke via thrombolysis,[105] or other novel applications. Currently, in clinical trials, the role of MRgFUS in brain tumors includes thermal ablation of malignant CNS neoplasms[14] and functional neurosurgery applications for neuropathic pain[15,16] and essential tremor.[17,18] Initial clinical tests with MRgFUS ablation for tremor and dyskinesia in patients with Parkinson's disease and obsessive compulsive disorder are also planned. The main current limitation of the technique is its limited "treatment envelope," which prevents high-intensity exposures close to the bone, meaning that superficial targets are not amenable to treatments (Figure 18.4). However, as we discuss toward the end of the chapter, new ways to expand the treatment envelope have recently emerged and are under intense investigation.

DRUG DELIVERY IN THE BRAIN: THE BLOOD-BRAIN BARRIER (BBB)

Drug delivery in the brain is very challenging due to the presence of the blood-brain barrier (BBB). The BBB consists of endothelial cells connected together by tight junctions, the basement membrane, and astrocytic endfeet. It is a specialized nonpermeable barrier that normally protects the brain from toxins, and helps maintain the homeostasis of the neuronal microenvironment. However, it also excludes 98% of

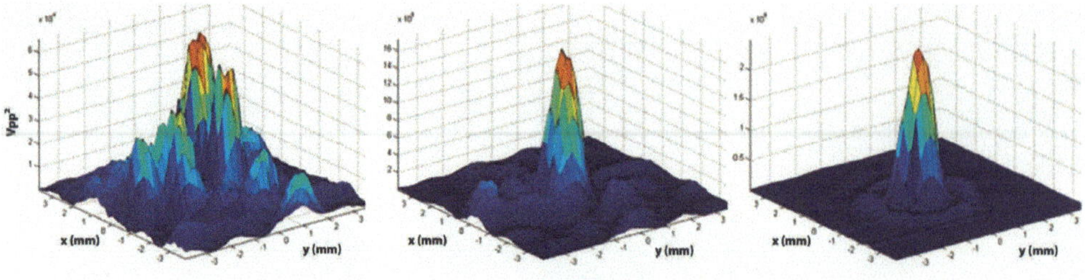

Figure 18.4 Estimated range in the brain in which the ExAblate 650 kHz Neuro system can currently target for thermal ablation. Only centrally located targets can be effectively ablated due to skull heating and other factors. The contours represent variability between skulls and target temperatures of 60°C. *(Figure courtesy InSightec, Haifa, Israel.)*

small-molecule drugs and approximately 100% of large-molecule neurotherapeutics from the brain parenchyma[106,107] allowing only small-molecule drugs with high lipid solubility and a molecular mass under 400–500 Da to cross it. Given the paucity of small-molecule drugs that are effective for CNS disorders, it is clear that the BBB is a significant limitation to the development and use of drugs in the brain. In addition to CNS disorders, brain tumors also face challenges for effective drug delivery. Despite their leaky vasculature, work in mice suggests that the blood–tumor barrier (BTB) is only partially compromised and that toxic concentrations of chemotherapy are only achieved in a small subset of tumors that are highly permeable.[108] Further, infiltrating cancer cells at the tumor margins and small metastatic seeds that may be protected by the BBB of surrounding normal tissue are also difficult to treat.[109]

Different methods that either bypass the CNS vascular barriers altogether, or facilitate passage across it via controlled exploitation of endogenous transport mechanisms have been explored for effective delivery of pharmaceuticals to the CNS (Table 18.1).[110–112] In this section we briefly describe the advantages and limitations of these drug delivery approaches in the brain and, then, we elaborate on the potential of FUS to overcome BBB/BTB and deliver pharmacologically significant amounts of drugs in the brain.

Current methods for drug delivery in the brain

Numerous approaches to deliver drugs to the brain have been developed. These include transvascular approaches that aim to develop or modify drugs, not larger than 400–450 Da, so that they can cross the BBB.[111,113,114] An alternative approach utilizes the molecular targeting of endothelial-surface receptors, which is colloquially termed the "Trojan Horse," and involves the use of a targeting ligand that binds to its receptor to activate endocytosis of the complex, which can have large molecular weight, into a vesicle that is transported across to, and released from the opposite pole (i.e., transcytosis).[115–121] Utilizing the solute carrier proteins (SLCs) on the endothelial surface that specifically transport many essential polar and charged nutrients across

Table 18.1 Different methods investigated to get around the BBB to deliver drugs to the brain

Method	Advantages	Disadvantages
Direct injection, convection-enhanced delivery, implantable devices	High local drug concentrations can be achieved; systemic administration avoided.	Invasive; side effects; challenging to control; not readily repeatable.
Intrathecal, intraventricular injection	Effectively delivers drugs to subarachnoid space, brain surface.	Little drug penetration beyond brain surface; invasive.
Transnasal delivery	Noninvasive; easy to administer; repeatable.	Small volume of drug delivered; interindividual variability.
BBB disruption via arterial injection of osmotic solution or other agents	Effectively delivers drugs to large brain regions; large clinical experience.	Invasive; requires general anesthesia; side effects; not readily repeatable or easy to target.
Modification of drugs to cross barrier through endogenous transport mechanisms	Easily administered; delivered to whole brain.	Requires systemic administration; expensive; each drug requires new development; clinical data lacking.
BBB disruption via FUS and microbubbles	Noninvasive; readily repeatable; can target drug delivery to desired volumes; can control "magnitude" of disruption; can be combined with drug-loaded microbubbles or magnetic particles for additional targeting.	Requires systemic administration; currently technically challenging; large volume/whole brain disruption unproven; no clinical data.

the BBB, is yet another paradigm in drug transport across the BBB.[114,122,123] While the above methods are highly promising and offer the ability to easily administer drugs to the CNS as in other organs, they do require the expense and time of developing new agents, and they result in drugs being delivered to the entire brain, which may not always be desirable.

Convection-enhanced delivery of pharmaceuticals has been shown to overcome the above limitations and achieve high local drug concentrations of some drugs.[124–127] However, its therapeutic index is still limited due to (i) risks associated with its invasiveness, (ii) exponential decrease (in space) of drug concentrations from the injection or implantation site,[128] and (iii) preferential delivery of drugs along white matter tracts.[129] Others have investigated methods to introduce drugs into the cerebrospinal fluid (CSF) via intrathecal or intraventricular routes or transnasally from the submucus space into the olfactory CSF.[130] However, penetration into the brain parenchyma can be limited,[110] and in the case of nasal delivery, only small drug volumes can be delivered and there is substantial variation among subjects making dosing uncertain.[131]

A method to temporarily disrupt the BBB to enable CNS delivery of circulating agents is by using hyperosmotic solutions such as mannitol or bradykinin. Administration of these solutions causes shrinkage of endothelial cells and consequent stretching of tight junctions[132–135] through which drugs may pass. This method has been shown repeatedly to enhance delivery of therapeutic agents to brain tumors, and several promising clinical trials have been performed.[136–144] While such methods can be an effective means to deliver drugs to large brain regions, they are invasive procedures that require general anesthesia, and can have significant side effects.[135,136] The BBB disruption is also not spatially targeted.

Therefore, having a less-invasive and targeted way to achieve BBB disruption would be highly beneficial and could potentially improve the therapeutic index of a wide range of neuropharmaceuticals. As we describe in this section, the use of ultrasound, when combined with circulating microbubbles, offers a potential way to disrupt the BBB in a targeted, noninvasive, and repeatable manner to deliver a wide range of drugs to the brain and to brain tumors. While to date the technique has only been performed in animals, numerous studies from different laboratories demonstrate that it is very promising for the treatment of a wide range of CNS disorders and brain tumors. Following, we present the most salient features of this technique, explain how it is performed and elaborate on the biophysics behind it.

Targeted drug delivery in the brain via FUS-induced BBB disruption

When microbubbles interact with an ultrasound field, a range of biological effects have been observed.[74] While exploring these effects, it was found that FUS combined with intravenously administered microbubbles can safely and consistently open the BBB.[68] The use of microbubbles (1–10 μm in size) that were originally developed as ultrasound imaging contrast agents is critical for this phenomena, as they appear to concentrate the ultrasound effects to the blood vessel walls and cause or trigger BBB disruption through the widening of tight junctions and the activation of transcellular mechanisms,[145] with little effect on the surrounding parenchyma.

The ultrasound exposures ("sonications") are applied concurrently with the microbubble administration as a few short bursts at a low duty cycle at focal pressure amplitudes that induce forced microbubble oscillations, typically few hundred kPa. These exposure levels are orders of magnitude lower than those used for thermal ablation, enabling sonication through the intact skull without inducing skull heating. Numerous parametric studies have been performed and have shown that BBB

Table 18.2 Reported effects of different parameters on BBB disruption via FUS and microbubbles

Parameter	Range	Effect on BBB disruption
Pressure amplitude	100–600 kPa	Increase in BBB disruption magnitude as pressure amplitude increases; saturation at some point[146–148]; vascular damage produced at high pressure amplitudes.
Ultrasound frequency	0.028–8 MHz	Decrease in BBB disruption threshold as frequency decreases; some evidence of improved safety for lower frequencies.[149]
Burst length	Few cycles–100 ms	For burst lengths less than 10 ms, BBB disruption threshold increases and BBB disruption magnitude decreases as burst length is reduced[150–153]; little or no increase in disruption magnitude for longer bursts.[68,151,154]
Pulse repetition frequency	0.1 Hz–10 kHz	BBB disruption magnitude increases as repetition frequency increases up to a point.[151] Other works have observed no effect on BBB disruption magnitude.[150]
Ultrasound contrast agent dose	10 µg/ml	Magnitude of BBB disruption increases with dose[147,151,155,156]; other experiments have reported no effect.[150]
Sonication duration	few bursts–few mins	Longer durations[148] or repeated sonication[157,158] increase magnitude of BBB disruption; damage reported with excessive sonication.[148,158]
Microbubble diameter	1–8 µm	Threshold for BBB disruption lower for larger microbubbles; disruption magnitude increased with larger microbubbles.[159–161]
Ultrasound contrast agent		Similar outcomes reported for Optison® and Definity® microbubbles.[162)] Sonovue® microbubbles and research agents are also commonly used.

disruption can occur over a wide range of parameters, as shown in Table 18.2. Typical parameters used in our laboratory are 10 ms bursts applied at 1 Hz for one minute.

The BBB disruption can occur almost immediately with sonication[163] and appears to decay exponentially over several hours thereafter.[68,146,157,164] The amount of agent delivered across the barrier appears to be much larger in gray matter than in white matter, presumably due to differences in vascular density.[165] Several studies have found that the barrier appears to be largely restored after several hours.[68,146,157,159,164,166,167] The duration of the opening appears to be reduced for larger tracers.[164] This window in time where the barrier is open is thought to be good for the prospect of delivering even long-circulating drugs, but not so long as to produce concern of toxicity arising from chronic BBB breakdown.

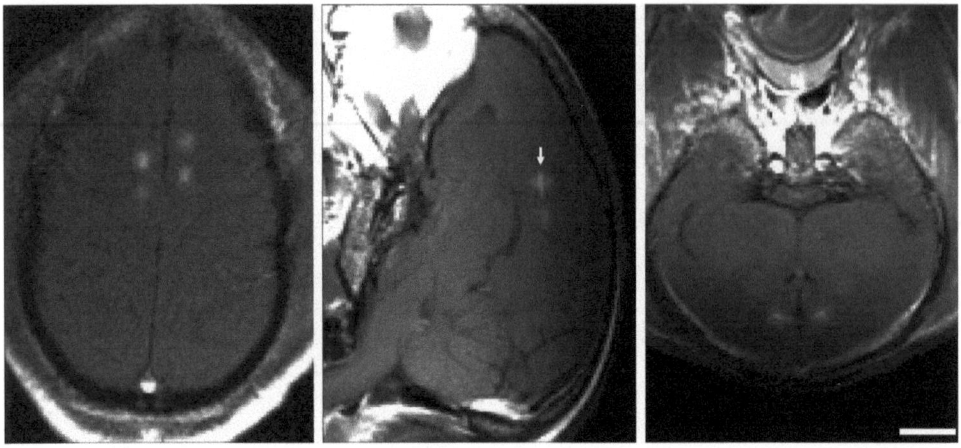

Figure 18.5 MR-evident BBB disruption after sonication of nonhuman primates. Contrast-enhanced T1-weighted MRI of the brain of a rhesus macaque showing enhancement at four focal targets in the cingulate cortex after injection of Gd-DTPA. (Left image: axial; middle: sagittal; right: coronal.) Enhancement of this MRI contrast agent, which normally does not extravasate into the brain, indicates areas with BBB disruption. Four targets were sonicated using a 220-kHz clinical MRI-guided FUS system along with an infusion of Definity microbubbles (Lantheus Med. Im. Inc., N. Billerica, MA). The dimensions of the disrupted spots were approximately 3 mm wide and 4 mm long. Note the leakage of contrast agent into the cingulate sulcus evident in the sagittal image (arrow). These images were obtained in a study evaluating passive acoustic mapping during microbubble-enhanced FUS (214) (bar: 1 cm). These images can also be acquired in the clinics to evaluate the treatment.

With a few simple modifications to enable low-intensity bursts, existing clinical brain FUS systems can be used for BBB disruption.[165] Clinical translation may also be possible using simpler FUS systems[168]; Figure 18.5 shows examples of targeted BBB disruption in a macaque using the clinical transcranial MRI-guided FUS system shown in Figure 18.3 (ExAblate, InSightec, Haifa, Israel). The device uses a hemispherical 1024-element phased array operating at 220 kHz, and is integrated with a 3 T MRI scanner. The focal region can be electronically steered to different locations using this array without physically moving the transducer. Volumes can be targeted by systematically steering the focal point to different targets, enabling one to deliver drugs to desired brain regions. Figure 18.6 shows an example of such "volumetric" FUS-induced BBB disruption. BBB disruption was evaluated using two MRI contrast agents that do not usually cross the BBB. As many as 100 targets (3 mm apart) per second can be sonicated using the parameters shown in Figure 18.6, resulting in BBB disruption in a volume as large as 50 cm^3 in 2–3 minutes. Also the use of low-frequency FUS can increase the range over which the beam can be steered, increasing the treatment envelope and making it possible to disrupt the BBB of large brain regions within 30–40 min.

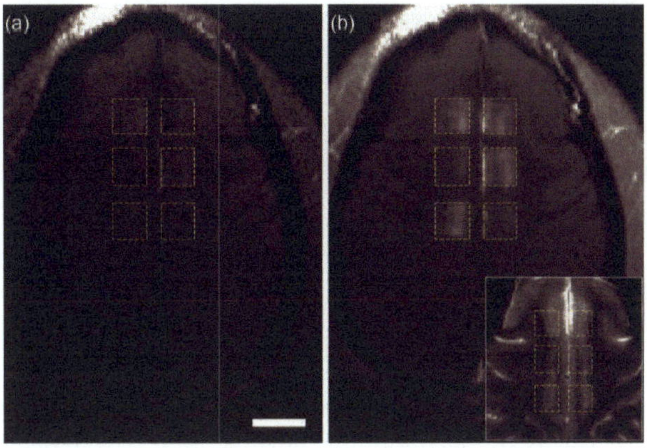

Figure 18.6 Demonstration of FUS-induced BBB disruption in nonhuman primates using contrast enhanced MRI. (a and b) Contrast-enhanced T1-weighted MRI after BBB disruption at six volumes (white squares) in the cingulate cortex. At each volume, the focal region was steered electronically in sequence to nine targets in a 3×3 grid using a phased array (the grid is not shown). (a) Low-level enhancement observed with gadofosveset trisodium, an MR contrast agent that binds to albumin in the blood (MW of albumin: $\sim$67 kDa); it was administered before sonication. (b) Enhancement after injection of Gd-DTPA (MW: 938 Da). The inset in (B) shows the same view in T2-weighted imaging. The enhancement patterns correspond to regions of cortical gray matter visible in T2-weighted imaging, demonstrating that gray matter can be targeted more easily. *Reprinted from Cancer Research, McDannold et al., 2012*[165]*; (©2012 American Association for Cancer Research.)*

Bioeffects induced in the brain by FUS and microbubbles

Even though numerous studies from different laboratories with different animal models have confirmed that FUS exposures combined with microbubbles can induce the disruption of the BBB, it remains elusive whether the FUS/microbubble interactions are physically modifying the vessel walls, or if they are triggering a physiological response that includes temporary BBB breakdown. While, as described below, a number of studies have investigated bioeffects occurring after FUS-mediated BBB disruption, the exact mechanism for the effect remains unknown.

Transmission electron microscopy (TEM) studies have shown delivery of tracers through widened tight junctions,[145,166] which could be consistent with a direct physical force pulling them apart, as well as active transport.[145,169] Other work has shown downregulation of the tight junctions proteins along with their mRNA, and recovery to normal levels at 12 hours postsonication.[170] Reorganization of connexin gap junction proteins have also been reported.[171] An increase of endothelial vesicles in normal[172] and tumor microvessels[173] have also been observed on TEM with an upregulation of caveolin proteins/mRNA, suggesting that caveolae-mediated transcytosis as a contributing mechanism for permeability. These researchers also found increased phosphorylation of Src and caveolin-1/2, noting that Src-induced phosphorylation of caveolins is a trigger for this transcytosis.[174] Further, Alonso et al. found increased protein ubiquitination in neurons but not glia, no increase in heat shock proteins, and only limited neuronal apoptosis at 24 hours in areas staining positive for extravasated albumin, after sonication.[175] Finally, intracellular signaling cascades in response to mechanical stimulation by FUS-induced BBB disruption is likely, but this effect has only recently been addressed.[176]

Although the aforementioned studies are not conclusive they do, however, support the hypothesis that the mechanical effects induced during the microbubble oscillations in the ultrasound field, which may include shear forces due to microstreaming or weak microbubble collapse and acoustic radiation forces that bring the oscillating bubbles in contact to the vessel wall, are responsible for the observed changes in protein expression and by extension to the BBB permeability. Further, the appearance of the brain after BBB disruption in light microscopy appears to be normal,[177] even after repeated weekly sessions.[165] The only major feature that has been observed in many studies is the presence of small clusters of extravassated red blood cells (petechiae).[178,179] It is thought that these petechiae are formed during inertial cavitation, and in experiments in which no wideband emissions (a signature for inertial cavitation) were observed, no such extravasations were found.[180] Therefore, it is evident that stable volumetric microbubble oscillations are mostly responsible for safe BBB disruption. As we describe below, during forced oscillations the microbubbles emit diverging pressure waves with characteristic spectral content that can be recorded, analyzed, quantified, and then used to fine-tune the sonication power for optimal BBB disruption. This ability may enable a level of control over drug delivery that is not available with other technologies. This flexibility, along with its noninvasiveness, lack of need for general anesthesia, and safety of repeated treatments make FUS a potentially transformative technology.

Targeted delivery of therapeutics

An additional advantage of this method for targeted drug delivery in the brain is that it appears to be "drug neutral"—that is, it appears that many agents with a wide range of properties can be successfully delivered across the BBB and/or the BTB. A large number of imaging tracers and therapeutic agents which normally don't cross the BBB have been delivered to the brain or to brain tumor models with FUS and microbubbles[181] (Figure 18.6). Both the amount of substance delivered and the distance from the blood vessels that it penetrates appears to depend on the drug size. Even though concentrated near the blood vessels, it has been shown that tracers up to 70 kDa can be delivered.[183] Beyond imaging tracers, a number of therapeutics for neurodegenerative diseases such as neuroprotective agents,[184,185] antibodies,[186,187] plasmid DNA,[188] and siRNA[189] have also been successfully delivered across the BBB using FUS and microbubbles. Other investigations have shown that circulating neural

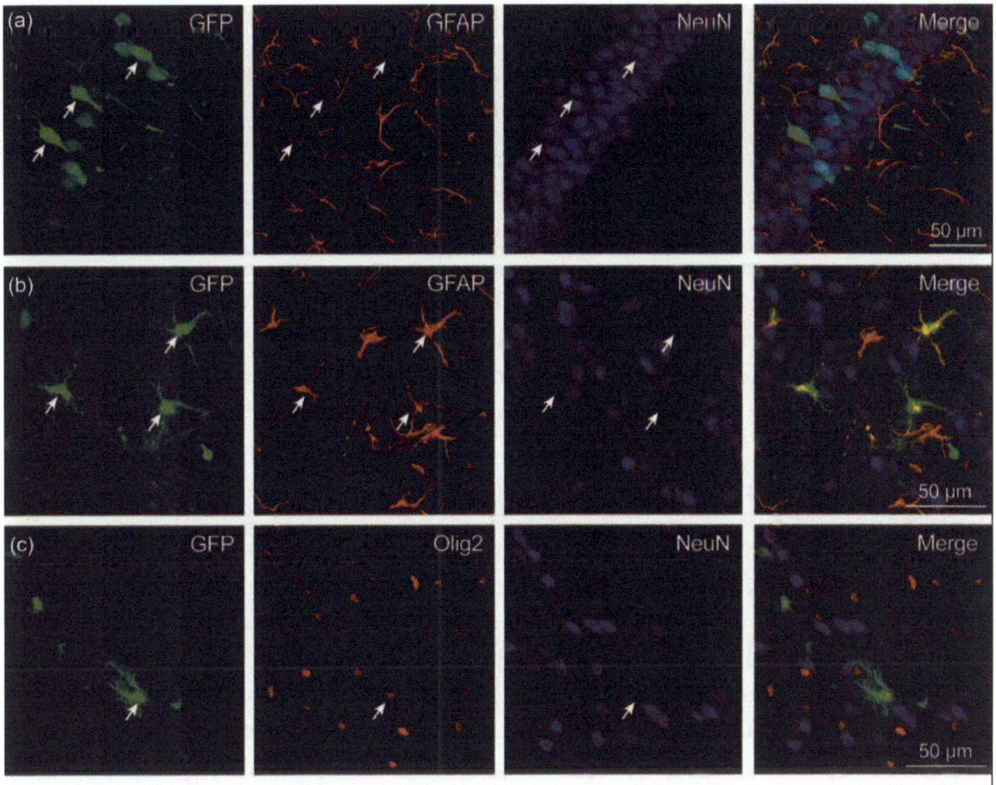

Figure 18.7 Transcranial FUS-induced gene delivery to neurons. Gene transfer to neurons, astrocytes, and oligodendrocytes after delivery of adeno-associated virus serotype 9 carrying the green fluorescent protein (GFP) to the mouse brain via FUS-induced BBB disruption. Immunohistochemistry was used to detect GFP expression in hippocampus for (a) NeuN-positive cells (neurons, white arrows), and striatum for (b) GFAP-positive cells (astrocytes, white arrows) and (c) Olig2-positive cells (oligodendrocytes, white arrow). *Modified from Thevenot et al.[182]; ©2012 Mary Ann Liebert, Inc.*

progenitor cells[190] and viral vectors for gene therapy[182,191,192] can be delivered to the sonicated regions after FUS-induced BBB disruption. An example of delivery of adeno-associated virus serotype 9 via FUS-induced BBB disruption to the different cell populations in the mouse brain is shown in Figure 18.7.

Further, chemotherapy agents,[157,193–196] chemotherapy packaged in liposomes,[147,197] targeted liposomes[198] and magnetic particles,[199–201] which allow for MRI-based tracking and enhanced delivery via magnetic targeting have also been delivered to the brain and to brain tumor models.[181] Others have loaded chemotherapy and other agents into the microbubbles used for the disruption,[184,188,201–203] which offers the possibility of achieving even higher local payload at the targeted region. In addition, Trastuzumab, an antibody-based agent used for HER2-positive breast cancer,[204,205] boronophenylalanine, which is used for boron neutron capture therapy, and natural killer cells, used for immunotherapy, can also be delivered to the brain and to brain tumor models.[206–208] The above studies have also shown that FUS enhancement of the BTB can slow tumor growth and/or improve survival in orthotopic murine models of primary or metastatic brain tumors.[193,196,197,199,200,205,209] While in some cases the response has been modest, several of these studies have seen substantial improvements. Using multiple treatments may be necessary to achieve a pronounced improvement.[197]

With so much accumulated evidence for the benefits of FUS-targeted drug delivery in the CNS, the clinical trials will start soon and hopefully will further support the benefits of this technology to the pharmacological treatment of brain tumors and CNS disorders.

IMAGE-GUIDED FUS

As described previously, therapeutic ultrasound utilizes the thermal and mechanical interactions between ultrasound and microbubbles oscillating in the ultrasound field. These interactions and the resulting bioeffects critically depend on the pressure amplitude at the focus (which is distorted after propagation through the skull), the acoustic parameters, the vascular density and other properties of the brain tissue, and the properties of the microbubbles. Other factors, such as vessel diameter may play a role.[210] Unfortunately, many of these parameters are difficult to predict and are expected to vary significantly for different patients, particularly in tumors or other pathologies. To overcome these challenges and extract accurate estimates of these parameters, FUS systems have been combined with numerous medical imaging modalities, most notably MRI and ultrasound imaging. These approaches have enabled (i) accurate target identification and determination of parameters to use (treatment planning), (ii) refinement during sonication to ensure the desired effect without overexposure (treatment monitoring), and (iii) evaluation of the treatment effects (treatment evaluation). In the following sections these approaches along with their benefits and limitations are described.

Treatment planning

The central questions in planning noninvasive or minimally invasive procedures with transcranial FUS are related to (i) target identification, (ii) FUS beam localization and targeting, and (iii) accurate focal pressure or power estimates. While for small animal experimentation a simple stereotactic system combined with *ex vivo* measurements might be adequate for accomplishing these tasks, for clinical implementation they can be very challenging to address. This challenge arises largely because in humans the thicker skull is complex (a layer of trabecular bone surrounded by layers of cortical bone), can vary substantially between individuals (3.5−9.5 mm[211]), and has significantly higher acoustic impedance than soft tissue and water that is used as coupling medium. As a result, the fraction the ultrasound beam that is transmitted depends strongly on the angle between the bone and the face of the transducer,[212] while skull's irregular shape, thickness and complexity can lead to strong deflection and distortions of the beam. These effects also make accurate focal pressure estimations challenging.

To address these challenges and ensure accurate treatment planning, numerous approaches have been developed and tested under clinical conditions so far. In particular, to correct for beam aberrations introduced by thick skulls, phased arrays composed of more than 1000 elements combined with skull aberration correction algorithms that utilize CT data are employed.[9,10] These approaches use acoustic simulations and CT scans to adjust the phase of each element in order to restore the focusing of the array after transmission through the skull.

Clinical experience with thermal ablation has shown that these aberration correction methods offer ∼1−2 mm accuracy in targeting.[14−18] MR temperature mapping[101] can be used to visualize low-level (1−2°C) focal heating and ensure accurate targeting before therapeutic exposure levels are used.[102] Other methods that can map small tissue displacements of a few microns induced by radiation force have also been developed for this purpose.[213,214] These MRI methods might also be used to predict the focal pressure amplitude[164] or to improve aberration correction[215] providing an additional method for comparison or cross validation of the modeling methods.

Treatment monitoring and control

Given the challenges in predicting the focal pressure amplitude when sonicating transcranially, effective monitoring of the procedures is essential for safe and effective outcomes. Depending on the specific interaction that it is exploited (thermal or mechanical), the methods to monitor the various procedures can be fundamentally different. For example, for thermal ablation, accurate measurement of focal temperature elevation is needed, while, for BBB disruption, monitoring of the microbubbles' emissions with high sensitivity is essential. In this section the two most widely available methods to accomplish these tasks transcranially and in real time are presented.

MR temperature imaging

Currently, the only widely available method for quantitative mapping of temperature changes in vivo uses MRI and the temperature sensitivity of the proton resonant frequency (PRF) of water, which is linear over the range of temperatures used for FUS thermal therapies and is independent of tissue type and thermal history.[216] While

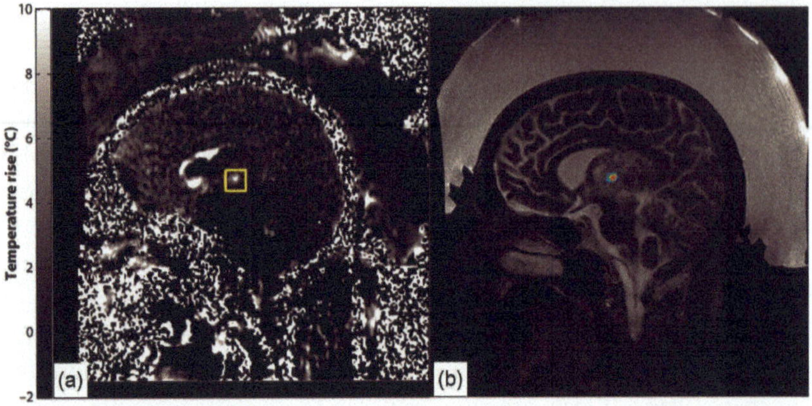

Figure 18.8 MR temperature imaging guidance of FUS ablation in humans. (a) MR temperature image during TcMRgFUS in a patient with recurrent glioma showing the feasibility transcranial focusing. Focal heating is evident in the boxed area (artifacts in the images were evident in the brain ventricles). (b) focal heating measured with MR temperature imaging superimposed on an anatomical MRI acquired before treatment.

MRTI has limitations (insensitivity in fat, for example), it ensures that the desired thermal dose has been reached and, as described above, the targeting was accurate. It also enables one to have closed-loop control over FUS thermal therapies.[217] Figure 18.8 shows example MRTI obtained during the first MRgFUS tests in patients with brain tumors.[14]

The temperature sensitivity of the PRF in water, which varies at a rate of about -0.01 ppm/°C, arises from heat-induced changes in hydrogen bonds and consequent changes in electron screening of the hydrogen nucleus.[218] These changes, which modify the magnetic field at the nucleus, result in a shift in PRF through the Larmor relationship. These changes in PRF can be estimated using phase maps of a gradient echo sequence referenced to either a pre-heating image or to surrounding tissue regions that are assumed to be unheated.[101] In a research setting, the spatial, temporal, and temperature resolution of MRTI can be better than 1 mm^3, one frame per second, and $\pm$ 1°C, respectively, and one can obtain multiple image slices. However, this performance is not possible at the current time for MRgFUS in the brain, where the introduction of the large-scale FUS phased array makes imaging challenging. A number of groups are developing MRI coils that can be integrated into the brain MRgFUS system.

Passive acoustic emissions monitoring and mapping

Currently, no MRI-based methods are available for characterizing and/or visualizing microbubble oscillations during the sonications. The inherently nonlinear microbubble oscillations combined with the micrometer and microsecond scales involved in these interactions provide very strict criteria for real-time monitoring and control. The majority of these criteria can be met using acoustic methods to analyze the sound emitted from the microbubbles when exposed to the FUS beam. Unlike standard ultrasound imaging, these methods use piezoelectric receivers operated in passive mode (i.e., only listening) to record and analyze the diverging pressure waves (i.e., acoustic emissions) emitted by oscillating microbubbles during sonication, such as FUS-induced BBB disruption.[180,219–221] The spectral content and strength of the recorded emissions is sufficient to characterize and subsequently control the microbubble oscillations.

Inertial cavitation is manifested in the frequency domain of the acoustic emissions as a broadband signal,[6] and has generally been associated with the production of vascular damage during BBB disruption.[180] Harmonic and/or sub- and ultraharmonic acoustic emissions in the absence of broadband signal are indicative of stable volumetric oscillations, which consistently have been associated with safe BBB disruption.[180,219,221] Therefore, depending on the spectral content and strength of the emissions the output of the device can be increased until strong harmonic, subharmonic, or ultraharmonic emissions are observed, and decreased if broadband emissions are detected.[180,221,223] The temporal resolution of these methods is currently less than a few msec.

An example of the good correlation observed during transcranial BBB disruption using a clinical FUS system in macaques between the MRI signal enhancement with the strength of the harmonic emissions is shown in Figure 18.9A. Integration of a large number of receivers (typically >64) into the FUS system with passive reconstruction methods[224,225] can be used to create two- or even three-dimensional maps of the microbubble activity to ensure that it is occurring at the expected location. Several studies have investigated passive imaging for brain applications.[222,226–230] Examples of passive acoustic mapping (PAM) from experiments in macaques using a linear receiver array integrated into a clinical brain MRgFUS system are shown in Figure 18.9b and 18.9c. We anticipate these methods, along with their integration into MRgFUS systems[230] could provide unprecedented control over the procedures in the brain and facilitate the development of other cavitation-based therapies, as we describe below.

Treatment evaluation

As described above, therapeutic ultrasound utilizes the thermal and mechanical interactions between ultrasound and microbubbles with tissues and the vasculature. MRI, which offers superior soft tissue contrast and precise perfusion measurements, is the most suitable imaging modality to assess these effects. Indeed, T2-weighted imaging demonstrating edema is used to identify successful thermal ablation of thalamic targets during FUS neurosurgery, while diffusion-weighted imaging (DWI), shows restriction of water diffusion induced by the treatment within 24 hours.[18] In addition, contrast-enhanced MRI and T2*- or susceptibility-weighted imaging can be used to verify

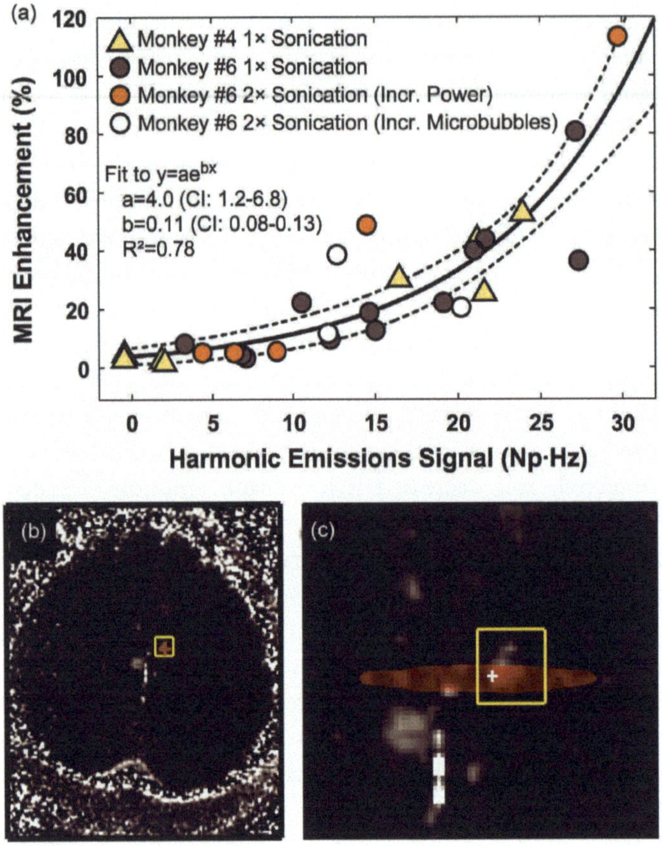

Figure 18.9 Comparison of MRI signal enhancement with microbubble acoustic emissions. (a) MRI signal enhancement after Gd-DTPA injection plotted as a function of the harmonic emissions signal strength measured with single-element detectors. *Reprinted from © 2012, Arvanitis et al.*[221] (b–c) Comparison between MRI signal enhancement and passive cavitation mapping. (b) Map showing the enhancement relative to a precontrast image. (c) Fusion of passive acoustic map with the T1-weighted MRI from (b). The red region shows the pixels in the cavitation map within 95% of the maximum value. This region overlapped with the contrast enhancement. The pixel with the maximum cavitation activity is noted with a " + ." This breakthrough work may enable the translation of microbubble based therapies in the brain to the clinics. *Modified from Arvanitis et al.,*[227] *©2013, Institute of Physics and Engineering in Medicine.*

that BBB disruption has occurred and whether vascular damage has occurred, respectively. For tumors, it may be necessary to compare the signal enhancement after contrast injection to measurements obtained before FUS.

For drug delivery, direct labeling of the therapeutic agent with a contrast agent for MRI or other modality might be useful to estimate drug uptake and penetration in the brain during or after FUS–BBB disruption.[201] Numerous studies have also used a

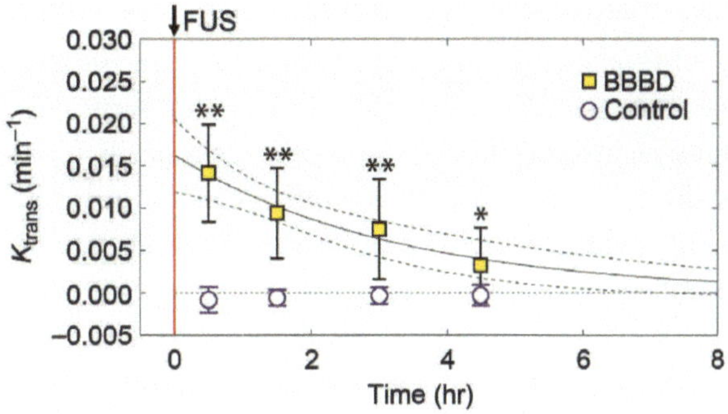

Figure 18.10 Dynamic and noninvasive assessment of FUS-induced BBB disruption. Dynamic contrast-enhanced MRI (DCE-MRI) can evaluate FUS-induced BBB disruption. A time series of T1-weighted MRI was acquired before, during, and after the injection of an MRI contrast agent. The change in signal intensity as a function of time was fit to a pharmacokinetic model[234] to estimate the transfer coefficient, K_{trans}, which describes transport of the contrast agent from the blood vessels to the brain. The mean K_{trans} values measured in regions of interest at sonicated locations and in corresponding nonsonicated structures in the contralateral hemisphere are plotted as a function of time. The decay of K_{trans}, which occurred due to restoration of the BBB, was fit to an exponential decay (solid line; dotted lines: 95% confidence intervals (CI)), yielding an estimated half-life of 2.2 hrs (95% CI: 1.1−3.4 hrs). These data are essential for determining the dosing of different drugs into the brain. *Modified from Park et al.,[157] ©2012 Elsevier B.V.*

standard contrast agent as a surrogate measurement of tissue drug concentrations.[147,204,231] Finally, estimation of the vascular transfer coefficients via analysis of dynamic contrast-enhanced MRI (DCE-MRI)[232] have been used to predict the resulting payload of drugs to the brain[157] and tumors.[233] Examples showing DCE-MRI evaluation of BBB disruption and its subsequent restoration over time are shown in Figure 18.10.

FUTURE DEVELOPMENTS AND PERSPECTIVES

With the ability to noninvasively focus acoustic energy to precise targets in the brain, FUS offers a wide range of therapies that can reduce morbidity, reduce the need for ionizing radiation, and enable the targeted delivery of seemingly any therapeutic to precisely targeted spots or volumes. With the development of simple methods to provide closed-loop control over the procedure, one can envision FUS becoming a truly transformative technology.

Most work to date on FUS-induced BBB disruption has focused on the delivery of therapeutic agents, and on standard histological methods to confirm that the procedure is safe. Recent work suggests that the procedure itself, or perhaps the body's response to it, may have therapeutic potential. For example, while studying the delivery of antibodies targeted to amyloid plaques in Alzheimer's disease model mice,[187] it was found that FUS-induced BBB disruption *alone* can reduce the size of the plaques, perhaps through the delivery of endogenous antibodies.[235] The same group also recently published work suggesting that the procedure can induce neurogenesis.[236] It may also be possible that a similar phenomenon triggered stem cell homing in experiments where neural precursor cells were injected systemically.[237] It is clear that there is much work to do to understand the subtle changes induced (or triggered) by mechanical stimulation of the microvasculature that occurs during FUS when combined with microbubbles. Carefully assessment of any detrimental effects that might emerge in the presence of these diseases is also important.

There are a number of other exciting uses of therapeutic ultrasound in the brain beyond thermal ablation and BBB disruption. For example, numerous studies over the years have shown that ultrasound, ranging in frequency from 20 kHz to 2 MHz, can aid in thrombolysis.[58,105,226,238–247] These studies have been conducted in vitro, in vivo and in clinical settings, and most efforts have focused upon enhancing the activity of thrombolytic agents. In particular, several experiments have demonstrated that the effectiveness of thrombolytic agents are increased with the use of intravascular ultrasound devices, which have applications in coronary arteries, the peripheral vasculature, and stroke.[105,245] Of particular interest are recent studies that demonstrate that even exposure to 2 MHz transcranial Doppler ultrasound bursts while infusing thrombolytic agents can result in increased thrombolysis in a clinical setting.[105] The clinical studies have shown a clear benefit for example in the multi-institutional CLOTBUST trial complete recanalization of vessels with tPA alone was 8% and 25% with tPA and ultrasound.[105] Other studies have shown that the introduction of microbubbles in conjunction with tPA and 2 MHz transcranial Doppler may improve the activity of ultrasound further,[248–252] although adverse events observed in a clinical trial with microbubbles[253] suggest more work on control and other factors is needed.

The use of microbubbles might also greatly expand the targetable areas in the brain where FUS ablation can be safely employed. As described above, the current systems cannot sonicate close to the skull without overheating the bone. To enable such sonications, the acoustic intensity needs to be greatly reduced. A way to achieve this reduction is to combine FUS with an intravenously injected microbubble-based ultrasound contrast agent (USCA). As described below, we have been investigating the use of such agents to reduce the power needed for FUS ablation in the brain. That work, along with studies by others, has shown that the power required for FUS ablation in the brain can be reduced by orders of magnitude with the use of an USCA.[254–256] With such a reduction, the heating produced in the skull bone can be decreased to a level below the threshold for thermal damage so targets directly adjacent to the bone can be ablated.

The combination of transcranial FUS and microbubbles may also be used to sensitize tumors to radiation. Using exposure levels above what are used for BBB disruption, a recent series of papers out of Sunnybrook and Women's Hospital in Toronto have demonstrated that microbubble-enhanced sonications is an effective radiosensitizer for a number of different tumor models.[257–262] While the method has not been tested in brain tumors to our knowledge, this method appears promising for enhancing the effectiveness of radiation therapy and radiosurgery. It might be possible that the method could also reduce the dose needed to achieve a therapeutic response.

FUS also has the potential to reversibly suppress or stimulate neuronal functions. It has been known for decades that FUS can reversibly suppress brain activity and modulate evoked potentials,[263–266] and a number of studies in small animals suggests that it may be capable of stimulation as well.[70,267–272] Reports this year have reported similar effects in a monkey[273] and in human volunteers.[274] If this technique can be translated to the clinic, it could provide a noninvasive method to modulate brain activity which, unlike transcranial magnetic or direct current stimulation, is highly localized and can be applied to deep targets. This approach could provide improved methods for surgical planning, functional brain mapping, or for new therapies.

Finally, additional benefit may be achieved by combining these different FUS therapies. For example, one might envision a therapy where one uses thermal ablation or microbubble-enhanced FUS ablation to destroy the hypervascular portion of the tumor, and then repeated sessions with BBB disruption in the surrounding brain to target the delivery of chemotherapy to infiltrating cancer cells.

CONCLUSION

FUS is uniquely capable of producing a broad spectrum of bioeffects that can be used for treatment of potentially extensive range of CNS diseases and disorders. It is a noninvasive, targeted, and repeatable method. It can be performed within an MRI unit, enabling precise anatomical and functional guidance along with real-time monitoring of its effects unlike any other technology. Major efforts have been undertaken to use the method to noninvasively ablate tissue volumes and deliver drugs to targeted brain regions through a temporary disruption of the BBB. Other capabilities of FUS include the ability to enhance thrombolysis, sensitize tissue to radiation, and reversibly modulate neuronal function. In recent years, with the development of devices capable of focusing ultrasound through the human skull, the demonstration of feasibility in humans, and the large number of preclinical studies that have been published, it has become clear that FUS has a real potential to transform the treatment of CNS disorders.

REFERENCES

1. Neuroinsights. *The Neurotechnology Industry 2008*. Report published by NeuroInsights; 2008.
2. Neuwelt EA, Bauer B, Fahlke C, et al. Engaging neuroscience to advance translational research in brain barrier biology. *Nat Rev Neurosci*. 2011;12(3):169–182.
3. Lynn JG, Zwemer RL, Chick AJ, Miller AE. A new method for the generation and use of focused ultrasound in experimental biology. *J Gen Physiol*. 1942;26:179–193.

4. Fry WJ. Intense ultrasound in investigations of the central nervous system. *Adv Biol Med Phys.* 1958;6:281–348.
5. Ballantine HT, Bell E, Manlapaz J. Progress and problems in the neurological applications of focused ultrasound. *J Neurosurg.* 1960;17:858–876.
6. Lele PP. Effects of ultrasound on "solid" mammalian tissues and tumors in vivo. In: Repacholi MH, Grondolfo M, Rindi A, eds. *Ultrasound: Medical Applications, Biological Effects and Hazard Potential.* New York: Plenum Pub. Corp.; 1987:275–306.
7. Vykhodtseva NI, Hynynen K, Damianou C. Histologic effects of high intensity pulsed ultrasound exposure with subharmonic emission in rabbit brain in vivo. *Ultrasound Med Biol.* 1995;21(7):969–979.
8. Jagannathan J, Sanghvi NT, Crum LA, et al. High-intensity focused ultrasound surgery of the brain: part 1—A historical perspective with modern applications. *Neurosurgery.* 2009;64(2):201–210.
9. Clement GT, Hynynen K. A non-invasive method for focusing ultrasound through the human skull. *Phys Med Biol.* 2002;47(8):1219–1236.
10. Aubry JF, Tanter M, Pernot M, Thomas JL, Fink M. Experimental demonstration of noninvasive transskull adaptive focusing based on prior computed tomography scans. *J Acoust Soc Am.* 2003;113(1):84–93.
11. Hynynen K, Clement GT, McDannold N, et al. 500-element ultrasound phased array system for noninvasive focal surgery of the brain: A preliminary rabbit study with ex vivo human skulls. *Magn Reson Med.* 2004;52(1):100–107.
12. Hynynen K, McDannold N, Clement G, et al. Pre-clinical testing of a phased array ultrasound system for MRI-guided noninvasive surgery of the brain: a primate study. *Eur J Radiol.* 2006;59(2):149–156.
13. Clement GT, White PJ, King RL, McDannold N, Hynynen K. A magnetic resonance imaging-compatible, large-scale array for trans-skull ultrasound surgery and therapy. *J Ultrasound Med.* 2005;24(8):1117–1125.
14. McDannold N, Clement GT, Black P, Jolesz F, Hynynen K. Transcranial magnetic resonance imaging-guided focused ultrasound surgery of brain tumors: initial findings in 3 patients. *Neurosurgery.* 2010;66(2):323–332.
15. Martin E, Jeanmonod D, Morel A, Zadicario E, Werner B. High-intensity focused ultrasound for noninvasive functional neurosurgery. *Ann Neurol.* 2009;66(6):858–861.
16. Jeanmonod D, Werner B, Morel A, et al. Transcranial magnetic resonance imaging-guided focused ultrasound: noninvasive central lateral thalamotomy for chronic neuropathic pain. *Neurosurg Focus.* 2012;32(1):E1.
17. Lipsman N, Schwartz ML, Huang Y, et al. MR-guided focused ultrasound thalamotomy for essential tremor: a proof-of-concept study. *Lancet Neurol.* 2013;12(5):462–468.
18. Elias WJ, Huss D, Voss T, et al. A pilot study of focused ultrasound thalamotomy for essential tremor. *N Engl J Med.* 2013;369(7):640–648.
19. Hynynen K, Davis KL. Small cylindrical ultrasound sources for induction of hyperthermia via body cavities or interstitial implants. *Int J Hyperthermia.* 1993;9(2):263–274.
20. Diederich CJ, Nau WH, Burdette EC, Bustany IS, Deardorff DL, Stauffer PR. Combination of transurethral and interstitial ultrasound applicators for high-temperature prostate thermal therapy. *Int J Hyperthermia.* 2000;16(5):385–403.
21. Lafon C, Chapelon JY, Prat F, et al. Design and preliminary results of an ultrasound applicator for interstitial thermal coagulation. *Ultrasound Med Biol.* 1998;24(1):113–122.
22. Foster RS, Bihrle R, Sanghvi NT, Fry FJ, Donohue JP. High-intensity focused ultrasound in the treatment of prostatic disease. *Eur Urol.* 1993;23(Suppl 1):29–33.
23. Hutchinson EB, Hynynen K. Intracavitary ultrasound phased arrays for noninvasive prostate surgery. *IEEE Trans Ultrason Ferroelectr Freq Contr.* 1996;43(6):1032.
24. Zimmer JE, Hynynen K, He DS, Marcus F. The feasibility of using ultrasound for cardiac ablation. *IEEE Trans Biomed Eng.* 1995;42(9):891–897.
25. Hynynen K, Dennie J, Zimmer JE, et al. Cylindrical ultrasonic transducers for cardiac catheter ablation. *IEEE Trans Biomed Eng.* 1997;44(2):144–151.

26. Lalonde R. Variable frequency field conjugate lenses for ultrasound hyperthermia. *IEEE Trans Ultrason Ferroelectr Freq Contr.* 1995;42(5):825.
27. Fjield T, Sorrentino V, Cline H, Hynynen K. Design and experimental verification of thin acoustic lenses for the coagulation of large tissue volumes. *Phys Med Biol.* 1997;42(12):2341–2354.
28. Cain CA, Umemura S-I. Concentric-ring and sector-vortex phased-array applicators for ultrasound hyperthermia. *IEEE Trans Microw Theory Tech.* 1986;34(5):542–551.
29. Ebbini ES, Umemura SI, Ibbini MS, Cain CA. A cylindrical-section ultrasound phased-array applicator for hyperthermia cancer therapy. *IEEE Trans Ultrason Ferroelectr Freq Contr.* 1988;35(5):561.
30. Fan X, Hynynen K. Control of the necrosed tissue volume during noninvasive ultrasound surgery using a 16-element phased array. *Med Phys.* 1995;22(3):297–306.
31. Hynynen K, Chung A, Fjield T, et al. Feasibility of using ultrasound phased arrays for MRI monitored noninvasive surgery. *IEEE Trans Ultrason Ferroelectr Freq Contr.* 1996;43(6):1043.
32. Daum DR, Smith NB, King R, Hynynen K. In vivo demonstration of noninvasive thermal surgery of the liver and kidney using an ultrasonic phased array. *Ultrasound Med Biol.* 1999;25(7):1087–1098.
33. Curiel L. 1.5D multi-elements phased array applied to high intensity focused ultrasound: 1999. Conference: Ultrasonics Symposium, 1999. Proceedings.1999 IEEE, Vol 2.
34. Clement GT, Sun J, Giesecke T, Hynynen K. A hemisphere array for non-invasive ultrasound brain therapy and surgery. *Phys Med Biol.* 2000;45(12):3707–3719.
35. Carstensen EL, Becroft SA, Law WK, Barber DB. Finite amplitude effects on thresholds for lesion production in tissues by unfocussed ultrasound. *J Acoust Soc Am.* 1981;70:302–309.
36. Swindell W. A theoretical study of nonlinear effects with focused ultrasound in tissues: an 'acoustic Bragg peak'. *Ultrasound Med Biol.* 1985;11(1):121.
37. Hynynen K. The role of nonlinear ultrasound propagation during hyperthermia treatments. *Med Phys.* 1991;18(6):1156–1163.
38. Fry FJ, Johnson LK. Tumor irradiation with intense ultrasound. *Ultrasound Med Biol.* 1978;4(4): 337–341.
39. Goss SA. The effect of high intensity ultrasonic irradiation on tumor growth. *IEEE Transactions on Sonics and Ultrasonics.* 1984;SU–31(5):491.
40. Yang R, Reilly CR, Rescorla FJ, et al. High-intensity focused ultrasound in the treatment of experimental liver cancer. *Arch Surg.* 1991;126(8):1002–1009.
41. Sibille A, Prat F, Chapelon JY, et al. Extracorporeal ablation of liver tissue by high-intensity focused ultrasound. *Oncology.* 1993;50(5):375–379.
42. Yang R, Reilly CR, Rescorla FJ, et al. Effects of high-intensity focused ultrasound in the treatment of experimental neuroblastoma. *J Pediatr Surg.* 1992;27(2):246–250.
43. Yang R, Sanghvi NT, Rescorla FJ, Kopecky KK, Grosfeld JL. Liver cancer ablation with extracorporeal high-intensity focused ultrasound. *Eur Urol.* 1993;23(Suppl 1):17–22.
44. Chen L, Rivens IH, ter Haar GR, Riddler S, Hill CR, Bensted JP. Histological changes in rat liver tumours treated with high-intensity focused ultrasound. *Ultrasound Med Biol.* 1993;19(1):67–74.
45. Prat F, Centarti M, Sibille A, et al. Extracorporeal high-intensity focused ultrasound for VX2 liver tumors in the rabbit. *Hepatology.* 1995;21(3):832–836.
46. Adams JB, Moore RG, Anderson JH, Strandberg JD, Marshall FF, Davoussi LR. High-intensity focused ultrasound ablation of rabbit kidney tumors. *J Endourol.* 1996;10(1):71–75.
47. Rowland IJ, Rivens IH, Chen L, et al. MRI study of hepatic tumours following high intensity focused ultrasound surgery. *Br J Radiol.* 1997;70:144–153.
48. Chen L, ter Haar HG, Hill CR, Eccles SA, Box G. Treatment of implanted liver tumors with focused ultrasound. *Ultrasound Med Biol.* 1998;24(9):1475–1488.
49. ter Haar GR, Rivens IH, Chen L, Riddler S. High intensity focused ultrasound for the treatment of rat tumours. *Phys Med Biol.* 1991;36(11):1495–1501.
50. Chen L, ter Haar GR, Robertson D, Bensted JP, Hill CR. Histological study of normal and tumor-bearing liver treated with focused ultrasound. *Ultrasound Med Biol.* 1999;25(5):847–856.
51. Vaezy S, Fujimoto VY, Walker C, Martin RW, Chi EY, Crum LA. Treatment of uterine fibroid tumors in a nude mouse model using high-intensity focused ultrasound. *Am J Obstet Gynecol.* 2000; 183(1):6–11.

52. Delon-Martin C, Vogt C, Chignier E, Guers C, Chapelon JY, Cathignol D. Venous thrombosis generation by means of high-intensity focused ultrasound. *Ultrasound Med Biol.* 1995;21(1):113–119.
53. Hynynen K, Colucci V, Chung A, Jolesz FA. Noninvasive arterial occlusion using MRI-guided focused ultrasound. *Ultrasound Med Biol.* 1996;22(8):1071–1077.
54. Rivens IH, Rowland IJ, Denbow M, Fisk NM, ter Haar GR, Leach MO. Vascular occlusion using focused ultrasound surgery for use in fetal medicine. *Eur J Ultrasound.* 1999;9(1):89–97.
55. Vaezy S, Martin R, Schmiedl U, et al. Liver hemostasis using high-intensity focused ultrasound. *Ultrasound Med Biol.* 1997;23(9):1413–1420.
56. Vaezy S, Martin R, Yaziji H, et al. Hemostasis of punctured blood vessels using high-intensity focused ultrasound. *Ultrasound Med Biol.* 1998;24(6):903–910.
57. Francis CW, Onundarson PT, Carstensen EL, et al. Enhancement of fibrinolysis in vitro by ultrasound. *J Clin Invest.* 1992;90(5):2063–2068.
58. Porter TR, LeVeen RF, Fox R, Kricsfeld A, Xie F. Thrombolytic enhancement with perfluorocarbon-exposed sonicated dextrose albumin microbubbles. *Am Heart J.* 1996;132(5): 964–968.
59. Unger EC, McCreery TP, Sweitzer RH, Caldwell VE, Wu Y. Acoustically active lipospheres containing paclitaxel: a new therapeutic ultrasound contrast agent. *Invest Radiol.* 1998;33(12):886–892.
60. Kim HJ, Greenleaf JF, Kinnick RR, Bronk JT, Bolander ME. Ultrasound-mediated transfection of mammalian cells. *Hum Gene Ther.* 1996;7(11):1339–1346.
61. Unger EC, McCreery TP, Sweitzer RH. Ultrasound enhances gene expression of liposomal transfection. *Invest Radiol.* 1997;32(12):723–727.
62. Greenleaf WJ, Bolander ME, Sarkar G, Goldring MB, Greenleaf JF. Artificial cavitation nuclei significantly enhance acoustically induced cell transfection. *Ultrasound Med Biol.* 1998;24(4):587–595.
63. Madio DP, van Gelderen P, DesPres D, et al. On the feasibility of MRI-guided focused ultrasound for local induction of gene expression. *J Magn Reson Imaging.* 1998;8(1):101–104.
64. Miller DL, Bao S, Gies RA, Thrall BD. Ultrasonic enhancement of gene transfection in murine melanoma tumors. *Ultrasound Med Biol.* 1999;25(9):1425–1430.
65. Huber PE, Pfisterer P. In vitro and in vivo transfection of plasmid DNA in the Dunning prostate tumor R3327-AT1 is enhanced by focused ultrasound. *Gene Ther.* 2000;7(17):1516–1525.
66. Lawrie A, Brisken AF, Francis SE, Cumberland DC, Crossman DC, Newman CM. Microbubble-enhanced ultrasound for vascular gene delivery. *Gene Ther.* 2000;7(23):2023–2027.
67. Vykhodtseva NI, Hynynen K, Damianou C. Pulse duration and peak intensity during focused ultrasound surgery: theoretical and experimental effects in rabbit brain in vivo. *Ultrasound Med Biol.* 1994;20(9):987–1000.
68. Hynynen K, McDannold N, Vykhodtseva N, Jolesz FA. Noninvasive MR imaging-guided focal opening of the blood-brain barrier in rabbits. *Radiology.* 2001;220(3):640–646.
69. Gavrilov LR, Tsirulnikov EM, Davies IA. Application of focused ultrasound for the stimulation of neural structures. *Ultrasound Med Biol.* 1996;22(2):179–192.
70. Tufail Y, Matyushov A, Baldwin N, et al. Transcranial pulsed ultrasound stimulates intact brain circuits. *Neuron.* 2010;66(5):681–694.
71. Yoo SS, Bystritsky A, Lee JH, et al. Focused ultrasound modulates region-specific brain activity. *Neuroimage.* 2011;56(3):1267–1275.
72. Prat F, Chapelon JY, Abou eF, et al. Focused liver ablation by cavitation in the rabbit: a potential new method of extracorporeal treatment. *Gut.* 1994;35(3):395–400.
73. Miller MW. Gene transfection and drug delivery. *Ultrasound Med Biol.* 2000;26(Suppl 1):S59–S62.
74. Nyborg WL, Carson PL, Carstensen EL, et al. *Exposure criteria for medical diagnostic ultrasound: II. Criteria based on all known mechanisms (NCRP Report No. 140).* Bethesda, MD: National Council on Radiation Protection and Measurements; 2002.
75. Crum LA, Hansen GM. Growth of air bubbles in tissue by rectified diffusion. *Phys Med Biol.* 1982; 27(3):413–417.
76. Leighton TG. *The Acoustic Bubble.* San Diego, CA: Academic Press Limited; 1994.
77. Edmonds PD, Sancier KM. Evidence for free radical production by ultrasonic cavitation in biological media. *Ultrasound Med Biol.* 1983;9:635–639.

78. Flynn HG. Generation of transient cavities in liquids by microsecond pulses of ultrasound. *J Acoust Soc Am*. 1982;72:1926–1932.
79. Apfel RE. Acoustic cavitation: a possible consequence of biomedical uses of ultrasound. *Br J Cancer Suppl*. 1982;45(5):140–146.
80. Miller DL. Particle gathering and microstreaming near ultrasonically activated gas-filled micropores. *J Acoust Soc Am*. 1988;84(4):1378–1387.
81. Fry WJ, Wulff VJ, Tucker D, Fry FJ. Physical factors involved in ultrasound induced changes in living systems: I. Identification of non-temperature effects. *J Acoust Soc Am*. 1950;22: 867–871.
82. Fry WJ, Fry RB. Determination of absolute sound levels and acoustic absorption coefficients by thermocouple probes: theory. *J Acoust Soc Am*. 1954;26:294–310.
83. Fry WJ, Barnard JW, Fry FJ, Krumins RF, Brennan JF. Ultrasonic lesions in the mammalian central nervous system. *Science*. 1955;122:517–518.
84. Fry WJ, Barnard JW, Fry FJ. Ultrasonically produced localized selective lesions in the central nervous system. *Am J Phys Med*. 1955;34:413–423.
85. Barnard JW, Fry WJ, Fry FJ, Brennan JF. Small localized ultrasonic lesions in the white and gray matter of the cat brain. *Archives of Neurological Psychiatry*. 1956;75:15–35.
86. Fry WJ, Brennan JF, Barnard JW. Histological study of changes produced by ultrasound in the gray and white matter of the central nervous system. *Ultrasound Med Biol*. 1957;3:110–130.
87. Lele PP. A simple method for production of trackless focal lesions with focused ultrasound: physical factors. *J Physiol*. 1962;160:494–512.
88. Basauri L, Lele PP. A simple method for production of trackless focal lesions with focused ultrasound: statistical evaluation of the effects of irradiation on the central nervous system of the cat. *J Physiol*. 1962;160:513–534.
89. Lele PP. Production of deep focal lesions by focused ultrasound—current status. *Ultrasonics*. 1967; 5:105–112.
90. Lele PP, Pierce AD. The thermal hypothesis of the mechanism of ultrasonic focal destruction in organised tissues.Interaction of ultrasound and biological tissues. In: Reid JM, Sikov MR, eds. *FDA 73-8008 BRH/DBE*. Washington, DC: Bureau of Radiological Health; 1973:121–128.
91. Lele PP. Ultrasound in surgery. In: Michaelson SM, Magin R, Carstensen EL, Miller MW, eds. *Fundamental and Applied Aspects of Nonionizing Radiation*. New York: Plenum Press; 1975:325–340.
92. Meyers R, Fry WJ, Fry FJ, Dreyer LL, Shultz DF, Noyes RF. Early experiences with ultrasonic irradiation of the pallidfugal and nigral complexes in hyperkinetic and hypertonic disorders. *J Neurosurg*. 1959;16:32–54.
93. Fry WJ, Fry FJ. Fundamental neurological research and human neurosurgery using intense ultrasound. *IRE Trans Med Electron*. 1960;ME-7:166–181.
94. Oka M, Okumura T, Yokoi H, et al. Surgical application of high intensity focused ultrasound. *Med J Osaka Univ*. 1960;10(3–4):427–442.
95. Heimburger RF. Ultrasound augmentation of central nervous system tumor therapy. *Indiana Med*. 1985;78(6):469–476.
96. Guthkelch AN, Carter LP, Cassady JR, et al. Treatment of malignant brain tumors with focused ultrasound hyperthermia and radiation: results of a phase I trial. *J Neurooncol*. 1991;10(3):271–284.
97. Park J, Jung S, Jung T-Y, Lee M-C. Focused ultrasound surgery for the treatment of recurrent anaplastic astrocytoma: A preliminary report: 2006. In: Clement GT, McDannold NJ, Hynynen K. Therapeutic Ultrasound: 5th International Symposium held 27–29 October 2005 in Boston, Massachusetts. Conference Proceedings, Vol. 829. Melville, NY: American Institute of Physics, 2006;238–240.
98. Ram Z, Cohen ZR, Harnof S, et al. Magnetic resonance imaging-guided, high-intensity focused ultrasound for brain tumor therapy. *Neurosurgery*. 2006;59(5):949–955.
99. Hynynen K, Jolesz FA. Demonstration of potential noninvasive ultrasound brain therapy through an intact skull. *Ultrasound Med Biol*. 1998;24(2):275–283.
100. Thomas J, Fink MA. Ultrasonic beam focusing through tissue inhomogeneities with a time reversal mirror: application to transskull therapy. *IEEE Trans Ultrason Ferroelectr Freq Contr*. 1996;43(6): 1122–1129.

101. Ishihara Y, Calderon A, Watanabe H, Okamoto K, Suzuki Y, Kuroda K. A precise and fast temperature mapping using water proton chemical shift. *Magn Reson Med.* 1995;34(6):814–823.
102. Hynynen K, Vykhodtseva NI, Chung AH, Sorrentino V, Colucci V, Jolesz FA. Thermal effects of focused ultrasound on the brain: determination with MR imaging. *Radiology.* 1997;204(1):247–253.
103. Vykhodtseva NI, Sorrentino V, Jolesz FA, Bronson RT, Hynynen K. MRI detection of the thermal effects of focused ultrasound on the brain. *Ultrasound Med Biol.* 2000;26(5):871–880.
104. Pernot M, Aubry JF, Tanter M, et al. In vivo transcranial brain surgery with an ultrasonic time reversal mirror. *J Neurosurgery.* 2007;106:1061–1066.
105. Alexandrov AV, Molina CA, Grotta JC, et al. Ultrasound-enhanced systemic thrombolysis for acute ischemic stroke. *N Engl J Med.* 2004;351(21):2170–2178.
106. Abbott NJ, Romero IA. Transporting therapeutics across the blood-brain barrier. *Mol Med Today.* 1996;2(3):106–113.
107. Pardridge WM. Molecular biology of the blood-brain barrier. *Mol Biotechnol.* 2005;30(1):57–70.
108. Lockman PR, Mittapalli RK, Taskar KS, et al. Heterogeneous blood-tumor barrier permeability determines drug efficacy in experimental brain metastases of breast cancer. *Clin Cancer Res.* 2010;16(23):5664–5678.
109. Eichler AF, Chung E, Kodack DP, Loeffler JS, Fukumura D, Jain RK. The biology of brain metastases: translation to new therapies. *Nat Rev Clin Oncol.* 2011;8(6):344–356.
110. Groothuis DR. The blood-brain and blood-tumor barriers: a review of strategies for increasing drug delivery. *Neuro-oncol.* 2000;2(1):45–59.
111. Pardridge WM. Blood-brain barrier delivery. *Drug Discov Today.* 2007;12(1–2):54–61.
112. Blumling III JP, Silva GA. Targeting the brain: advances in drug delivery. *Curr Pharm Biotechnol.* 2012;13(12):2417–2426.
113. Oldendorf WH, Hyman S, Braun L, Oldendorf SZ. Blood-brain barrier: penetration of morphine, codeine, heroin, and methadone after carotid injection. *Science.* 1972;178(4064):984–986.
114. Abbott NJ, Patabendige AA, Dolman DE, Yusof SR, Begley DJ. Structure and function of the blood-brain barrier. *Neurobiol Dis.* 2010;37(1):13–25.
115. Pardridge WM. Drug targeting to the brain. *Pharm Res.* 2007;24(9):1733–1744.
116. Blasi P, Giovagnoli S, Schoubben A, Ricci M, Rossi C. Solid lipid nanoparticles for targeted brain drug delivery. *Adv Drug Deliv Rev.* 2007;59(6):454–477.
117. Olivier JC. Drug transport to brain with targeted nanoparticles. *NeuroRx.* 2005;2(1):108–119.
118. Schnyder A, Huwyler J. Drug transport to brain with targeted liposomes. *NeuroRx.* 2005;2(1):99–107.
119. Broadwell RD, Balin BJ, Salcman M. Transcytotic pathway for blood-borne protein through the blood-brain barrier. *Proc Natl Acad Sci USA.* 1988;85(2):632–636.
120. Mellman I. Endocytosis and molecular sorting. *Annu Rev Cell Dev Biol.* 1996;12:575–625.
121. Mukherjee S, Ghosh RN, Maxfield FR. Endocytosis. *Physiol Rev.* 1997;77(3):759–803.
122. Zhang EY, Knipp GT, Ekins S, Swaan PW. Structural biology and function of solute transporters: implications for identifying and designing substrates. *Drug Metab Rev.* 2002;34(4):709–750.
123. Pardridge WM. Blood-brain barrier drug targeting: the future of brain drug development. *Mol Interv.* 2003;3(2):90–105:51.
124. Bobo RH, Laske DW, Akbasak A, Morrison PF, Dedrick RL, Oldfield EH. Convection-enhanced delivery of macromolecules in the brain. *Proc Natl Acad Sci USA.* 1994;91(6):2076–2080.
125. Guerin C, Olivi A, Weingart JD, Lawson HC, Brem H. Recent advances in brain tumor therapy: local intracerebral drug delivery by polymers. *Invest New Drugs.* 2004;22(1):27–37.
126. Brem H, Gabikian P. Biodegradable polymer implants to treat brain tumors. *J Control Release.* 2001;74(1–3):63–67.
127. Tomita T. Interstitial chemotherapy for brain tumors: review. *J Neurooncol.* 1991;10(1):57–74.
128. Fung LK, Shin M, Tyler B, Brem H, Saltzman WM. Chemotherapeutic drugs released from polymers: distribution of 1,3-bis(2-chloroethyl)-1-nitrosourea in the rat brain. *Pharm Res.* 1996;13(5):671–682.

129. Voges J, Reszka R, Gossmann A, et al. Imaging-guided convection-enhanced delivery and gene therapy of glioblastoma. *Ann Neurol.* 2003;54(4):479–487.
130. Illum L. Nasal drug delivery—possibilities, problems and solutions. *J Control Release.* 2003;87 (1–3):187–198.
131. Pires A, Fortuna A, Alves G, Falcao A. Intranasal drug delivery: how, why and what for? *J Pharm Pharm Sci.* 2009;12(3):288–311.
132. Rapoport SI. Effect of concentrated solutions on blood-brain barrier. *Am J Physiol.* 1970;219(1): 270–274.
133. Rapoport SI. Advances in osmotic opening of the blood-brain barrier to enhance CNS chemotherapy. *Expert Opin Investig Drugs.* 2001;10(10):1809–1818.
134. Kroll RA, Neuwelt EA. Outwitting the blood-brain barrier for therapeutic purposes: osmotic opening and other means. *Neurosurgery.* 1998;42(5):1083–1099.
135. Bellavance MA, Blanchette M, Fortin D. Recent advances in blood-brain barrier disruption as a CNS delivery strategy. *AAPS J.* 2008;10(1):166–177.
136. Doolittle ND, Miner ME, Hall WA, Siegal T, Jerome E, Osztie E, et al. Safety and efficacy of a multicenter study using intraarterial chemotherapy in conjunction with osmotic opening of the blood-brain barrier for the treatment of patients with malignant brain tumors. *Cancer.* 2000;88(3): 637–647.
137. Hall WA, Doolittle ND, Daman M, et al. Osmotic blood-brain barrier disruption chemotherapy for diffuse pontine gliomas. *J Neurooncol.* 2006;77(3):279–284.
138. Jahnke K, Kraemer DF, Knight KR, et al. Intraarterial chemotherapy and osmotic blood-brain barrier disruption for patients with embryonal and germ cell tumors of the central nervous system. *Cancer.* 2008;112(3):581–588.
139. Angelov L, Doolittle ND, Kraemer DF, et al. Blood-brain barrier disruption and intra-arterial methotrexate-based therapy for newly diagnosed primary CNS lymphoma: a multi-institutional experience. *J Clin Oncol.* 2009;27(21):3503–3509.
140. Boockvar JA, Tsiouris AJ, Hofstetter CP, et al. Safety and maximum tolerated dose of superselective intraarterial cerebral infusion of bevacizumab after osmotic blood-brain barrier disruption for recurrent malignant glioma. Clinical article. *J Neurosurg.* 2011;114(3):624–632.
141. Guillaume DJ, Doolittle ND, Gahramanov S, Hedrick NA, Delashaw JB, Neuwelt EA. Intraarterial chemotherapy with osmotic blood-brain barrier disruption for aggressive oligodendroglial tumors: results of a phase I study. *Neurosurgery.* 2010;66(1):48–58.
142. Gutman M, Laufer R, Eisenthal A, et al. Increased microvascular permeability induced by prolonged interleukin-2 administration is attenuated by the oxygen-free-radical scavenger dimethylthiourea. *Cancer Immunol Immunother.* 1996;43(4):240–244.
143. Black KL, Chio CC. Increased opening of blood-tumour barrier by leukotriene C4 is dependent on size of molecules. *Neurol Res.* 1992;14(5):402–404.
144. de Vries HE, Blom-Roosemalen MC, van Oosten M, et al. The influence of cytokines on the integrity of the blood-brain barrier in vitro. *J Neuroimmunol.* 1996;64(1):37–43.
145. Sheikov N, McDannold N, Vykhodtseva N, Jolesz F, Hynynen K. Cellular mechanisms of the blood-brain barrier opening induced by ultrasound in presence of microbubbles. *Ultrasound Med Biol.* 2004;30(7):979–989.
146. Hynynen K, McDannold N, Sheikov NA, Jolesz FA, Vykhodtseva N. Local and reversible blood-brain barrier disruption by noninvasive focused ultrasound at frequencies suitable for trans-skull sonications. *Neuroimage.* 2005;24(1):12–20.
147. Treat LH, McDannold N, Zhang Y, Vykhodtseva N, Hynynen K. Targeted delivery of doxorubicin to the rat brain at therapeutic levels using MRI-guided focused ultrasound. *Int J Cancer.* 2007;121(4): 901–907.
148. Chopra R, Vykhodtseva N, Hynynen K. Influence of exposure time and pressure amplitude on blood-brain-barrier opening using transcranial ultrasound exposures. *ACS Chem Neurosci.* 2010;1(5): 391–398.

149. McDannold N, Vykhodtseva N, Hynynen K. Blood-brain barrier disruption induced by focused ultrasound and circulating preformed microbubbles appears to be characterized by the mechanical index. *Ultrasound Med Biol.* 2008;34(5):834–840.
150. McDannold N, Vykhodtseva N, Hynynen K. Effects of acoustic parameters and ultrasound contrast agent dose on focused-ultrasound induced blood-brain barrier disruption. *Ultrasound Med Biol.* 2008;34(6):930–937.
151. Choi JJ, Selert K, Gao Z, Samiotaki G, Baseri B, Konofagou EE. Noninvasive and localized blood-brain barrier disruption using focused ultrasound can be achieved at short pulse lengths and low pulse repetition frequencies. *J Cereb Blood Flow Metab.* 2011;31(2):725–737.
152. Choi JJ, Selert K, Vlachos F, Wong A, Konofagou EE. Noninvasive and localized neuronal delivery using short ultrasonic pulses and microbubbles. *Proc Natl Acad Sci USA.* 2011;108(40): 16539–16544.
153. Bing KF, Howles GP, Qi Y, Palmeri ML, Nightingale KR. Blood-brain barrier (BBB) disruption using a diagnostic ultrasound scanner and definity in mice. *Ultrasound Med Biol.* 2009;35(8): 1298–1308.
154. O'Reilly MA, Huang Y, Hynynen K. The impact of standing wave effects on transcranial focused ultrasound disruption of the blood-brain barrier in a rat model. *Phys Med Biol.* 2010;55(18): 5251–5267.
155. Yang FY, Fu WM, Chen WS, Yeh WL, Lin WL. Quantitative evaluation of the use of microbubbles with transcranial focused ultrasound on blood-brain-barrier disruption. *Ultrason Sonochem.* 2008; 15(4):636–643.
156. Weng JC, Wu SK, Lin WL, Tseng WY. Detecting blood-brain barrier disruption within minimal hemorrhage following transcranial focused ultrasound: a correlation study with contrast-enhanced MRI. *Magn Reson Med.* 2011;65(3):802–811.
157. Park J, Zhang Y, Vykhodtseva N, Jolesz FA, McDannold NJ. The kinetics of blood brain barrier permeability and targeted doxorubicin delivery into brain induced by focused ultrasound. *J Control Release.* 2012;162(1):134–142.
158. Yang FY, Lin YS, Kang KH, Chao TK. Reversible blood-brain barrier disruption by repeated transcranial focused ultrasound allows enhanced extravasation. *J Control Release.* 2011;150(1):111–116.
159. Samiotaki G, Vlachos F, Tung YS, Konofagou EE. A quantitative pressure and microbubble-size dependence study of focused ultrasound-induced blood-brain barrier opening reversibility in vivo using MRI. *Magn Reson Med.* 2012;67(3):769–777.
160. Choi JJ, Feshitan JA, Baseri B, et al. Microbubble-size dependence of focused ultrasound-induced blood-brain barrier opening in mice in vivo. *IEEE Trans Biomed Eng.* 2010;57(1):145–154.
161. Vlachos F, Tung YS, Konofagou E. Permeability dependence study of the focused ultrasound-induced blood-brain barrier opening at distinct pressures and microbubble diameters using DCE-MRI. *Magn Reson Med.* 2011;66(3):821–830.
162. McDannold N, Vykhodtseva N, Hynynen K. Use of ultrasound pulses combined with Definity for targeted blood-brain barrier disruption: A feasibility study. *Ultrasound Med Biol.* 2007;33(4): 584–590.
163. Cho EE, Drazic J, Ganguly M, Stefanovic B, Hynynen K. Two-photon fluorescence microscopy study of cerebrovascular dynamics in ultrasound-induced blood-brain barrier opening. *J Cereb Blood Flow Metab.* 2011;31(9):1852–1862.
164. Marty B, Larrat B, Van Landeghem M, et al. Dynamic study of blood-brain barrier closure after its disruption using ultrasound: a quantitative analysis. *J Cereb Blood Flow Metab.* 2012;32(10): 1948–1958.
165. McDannold N, Arvanitis CD, Vykhodtseva N, Livingstone MS. Temporary disruption of the blood-brain barrier by use of ultrasound and microbubbles: safety and efficacy evaluation in rhesus macaques. *Cancer Res.* 2012;72(14):3652–3663.
166. Sheikov N, McDannold N, Sharma S, Hynynen K. Effect of focused ultrasound applied with an ultrasound contrast agent on the tight junctional integrity of the brain microvascular endothelium. *Ultrasound Med Biol.* 2008;34(7):1093–1104.

167. Howles GP, Bing KF, Qi Y, Rosenzweig SJ, Nightingale KR, Johnson GA. Contrast-enhanced in vivo magnetic resonance microscopy of the mouse brain enabled by noninvasive opening of the blood-brain barrier with ultrasound. *Magn Reson Med.* 2010;64(4):995–1004.
168. Marquet F, Tung YS, Teichert T, Ferrera VP, Konofagou EE. Noninvasive, transient and selective blood-brain barrier opening in non-human primates in vivo. *PLoS One.* 2011;6(7):e22598.
169. Sheikov N, McDannold N, Jolesz F, Zhang YZ, Tam K, Hynynen K. Brain arterioles show more active vesicular transport of blood-borne tracer molecules than capillaries and venules after focused ultrasound-evoked opening of the blood-brain barrier. *Ultrasound Med Biol.* 2006;32(9):1399–1409.
170. Shang X, Wang P, Liu Y, Zhang Z, Xue Y. Mechanism of low-frequency ultrasound in opening blood-tumor barrier by tight junction. *J Mol Neurosci.* 2011;43(3):364–369.
171. Alonso A, Reinz E, Jenne JW, Fatar M, Schmidt-Glenewinkel H, Hennerici MG, et al. Reorganization of gap junctions after focused ultrasound blood-brain barrier opening in the rat brain. *J Cereb Blood Flow Metab.* 2010;30(7):1394–1402.
172. Deng J, Huang Q, Wang F, et al. The role of caveolin-1 in blood-brain barrier disruption induced by focused ultrasound combined with microbubbles. *J Mol Neurosci.* 2012;46(3):677–687.
173. Xia CY, Liu YH, Wang P, Xue YX. Low-frequency ultrasound irradiation increases blood-tumor barrier permeability by transcellular pathway in a rat glioma model. *J Mol Neurosci.* 2012;48(1): 281–290.
174. Shajahan AN, Tiruppathi C, Smrcka AV, Malik AB, Minshall RD. Gbetagamma activation of Src induces caveolae-mediated endocytosis in endothelial cells. *J Biol Chem.* 2004;279(46): 48055–48062.
175. Alonso A, Reinz E, Fatar M, Jenne J, Hennerici MG, Meairs S. Neurons but not glial cells overexpress ubiquitin in the rat brain following focused ultrasound-induced opening of the blood-brain barrier. *Neuroscience.* 2010;169(1):116–124.
176. Jalali S, Huang Y, Dumont DJ, Hynynen K. Focused ultrasound-mediated BBB disruption is associated with an increase in activation of AKT: experimental study in rats. *BMC Neurol.* 2010; 10:114.
177. Baseri B, Choi JJ, Tung YS, Konofagou EE. Multi-modality safety assessment of blood-brain barrier opening using focused ultrasound and Definity microbubbles: a short-term study. *Ultrasound Med Biol.* 2010;36(9):1445–1459.
178. McDannold N, Vykhodtseva N, Raymond S, Jolesz FA, Hynynen K. MRI-guided targeted blood-brain barrier disruption with focused ultrasound: histological findings in rabbits. *Ultrasound Med Biol.* 2005;31(11):1527–1537.
179. Liu HL, Wai YY, Chen WS, et al. Hemorrhage detection during focused-ultrasound induced blood-brain-barrier opening by using susceptibility-weighted magnetic resonance imaging. *Ultrasound Med Biol.* 2008;34(4):598–606.
180. McDannold N, Vykhodtseva N, Hynynen K. Targeted disruption of the blood-brain barrier with focused ultrasound: association with cavitation activity. *Phys Med Biol.* 2006;51(4):793–807.
181. Aryal M, Arvanitis CD, Alexander PM, McDannold N. Ultrasound-mediated blood-brain barrier disruption for targeted drug delivery in the central nervous system. *Adv Drug Deliv Rev.* 2014; 72:94–109.
182. Thevenot E, Jordao JF, O'Reilly MA, et al. Targeted delivery of self-complementary adeno-associated virus serotype 9 to the brain, using magnetic resonance imaging-guided focused ultrasound. *Hum Gene Ther.* 2012;23(11):1144–1155.
183. Choi JJ, Wang S, Tung YS, Morrison III B, Konofagou EE. Molecules of various pharmacologically-relevant sizes can cross the ultrasound-induced blood-brain barrier opening in vivo. *Ultrasound Med Biol.* 2010;36(1):58–67.
184. Wang F, Shi Y, Lu L, et al. Targeted delivery of GDNF through the blood-brain barrier by MRI-guided focused ultrasound. *PLoS One.* 2012;7(12):e52925.
185. Baseri B, Choi JJ, Deffieux T, et al. Activation of signaling pathways following localized delivery of systemically administered neurotrophic factors across the blood-brain barrier using focused ultrasound and microbubbles. *Phys Med Biol.* 2012;57(7):N65–N81.

186. Raymond SB, Treat LH, Dewey JD, McDannold N, Hynynen K, Bacskai BJ. Ultrasound enhanced delivery of molecular imaging and therapeutic agents in Alzheimer's disease mouse models. *PLoS One*. 2008;3(5):e2175.
187. Jordao JF, Ayala-Grosso CA, Markham K, et al. Antibodies targeted to the brain with image-guided focused ultrasound reduces amyloid-beta plaque load in the TgCRND8 mouse model of Alzheimer's disease. *PLoS One*. 2010;5(5):e10549.
188. Huang Q, Deng J, Wang F, et al. Targeted gene delivery to the mouse brain by MRI-guided focused ultrasound-induced blood-brain barrier disruption. *Exp Neurol*. 2012;233(1):350–356.
189. Burgess A, Huang Y, Querbes W, Sah DW, Hynynen K. Focused ultrasound for targeted delivery of siRNA and efficient knockdown of Htt expression. *J Control Release*. 2012;163(2):125–129.
190. Burgess A, Ayala-Grosso CA, Ganguly M, Jordao JF, Aubert I, Hynynen K. Targeted delivery of neural stem cells to the brain using MRI-guided focused ultrasound to disrupt the blood-brain barrier. *PLoS One*. 2011;6(11):e27877.
191. Alonso A, Reinz E, Leuchs B, et al. Focal delivery of AAV2/1-transgenes into the rat brain by localized ultrasound-induced BBB opening. *Mol Ther Nucleic Acids*. 2013;2:e73.
192. Hsu PH, Wei KC, Huang CY, et al. Noninvasive and targeted gene delivery into the brain using microbubble-facilitated focused ultrasound. *PLoS One*. 2013;8(2):e57682.
193. Liu HL, Hua MY, Chen PY, et al. Blood-brain barrier disruption with focused ultrasound enhances delivery of chemotherapeutic drugs for glioblastoma treatment. *Radiology*. 2010;255(2):415–425.
194. Mei J, Cheng Y, Song Y, et al. Experimental study on targeted methotrexate delivery to the rabbit brain via magnetic resonance imaging-guided focused ultrasound. *J Ultrasound Med*. 2009;28(7): 871–880.
195. Zeng HQ, Lu L, Wang F, Luo Y, Lou SF. Focused ultrasound-induced blood-brain barrier disruption enhances the delivery of cytarabine to the rat brain. *J Chemother*. 2012;24(6):358–363.
196. Wei KC, Chu PC, Wang HY, et al. Focused ultrasound-induced blood-brain barrier opening to enhance temozolomide delivery for glioblastoma treatment: a preclinical study. *PLoS One*. 2013;8(3):e58995.
197. Aryal M, Vykhodtseva N, Zhang YZ, Park J, McDannold N. Multiple treatments with liposomal doxorubicin and ultrasound-induced disruption of blood-tumor and blood-brain barriers improve outcomes in a rat glioma model. *J Control Release*. 2013;169(1–2):103–111.
198. Yang FY, Wong TT, Teng MC, et al. Focused ultrasound and interleukin-4 receptor-targeted liposomal doxorubicin for enhanced targeted drug delivery and antitumor effect in glioblastoma multiforme. *J Control Release*. 2012;160(3):652–658.
199. Liu HL, Hua MY, Yang HW, et al. Magnetic resonance monitoring of focused ultrasound/magnetic nanoparticle targeting delivery of therapeutic agents to the brain. *Proc Natl Acad Sci USA*. 2010;107(34):15205–15210.
200. Chen PY, Liu HL, Hua MY, et al. Novel magnetic/ultrasound focusing system enhances nanoparticle drug delivery for glioma treatment. *Neuro Oncol*. 2010;12(10):1050–1060.
201. Fan CH, Ting CY, Lin HJ, et al. SPIO-conjugated, doxorubicin-loaded microbubbles for concurrent MRI and focused-ultrasound enhanced brain-tumor drug delivery. *Biomaterials*. 2013;34(14): 3706–3715.
202. Ting CY, Fan CH, Liu HL, et al. Concurrent blood-brain barrier opening and local drug delivery using drug-carrying microbubbles and focused ultrasound for brain glioma treatment. *Biomaterials*. 2012;33(2):704–712.
203. Fan CH, Ting CY, Liu HL, et al. Antiangiogenic-targeting drug-loaded microbubbles combined with focused ultrasound for glioma treatment. *Biomaterials*. 2013;34(8):2142–2155.
204. Kinoshita M, McDannold N, Jolesz FA, Hynynen K. Noninvasive localized delivery of Herceptin to the mouse brain by MRI-guided focused ultrasound-induced blood-brain barrier disruption. *Proc Natl Acad Sci USA*. 2006;103(31):11719–11723.
205. Park EJ, Zhang YZ, Vykhodtseva N, McDannold N. Ultrasound-mediated blood-brain/blood-tumor barrier disruption improves outcomes with trastuzumab in a breast cancer brain metastasis model. *J Control Release*. 2012;163(3):277–284.

206. Yang FY, Chen YW, Chou FI, Yen SH, Lin YL, Wong TT. Boron neutron capture therapy for glioblastoma multiforme: enhanced drug delivery and antitumor effect following blood-brain barrier disruption induced by focused ultrasound. *Future Oncol.* 2012;8(10):1361–1369.
207. Alkins RD, Brodersen PM, Sodhi RN, Hynynen K. Enhancing drug delivery for boron neutron capture therapy of brain tumors with focused ultrasound. *Neuro Oncol.* 2013;15(9):1225–1235.
208. Alkins R, Burgess A, Ganguly M, et al. Focused ultrasound delivers targeted immune cells to metastatic brain tumors. *Cancer Res.* 2013;73(6):1892–1899.
209. Treat LH, McDannold N, Zhang Y, Vykhodtseva N, Hynynen K. Improved anti-tumor effect of liposomal doxorubicin after targeted blood-brain barrier disruption by MRI-guided focused ultrasound in rat glioma. *Ultrasound Med Biol.* 2012;38(10):1716–1725.
210. Sassaroli E, Hynynen K. Resonance frequency of microbubbles in small blood vessels: a numerical study. *Phys Med Biol.* 2005;50(22):5293–5305.
211. Pichardo S, Sin VW, Hynynen K. Multi-frequency characterization of the speed of sound and attenuation coefficient for longitudinal transmission of freshly excised human skulls. *Phys Med Biol.* 2011;56(1):219–250.
212. White PJ, Clement GT, Hynynen K. Longitudinal and shear mode ultrasound propagation in human skull bone. *Ultrasound Med Biol.* 2006;32(7):1085–1096.
213. McDannold N, Maier SE. Magnetic resonance acoustic radiation force imaging. *Med Phys.* 2008;35 (8):3748–3758.
214. Radicke M, Engelbertz A, Habenstein B, et al. New image contrast method in magnetic resonance imaging via ultrasound. *Hyperfine Interact.* 2008;181:21–26.
215. Vyas U, Kaye E, Pauly KB. Transcranial phase aberration correction using beam simulations and MR-ARFI. *Med Phys.* 2014;41(3):032901.
216. Peters RD, Hinks RS, Henkelman RM. Ex vivo tissue-type independence in proton-resonance frequency shift MR thermometry. *Magn Reson Med.* 1998;40(3):454–459.
217. Mougenot C, Salomir R, Palussiere J, Grenier N, Moonen CT. Automatic spatial and temporal temperature control for MR-guided focused ultrasound using fast 3D MR thermometry and multispiral trajectory of the focal point. *Magn Reson Med.* 2004;52:1005–1015.
218. Hindman JC. Proton resonance shift of water in the gas and liquid states. *J Chem Phys.* 1966;44: 4582–4592.
219. Tung YS, Vlachos F, Choi JJ, Deffieux T, Selert K, Konofagou EE. In vivo transcranial cavitation threshold detection during ultrasound-induced blood-brain barrier opening in mice. *Phys Med Biol.* 2010;55(20):6141–6155.
220. Tung YS, Marquet F, Teichert T, Ferrera V, Konofagou EE. Feasibility of noninvasive cavitation-guided blood-brain barrier opening using focused ultrasound and microbubbles in nonhuman primates. *Appl Phys Lett.* 2011;98(16):163704.
221. Arvanitis CD, Livingstone MS, Vykhodtseva N, McDannold N. Controlled ultrasound-induced blood-brain barrier disruption using passive acoustic emissions monitoring. *PLoS One.* 2012;7(9): e45783.
222. O'Reilly MA, Hynynen K. A super-resolution ultrasound method for brain vascular mapping. *Med Phys.* 2013;40(11):110701.
223. O'Reilly MA, Hynynen K. Blood-brain barrier: real-time feedback-controlled focused ultrasound disruption by using an acoustic emissions-based controller. *Radiology.* 2012;263(1):96–106.
224. Salgaonkar VA, Datta S, Holland CK, Mast TD. Passive cavitation imaging with ultrasound arrays. *J Acoust Soc Am.* 2009;126(6):3071–3083.
225. Gyongy M, Coussios CC. Passive cavitation mapping for localization and tracking of bubble dynamics. *J Acoust Soc Am.* 2010;128(4):EL175–EL180.
226. Gateau J, Aubry JF, Chauvet D, Boch AL, Fink M, Tanter M. In vivo bubble nucleation probability in sheep brain tissue. *Phys Med Biol.* 2011;56(22):7001–7015.
227. Arvanitis CD, Livingstone MS, McDannold N. Combined ultrasound and MR imaging to guide focused ultrasound therapies in the brain. *Phys Med Biol.* 2013;58(14):4749–4761.

228. Arvanitis CD, McDannold N. Integrated ultrasound and magnetic resonance imaging for simultaneous temperature and cavitation monitoring during focused ultrasound therapies. *Med Phys*. 2013;40(11):112901.
229. Jones RM, O'Reilly MA, Hynynen K. Transcranial passive acoustic mapping with hemispherical sparse arrays using CT-based skull-specific aberration corrections: a simulation study. *Phys Med Biol*. 2013;58(14):4981–5005.
230. O'Reilly MA, Jones RM, Hynynen K. Three-dimensional transcranial ultrasound imaging of microbubble clouds using a sparse hemispherical array. *IEEE Trans Biomed Eng*. 2014;61(4):1285–1294.
231. Yang FY, Horng SC, Lin YS, Kao YH. Association between contrast-enhanced MR images and blood-brain barrier disruption following transcranial focused ultrasound. *J Magn Reson Imaging*. 2010;32(3):593–599.
232. Vlachos F, Tung YS, Konofagou EE. Permeability assessment of the focused ultrasound-induced blood-brain barrier opening using dynamic contrast-enhanced MRI. *Phys Med Biol*. 2010;55(18):5451–5466.
233. Chu PC, Chai WY, Hsieh HY, et al. Pharmacodynamic analysis of magnetic resonance imaging-monitored focused ultrasound-induced blood-brain barrier opening for drug delivery to brain tumors. *Biomed Res Int*. 2013;2013:627496.
234. Tofts PS, Kermode AG. Measurement of the blood-brain barrier permeability and leakage space using dynamic MR imaging. 1. Fundamental concepts. *Magn Reson Med*. 1991;17(2):357–367.
235. Jordao JF, Thevenot E, Markham-Coultes K, et al. Amyloid-beta plaque reduction, endogenous antibody delivery and glial activation by brain-targeted, transcranial focused ultrasound. *Exp Neurol*. 2013.
236. Scarcelli T, Jordao JF, O'Reilly MA, Ellens N, Hynynen K, Aubert I. Stimulation of hippocampal neurogenesis by transcranial focused ultrasound and microbubbles in adult mice. *Brain Stimul*. 2014;7(2):304–307.
237. Burgess A, Ayala-Grosso C, Ganguly M, Jordao J, Aubert I, Hynynen K. Targeted delivery of neural stem cells using MRI-guided focused ultrasound. Ultrasonics Proceedings 2011.
238. Francis CW. Ultrasound-enhanced thrombolysis. *Echocardiography*. 2001;18(3):239–246.
239. Harpaz D, Chen X, Francis CW, Meltzer RS. Ultrasound accelerates urokinase-induced thrombolysis and reperfusion. *Am Heart J*. 1994;127(5):1211–1219.
240. Frenkel V, Oberoi J, Stone MJ, et al. Pulsed high-intensity focused ultrasound enhances thrombolysis in an in vitro model. *Radiology*. 2006;239(1):86–93.
241. Everbach EC, Francis CW. Cavitational mechanisms in ultrasound-accelerated thrombolysis at 1 MHz. *Ultrasound Med Biol*. 2000;26(7):1153–1160.
242. Riggs PN, Francis CW, Bartos SR, Penney DP. Ultrasound enhancement of rabbit femoral artery thrombolysis. *Cardiovasc Surg*. 1997;5(2):201–207.
243. Unger EC, Matsunaga TO, McCreery T, Schumann P, Sweitzer R, Quigley R. Therapeutic applications of microbubbles. *Eur J Radiol*. 2002;42(2):160–168.
244. Stone MJ, Frenkel V, Dromi S, et al. Pulsed-high intensity focused ultrasound enhanced tPA mediated thrombolysis in a novel in vivo clot model, a pilot study. *Thromb Res*. 2007.
245. Tachibana K, Tachibana S. Prototype therapeutic ultrasound emitting catheter for accelerating thrombolysis. *J Ultrasound Med*. 1997;16(8):529–535.
246. Atar S, Luo H, Birnbaum Y, Hansmann D, Siegel RJ. The use of transducer-tipped ultrasound catheter for recanalization of thrombotic arterial occlusions. *Echocardiography*. 2001;18(3):233–237.
247. Tiukinhoy-Laing SD, Huang S, Klegerman M, Holland CK, McPherson DD. Ultrasound-facilitated thrombolysis using tissue-plasminogen activator-loaded echogenic liposomes. *Thromb Res*. 2007;119(6):777–784.
248. Reinhard M, Hetzel A, Kruger S, Kretzer S, Talazko J, Ziyeh S, et al. Blood-brain barrier disruption by low-frequency ultrasound. *Stroke*. 2006;37(6):1546–1548.

249. Culp WC, Porter TR, Lowery J, Xie F, Roberson PK, Marky L. Intracranial clot lysis with intravenous microbubbles and transcranial ultrasound in swine. *Stroke.* 2004;35(10):2407–2411.
250. Porter TR, Kricsfeld D, Lof J, Everbach EC, Xie F. Effectiveness of transcranial and transthoracic ultrasound and microbubbles in dissolving intravascular thrombi. *J Ultrasound Med.* 2001;20(12):1313–1325.
251. Molina CA, Ribo M, Rubiera M, et al. Microbubble administration accelerates clot lysis during continuous 2-MHz ultrasound monitoring in stroke patients treated with intravenous tissue plasminogen activator. *Stroke.* 2006;37(2):425–429.
252. Xie F, Tsutsui JM, Lof J, et al. Effectiveness of lipid microbubbles and ultrasound in declotting thrombosis. *Ultrasound Med Biol.* 2005;31(7):979–985.
253. Daffertshofer M, Gass A, Ringleb P, et al. Transcranial low-frequency ultrasound-mediated thrombolysis in brain ischemia: increased risk of hemorrhage with combined ultrasound and tissue plasminogen activator: results of a phase II clinical trial. *Stroke.* 2005;36(7):1441–1446.
254. McDannold NJ, Vykhodtseva NI, Hynynen K. Microbubble contrast agent with focused ultrasound to create brain lesions at low power levels: MR imaging and histologic study in rabbits. *Radiology.* 2006;241(1):95–106.
255. Vykhodtseva N, McDannold N, Hynynen K. Induction of apoptosis in vivo in the rabbit brain with focused ultrasound and Optison. *Ultrasound Med Biol.* 2006;32(12):1923–1929.
256. McDannold N, Zhang YZ, Power C, Jolesz F, Vykhodtseva N. Nonthermal ablation with microbubble-enhanced focused ultrasound close to the optic tract without affecting nerve function. *J Neurosurg.* 2013;119(5):1208–1220.
257. Al Mahrouki AA, Karshafian R, Giles A, Czarnota GJ. Bioeffects of ultrasound-stimulated microbubbles on endothelial cells: gene expression changes associated with radiation enhancement in vitro. *Ultrasound Med Biol.* 2012;38(11):1958–1969.
258. Czarnota GJ, Karshafian R, Burns PN, et al. Tumor radiation response enhancement by acoustical stimulation of the vasculature. *Proc Natl Acad Sci USA.* 2012;109(30):E2033–E2041.
259. Nofiele JT, Karshafian R, Furukawa M, et al. Ultrasound-activated microbubble cancer therapy: ceramide production leading to enhanced radiation effect in vitro. *Technol Cancer Res Treat.* 2013;12 (1):53–60:Epub 2012 Aug 10.
260. Tran WT, Iradji S, Sofroni E, Giles A, Eddy D, Czarnota GJ. Microbubble and ultrasound radioenhancement of bladder cancer. *Br J Cancer.* 2012;107(3):469–476.
261. El Kaffas A, Nofiele J, Giles A, Cho S, Liu SK, Czarnota GJ. Dll4-notch signalling blockade synergizes combined ultrasound-stimulated microbubble and radiation therapy in human colon cancer xenografts. *PLoS ONE.* 2014;9(4):e93888.
262. Kwok SJ, El Kaffas A, Lai P, et al. Ultrasound-mediated microbubble enhancement of radiation therapy studied using three-dimensional high-frequency power Doppler ultrasound. *Ultrasound Med Biol.* 2013;39(11):1983–1990.
263. Fry FJ, Ades HW, Fry WJ. Production of reversible changes in the central nervous system by ultrasound. *Science.* 1958;127:83–84.
264. Adrianov OS, Vykhodtseva NI, Fokin VF, Uranova NA, Avirom VM. [Reversible functional shutdown of the optic tract on exposure to focused ultrasound]. *Biull Eksp Biol Med.* 1984;97(6): 760–762.
265. Bachtold MR, Rinaldi PC, Jones JP, Reines F, Price LR. Focused ultrasound modifications of neural circuit activity in a mammalian brain. *Ultrasound Med Biol.* 1998;24(4):557–565.
266. Min BK, Bystritsky A, Jung KI, Fischer K, Zhang Y, Maeng LS, et al. Focused ultrasound-mediated suppression of chemically-induced acute epileptic EEG activity. *BMC Neurosci.* 2011;12:23.
267. Tyler WJ. Noninvasive neuromodulation with ultrasound? A continuum mechanics hypothesis. *Neuroscientist.* 2011;17(1):25–36.
268. Kim H, Taghados SJ, Fischer K, Maeng LS, Park S, Yoo SS. Noninvasive transcranial stimulation of rat abducens nerve by focused ultrasound. *Ultrasound Med Biol.* 2012;38(9):1568–1575.
269. Yoo SS, Bystritsky A, Lee JH, Zhang Y, Fischer K, Min BK, et al. Focused ultrasound modulates region-specific brain activity. *Neuroimage.* 2011;56(3):1267–1275.

270. King RL, Brown JR, Pauly KB. Localization of ultrasound-induced in vivo neurostimulation in the mouse model. *Ultrasound Med Biol.* 2014;40(7):1512–1522.
271. Menz MD, Oralkan O, Khuri-Yakub PT, Baccus SA. Precise neural stimulation in the retina using focused ultrasound. *J Neurosci.* 2013;33(10):4550–4560.
272. Mehic E, Xu JM, Caler CJ, Coulson NK, Moritz CT, Mourad PD. Increased anatomical specificity of neuromodulation via modulated focused ultrasound. *PLoS One.* 2014;9(2):e86939.
273. Deffieux T, Younan Y, Wattiez N, Tanter M, Pouget P, Aubry JF. Low-intensity focused ultrasound modulates monkey visuomotor behavior. *Curr Biol.* 2013;23(23):2430–2433.
274. Legon W, Sato TF, Opitz A, Mueller J, Barbour A, Williams A, et al. Transcranial focused ultrasound modulates the activity of primary somatosensory cortex in humans. *Nat Neurosci.* 2014;17(2): 322–329.

CHAPTER 19

Robotics for Image-Guided Neurosurgery

Sanju Lama and Garnette R. Sutherland
Department of Clinical Neurosciences and Hotchkiss Brain Institute, University of Calgary, Calgary, Alberta, Canada

INTRODUCTION

Image guidance has remained a pivotal part of neurosurgical advances as modern neurosurgery has heavily depended on, and over time paralleled, the emergence of newer imaging techniques and modalities. As early as the discovery of the X-ray in 1895 by Wilhelm Roentgen,[1,2] neurosurgery has adapted its reliance on brain imaging to guide lesion localization and visualization.[3–6] Ever since, neurosurgery has made significant progress with advanced imaging techniques, parallel technological innovation, and the integration of both into the neurosurgical operating room. Of the several advanced surgical adjuncts being translated to the operating room, it is the merging of machine technology such as image-guided robotics into neurosurgery that may well transform the standards, principles, and practices of neurosurgery (Figure 19.1).

Progress in the realm of medical imaging is firmly rooted in the quest for improved diagnostic accuracy, lesion localization, and disease surveillance. Coupling existing X-ray imaging technology to the rapidly evolving microprocessor, allowing for computed tomography (CT) and related radiation-free magnetic resonance (MR) imaging (MRI) platforms, created the modern era of medical imaging[8]. The advent of CT brain imaging in 1971[9,10] and MR brain imaging in the early 1980s[11–13] revolutionized the diagnosis and treatment of neurological diseases. A logical progression in the imaging of anatomy and chemical processes within the human body, contemporary techniques, such as ultrasonography (USG),[14] positron emission tomography (PET),[15] MR spectroscopy,[16,17] and Raman spectroscopy,[18] have all been deployed for improved image guidance and resection control during surgery.

Many of these imaging technologies have been translated into the neurosurgical operating room for therapeutic applications. Advanced brain mapping in the operating room led to the generation of three-dimensional (3D) preoperative image data based on Cartesian coordinates for registration and localization.[19] This has formed an integral part of neuro-navigation technology where image fusion and display are integrated with real-time cranial navigation, increasing surgical accuracy (Figure 19.2).[20,21] These technologies have allowed surgery to progress toward the ultimate goal of operating within an image in real time—i.e., an intraoperative

A. Golby (Ed): Image-Guided Neurosurgery
DOI: http://dx.doi.org/10.1016/B978-0-12-800870-6.00019-4

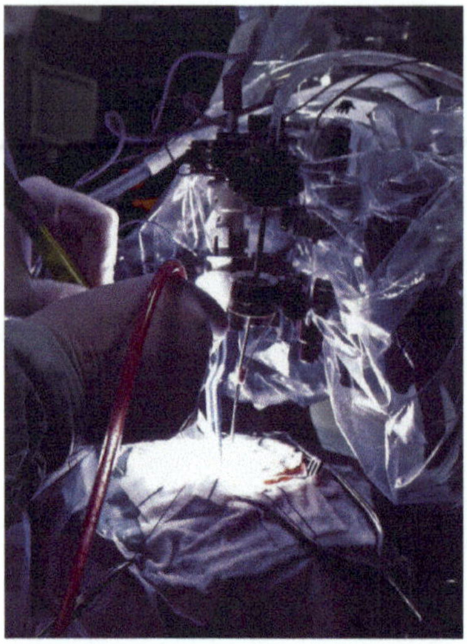

Figure 19.1 Image-guided robot, neuroArm, merged with microsurgery. *(Reproduced with permission from* Journal of Neurosurgery, *Sutherland et al.*[7]*)*

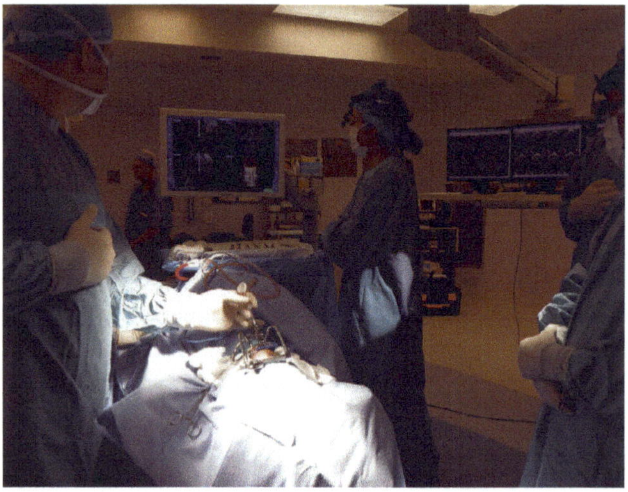

Figure 19.2 Continuous electromagnetic navigation during transoral resection of clival chordoma.

imaging platform that reveals brain shift and other effects of surgical manipulation while surgery is being performed. The concept consists of an image-guided robotic platform capable of both microsurgery and stereotaxy, which conducts telesurgery with the patient within the bore of an MRI scanner. Based on a master–slave paradigm, the system may include a pair of robotic manipulators operating in the magnet bore, whereas a surgeon commands the robot from a sensory immersive workstation that recreates the sight, sound, and touch of surgery. A common denominator underlying this achievement—i.e., the seamless integration of advanced imaging and machine *robotic* technologies into the operating room—is the rapidly advancing computational power in both hardware and software.

CONVERGENCE OF TECHNOLOGIES AND IMAGING IN NEUROSURGERY

Modern neurosurgery has evolved through the invention and integration of numerous technologies into the operating room, starting with the advent of electrocautery for hemostasis in 1926 emerging from the partnership between surgery and engineering, with contributions from Harvey Cushing and William Bovie, respectively.[22]

As the beginning of the last century set the pace for multiple inventions and interventions in neurosurgery, it has only continued to evolve. Accordingly, advances in imaging modalities, paired with the innovative refinement of machine technology, have refigured the neurosurgical operating room.[23–25] As much as surgeons have relied on supportive preoperative diagnostic, and more recently, interdissection brain imaging data to improve surgical resections and outcomes, they have also heavily relied on the design and invention of innovative tool sets and surgical adjuncts to aid in the actual performance of surgery.[26–29] As the world of engineering and industry advanced toward robotics and automation in the middle of the twentieth century, neurosurgeons witnessed—and participated in—this progression toward technologies that could potentially revolutionize the norms of the operating rooms worldwide. The latest technological breakthrough or surgical adjunct to make it to the list has been a robot capable of performing neurosurgery while being controlled by a surgeon at a sensory immersive workstation.[7,24,30–32] The workstation provides a responsive human–machine interface that converges the imaging data, 3D view of the operating site, MR-ID display of the robotic manipulators in space, and specialized haptic hand controllers that employ the robotic manipulators in multiple degrees of freedom (DOFs) (Figure 19.3). Integrating a navigation system for lesion localization in 3D space and precise surgical planning and execution accentuates the paradigm of minimalistic surgery. In principle, the approach steers surgeons toward performing beyond their physical limits of precision and accuracy, enabling them to navigate an increasingly narrow surgical corridor.[33] Basing the practice toward optimal resection control, the technology forms the ideal combination as neurosurgery begins to achieve a minimalistic approach with advanced intraoperative image guidance, coupled with high-fidelity robotic precision and accuracy.

IMAGE-GUIDED NEUROSURGERY AND ROBOTICS

Intraoperative imaging technologies in neurosurgery made their entry into the operating room in the early 1990s in the form of intraoperative CT (iCT)[34,35] and

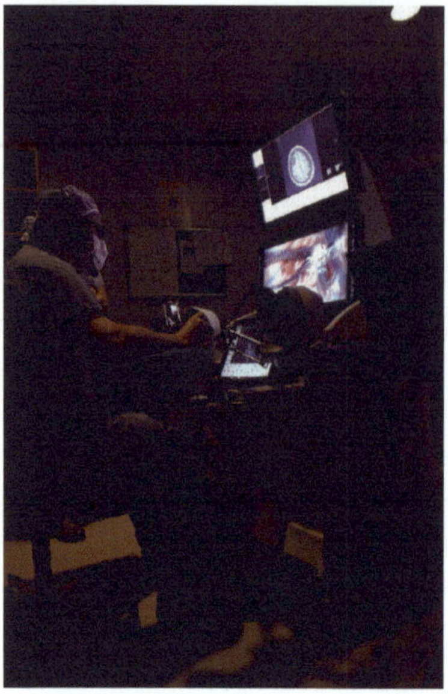

Figure 19.3 Sensory immersive workstation: neuroArm, where the sight, sound, and touch of surgery are recreated together with 3D MRI display.

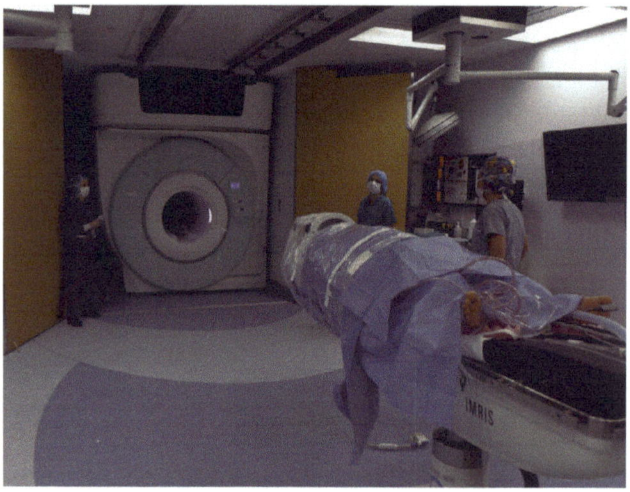

Figure 19.4 Intraoperative surgical planning MRI using an iMRI system based on a movable ceiling-mounted 3.0 T magnet.

intraoperative MRI (iMRI),[36–38] joining existing tools such as intraoperative X-ray fluoroscopy and angiography.[39] The idea of iMRI was developed with the goal of obtaining maximal resection of lesion through identification of residual tumor intraoperatively (Figure 19.4). This would allow additional surgery to be performed if

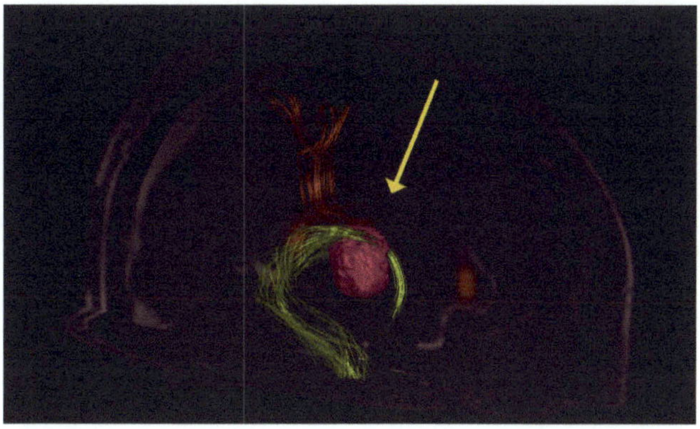

Figure 19.5 Intraoperative surgical planning MRI with DTI tractography, showing relationship of planned surgical corridor to connecting motor pathways and fornix.

necessary—in the same sitting, thus preventing redo surgery weeks or months later. Furthermore, iMRI can also demonstrate intraoperative brain shift by providing an updated MRI for neuronavigation when surgery has rendered registration based on preoperative imaging data that are no longer accurate. Advanced imaging sequences such as functional MRI (fMRI) and diffusion tensor imaging-tractography (DTI-DTT) continue to add extensive information on the functional cortex and white matter components relative to the lesion, leading to safer and more precise surgery (Figure 19.5).[40–43] Furthermore, for procedures in the eloquent regions of the brain, it is imperative that the lesion be cautiously approached using the full spectrum of imaging and surgical adjuncts, including brain mapping for neurophysiological monitoring and robotics—if they are available.[44]

The role of a robot in neurosurgery would be to perform complex neurosurgical maneuvers with the accuracy, precision, and dexterity of a machine, where human capabilities meet the limits of narrower surgical corridors. It is only logical to seek robotic assistance for the superb execution of procedures that are not physically possible. With robust patient safety mechanisms prioritized in such a system, its entry into the operating room is well warranted for the intended robot-assisted procedures. Thus, the judicious assimilation of microsurgical finesse with adept tool technologies, together with advanced imaging, neuronavigation, and state-of-the-art robotic human machine interface, has once again offered the ultimate solution to complex neurosurgical problems in the operating room.

Robots in neurosurgery?

In a time where reliance on rapidly updating computer intelligence inherently drives an era into the next generation, robotics is an intriguing aspect that has been introduced to mankind with the intent to assist in various tasks, e.g., precise repetitive action in industrial sector and/or space exploration.[45–47] The programmable precision and strength of a machine technology enable surgeries to unfold independent of factors such as fatigue and time limitations. The idea of incorporating such a machine into neurosurgery, while interesting, is not only a formidable proposition, but also something that is accompanied by considerable risk to an already complex operating room environment. However, the notion of having such an option where performing complex tasks with a machine-based accuracy and semiautomated certainty appealed to the surgeons, particularly if such a system could be used to aid in the performance of surgery. The fundamental questions that need to be addressed in advance are those associated with patient safety, surgical sterility, regulatory approvals, ethical standards, financial cost, dedicated engineering proficiency, seamless integration of the tools into surgeon's training, and execution of surgical procedures at par or better than in a conventional neurosurgery.[33] For the initial part, this goal is thus far attainable by merging the executive capacity of the human brain with the predefined precision and accuracy of a robot—i.e., a master surgeon in the loop, which combines the best of human and machine technology, interfacing the patient pathology and its treatment.

With image guidance forming an integral part of this technological epitome, material selection as it relates to MR safety, payload, swift motion, hardware and software compatibility, together with incorporation of image data for neuronavigation–lesion localization will streamline the ultimate execution of minimally invasive, precise technical delicacy in the performance of surgery.

Neurosurgical robot design aspects

Based on the fundamental principle of human–machine interaction, surgical robots can be broadly categorized into three main design configurations[48,49] (Figure 19.6): (a) *Supervisory controlled system,* in which specific and required motions of the robot are predetermined and specified by the surgeon; based on these specifications, the robot operates autonomously.[52] The surgeon's role then transitions to overseeing the robotic performance, ensuring no errors occur in the process. However, due to time lag between the robotic operating site and the relay of its action to the computer, the system requires a robust mechanism, i.e., an emergency stop command, to halt the procedure should an unintended motion occur, which is a main risk of any machine-operated system. (b) *Telesurgical system,* where the robotic end effectors are controlled and manipulated by the surgeon in real time, through a remotely controlled workstation interface.[53,54] The procedure occurs in real time, with opportunities for sensory feedback mechanisms to be incorporated in the robot through sensor technologies.[32] (c) *Shared control system,* where the surgeon is manually in control of a robotic end effector that augments the hand steadiness of the operating surgeon. In such a system,

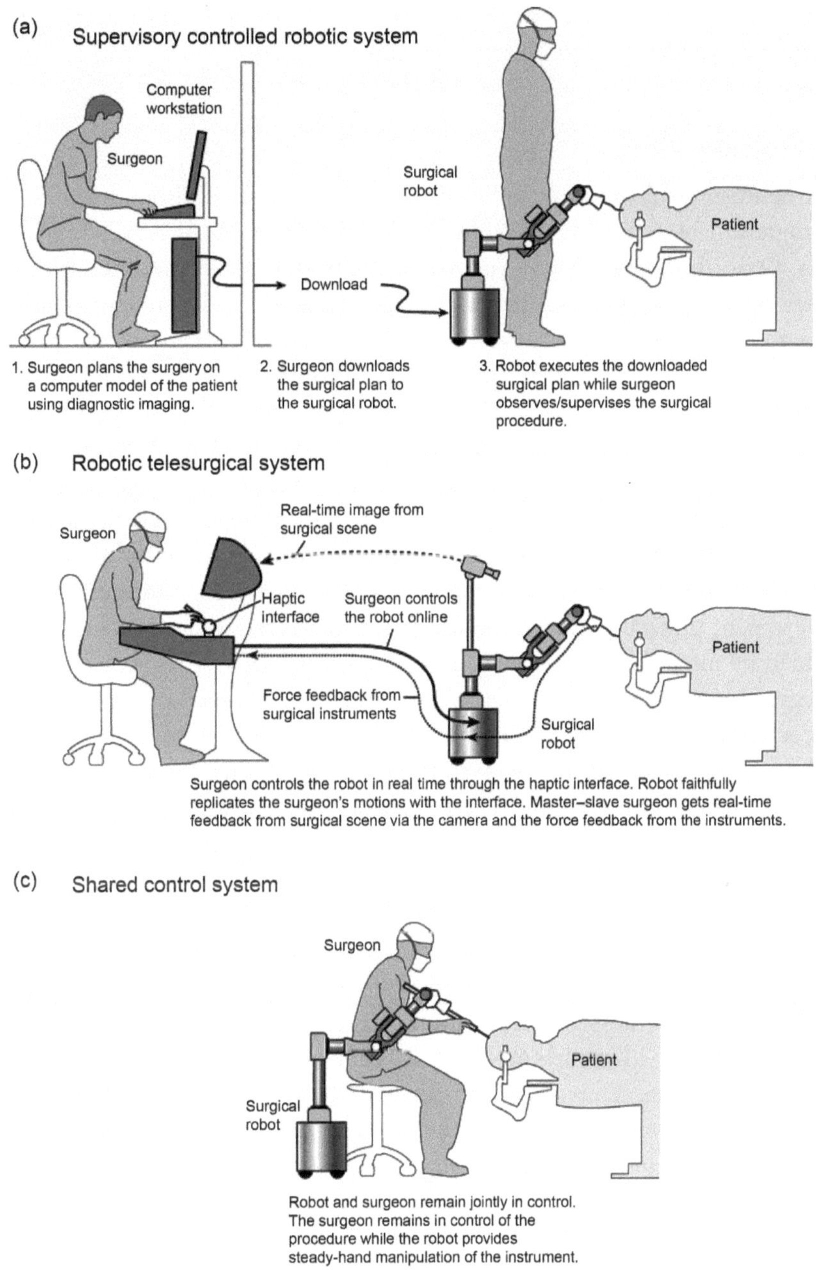

Figure 19.6 Three robotic systems proposed by Nathoo in 2005.[50,51]

the dexterity is enhanced via mechanical solutions that reduce or filter physiologic tremor, thereby producing more precise motions.[55]

Neurosurgical robots

The first neurosurgical robot, introduced in 1988, was a supervisory controlled system called Programmable Universal Machine for Assembly (PUMA, Advanced Search and Robotics, Oxford, Connecticut). This was an industrial robot modified for neurosurgery, interfaced with a CT scanner for image guidance and equipped with a probe guide for tool placement in 3D space at its end effector.[56,57] One of the first applications was CT-guided brain tumor biopsy, which eliminated the need for a bulky head frame for stereotaxy. This was closely followed by the first Food and Drug Administration (FDA)-approved robotic system NeuroMate (Integrated Surgical Systems, Inc., Davis, CA, USA), an image-guided, computer-controlled robotic system specifically designed for neurosurgery.[58,59] The system allowed precise image-based surgical planning and visualization of multiple trajectories and included an arm assembly with five DOFs together with a computer-based kinematical positioning software system. The system utilized a passive robotic arm that moved in a preprogrammed direction defined by integrated neuronavigation systems for stereotactic biopsy or functional neurosurgical applications. Although these systems introduced a transformation in robotic technology utilization in the operating room, the lack of true image guidance in real time, taking into account brain shift, positional accuracy, and dexterity in multiple DOFs, remained a limitation. These, together with the need for incorporating a sensory data feedback loop, drove the need for refinement of such technology.

As time progressed, multiple robotic systems entered neurosurgery, with a few reaching commercialization and others remaining at an experimental stage (Figure 19.7). A most successful robotic surgical system, da Vinci (Intuitive Surgical, Sunnyvale, CA), based on an endoscopic platform and initially developed for cardiac surgery, has enjoyed widespread application to general surgery, urology, obstetrics, and gynecology. Its use in neurosurgery has been confined to selected skull base procedures.[60] Minerva (University of Lausanne, Lausanne, Switzerland), designed to guide and direct tools held by a robot under direct and real-time CT guidance, faced limitations of single-dimensional incursions and need for real-time CT.[61] A current generation NeuroMate (Integrate Surgical Systems, Inc. Davis, California) has been upgraded to perform—in addition to stereotactic biopsy—electrode placement for deep brain recording or stimulation, and, when combined with other technologies, it can perform stereotactic encephalography, transcranial magnetic stimulation, radiosurgery, and neuroendoscopy.[50] Another FDA-approved system PathFinder (Prosurgics, Inc., Cupertino, CA, USA) allows the surgeon to specify a target and trajectory based on preoperative images, which are utilized by the robot to guide instruments with

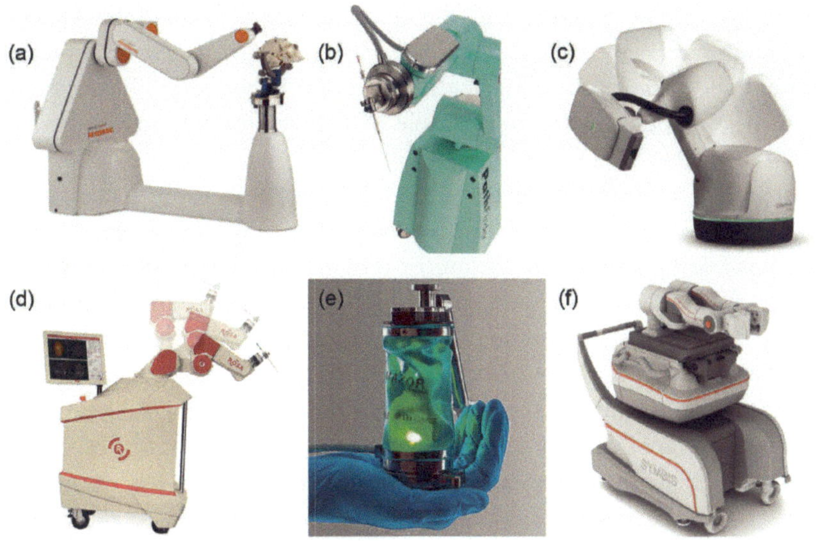

Figure 19.7 Commercially available robotic systems applicable to neurosurgery: (a) NeuroMate – Integrate Surgical Systems, Inc. Davis, CA, USA; (b) PathFinder – Prosurgics Inc., Cupertino, CA, USA; (c) Cyberknife – Accuray, Sunnyvale, CA, USA; (d) Rosa – Medtech, Castelnau Le Lez, France; (e) SpineAssist – Mazor Robotics, Amelia Island, FL, USA; (f) SYMBIS – IMRIS Inc., Minnetonka, MN, USA.

submillimeter accuracy, in procedures such as needle biopsy and surface identification, and guiding drills for burr hole placement.[51] A small robot called Renaissance (Mazor Robotics, Amelia Island, FL, USA), initial version named SpineAssist, made entry in the area of spine surgery in the form of a system that can be mounted onto the exact site of a surgery (spine) to provide tool guidance.[62] The system includes a robust software system for multiple spine procedures, as well as an add-on fluoroscopy C-arms for 3D image verification for procedures such as implant placement. A system marketed by Medtech (Montpellier, France), Rosa, has been gaining acceptance for stereotactic neurosurgical procedures, including brain tumor biopsies and electrode placement, and incorporates endoscopic platforms and laser technology for minimally invasive procedures.[63] In sum, integrating these disruptive technologies into the operating room allowed the opportunity to push boundaries for increased precision and accuracy toward increased safety and surgical minimalism. To establish their superiority, or otherwise, would require more widespread adoption of such technology providing a basis for randomized controlled trials.

As surgeons pursue real-time image guidance for improved resection control, reliance on image-guided robotics is increasingly required. Minerva, a CT-guided robot, is capable of operating within the scanner, and several others were designed to operate within an MRI machine.[64] Examples include the MR-compatible robot capable of stereotaxy while mounted outside the magnet, which was developed at the Brigham and Women's

Hospital, Boston, Massachusetts, and the Surgical Assist Technology Group of the Agency of Industrial Science and Technology, Tsukuba, Japan.[65] This was preceded by a similar system, also developed in Japan (Mechatronics Laboratory, University of Tokyo, Tokyo, Japan), a robotic manipulator capable of stereotaxy while remaining within the aperture of the magnet.[66]

As the field of robotics advanced and neurosurgeons became more familiar with the technology, the reality of robotic microsurgery began to take shape. This evolution was made possible by adding the feature of tool actuation and micromanipulation of microsurgical instruments at the end effector. It was necessary that the tools be capable of moving in multiple directions and performing tasks that mimic conventional neurosurgery. The solution improved dexterity and haptics through mechanisms of shared control (SteadyHand) and telesurgical robotic platforms.[55,67–69] These systems are capable of microsurgery, as they allow the master surgeon in control of the operation to perform the preplanned surgical procedure and to bring in the element of judgment based on past experience while executing the command to the robot. These robotic platforms include important design specifications crucial for micromanipulation, which allow mobility in multiple DOFs and the capability of motion and force scaling, tremor filtering, and force feedback or haptics. An advanced example of such a technology that includes all the pertinent design specifications is the neuroArm, an MR-compatible robot capable of both microsurgery and stereotaxy, designed and developed at the University of Calgary, Calgary, Alberta, Canada. For microsurgery, a digitizing arm colocalizes the neuroArm manipulators to the radiofrequency coil and hence to the intraoperatively acquired images, whereas for stereotaxy, the manipulator is positioned on a platform within the magnet registered to the magnet isocenter. The system, built in collaboration with Macdonald, Dettwiler and Associates (MDA, Brampton, Ontario, Canada), draws from the technology developed for space robotic ventures, namely, Canadarm and Dextre. Insights gained from space robotics included material selection, multiple degrees of motion, and payload. Based on a telerobotic platform, the system has conferred significant attention to both the design of the robot for improved dexterity and freedom of end effector mobility, as well as an equally robust and information-savvy human–machine interface—in other words, a sensory-immersive workstation.[30] neuroArm combines the robust design standards of the aerospace industry with stringent health and safety regulations; features include a microsensor technology for augmented haptics feedback (ATI, Apex NC, USA) (Figure 19.8) and a robust safety mechanism for unintended motion and hazard prevention. With a high-powered dedicated engineering expertise at hand, the operator is responsible for tool manipulation during surgery, ensuring collision avoidance and on-site judgment while being immersed in the sensory environment that recreates the sight (3D stereoscopic view of surgical site on high-definition monitors), sound (wireless headsets and mouthpiece for acoustics and communication) and touch (haptics

Figure 19.8 neuroArm end effector tool holder with ATI Nano17 force sensors (arrows), beneath the sterile drape in contact with the tool holder. Forces of tool tissue interaction are transferred to the haptic hand controller at the workstation.

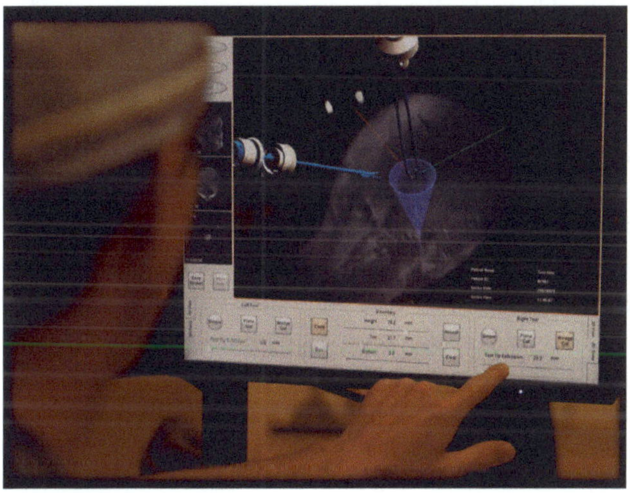

Figure 19.9 neuroArm MR-ID display showing an electronic highway through which tools may be inserted, creating no-go zones.

feedback and end effector-based sensor mechanism) of surgery. In addition, drawing on the ingenuity of aerospace industry design specifications and training mechanism, neuroArm is equipped with a simulator paradigm that not only allows the practice of intended tasks in a virtual reality simulation platform, but also sets the safety standards of electronic highways (no-go zones) that can be preregistered for ensuring safety while approaching a given target (Figure 19.9). Coupled with the neuroArm's intrinsic navigation machinery for patient registration and lesion localization, patient safety could be markedly improved in this robotic platform. In addition, the safety feature of a foot switch *Live Man's Switch*, which bypasses the computer and directly commands the robot to stop, provides a fail-safe smart conduit to stop inadvertent motion instantaneously. This feature was incorporated into the neuroArm system upon rigorous safety review following an inadvertent motion of the robotic arm in response to, but contrasting (against) the surgeon's command from the remote workstation.[7] With clinical experience being incrementally accumulated and commercial version in place (SYMBIS Surgical System, IMRIS, Inc., Minnetonka, Minnesota), robotic neurosurgery is poised to keep up with the changing technology of the future. For the reader's reference and comparison, a case presentation of right frontal grade II oligoastrocytoma in the following section presents the intricacies of merging an image-guided robot onto the neurosurgical platform that sets it apart from a conventional microsurgery. It is also important to note the value of intraoperative interdissection MRI in this particular case, where following identification of residual tumor, the decision was made to further resect the lesion.

From the above, it seems clear that many elite neurosurgical centers in the world have shown considerable enthusiasm for designing or integrating robotic technologies into their surgeries. Accordingly, many robotic designs and developments have made it to the list of surgical armamentarium. However, only a few are successfully commercialized for global adoption. This issue may relate to the complexity of building a fail-safe technology that mimics and, in some respect, surpasses human capabilities in performing brain surgery, thus limiting the long and strenuous journey through regulatory exercise and translational pathway. As a summary outline, Table 19.1 presents an overview of the selected commercial robots, highlighting their differences and applications.

neuroArm case illustration

A 30-year-old right-handed man presented to the emergency department with a generalized tonic–clonic seizure. Investigations with CT and MR brain imaging showed a nonenhancing right frontal intra-axial lesion consistent with low-grade glioma. Following treatment with antiseizure medication, the decision was made to proceed to surgical removal of the right frontal lesion, augmented with iMRI, neuronavigation and the neuroArm robot for microsurgery.

Following general anesthesia, the patient was positioned supine on the operating (OR) table with his head held in a three-pin head holder. The intraoperative magnet (IMRIS, Inc., Minnetonka, MN, USA) was brought to the operating site for surgical planning MRI (Figure 19.10, upper panel). With imaging data acquired, the magnet was returned to its alcove, and the images registered to a neurosurgical navigation system (Medtronic, Minneapolis, MN, USA) for surgical planning and lesion

Table 19.1 Selected neurosurgical robots—commercial

Year	System	DOF	#EE	Navigation	Imaging	Purpose	Use	Regulatory	Limitations
2003	NeuroMate (Renishaw)	5	1	Frame based or frame less	CT	Stereotaxy Lesion localization	Stereotaxy Functional neurosurgery Skull base drilling	First FDA approved system	No microsurgery No tool actuation
2006	PathFinder (Prosurgics)	6	1	Frameless	CT	Stereotaxy	Epilepsy surgery	Accurate navigation	CT dependent Surgical ergonomics
2006	SpineAssist (Mazor Robotics)	6	1	Frame based	CT	Spinal instrumentation	Guide for tool positioning and placement	FDA approved	Limited to spine
2011	Rosa (Medtech)	6	1	Frameless	CT/MRI	Stereotaxy	Stereotactic biopsy Functional neurosurgery	FDA approved Endoscopy/laser? haptics	No microsurgery
2014	SYMBIS (IMRIS)	7	2	Frameless	CT/MRI	Stereotaxy Microsurgery	Intracranial pathology	In review	Cost?

Abbreviations: DOF: degrees of freedom; #EE: number of end effectors; PUMA: programmable universal machine for assembly; CT: computed tomography; FDA: Food and Drug Administration; OR: operating room; MRI: magnetic resonance imaging.

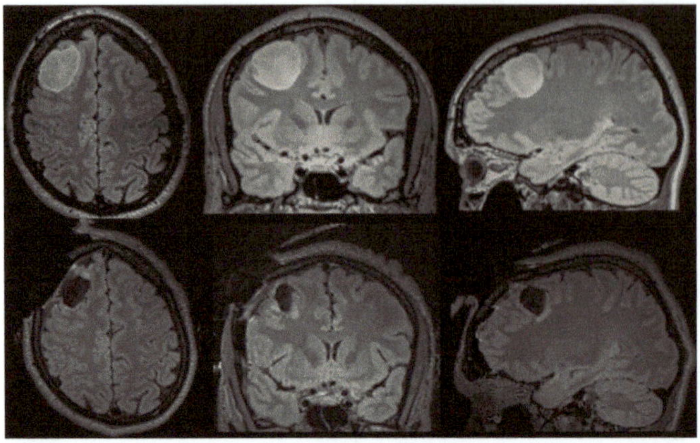

Figure 19.10 Surgical planning (upper panel) and interdissection (lower panel) fluid-attenuated inversion recovery MR images: axial, coronal, and sagittal, obtained from a patient with a right frontal oligoastrocytoma (WHO grade II).

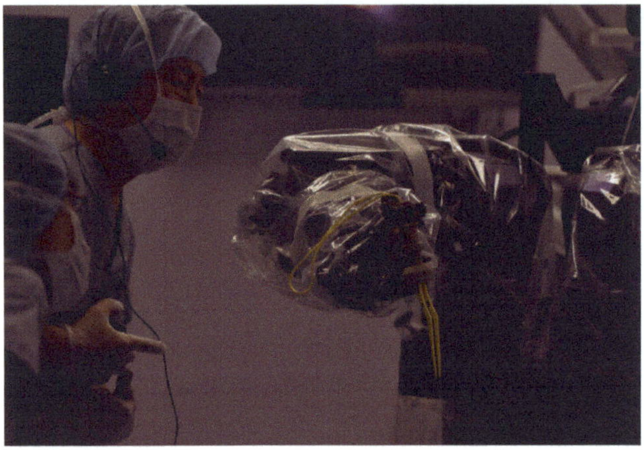

Figure 19.11 Testing the manipulators after sterile draping and initial tool placement.

localization. Following mapping of the lesion for craniotomy placement, a right frontotemporal craniotomy was performed in a conventional fashion. During the craniotomy, brain decompression was enhanced through the administration of mannitol (1 g/kg), controlled hyperventilation and by maintaining the patient in a relative head-up position. A cruciate dural incision centered on the intra-axial lesion was performed. As expected, an expanded right frontal gyrus was defined and cross-checked with the navigation probe. The pia/arachnoid overlying the gyrus was coagulated and the tumor entered, allowing sample acquisition, which was consistent with low-grade glioma. As the case proceeded, neuroArm was brought to the operating room sited at a distance from the OR table while the microscopic vision apparatus

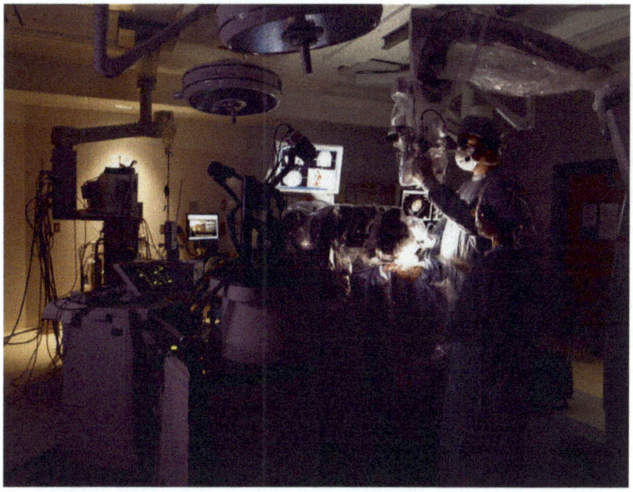

Figure 19.12 Robot-assisted microsurgery: neuroArm replaces the primary surgeon who moves to a sensory immersive workstation in an adjacent room. While at the workstation, the primary surgeon controls the robotic manipulators (arms) and communicates with the surgical staff via wireless communication device/headsets.

and neuroArm manipulator functionality were cross-checked for operation by engineers communicating through a wireless communication device from the workstation (Figure 19.11). The scrub nurse draped the robot, attaching the sterile bipolar tool on the right manipulator and suction in the left.

Following further tumor decompression, neuroArm was brought to the surgical site and positioned as the primary surgeon with the assistant surgeon on the opposite side. Accordingly the microscope (Zeiss, OPMI Pentero) was positioned and adjusted to accommodate the assistant surgeon's surgical angle and orientation (Figure 19.12). The team, which included the primary surgeon, the assistant surgeon, the scrub nurse, the circulating nurse, the robot engineer, and the robot technician, all donned communicating headset devices for enhanced two-way communication between the operating team and the primary surgeon at the workstation.

With the robot at the surgical site, microsurgery continued. The relative plane between tumor and adjacent brain was defined and dissected. The anterior, posterior, lateral, and medial tumor–brain interfaces were defined and dissection proceeded downward into the white matter. In this manner, a circumferential dissection of the tumor was achieved. In a well-coordinated master–slave paradigm, the robotic manipulators maneuvered as per the surgeon's command, while the assistant surgeon assisted in the operation with constant command feedback from and to the surgeon at the workstation. As per protocol, the team was asked to inform the primary surgeon to stop any time the robotic arms or draped cables could potentially hit the patient or OR table or anything of concern in or around the surgical site.

Accordingly, with the tumor removed, the tumor cavity was inspected and hemostasis achieved with diathermy. Once the procedure was deemed complete, the robot was removed from the surgical site and the wound draped in preparation for interdissection MR images.

Interdissection MR images showed an unexpected residual tumor localized to the medial aspect of the resection cavity (Figure 19.11, lower panel); thus, the decision was made to reoperate. The wound was reexposed and a tumor underlying the medial aspect of the cavity was defined and removed. With the surgical objective accomplished, the microscope was removed from the operative site, and the wound closed in the usual fashion. Postoperatively, the patient was transferred to the recovery room and, when examined, was clinically stable and neurologically intact.

Follow-up MRI showed >95% of tumor resection with minimal fluid-attenuated inversion recovery abnormality. Final pathology was consistent with oligoastrocytoma grade II. The patient is currently undergoing follow-up with serial MR brain imaging studies.

TOWARD AN IDEAL NEUROSURGICAL ROBOT

Time is the only variable that remains constant and is unidirectional, driving a natural evolution of technologies. As the world moves toward miniaturized mimicry of human capabilities via machines, near-real transformation of artificial intelligence becomes a distinct possibility.[70] To achieve this, it may be postulated that an MR-compatible neurosurgical robot fluid in its action of articulated wrists for microsurgery and stereotaxy, although MR imaging is acquired in real time, can be built for neurosurgery. The groundwork for such is already laid with prototype systems, by providing knowledge and information for improvement of the technology. However, an ideal robotic platform that operates in a semiautomated fashion, its microprocessors propelling information at a faster rate such that a continuous data is acquired, is still further away from reality. The following may mark some of the features desirable of a future neurosurgical robot that will surpass its predecessors as it relates to humanoid performance and skill sets, with the capability to perform based on memory and experience.

i. Better mechanics of end effector manipulators that reflects a specialized articulated wrist for tool attachment and application: Current robotics systems are somewhat bulky in their manipulators, thus making actuation and manipulation devoid of fluidity and smoothness, i.e., motions may be somewhat coarse and jerky. The solution would be increased DOFs, with increased use of joint links, gears and motors within the wrist, and elbow and shoulder joints of the manipulator for flexible and suave operation.

ii. Superior sensor technology: The neuroArm system signifies the importance of built-in strain gauge sensors with its force-sensing capacity that transmits the forces of dissection and tool tissue interaction back to the surgeon's fingertips maneuvering the hand controller for tool actuation. However, this could be improved and the sense of touch further augmented through the use of advanced and increasingly miniaturized microprocessor sensors and encoders that will compound the effect and provision of continuous data feed onto the workstation. This will enable the surgeon together with the robot to eliminate any time lag in information processing and better modulate the pace of surgery and decision making in the process.

Silicon-based microprocessor sensors made their entry into the engineering mechanics in the early 1960s, thanks to Art Zias and his development of piezoresistive accelerometers and pressure sensors for the aerospace, industrial, and microswitch divisions.[71] Although the concept of remote sensing had begun for radar-based systematic aerial photography and surveillance during World War I, the technology became increasingly refined as air defense, space exploration, and automotive and electronic communication technology rapidly advanced the field. Present-day research and development on nanoscale sensor technology ensued, be it chemical, mechanical, or biological.[72,73] As to how the technology would benefit a neurosurgical robot of the future, the scope perhaps is unlimited. Chemical and biological nanosensors if incorporated into the body environment with specific cell molecule-based targets which the robotic sensors counterpart recognizes, coupling such target recognition and execution of operation would significantly alter/improve the performance of surgery and outcomes. This, in part, relates to and expands upon the philosophy of molecular imaging and therapy that combines cell-based targeted contrast agents-nanoparticles for visualization of pathology such that surgery now transforms from an organ to a cellular level. Alongside, the theme remains of an automated robotic surgery (cautiously) entering the operating room environment, enabling a programmable mechanism by which the robot operates on the targeted cells.

iii. Haptic hand controller for superior human-machine interface: Present teleoperated systems have invariably relied on commercial hand controllers that form the cornerstone of the human–machine interface. It is important that this interface be ideal such that commanding the robot and executing superior result/surgical performance become a seamlessly orchestrated action. Toward this, compatible mechanics hardware and software dynamics must exist. Not only a well-defined workspace and ergonomic haptic hand piece are desired, a robust software platform that allows exceptional command and control of the system where error or interruption is no more an option. Furthermore, an intricately related element of an ideal hand controller is its detection and transmission analog of haptic interface that it receives from the sensors and encoders installed on the robotic end effectors bringing in the realism of tool tissue interaction onto the surgeon's hands.

iv. Simulation platforms: A related and natural component of any robotic platform is simulation in a virtual environment. The image-guided neurosurgical robot for microsurgery and stereotaxy, neuroArm comes with a built-in, customized simulator for rehearsal of teleoperation in a virtual environment. An advanced simulation platform in robotic surgery always permits an artificial environment where the possibility of case rehearsal, play forward, and playback allows faster surgeon training with shorter and flatter learning curves. Furthermore, derived from aerospace industry where simulation has a widespread and well-recognized applicability in training of pilots, the system pushes present-day surgical education toward more standardized and streamlined practice and principles. In turn, the principle could be expanded to more specific tasks relative to case-specific simulation-based planning and rehearsal.

v. Improved vision, optics, and wireless neuronavigation: As technology advances, cameras and monitors make leaps of progress in terms of their ability to provide high-definition 3D imagery of objects. Present robotic systems do rely on the operating microscope for field vision that works in tandem with the assistant surgeon. However, as surgery moves into the magnet bore, with or without the assistant surgeon, it is pertinent that a high-definition robotic camera strategically be built onto the robot relay 3D stereoscopic view of the surgical site back to the workstation where the operator is located. This development corresponds to the contemporary development of high-definition monitors for image display, the quality of which cannot be compromised. In addition, the importance of a wireless navigation system mounted onto this robot cannot be emphasized more. With the patient position fixed in space, a robust robot-based navigation system for patient registration and surgical planning adds a dimension of consistency and standardized care into the equation.

vi. Interactive robotic platform: In the field of robotics, the thought of an interactive robot has been of much interest and the object of significant work. Artificial intelligence is a topic that has intrigued mankind for several decades, and envisioning a robot capable of interacting and guessing will only be a benefit to what can complement human intelligence. Although challenging, this is a field pursued and closely followed by anyone dedicated to robotic research and development. This would also mean the creation of a highly advanced software capacity that recreates an artificial human neuronal network and connectivity, perhaps a far-reaching hypothesis at this stage but not entirely unimaginable.

vii. Digital data matrix: As stated, a robotic platform only directs the world toward what is coined as information age surgery. It is possible that in such a world, information overload could lead to system failure and crashing—encryption of data. Thus, streamlining and amalgamating the volume of digital data, both mechanical and biological recording of forces of dissection, audio and visual rendering, and so on, becomes an arduous endeavor on its own. This can be overcome with dedicated and robust hardware and software design with full-capacity continuous data storage drives. Furthermore, incorporating the micro sensor technology onto the biological/molecular environment, together with reciprocal sensors in robotic tool technology, would perhaps provide a whole new dimension to lesion tracking and localization in cranial space. A hypothetical micro-apparatus where recognition of molecular signals indicates effectively treated tissue would help characterize disease biology as surgery is being performed. This would add to an incremental volume of biological data in a digital recordable matrix for diagnostics and treatment paradigms.

viii. Parallel computing: As computers became integral parts of life on earth, the way humans think and perform and conduct their day-to-day activities may have been revolutionized by this machine technology. However, when it comes to robotic and virtual reality platforms, an ideal computer and its performance in surgery may still be far from reality. Whether it is the time delay and its relationship to the mechanical aspects of the robot, the software code and processing of data still pose a significant challenge in the seamless integration of a robotic system onto the performance of surgery. The reason is that a computer cannot work like a human brain: i.e., computers are built on vertical processing, whereas the human brain is capable of parallel thinking, allowing the ability to multitask and guess based on past experience. This fundamental gap between innate and artificial intelligence, if filled in, will create an era of robots and humans existing in unison. Via its dependency on computers today, the world continues to remain vigil to the early contributions of Alan Turing, an ageless computer scientist in his own right, whose timeless contributions include the design of the first detailed stored program computer and his historical role in ending World War II through his ability to read and decipher the cryptic wartime secret codes.[74]

Incorporating the above advances into the design of a modern neurosurgical robot will minimize the limitations of the present-day robot toward one with more fluid and robust performance matrices. With miniaturized sensors, superior haptics, auditory and visual feedback and, of paramount importance, faster, more efficient computing power, the vision of building a more proficient neurosurgical robot remains a close possibility. Furthermore, including an innate computing capability of a system that is able to perform based on a prior command or memory, the realization of a robot that is able to guess becomes achievable.

In conclusion, the direction of image-guided neurosurgery and robotics is an ever-changing and rapidly expanding specialty. Heavy reliance on numerous technologies poses tremendous challenges in making such a system fail-safe. Nevertheless, deriving assistance from robots for achieving what is not possible otherwise continues to entice the neurosurgical community. The fundamental and accessory components of robotics gain access to young minds as children are drawn to electronic (video) games and scenarios using hand controllers and robotic platforms. As engineering, industry, and medicine cross paths, the possibility of realizing a robot as an assistant and surgical adjunct in the operating rooms becomes a conceivable proposition. The exponential proliferation of global innovation and cross-fertilization of disciplines will only make the vision of having an ideal robot in the neurosurgical operating rooms a reality one day. True as it remains to date, da Vinci's quote, "Everything changes ... nothing is certain," the unidirectional forward motion of time can only define the limitless possibility the future holds.

DISCLOSURE

Garnette R. Sutherland holds shares in IMRIS, the company manufacturing and distributing both intraoperative MRI and neuroArm technology. His name is listed on many of the founding patents of IMRIS.

This work was supported by grants from the Canada Foundation for Innovation, Western Economic Diversification (Canada), and Alberta Advanced Education and Technology.

REFERENCES

1. On a new kind of rays. By W.C. Rontgen. Translated by Arthur Stanton from the Sitzungsberichte der Wurzburger Physic-medic. Gesellschaft, 1895. Nature, January 23, 1896. Radiography 1970 Aug;36(428):185–188.
2. Tubiana M. Wilhelm Conrad Rontgen and the discovery of X-rays. *Bull Acad Natl Med*. 1996;180 (1):97–108.
3. Seibert JA. One hundred years of medical diagnostic imaging technology. *Health Phys*. 1995;69 (5):695–720.
4. Young BR, Funch RB, MacMoran JW, Parker JA, Bernhard RA. Radiology of the skull and central nervous system. *Prog Neurol Psychiatry*. 1966;21:374–390.
5. Kanal E, Maravilla K, Rowley HA. Gadolinium contrast agents for CNS imaging: current concepts and clinical evidence. *Am J Neuroradiol*. 2014;35(12):2215–2226.
6. McLaren JR. A method for localizing the pineal gland on anteroposterior projections. *Am J Roentgenol Radium Ther Nucl Med*. 1959;81(6):945–952.
7. Sutherland GR, Lama S, Gan LS, Wolfsberger S, Zareinia K. Merging machines with microsurgery: clinical experience with neuroArm. *J Neurosurg*. 2013;118(3):521–529.

8. O'Connor JP, Tofts PS, Miles KA, Parkes LM, Thompson G, Jackson A. Dynamic contrast-enhanced imaging techniques: CT and MRI. *Br J Radiol.* 2011;84(Spec No 2):S112–S120.
9. Ambrose J, Hounsfield G. Computerized transverse axial tomography. *Br J Radiol.* 1972;46:148–149.
10. Ambrose J, Hounsfield G. Computerized transverse axial scanning (tomography). Part 2: clinical applications. *Br J Radiol.* 1972;46:1923–1947.
11. Lauterbur PC. Image formation by induced local interactions. Examples employing nuclear magnetic resonance. 1973. *Nature.* 1973;242:190–191.
12. Mansfield P, Maudsley AA. Line scan proton spin imaging in biological structures by NMR. *Phys Med Biol.* 1976;21(5):847–852.
13. Mansfield P. Multi-planar image formation using NMR spin echoes. *J Phys C: Solid State Phys.* 1977;10:L55–L58.
14. Ruggiero M, Magherini S, Fiore MG, et al. Transcranial sonography: a technique for the study of the temporal lobes of the human and non-human primate brain. *Ital J Anat Embryol.* 2013;118(3):241–255.
15. Nasrallah I, Dubroff J. An overview of PET neuroimaging. *Semin Nucl Med.* 2013;43(6):449–461.
16. Kaibara T, Tyson RL, Sutherland GR. Human cerebral neoplasms studied using MR spectroscopy: a review. *Biochem Cell Biol.* 1998;76(2–3):477–486.
17. Peeling J, Sutherland G. High-resolution 1 H NMR spectroscopy studies of extracts of human cerebral neoplasms. *Magn Reson Med.* 1992;24(1):123–136.
18. Krafft C, Steiner G, Beleites C, Salzer R. Disease recognition by infrared and Raman spectroscopy. *J Biophotonics.* 2009;2(1–2):13–28.
19. Schulz C, Waldeck S, Mauer UM. Intraoperative image guidance in neurosurgery: development, current indications, and future trends. *Radiol Res Pract.* 2012;2012:197364.
20. Mert A, Gan LS, Knosp E, Sutherland GR, Wolfsberger S. Advanced cranial navigation. *Neurosurgery.* 2013;72(Suppl 1):43–53.
21. Nimsky C, von Keller B, Schlaffer S, et al. Updating navigation with intraoperative image data. *Top Magn Reson Imaging.* 2009;19(4):197–204.
22. Cushing H, Bovie T. Electrosurgery as an aid to the removal of intracranial tumors, with a preliminary note on a new surgical-current generator by W.T. Bovie. *Surg Gynecol Obstet.* 1928;27:751–785.
23. Sutherland GR, Kaibara T, Louw D, Hoult DI, Tomanek B, Saunders J. A mobile high–field magnetic resonance system for neurosurgery. *J Neurosurg.* 1999;91(5):804–813.
24. Sutherland GR, Latour I, Greer AD. Integrating an image-guided robot with intraoperative MRI: a review of the design and construction of neuroArm. *IEEE Eng Med Biol Mag.* 2008;27(3):59–65.
25. Jolesz FA. Intraoperative imaging in neurosurgery: where will the future take us? *Acta Neurochir Suppl.* 2011;109:21–25.
26. Rhoton Jr. AL. Operative techniques and instrumentation for neurosurgery. *Neurosurgery.* 2003;53(4):907–934:discussion 934.
27. Yasargil MG. From the microsurgical laboratory to the operating theatre. *Acta Neurochir (Wien)* 2005;147(5):465–468.
28. Blackie RA, Gordon A. Histological appearances of intracranial biopsies obtained using the Cavitron ultrasonic surgical aspirator. *J Clin Pathol.* 1984;37(10):1101–1104.
29. Kirkup J. The history and evolution of surgical instruments. VI. The surgical blade: from finger nail to ultrasound. *Ann R Coll Surg Engl.* 1995;77(5):380–388.
30. Greer AD, Newhook P, Sutherland GR. Human–machine interface for robotic surgery and stereotaxy. *IEEE/ASME Trans MRI Compat Mechatronic Syst.* 2008;13:355–361.
31. Sutherland GR, Latour I, Greer AD, Fielding T, Feil G, Newhook P. An image-guided magnetic resonance-compatible surgical robot. *Neurosurgery.* 2008;62(2):286–292.
32. Pandya S, Motkoski JW, Serrano-Almeida C, Greer AD, Latour I, Sutherland GR. Advancing neurosurgery with image-guided robotics. *J Neurosurg.* 2009;111(6):1141–1149.
33. Motkoski JW, Sutherland GR. Progress in neurosurgical robotics. In: Jolesz FA, ed. *Intraoperative Imaging and Image-Guided Therapy.* New York, NY: Springer; 2014:601–612.

34. Lunsford LD, Martinez AJ, Latchaw RE. Stereotaxic surgery with a magnetic resonance- and computerized tomography-compatible system. *J Neurosurg*. 1986;64(6):872–878.
35. Lunsford LD, Kondziolka D, Bissonette DJ. Intraoperative imaging of the brain. *Stereotact Funct Neurosurg*. 1996;66(1–3):58–64.
36. Black PM, Moriarty T, Alexander 3rd E, et al. Development and implementation of intraoperative magnetic resonance imaging and its neurosurgical applications. *Neurosurgery*. 1997;41(4):831–842.
37. Kaibara T, Saunders JK, Sutherland GR. Advances in mobile intraoperative magnetic resonance imaging. *Neurosurgery*. 2000;47(1):131–137:discussion 137–138.
38. Hadani M, Spiegelman R, Feldman Z, Berkenstadt H, Ram Z. Novel, compact, intraoperative magnetic resonance imaging-guided system for conventional neurosurgical operating rooms. *Neurosurgery*. 2001;48(4):799–807.
39. Smith RW. Intraoperative intracranial angiography. *Neurosurgery*. 1977;1(2):107–110.
40. Nimsky C. Fiber tracking—a reliable tool for neurosurgery? *World Neurosurg*. 2010;74(1):105–106.
41. Mamata H, Mamata Y, Westin CF, et al. High-resolution line scan diffusion tensor MR imaging of white matter fiber tract anatomy. *Am J Neuroradiol*. 2002;23(1):67–75.
42. Golby AJ, Kindlmann G, Norton I, Yarmarkovich A, Pieper S, Kikinis R. Interactive diffusion tensor tractography visualization for neurosurgical planning. *Neurosurgery*. 2011;68(2):496–505.
43. Lurito JT, Lowe MJ, Sartorius C, Mathews VP. Comparison of fMRI and intraoperative direct cortical stimulation in localization of receptive language areas. *J Comput Assist Tomogr*. 2000;24 (1):99–105.
44. Lama S, Wolfsberger S, Sutherland GR. Optimal resection of insular glioma with image-guided technologies. Controversies in Neurosurgery II Ed. Ossama Al-Mefty 2013:182–188.
45. *An Automatic Block-Setting Crane*. Meccano Magazine Liverpool UK: Meccano; 1938;23:172.
46. *History of Industrial Robots*. Frankfurt, Germany: International Federation of Robotics; 2012.
47. Cole T. Solar system exploration, technology development, and neurosurgery. *Neurosurgery*. 1997;41 (6):1440–1441.
48. Nathoo N, Cavusoglu MC, Vogelbaum MA, Barnett GH. In touch with robotics: neurosurgery for the future. *Neurosurgery*. 2005;56(3):421–433.
49. Cossetto TL, Zareinia K, Sutherland GR. Robotics for neurosurgery. In: Gomes P, ed. *Medical Robotics Minimally Invasive Surgery*. Cambridge Woodhead Publishing Ltd; 2012:59–77.
50. Beasley RA. Medical robots: current systems and research directions. *J Rob*. 2012:14:401613.
51. Morgan P, Carter TJ, Davis S, et al. The application accuracy of the PathFinder neurosurgical robot. In: Proceedings of CARS 2003;1256:561–567.
52. Jiang B, Veeravagu A, Feroze AH, et al. CyberKnife radiosurgery for the management of skull base and spinal chondrosarcomas. *J Neurooncol*. 2013;114(2):209–218.
53. Kappert U, Cichon R, Schneider J, et al. Robotic coronary artery surgery—the evolution of a new minimally invasive approach in coronary artery surgery. *Thorac Cardiovasc Surg*. 2000;48(4):193–197.
54. Abbou CC, Hoznek A, Salomon L, et al. Laparoscopic radical prostatectomy with a remote controlled robot. *J Urol*. 2001;165(6 Pt 1):1964–1966.
55. Taylor RH, Jensen P, Whitcomb L, et al. A steady-hand robotic system for microsurgical augmentation. Johns Hopkins University. 1999;18(12):1201–1210. Available at: <http://urobotics.urology.jhu.edu/pub/1999-taylor-ijrr.pdf>.
56. Kwoh YS, Hou J, Jonckheere EA, Hayati S. A robot with improved absolute positioning accuracy for CT guided stereotactic brain surgery. *IEEE Trans Biomed Imaging*. 1988;35:153–160.
57. Drake JM, Joy M, Goldenberg A, Kreindler D. Computer- and robot-assisted resection of thalamic astrocytomas in children. *Neurosurgery*. 1991;29(1):27–33.
58. Li QH, Zamorano L, Pandya A, Perez R, Gong J, Diaz F. The application accuracy of the NeuroMate robot—a quantitative comparison with frameless and frame-based surgical localization systems. *Comput Aided Surg*. 2002;7(2):90–98.
59. Varma TR, Eldridge PR, Forster A, et al. Use of the NeuroMate stereotactic robot in a frameless mode for movement disorder surgery. *Stereotact Funct Neurosurg*. 2003;80(1–4):132–135.
60. Lee JY, Lega B, Bhowmick D, et al. Da Vinci robot-assisted transoral odontoidectomy for basilar invagination. *J Otorhinolaryngol Relat Spec*. 2010;72(2):91–95.

61. Glauser D, Fankhauser H, Epitaux M, Hefti JL, Jaccottet A. Neurosurgical robot Minerva: first results and current developments. *J Image Guid Surg*. 1995;1(5):266–272.
62. Sukovich W, Brink-Danan S, Hardenbrook M. Miniature robotic guidance for pedicle screw placement in posterior spinal fusion: early clinical experience with the SpineAssist. *Int J Med Rob*. 2006;2 (2):114–122.
63. Gonzalez-Martinez J, Vadera S, Mullin J, et al. Robot-assisted stereotactic laser ablation in medically intractable epilepsy: operative technique. *Neurosurgery*. 2014;10:167–173.
64. Fankhauser H, Glauser D, Flury P, et al. Robot for CT-guided stereotactic neurosurgery. *Stereotact Funct Neurosurg*. 1994;63(1–4):93–98.
65. Chinzei K, Miller K. Towards MRI guided surgical manipulator. *Med Sci Monit*. 2001;7:153–163.
66. Masamune K, Kobayashi E, Masutani Y, et al. Development of an MRI-compatible needle insertion manipulator for stereotactic neurosurgery. *J Image Guid Surg*. 1995;1(4):242–248.
67. Hongo K, Kobayashi S, Kakizawa Y, et al. NeuRobot: telecontrolled micromanipulator system for minimally invasive microneurosurgery—preliminary results. *Neurosurgery*. 2002;51:985–988.
68. Le Roux PD, Das H, Esquenazi S, Kelly PJ. Robot-assisted microsurgery: a feasibility study in the rat. *Neurosurgery*. 2001;48(3):584–589.
69. Das H, Zak H, Johnson J, Crouch J, Frambach D. Evaluation of a telerobotic system to assist surgeons in microsurgery. *Comput Aided Surg*. 1999;4(1):15–25.
70. Attenello FJ, Lee B, Yu C, Liu CY, Apuzzo ML. Supplementing the neurosurgical virtuoso: evolution of automation from mythology to operating room adjunct. *World Neurosurg*. 2014;81 (5–6):719–729.
71. Sutherland GR. Neurorobotics: driving the paradigm shift. *World Neurosurg*. 2014;81(5–6):668.
72. Giouroudi I, Keplinger F. Microfluidic biosensing systems using magnetic nanoparticles. *Int J Mol Sci*. 2013;14(9):18535–18556.
73. Batra AK, Chilvery AK, Guggilla P, Aggarwal M, Currie Jr. JR. Micro- and nano-structured metal oxides based chemical sensors: an overview. *J Nanosci Nanotechnol*. 2014;14(2):2065–2085.
74. Dyson G. Turing centenary: the dawn of computing. *Nature*. 2012;482(7386):459–460.

INDEX

Note: Page numbers followed by "*f*" and "*t*" refer to figures and tables, respectively.

C

D

E

G

H

I

S

FSC
www.fsc.org
MIX
Papier | Fördert gute Waldnutzung
FSC® C083411

Kolibri 360 GmbH
Ferdinand-Jühlke-Straße 7
99095 Erfurt, Deutschland
produktsicherheit@kolibri360.de